Annals of Mathematics Studies
Number 173

On the cohomology of certain noncompact Shimura varieties

Sophie Morel

with an appendix by Robert Kottwitz

PRINCETON UNIVERSITY PRESS

PRINCETON AND OXFORD

Published by Princeton University Press,
41 William Street, Princeton, New Jersey 08540

In the United Kingdom: Princeton University Press,

6 Oxford Street, Woodstock, Oxfordshire OX20 1TW

Library of Congress Cataloging-in-Publication Data
Morel, Sophie, 1979 -
On the cohomology of certain noncompact Shimura varieties/Sophie Morel;
with an appendix by Robert Kottwitz.
p. cm. – (Annals of mathematics ; 173)
Includes bibliographical references and index.
ISBN 978-0-691-14292-0 (hardcover : alk. paper)
ISBN 978-0-691-14293-7 (paperback : alk. paper).
Shimura varieties.
2. Homology theory. I. Title.
QA242.5.M67 2010
516.3'52–dc22

2009034338

British Library Cataloging-in-Publication Data is available

Printed on acid-free paper. ∞

press.princeton.edu

Typset by SR Nova Pvt Ltd, Bangalore, India

Printed in the United States of America

10 9 8 7 6 5 4 3 2 1

Contents

Preface

The goal of this text is to calculate the trace of a Hecke correspondence composed with a (big enough) power of the Frobenius automorphism at a good place on the intersection cohomology of the Baily-Borel compactification of certain Shimura varieties, and then to stabilize the result for the Shimura varieties associated to unitary groups over $\mathbb{Q}$.

The main result is theorem 8.4.3. It expresses the above trace in terms of the twisted trace formula on products of general linear groups, for well-chosen test functions.

Here are two applications of this result. The first (corollary 8.4.5) is about the calculation of the L-function of the intersection complex of the Baily-Borel compactification.

Theorem A *Let E be a quadratic imaginary extension of $\mathbb{Q}$, $\mathbf{G} = \mathbf{GU}(p,q)$ one of the unitary groups defined by using E (cf. 2.1), K a neat open compact subgroup of $\mathbf{G}(\mathbb{A}_f)$, $M^{\mathrm{K}}(\mathbf{G},\mathcal{X})$ the associated Shimura variety (cf. 2.1 and 1.1), and V an irreducible algebraic representation of $\mathbf{G}$. Denote by $IC^{\mathrm{K}}V$ the intersection complex of the Baily-Borel compactification of $M^{\mathrm{K}}(\mathbf{G},\mathcal{X})$ with coefficients in V. Let $\mathcal{E}_G$ be the set of elliptic endoscopic groups $\mathbf{G}(\mathbf{U}^*(n_1)\times\mathbf{U}^*(n_2))$ of $\mathbf{G}$, where $n_1, n_2 \in \mathbb{N}$ are such that $n_1+n_2 = p+q$ and n_2 is even. For every $\mathbf{H}\in\mathcal{E}_G$, let Π_H be the set of equivalence classes of automorphic representations of $\mathbf{H}(\mathbb{A}_E)$.*

Assume that K is small enough. Then there exist, for every $\mathbf{H}\in\mathcal{E}_G$, an explicit finite set R_H of algebraic representations of ${}^L\mathbf{H}_E$ and a family of complex numbers $(c_H(\pi_H, r_H))_{\pi_H\in\Pi_H, r_H\in R_H}$, almost all zero, such that, for every finite place $\wp$ of E above a prime number where K is hyperspecial,

$$\log L_\wp(s, IC^{\mathrm{K}}V) = \sum_{\mathbf{H}\in\mathcal{E}_G}\sum_{\pi_H\in\Pi_H}\sum_{r_H\in R_H} c_H(\pi_H, r_H)\log L\left(s-\frac{d}{2}, \pi_{H,\wp}, r_H\right),$$

where $d = pq$ is the dimension of $M^{\mathrm{K}}(\mathbf{G},\mathcal{X})$.

See the statement of corollary 8.4.5 for more details. The second application is corollary 8.4.9. We give a simplified statement of this corollary and refer to 8.4 for the definitions.

Theorem B *Let n be a positive integer that is not divisible by 4 and E an imaginary quadratic extension of $\mathbb{Q}$. Denote by θ the automorphism $g \longmapsto {}^t\overline{g}^{-1}$ of $R_{E/\mathbb{Q}}\mathbf{GL}_{n,E}$. Let π be a θ-stable cuspidal automorphic representation of $\mathbf{GL}_n(\mathbb{A}_E)$ that is regular algebraic. Let S be the union of the set of prime numbers that ramify* *in E and the set of prime numbers under finite places of E where π is ramified. Then there exist a number field K, a positive integer N and, for every finite place λ of K, a continuous finite-dimensional representation σ_λ of $\mathrm{Gal}(\overline{\mathbb{Q}}/E)$ with coefficients in K_λ, exists such that:*

(i) The representation σ_λ is unramified outside $S \cup \{\ell\}$, where ℓ is the prime number under λ, and pure of weight $1-n$.

(ii) For every place $\wp$ of E above a prime number $p \notin S$, for every finite place $\lambda \not| p$ of K,

$$\log L_\wp(s, \sigma_\lambda) = N \log L\left(s + \frac{n-1}{2}, \pi_\wp\right).$$

In particular, π satisfies the Ramanujan-Petersson conjecture at every finite unramified place.

There is also a result for n divisible by 4, but it is weaker and its statement is longer. See also chapter 7 for applications of the stabilized fixed point formula (corollary 6.3.2) that do not use base change to $\mathbf{GL}_n$.

The method used in this text is the one developed by Ihara, Langlands, and Kottwitz: comparison of the fixed point formula of Grothendieck-Lefschetz and of the trace formula of Arthur-Selberg. In the case where the Shimura variety is compact and the group has no endoscopy, this method is explained in the article [K10] of Kottwitz. Using base change from unitary groups to $\mathbf{GL}_n$, Clozel deduced from this a version of theorem B with supplementary conditions on the automorphic representation at a finite place (cf. [Cl5] and the article [CL] of Clozel and Labesse). The case of a compact Shimura variety (more generally, of the cohomology with compact support) and of a group that might have nontrivial endoscopy is treated by Kottwitz in the articles [K11] and [K9], modulo the fundamental lemma. For unitary groups, the fundamental lemma (and the twisted version that is used in the stabilization of the fixed point formula) is now known thanks to the work of Laumon-Ngo ([LN]), Hales ([H2]) and Waldspurger ([Wa1], [Wa2], [Wa3]); note that the fundamental lemma is even known in general thanks to the recent article of Ngo ([Ng]).

The case of $\mathbf{GL}_2$ over $\mathbb{Q}$ (i.e., of modular curves) has been treated in the book [DK], and the case of $\mathbf{GL}_2$ over a totally real number field in the article [BL] of Brylinski and Labesse. In these two cases, the Shimura variety is noncompact (but its Baily-Borel compactification is not too complicated) and the group has no endoscopy.

One of the simplest cases where the Shimura variety is noncompact and the group has nontrivial endoscopy is that of the unitary group $\mathbf{GU}(2,1)$. This case has been studied in the book [LR]. For groups of semisimple rank 1, Rapoport proved in [Ra] the (unstabilized) fixed point formula in the case where the Hecke correspondance is trivial. The stabilized fixed point formula for the symplectic groups $\mathbf{GSp}_{2n}$ is proved in [M3] (and some applications to $\mathbf{GSp}_4$ and $\mathbf{GSp}_6$, similar to theorem A, are given). In [Lau1] and [Lau2], Laumon obtained results similar to theorem B for the group $\mathbf{GSp}_4$ by using the cohomology with compact support (instead of the intersection cohomology).

Finally, note that Shin recently obtained results analogous to theorem B, and also results about ramified places, by studying the cohomology of Igusa varieties attached to compact unitary Shimura varieties (cf. [Shi1], [Shi2], [Shi3]). This builds on previous work of Harris and Taylor ([HT]), Mantovan ([Ma1], [Ma2]) and Fargues ([Fa]).

We give a quick description of the chapters.

Chapter 1 contains "known facts" about the fixed point formula. When the Shimura variety is associated to a unitary group over $\mathbb{Q}$ and the Hecke correspondence is trivial, the fixed point formula has been proved in [M1] (theorem 5.3.3.1). The article [M2] contains the theoretical tools needed to treat the case of nontrivial Hecke correspondences for Siegel modular varieties (proposition 5.1.5 and theorem 5.2.2), but does not finish the calculation. We generalize here the results of [M2] under certain conditions on the group (that are satisfied by unitary groups over $\mathbb{Q}$ and by symplectic groups), then use them to calculate the trace on the intersection cohomology of a Hecke correspondence twisted by a high enough power of Frobenius in the case when the Shimura variety and the boundary strata of its Baily-Borel compactification are of the type considered by Kottwitz in his article [K11] (i.e., PEL of type A or C). The result is given in theorem 1.7.1.

Chapters 2 to 6 treat the stabilization of the fixed point formula for unitary groups over $\mathbb{Q}$. We prove conjecture (10.1) of the article [K9] (corollary 6.3.2). Kottwitz stabilized the elliptic part of the fixed point formula in [K9], and the method of this book to stabilize the terms coming from the boundary is inspired by his method. The most complicated calculations are at the infinite place, where we need to show a formula for the values at certain elements of the stable characters of discrete series (proposition 3.4.1). This formula looks a little like the formulas established by Goresky, Kottwitz et MacPherson ([GKM] theorems 5.1 and 5.2), though it has fewer terms. This result is special to unitary groups: the analogous formula for symplectic groups (cf. section 4 of [M3]) is much more complicated, and more different from the formulas of [GKM].

In chapter 2, we define the unitary groups over $\mathbb{Q}$ that we will study, as well as their Shimura data, and we recall some facts about their endoscopy.

Chapter 3 contains the calculations at the infinite place.

Chapter 4 contains explicit calculations, at an unramified place of the group, of the Satake transform, the base change map, the transfer map, and the twisted transfer map, and a compatibility result for the twisted transfer and constant term maps.

In chapter 5, we recall the stabilization by Kottwitz of the geometric side of the invariant trace formula when the test function is stable cuspidal at infinity (cf. [K13]). This stabilization relies on the calculation by Arthur of the geometric side of the invariant trace formula for a function that is stable cuspidal at infinity (cf. [A6] formula (3.5) and theorem 6.1), and uses only the fundamental lemma (it does not require the weighted fundamental lemma). Unfortunately, this result is unpublished. Chapter 5 also contains the normalization of the Haar measures and of the transfer factors, the statement of the fundamental lemmas we use, and a summary of the results that are known about these fundamental lemmas.

In chapter 6, we put the results of chapters 2, 3, and 4 together and stabilize the fixed point formula.

Chapter 7 gives applications of the stabilized fixed point formula that do not use base change to $\mathbf{GL}_n$. First, in section 7.1, we show how to make the results of this chapter formally independent from Kottwitz's unpublished article [K13] (this is merely a formal game, because of course a large part of this book was inspired by [K13] in the first place). In section 7.2, we express the logarithm of the L-function of the intersection complex at a finite place above a big enough prime number as a sum (a priori with complex coefficients) of logarithms of local L-functions of automorphic representations of unitary groups. We also give, in section 7.3, an application to the Ramanujan-Petersson conjecture (at unramified places) for certain discrete automorphic representations of unitary groups.

Chapter 8 gives applications of the stabilized fixed point formula that use base change to $\mathbf{GL}_n$. In section 8.1, we recall some facts about nonconnected groups. In sections 8.2 and 8.3, we study the twisted trace formula for certain test functions. We give applications of this in 8.4; in particular, we obtain another formula for the L-function of the intersection complex, this time in terms of local L-functions of automorphic representations of general linear groups. The simple twisted trace formula proved in this chapter implies some weak base change results; these have been worked out in section 8.5.

In chapter 9, we prove the particular case of the twisted fundamental lemma that is used in the stabilization of the fixed point formula in the article [K9] of Kottwitz, the articles [Lau1] and [Lau2] of Laumon and chapter 6. The methods of this chapter are not new, and no attempt is made to obtain the most general result possible. Waldspurger showed in [Wa3] that the twisted fundamental lemma for the unit of the Hecke algebra is a consequence of the ordinary fundamental lemma (and, in the general case, of the nonstandard fundamental lemma). We show that, in the particular case that we need, the twisted fundamental lemma for the unit of the Hecke algebra implies the twisted fundamental lemma for all the functions of the Hecke algebra. The method is the same as in the article [H2] of Hales (i.e., it is the method inspired by the article [Cl3] of Clozel, and by the remark of the referee of that article).

The appendix by Robert Kottwitz contains a comparison theorem between the twisted transfer factors of [KS] and [K9]. This result is needed to use the twisted fundamental lemma in the stabilization of the fixed point formula.

It is a great pleasure for me to thank Robert Kottwitz and Gérard Laumon. Robert Kottwitz very kindly allowed me to read his unpublished manuscript [K13], that has been extremely helpful to me in writing this text. He also helped me fix a problem in the proof of proposition 8.2.3, pointed out several mistakes in chapter 9 and agreed to write his proof of the comparison of twisted transfer factors as an appendix of this book. Gérard Laumon suggested that I study the intersection cohomology of noncompact Shimura varieties and has spent countless hours patiently explaining the subject to me. I also thank Jean-Loup Waldspurger for sending me a complete version of his manuscript [Wa3] on twisted endoscopy before it was published.

I am grateful to the other mathematicians who have answered my questions or pointed out simpler arguments to me, in particular, Pierre-Henri Chaudouard, Laurent Fargues, Günter Harder, Colette Moeglin, Bao Chau Ngo, Sug Woo Shin, and Marie-France Vignéras (I am especially grateful to Sug Woo Shin for repeatedly correcting my misconceptions about the spectral side of the twisted trace formula).

Finally, I would like to express my gratitude to the anonymous referee for finding several mistakes and inaccuracies in the first version of this text.

This text was written entirely while I was a Clay Research Fellow of the Clay Mathematics Institute, and worked as a member at the Institute for Advanced Study in Princeton. Moreover, I have been partially supported by the National Science Foundation under agreements number DMS-0111298 and DMS-0635607.

Chapter One

The fixed point formula

1.1 SHIMURA VARIETIES

The reference for this section is [P2] §3.

Let $\mathbb{S} = R_{\mathbb{C}/\mathbb{R}}\mathbb{G}_{m,\mathbb{R}}$. Identify $\mathbb{S}(\mathbb{C}) = (\mathbb{C} \otimes_{\mathbb{R}} \mathbb{C})^\times$ and $\mathbb{C}^\times \times \mathbb{C}^\times$ using the morphism $a \otimes 1 + b \otimes i \longmapsto (a + ib, a - ib)$, and write $\mu_0 : \mathbb{G}_{m,\mathbb{C}} \longrightarrow \mathbb{S}_{\mathbb{C}}$ for the morphism $z \longmapsto (z, 1)$.

The definition of (pure) Shimura data that will be used here is that of [P2] (3.1), up to condition (3.1.4). So a pure Shimura datum is a triple $(\mathbf{G}, \mathcal{X}, h)$ (that will often be written simply $(\mathbf{G}, \mathcal{X})$), where $\mathbf{G}$ is a connected reductive linear algebraic group over $\mathbb{Q}$, $\mathcal{X}$ is a set with a transitive action of $\mathbf{G}(\mathbb{R})$ and $h : \mathcal{X} \longrightarrow \mathrm{Hom}(\mathbb{S}, \mathbf{G}_{\mathbb{R}})$ is a $\mathbf{G}(\mathbb{R})$-equivariant morphism, satisfying conditions (3.1.1), (3.1.2), (3.1.3), and (3.1.5) of [P2], but not necessarily condition (3.1.4) (i.e., the group $\mathbf{G}^{\mathrm{ad}}$ may have a simple factor of compact type defined over $\mathbb{Q}$).

Let $(\mathbf{G}, \mathcal{X}, h)$ be a Shimura datum. The field of definition F of the conjugacy class of cocharacters $h_x \circ \mu_0 : \mathbb{G}_{m,\mathbb{C}} \longrightarrow \mathbf{G}_{\mathbb{C}}$, $x \in \mathcal{X}$, is called the *reflex field* of the datum. If K is an open compact subgroup of $\mathbf{G}(\mathbb{A}_f)$, there is an associated Shimura variety $M^{\mathrm{K}}(\mathbf{G}, \mathcal{X})$, which is a quasi-projective algebraic variety over F satisfying

$$M^{\mathrm{K}}(\mathbf{G}, \mathcal{X})(\mathbb{C}) = \mathbf{G}(\mathbb{Q}) \setminus (\mathcal{X} \times \mathbf{G}(\mathbb{A}_f)/\mathrm{K}).$$

If moreover K is *neat* (cf. [P1] 0.6), then $M^{\mathrm{K}}(\mathbf{G}, \mathcal{X})$ is smooth over F. Let $M(\mathbf{G}, \mathcal{X})$ be the inverse limit of the $M^{\mathrm{K}}(\mathbf{G}, \mathcal{X})$, taken over the set of open compact subgroups K of $\mathbf{G}(\mathbb{A}_f)$.

Let $g, g' \in \mathbf{G}(\mathbb{A}_f)$, and let K, K' be open compact subgroups of $\mathbf{G}(\mathbb{A}_f)$ such that $\mathrm{K}' \subset g\mathrm{K}g^{-1}$. Then there is a finite morphism

$$T_g : M^{\mathrm{K}'}(\mathbf{G}, \mathcal{X}) \longrightarrow M^{\mathrm{K}}(\mathbf{G}, \mathcal{X}),$$

which is given on complex points by

$$\begin{cases} \mathbf{G}(\mathbb{Q}) \setminus (\mathcal{X} \times \mathbf{G}(\mathbb{A}_f)/\mathrm{K}') & \longrightarrow \quad \mathbf{G}(\mathbb{Q}) \setminus (\mathcal{X} \times \mathbf{G}(\mathbb{A}_f)/\mathrm{K}), \\ \qquad \mathbf{G}(\mathbb{Q})(x, h\mathrm{K}') & \longmapsto \quad \mathbf{G}(\mathbb{Q})(x, hg\mathrm{K}). \end{cases}$$

If K is neat, then the morphism T_g is étale.

Fix K. The Shimura variety $M^{\mathrm{K}}(\mathbf{G}, \mathcal{X})$ is not projective over F in general, but it has a compactification $j : M^{\mathrm{K}}(\mathbf{G}, \mathcal{X}) \longrightarrow M^{\mathrm{K}}(\mathbf{G}, \mathcal{X})^*$, the Satake-Baily-Borel (or Baily-Borel, or minimal Satake, or minimal) compactification, such that $M^{\mathrm{K}}(\mathbf{G}, \mathcal{X})^*$ is a normal projective variety over F and $M^{\mathrm{K}}(\mathbf{G}, \mathcal{X})$ is open dense in $M^{\mathrm{K}}(\mathbf{G}, \mathcal{X})^*$. Note that $M^{\mathrm{K}}(\mathbf{G}, \mathcal{X})^*$ is not smooth in general (even when K is neat). The set of complex points of $M^{\mathrm{K}}(\mathbf{G}, \mathcal{X})^*$ is

$$M^{\mathrm{K}}(\mathbf{G}, \mathcal{X})^*(\mathbb{C}) = \mathbf{G}(\mathbb{Q}) \setminus (\mathcal{X}^* \times \mathbf{G}(\mathbb{A}_f)/\mathrm{K}),$$

where $\mathcal{X}^*$ is a topological space having $\mathcal{X}$ as an open dense subset and such that the $\mathbf{G}(\mathbb{Q})$-action on $\mathcal{X}$ extends to a continuous $\mathbf{G}(\mathbb{Q})$-action on $\mathcal{X}^*$. As a set, $\mathcal{X}^*$ is the disjoint union of $\mathcal{X}$ and of boundary components $\mathcal{X}_P$ indexed by the set of admissible parabolic subgroups of $\mathbf{G}$ (a parabolic subgroup of $\mathbf{G}$ is called *admissible* if it is not equal to $\mathbf{G}$ and if its image in every simple factor $\mathbf{G}'$ of $\mathbf{G}^{\mathrm{ad}}$ is equal to $\mathbf{G}'$ or to a maximal parabolic subgroup of $\mathbf{G}'$, cf. [P1] 4.5). If $\mathbf{P}$ is an admissible parabolic subgroup of $\mathbf{G}$, then $\mathbf{P}(\mathbb{Q}) = \mathrm{Stab}_{\mathbf{G}(\mathbb{Q})}(\mathcal{X}_P)$; the $\mathbf{P}(\mathbb{Q})$-action on $\mathcal{X}_P$ extends to a transitive $\mathbf{P}(\mathbb{R})$-action, and the unipotent radical of $\mathbf{P}$ acts trivially on $\mathcal{X}_P$.

For every g, K, K$'$ as above, there is a finite morphism $\overline{T}_g : M^{\mathrm{K}'}(\mathbf{G}, \mathcal{X})^* \longrightarrow M^{\mathrm{K}}(\mathbf{G}, \mathcal{X})^*$ extending the morphism T_g.

From now on, we will assume that $\mathbf{G}$ satisfies the following condition. Let $\mathbf{P}$ be an admissible parabolic subgroup of $\mathbf{G}$, $\mathbf{N}_P$ be its unipotent radical, $\mathbf{U}_P$ the center of $\mathbf{N}_P$, and $\mathbf{M}_P = \mathbf{P}/\mathbf{N}_P$ the Levi quotient. Then there exist two connected reductive subgroups $\mathbf{L}_P$ and $\mathbf{G}_P$ of $\mathbf{M}_P$ such that

- $\mathbf{M}_P$ is the direct product of $\mathbf{L}_P$ and $\mathbf{G}_P$;
- $\mathbf{G}_P$ contains $\mathbf{G}_1$, where $\mathbf{G}_1$ is the normal subgroup of $\mathbf{M}_P$ defined by Pink in [P2] (3.6), and the quotient $\mathbf{G}_P/\mathbf{G}_1 Z(\mathbf{G}_P)$ is $\mathbb{R}$-anisotropic;
- $\mathbf{L}_P \subset \mathrm{Cent}_{\mathbf{M}_P}(\mathbf{U}_P) \subset Z(\mathbf{M}_P)\mathbf{L}_P$;
- $\mathbf{G}_P(\mathbb{R})$ acts transitively on $\mathcal{X}_P$, and $\mathbf{L}_P(\mathbb{R})$ acts trivially on $\mathcal{X}_P$;
- for every neat open compact subgroup K_M of $\mathbf{M}_P(\mathbb{A}_f)$, $\mathrm{K}_M \cap \mathbf{L}_P(\mathbb{Q}) = \mathrm{K}_M \cap \mathrm{Cent}_{\mathbf{M}_P(\mathbb{Q})}(\mathcal{X}_P)$.

Denote by $\mathbf{Q}_P$ the inverse image of $\mathbf{G}_P$ in $\mathbf{P}$.

Remark 1.1.1 If $\mathbf{G}$ satisfies this condition, then, for every admissible parabolic subgroup $\mathbf{P}$ of $\mathbf{G}$, the group $\mathbf{G}_P$ satisfies the same condition.

Example 1.1.2 Any interior form of the general symplectic group $\mathbf{GSp}_{2n}$ or of the quasi-split unitary group $\mathbf{GU}^*(n)$ defined in section 2.1 satisfies the condition.

The boundary of $\mathbf{M}^{\mathrm{K}}(\mathbf{G}, \mathcal{X})^*$ has a natural stratification (this stratification exists in general, but its description is a little simpler when $\mathbf{G}$ satisfies the above condition). Let $\mathbf{P}$ be an admissible parabolic subgroup of $\mathbf{G}$. Pink has defined a morphism $\mathcal{X}_P \longrightarrow \mathrm{Hom}(\mathbb{S}, \mathbf{G}_{P,\mathbb{R}})$ ([P2] (3.6.1)) such that $(\mathbf{G}_P, \mathcal{X}_P)$ is a Shimura datum and the reflex field of $(\mathbf{G}_P, \mathcal{X}_P)$ is F. Let $g \in \mathbf{G}(\mathbb{A}_f)$. Let $\mathrm{H}_P = g\mathrm{K}g^{-1} \cap \mathbf{P}(\mathbb{Q})\mathbf{Q}_P(\mathbb{A}_f)$, $\mathrm{H}_L = g\mathrm{K}g^{-1} \cap \mathbf{L}_P(\mathbb{Q})\mathbf{N}_P(\mathbb{A}_f)$, $\mathrm{K}_Q = g\mathrm{K}g^{-1} \cap \mathbf{Q}_P(\mathbb{A}_f)$, and $\mathrm{K}_N = g\mathrm{K}g^{-1} \cap \mathbf{N}_P(\mathbb{A}_f)$. Then (cf. [P2] (3.7)) there is a morphism, finite over its image,

$$M^{\mathrm{K}_Q/\mathrm{K}_N}(\mathbf{G}_P, \mathcal{X}_P) \longrightarrow M^{\mathrm{K}}(\mathbf{G}, \mathcal{X})^* - M^{\mathrm{K}}(\mathbf{G}, \mathcal{X}).$$

The group H_P acts on the right on $M^{\mathrm{K}_Q/\mathrm{K}_N}(\mathbf{G}_P, \mathcal{X}_P)$, and this action factors through the finite group $\mathrm{H}_P/\mathrm{H}_L\mathrm{K}_Q$. Denote by $i_{P,g}$ the locally closed immersion

$$M^{\mathrm{K}_Q/\mathrm{K}_N}(\mathbf{G}_P, \mathcal{X}_P)/\mathrm{H}_P \longrightarrow M^{\mathrm{K}}(\mathbf{G}, \mathcal{X})^*.$$

This immersion extends to a finite morphism

$$\bar{i}_{P,g} : M^{\mathrm{K}_Q/\mathrm{K}_N}(\mathbf{G}_P, \mathcal{X}_P)^*/\mathrm{H}_P \longrightarrow M^{\mathrm{K}}(\mathbf{G}, \mathcal{X})^*$$

(this morphism is not a closed immersion in general). The boundary of $M^{\mathrm{K}}(\mathbf{G}, \mathcal{X})^*$ is the union of the images of the morphisms $i_{P,g}$, for $\mathbf{P}$ an admissible parabolic subgroup of $\mathbf{G}$ and $g \in \mathbf{G}(\mathbb{A}_f)$. If $\mathbf{P}'$ is another admissible parabolic subgroup of $\mathbf{G}$ and $g' \in \mathbf{G}(\mathbb{A}_f)$, then the images of the immersions $i_{P,g}$ and $i_{P',g'}$ are equal if and only if there exists $\gamma \in \mathbf{G}(\mathbb{Q})$ such that $\mathbf{P}' = \gamma \mathbf{P} \gamma^{-1}$ and $\mathbf{P}(\mathbb{Q})\mathbf{Q}_P(\mathbb{A}_f)g\mathrm{K} = \mathbf{P}(\mathbb{Q})\mathbf{Q}_P(\mathbb{A}_f)\gamma^{-1}g'\mathrm{K}$; if there is no such γ, then these images are disjoint. If K is neat, then $\mathrm{K}_Q/\mathrm{K}_N$ is also neat and the action of $\mathrm{H}_P/\mathrm{H}_L\mathrm{K}_N$ on $M^{\mathrm{K}_Q/\mathrm{K}_N}(\mathbf{G}_P, \mathcal{X}_P)$ is free (so $M^{\mathrm{K}_Q/\mathrm{K}_N}(\mathbf{G}_P, \mathcal{X}_P)/\mathrm{H}_P$ is smooth).

The images of the morphisms $i_{P,g}$, $g \in \mathbf{G}(\mathbb{A}_f)$, are the *boundary strata* of $M^{\mathrm{K}}(\mathbf{G}, \mathcal{X})^*$ associated to $\mathbf{P}$.

To simplify notation, assume from now on that $\mathbf{G}^{\mathrm{ad}}$ is simple. Fix a minimal parabolic subgroup $\mathbf{P}_0$ of $\mathbf{G}$. A parabolic subgroup of $\mathbf{G}$ is called *standard* if it contains $\mathbf{P}_0$. Let $\mathbf{P}_1, \dots, \mathbf{P}_n$ be the maximal standard parabolic subgroups of $\mathbf{G}$, with the numbering satisfying $r \leq s$ if and only if $\mathbf{U}_{P_r} \subset \mathbf{U}_{P_s}$ (cf. [GHM] (22.3)). Write $\mathbf{N}_r = \mathbf{N}_{P_r}$, $\mathbf{G}_r = \mathbf{G}_{P_r}$, $\mathbf{L}_r = \mathbf{L}_{P_r}$, $i_{r,g} = i_{P_r,g}$, etc.

Let $\mathbf{P}$ be a standard parabolic subgroup of $\mathbf{G}$. Write $\mathbf{P} = \mathbf{P}_{n_1} \cap \dots \cap \mathbf{P}_{n_r}$, with $n_1 < \dots < n_r$. The Levi quotient $\mathbf{M}_P = \mathbf{P}/\mathbf{N}_P$ is the direct product of $\mathbf{G}_{n_r}$ and of a Levi subgroup $\mathbf{L}_P$ of $\mathbf{L}_{n_r}$. Let $\mathcal{C}_P$ be the set of n-uples $(X_1, \dots, X_r)$, where

- X_1 is a boundary stratum of $M^{\mathrm{K}}(\mathbf{G}, \mathcal{X})^*$ associated to $\mathbf{P}_{n_1}$;
- for every $i \in \{1, \dots, r-1\}$, X_{i+1} is a boundary stratum of X_i associated to the maximal parabolic subgroup $(\mathbf{P}_{n_{i+1}} \cap \mathbf{Q}_{n_i})/\mathbf{N}_{n_i}$ of $\mathbf{G}_{n_i}$.

Let $\mathcal{C}_P^1$ be the quotient of $\mathbf{G}(\mathbb{A}_f) \times \mathbf{Q}_{n_1}(\mathbb{A}_f) \times \dots \times \mathbf{Q}_{n_{r-1}}(\mathbb{A}_f)$ by the following equivalence relation: $(g_1, \dots, g_r)$ is equivalent to $(g_1', \dots, g_r')$ if and only if, for every $i \in \{1, \dots, r\}$,

$$(\mathbf{P}_{n_1} \cap \dots \cap \mathbf{P}_{n_i})(\mathbb{Q})\mathbf{Q}_{n_i}(\mathbb{A}_f)g_i \dots g_1\mathrm{K} = (\mathbf{P}_{n_1} \cap \dots \cap \mathbf{P}_{n_i})(\mathbb{Q})\mathbf{Q}_{n_i}(\mathbb{A}_f)g_i' \dots g_1'\mathrm{K}.$$

Proposition 1.1.3 (i) *The map* $\mathbf{G}(\mathbb{A}_f) \longrightarrow \mathcal{C}_P^1$ *that sends* g *to the class of* $(g, 1, \dots, 1)$ *induces a bijection* $\mathbf{P}(\mathbb{Q})\mathbf{Q}_{n_r}(\mathbb{A}_f) \setminus \mathbf{G}(\mathbb{A}_f)/\mathrm{K} \xrightarrow{\sim} \mathcal{C}_P^1$.

(ii) *Define a map* $\varphi' : \mathcal{C}_P^1 \longrightarrow \mathcal{C}_P$ *in the following way. Let* $(g_1, \dots, g_r) \in \mathbf{G}(\mathbb{A}_f) \times \mathbf{Q}_{n_1}(\mathbb{A}_f) \times \dots \times \mathbf{Q}_{n_{r-1}}(\mathbb{A}_f)$. *For every* $i \in \{1, \dots, r\}$, *write*

$$\mathrm{H}_i = (g_i \dots g_1)\mathrm{K}(g_i \dots g_1)^{-1} \cap (\mathbf{P}_{n_1} \cap \dots \cap \mathbf{P}_{n_i})(\mathbb{Q})\mathbf{Q}_{n_i}(\mathbb{A}_f),$$

and let K_i *be the image of* $\mathrm{H}_i \cap \mathbf{Q}_{n_i}(\mathbb{A}_f)$ *by the obvious morphism* $\mathbf{Q}_{n_i}(\mathbb{A}_f) \longrightarrow \mathbf{G}_{n_i}(\mathbb{A}_f)$. *Then* φ' *sends the class of* $(g_1, \dots, g_r)$ *to the* n*-tuple* $(X_1, \dots, X_r)$, *where* $X_1 = \mathrm{Im}(i_{n_1,g_1}) = M^{\mathrm{K}_1}(\mathbf{G}_{n_1}, \mathcal{X}_{n_1})/\mathrm{H}_1$ *and, for every* $i \in \{1, \dots, r-1\}$, X_{i+1} *is the boundary stratum of* $X_i = M^{\mathrm{K}_i}(\mathbf{G}_{n_i}, \mathcal{X}_{n_i})/\mathrm{H}_i$ *image of the morphism* $i_{P',g}$, *with* $\mathbf{P}' = (\mathbf{P}_{n_{i+1}} \cap \mathbf{Q}_{n_i})/\mathbf{N}_{n_i}$ *(a maximal parabolic subgroup of* $\mathbf{G}_{n_i}$*) and* $g = g_{i+1}\mathbf{N}_{n_i}(\mathbb{A}_f) \in \mathbf{G}_{n_i}(\mathbb{A}_f)$.

Then this map $\mathcal{C}_P^1 \longrightarrow \mathcal{C}_P$ *is well defined and bijective.*

The proposition gives a bijection $\varphi_P : \mathbf{P}(\mathbb{Q})\mathbf{Q}_{n_r}(\mathbb{A}_f) \setminus \mathbf{G}(\mathbb{A}_f)/\mathrm{K} \xrightarrow{\sim} \mathcal{C}_P$. On the other hand, there is a map from $\mathcal{C}_P$ to the set of boundary strata of $M^{\mathrm{K}}(\mathbf{G}, \mathcal{X})^*$ associated to $\mathbf{P}_{n_r}$, defined by sending $(X_1, \dots, X_r)$ to the image of X_r in $M^{\mathrm{K}}(\mathbf{G}, \mathcal{X})^*$. After identifying $\mathcal{C}_P$ to $\mathbf{P}(\mathbb{Q})\mathbf{Q}_{n_r}(\mathbb{A}_f) \setminus \mathbf{G}(\mathbb{A}_f)/\mathrm{K}$ using φ_P and the second set to

$\mathbf{P}_{n_r}(\mathbb{Q})\mathbf{Q}_{n_r}(\mathbb{A}_f) \setminus \mathbf{G}(\mathbb{A}_f)/\mathrm{K}$ using $g \longmapsto \mathrm{Im}(i_{n_r,g})$, this map becomes the obvious projection $\mathbf{P}(\mathbb{Q})\mathbf{Q}_{n_r}(\mathbb{A}_f) \setminus \mathbf{G}(\mathbb{A}_f)/\mathrm{K} \longrightarrow \mathbf{P}_{n_r}(\mathbb{Q})\mathbf{Q}_{n_r}(\mathbb{A}_f) \setminus \mathbf{G}(\mathbb{A}_f)/\mathrm{K}$.

Proof.

(i) As $\mathbf{Q}_{n_r} \subset \mathbf{Q}_{n_{r-1}} \subset \cdots \subset \mathbf{Q}_{n_1}$, it is easy to see that, in the definition of $\mathcal{C}^1_P$, $(g_1, \ldots, g_r)$ is equivalent to $(g'_1, \ldots, g'_r)$ if and only if

$$(\mathbf{P}_{n_1} \cap \cdots \cap \mathbf{P}_{n_r})(\mathbb{Q})\mathbf{Q}_{n_r}(\mathbb{A}_f)g_r \ldots g_1\mathrm{K}$$
$$= (\mathbf{P}_{n_1} \cap \cdots \cap \mathbf{P}_{n_r})(\mathbb{Q})\mathbf{Q}_{n_r}(\mathbb{A}_f)g'_r \ldots g'_1\mathrm{K}.$$

The result now follows from the fact that $\mathbf{P} = \mathbf{P}_{n_1} \cap \cdots \cap \mathbf{P}_{n_r}$.

(ii) We first check that φ' is well defined. Let $i \in \{1, \ldots, r-1\}$. If $X_i = M^{\mathrm{K}_i}(\mathbf{G}_{n_i}, \mathcal{X}_{n_i})/\mathrm{H}_i$ and X_{i+1} is the boundary stratum $\mathrm{Im}(i_{P',g})$ of X_i, with $\mathbf{P}'$ and g as in the proposition, then $X_{i+1} = M^{\mathrm{K}'}(\mathbf{G}_{n_{i+1}}, \mathcal{X}_{n_{i+1}})/\mathrm{H}'$, where $\mathrm{H}' = g_{i+1}\mathrm{H}_i g_{i+1}^{-1} \cap \mathbf{P}_{n_{i+1}}(\mathbb{Q})\mathbf{Q}_{n_{i+1}}(\mathbb{A}_f)$ and K' is the image of $\mathrm{H}' \cap \mathbf{Q}_{n_{i+1}}(\mathbb{A}_f)$ in $\mathbf{G}_{n_{i+1}}(\mathbb{A}_f)$. As $g_{i+1} \in \mathbf{Q}_{n_i}(\mathbb{A}_f)$,

$$\mathrm{H}' = (g_{i+1} \ldots g_1)\mathrm{K}(g_{i+1} \ldots g_1)^{-1} \cap (\mathbf{P}_{n_1} \cap \cdots \cap \mathbf{P}_{n_i})(\mathbb{Q})\mathbf{Q}_{n_i}(\mathbb{A}_f)$$
$$\cap \mathbf{P}_{n_{i+1}}(\mathbb{Q})\mathbf{Q}_{n_{i+1}}(\mathbb{A}_f).$$

On the other hand, it is easy to see that

$$(\mathbf{P}_{n_1} \cap \cdots \cap \mathbf{P}_{n_i})(\mathbb{Q})\mathbf{Q}_{n_i}(\mathbb{A}_f) \cap \mathbf{P}_{n_{i+1}}(\mathbb{Q})\mathbf{Q}_{n_{i+1}}(\mathbb{A}_f)$$
$$= (\mathbf{P}_{n_1} \cap \cdots \cap \mathbf{P}_{n_{i+1}})(\mathbb{Q})\mathbf{Q}_{n_{i+1}}(\mathbb{A}_f).$$

Hence $\mathrm{H}' = \mathrm{H}_{i+1}$, and $X_{i+1} = M^{\mathrm{K}_{i+1}}(\mathbf{G}_{n_{i+1}}, \mathcal{X}_{n_{i+1}})/\mathrm{H}_{i+1}$. It is also clear that the n-tuple $(X_1, \ldots, X_r)$ defined in the proposition does not change if $(g_1, \ldots, g_r)$ is replaced by an equivalent r-tuple.

It is clear that φ' is surjective. We want to show that it is injective. Let $c, c' \in \mathcal{C}^1_P$; write $(X_1, \ldots, X_r) = \varphi'(c)$ and $(X'_1, \ldots, X'_r) = \varphi'(c')$, and suppose that $(X_1, \ldots, X_r) = (X'_1, \ldots, X'_r)$. Fix representatives $(g_1, \ldots, g_n)$ and $(g'_1, \ldots, g'_n)$ of c and c'. As before, write, for every $i \in \{1, \ldots, n\}$,

$$\mathrm{H}_i = (g_i \ldots g_1)\mathrm{K}(g_i \ldots g_1)^{-1} \cap (\mathbf{P}_{n_1} \cap \cdots \cap \mathbf{P}_{n_i})(\mathbb{Q})\mathbf{Q}_{n_i}(\mathbb{A}_f),$$
$$\mathrm{H}'_i = (g'_i \ldots g'_1)\mathrm{K}(g'_i \ldots g'_1)^{-1} \cap (\mathbf{P}_{n_1} \cap \cdots \cap \mathbf{P}_{n_i})(\mathbb{Q})\mathbf{Q}_{n_i}(\mathbb{A}_f).$$

Then the equality $X_1 = X'_1$ implies that $\mathbf{P}_{n_1}(\mathbb{Q})\mathbf{Q}_{n_1}(\mathbb{A}_f)g_1\mathrm{K} = \mathbf{P}_{n_1}(\mathbb{Q})\mathbf{Q}_{n_1}(\mathbb{A}_f)g'_1\mathrm{K}$ and, for every $i \in \{1, \ldots, r-1\}$, the equality $X_{i+1} = X'_{i+1}$ implies that

$$\mathbf{P}_{n_{i+1}}(\mathbb{Q})\mathbf{Q}_{n_{i+1}}(\mathbb{A}_f)g_{i+1}\mathrm{H}_i(g_i \ldots g_1) = \mathbf{P}_{n_{i+1}}(\mathbb{Q})\mathbf{Q}_{n_{i+1}}(\mathbb{A}_f)g'_{i+1}\mathrm{H}'_i(g'_i \ldots g'_1).$$

So $(g_1, \ldots, g_r)$ and $(g'_1, \ldots, g'_r)$ are equivalent, and $c = c'$. □

1.2 LOCAL SYSTEMS AND PINK'S THEOREM

Fix a number field K. If $\mathbf{G}$ is a linear algebraic group over $\mathbb{Q}$, let $\mathrm{Rep}_{\mathbf{G}}$ be the category of algebraic representations of $\mathbf{G}$ defined over K. Fix a prime number ℓ and a place λ of K over ℓ.

Let $\mathbf{M}$ be a connected reductive group over $\mathbb{Q}$, $\mathbf{L}$ and $\mathbf{G}$ connected reductive subgroups of $\mathbf{M}$ such that $\mathbf{M}$ is the direct product of $\mathbf{L}$ and $\mathbf{G}$, and $(\mathbf{G}, \mathcal{X})$ a Shimura datum. Extend the $\mathbf{G}(\mathbb{A}_f)$-action on $M(\mathbf{G}, \mathcal{X})$ to an $\mathbf{M}(\mathbb{A}_f)$-action by the obvious map $\mathbf{M}(\mathbb{A}_f) \longrightarrow \mathbf{G}(\mathbb{A}_f)$ (so $\mathbf{L}(\mathbb{A}_f)$ acts trivially). Let K_M be a neat open compact subgroup of $\mathbf{M}(\mathbb{A}_f)$. Write $\mathrm{H} = \mathrm{K}_M \cap \mathbf{L}(\mathbb{Q})\mathbf{G}(\mathbb{A}_f)$, $\mathrm{H}_L = \mathrm{K}_M \cap \mathbf{L}(\mathbb{Q})$ (an arithmetic subgroup of $\mathbf{L}(\mathbb{Q})$), and $\mathrm{K} = \mathrm{K}_M \cap \mathbf{G}(\mathbb{A}_f)$. The group H acts on the Shimura variety $M^{\mathrm{K}}(\mathbf{G}, \mathcal{X})$, and the quotient $M^{\mathrm{K}}(\mathbf{G}, \mathcal{X})/\mathrm{H}$ is equal to $M^{\mathrm{H}/\mathrm{H}_L}(\mathbf{G}, \mathcal{X})$ (H/H_L is a neat open compact subgroup of $\mathbf{G}(\mathbb{A}_f)$).

Remark 1.2.1 It is possible to generalize the morphisms T_g of section 1.1: If $m \in \mathbf{L}(\mathbb{Q})\mathbf{G}(\mathbb{A}_f)$ and K'_M is an open compact subgroup of $\mathbf{M}(\mathbb{A}_f)$ such that $\mathrm{K}'_M \cap \mathbf{L}(\mathbb{Q})\mathbf{G}(\mathbb{A}_f) \subset m\mathrm{H}m^{-1}$, then there is a morphism

$$T_m : M(\mathbf{G}, \mathcal{X})/\mathrm{H}' \longrightarrow M(\mathbf{G}, \mathcal{X})/\mathrm{H},$$

where $\mathrm{H}' = \mathrm{K}'_M \cap \mathbf{L}(\mathbb{Q})\mathbf{G}(\mathbb{A}_f)$ and $\mathrm{H} = \mathrm{K}_M \cap \mathbf{L}(\mathbb{Q})\mathbf{G}(\mathbb{A}_f)$. This morphism is simply the one induced by the injection $\mathrm{H}' \longrightarrow \mathrm{H}$, $h \longmapsto mhm^{-1}$ (equivalently, it is induced by the endomorphism $x \longmapsto xm$ of $M(\mathbf{G}, \mathcal{X})$).

There is an additive triangulated functor $V \longmapsto \mathcal{F}^{\mathrm{H}/\mathrm{H}_L} R\Gamma(\mathrm{H}_L, V)$ from the category $D^b(\mathrm{Rep}_{\mathbf{M}})$ to the category of λ-adic complexes on $M^{\mathrm{K}}(\mathbf{G}, \mathcal{X})/\mathrm{H}$,[1] constructed using the functors $\mu_{\Gamma,\varphi}$ of Pink (cf. [P1] (1.10)) for the profinite étale (and Galois of group H/H_L) covering $M(\mathbf{G}, \mathcal{X}) \longrightarrow M^{\mathrm{K}}(\mathbf{G}, \mathcal{X})/\mathrm{H}$ and the properties of the arithmetic subgroups of $\mathbf{L}(\mathbb{Q})$. This construction is explained in [M1] 2.1.4. For every $V \in \mathrm{Ob}\, D^b(\mathrm{Rep}_{\mathbf{M}})$ and $k \in \mathbb{Z}$, $\mathrm{H}^k \mathcal{F}^{\mathrm{H}/\mathrm{H}_L} R\Gamma(\mathrm{H}_L, V)$ is a lisse λ-adic sheaf on $M^{\mathrm{K}}(\mathbf{G}, \mathcal{X})/\mathrm{H}$, whose fiber is (noncanonically) isomorphic to

$$\bigoplus_{i+j=k} \mathrm{H}^i(\mathrm{H}_L, \mathrm{H}^j V).$$

Remark 1.2.2 If Γ is a neat arithmetic subgroup of $\mathbf{L}(\mathbb{Q})$ (e.g., $\Gamma = \mathrm{H}_L$), then it is possible to compute $R\Gamma(\Gamma, V)$ in the category $D^b(\mathrm{Rep}_{\mathbf{G}})$, because Γ is of type FL (cf. [BuW], theorem 3.14).

We will now state a theorem of Pink about the direct image of the complexes $\mathcal{F}^{\mathrm{H}/\mathrm{H}_L} R\Gamma(\mathrm{H}_L, V)$ by the open immersion $j : M^{\mathrm{K}}(\mathbf{G}, \mathcal{X})/\mathrm{H} \longrightarrow M^{\mathrm{K}}(\mathbf{G}, \mathcal{X})^*/\mathrm{H}$. Let $\mathbf{P}$ be an admissible parabolic subgroup of $\mathbf{G}$ and $g \in \mathbf{G}(\mathbb{A}_f)$. Write

$$\mathrm{H}_P = g\mathrm{H}g^{-1} \cap \mathbf{L}(\mathbb{Q})\mathbf{P}(\mathbb{Q})\mathbf{Q}_P(\mathbb{A}_f),$$

$$\mathrm{H}_{P,L} = g\mathrm{H}g^{-1} \cap \mathbf{L}(\mathbb{Q})\mathbf{L}_P(\mathbb{Q})\mathbf{N}_P(\mathbb{A}_f),$$

$$\mathrm{K}_N = g\mathrm{H}g^{-1} \cap \mathbf{N}_P(\mathbb{A}_f),$$

$$\mathrm{K}_G = (g\mathrm{H}g^{-1} \cap \mathbf{Q}_P(\mathbb{A}_f))/(g\mathrm{H}g^{-1} \cap \mathbf{N}_P(\mathbb{A}_f)),$$

and $i = i_{P,g} : M^{\mathrm{K}_G}(\mathbf{G}_P, \mathcal{X}_P)/\mathrm{H}_P \longrightarrow M^{\mathrm{K}}(\mathbf{G}, \mathcal{X})^*/\mathrm{H}$.

Then theorem 4.2.1 of [P2] implies the following result (cf. [M1] 2.2).

[1] Here, and in the rest of the book, the notation $R\Gamma$ will be used to denote the right derived functor of the functor H^0.

Theorem 1.2.3 *For every* $V \in \operatorname{Ob} D^b(\mathrm{Rep}_{\mathbf{M}})$*, there are canonical isomorphisms*

$$i^* Rj_* \mathcal{F}^{\mathrm{H}/\mathrm{H}_L} R\Gamma(\mathrm{H}_L, V) \simeq \mathcal{F}^{\mathrm{H}_P/\mathrm{H}_{P,L}} R\Gamma(\mathrm{H}_{P,L}, V)$$
$$\simeq \mathcal{F}^{\mathrm{H}_P/\mathrm{H}_{P,L}} R\Gamma(\mathrm{H}_{P,L}/\mathrm{K}_N, R\Gamma(\mathrm{Lie}(\mathbf{N}_P), V)).$$

The last isomorphism uses van Est's theorem, as stated (and proved) in [GHM] 24.

We will also use local systems on locally symmetric spaces that are not necessarily hermitian. We will need the following notation. Let $\mathbf{G}$ be a connected reductive group over $\mathbb{Q}$. Fix a maximal compact subgroup K_∞ of $\mathbf{G}(\mathbb{R})$. Let $\mathbf{A}_G$ be the maximal ($\mathbb{Q}$-)split torus of the center of $\mathbf{G}$, $\mathcal{X} = \mathbf{G}(\mathbb{R})/\mathrm{K}_\infty \mathbf{A}_G(\mathbb{R})^0$ and $q(\mathbf{G}) = \dim(\mathcal{X})/2 \in \frac{1}{2}\mathbb{Z}$. Write

$$M^{\mathrm{K}}(\mathbf{G}, \mathcal{X})(\mathbb{C}) = \mathbf{G}(\mathbb{Q}) \setminus (\mathcal{X} \times \mathbf{G}(\mathbb{A}_f)/\mathrm{K})$$

(even though $(\mathbf{G}, \mathcal{X})$ is not a Shimura datum in general, and $\mathbf{M}^{\mathrm{K}}(\mathbf{G}, \mathcal{X})(\mathbb{C})$ is not always the set of complex points of an algebraic variety). If K is small enough (e.g., neat), this quotient is a real analytic variety. There are morphisms T_g ($g \in \mathbf{G}(\mathbb{A}_f)$) defined exactly as in section 1.1.

Let $V \in \operatorname{Ob} \mathrm{Rep}_{\mathbf{G}}$. Let $\mathcal{F}^{\mathrm{K}} V$ be the sheaf of local sections of the morphism

$$\mathbf{G}(\mathbb{Q}) \setminus (V \times \mathcal{X} \times \mathbf{G}(\mathbb{A}_f)/\mathrm{K}) \longrightarrow \mathbf{G}(\mathbb{Q}) \setminus (\mathcal{X} \times \mathbf{G}(\mathbb{A}_f)/\mathrm{K})$$

(where $\mathbf{G}(\mathbb{Q})$ acts on $V \times \mathcal{X} \times \mathbf{G}(\mathbb{A}_f)/\mathrm{K}$ by $(\gamma, (v, x, g\mathrm{K})) \longmapsto (\gamma.v, \gamma.x, \gamma g\mathrm{K})$). As suggested by the notation, there is a connection between this sheaf and the local systems defined above: if $(\mathbf{G}, \mathcal{X})$ is a Shimura datum, then $\mathcal{F}^{\mathrm{K}} V \otimes K_\lambda$ is the inverse image on $M^{\mathrm{K}}(\mathbf{G}, \mathcal{X})(\mathbb{C})$ of the λ-adic sheaf $\mathcal{F}^{\mathrm{K}} V$ on $M^{\mathrm{K}}(\mathbf{G}, \mathcal{X})$ (cf. [L1], p. 38 or [M1] 2.1.4.1).

Let Γ be a neat arithmetic subgroup of $\mathbf{G}(\mathbb{Q})$. Then the quotient $\Gamma \setminus \mathcal{X}$ is a real analytic variety. For every $V \in \operatorname{Ob} \mathrm{Rep}_{\mathbf{G}}$, let $\mathcal{F}^\Gamma V$ be the sheaf of local sections of the morphism

$$\Gamma \setminus (V \times \mathcal{X}) \longrightarrow \Gamma \setminus \mathcal{X}$$

(where Γ acts on $V \times \mathcal{X}$ by $(\gamma, (v, x)) \longmapsto (\gamma.v, \gamma.x)$).

Let K be a neat open compact subgroup of $\mathbf{G}(\mathbb{A}_f)$, and let $(g_i)_{i \in I}$ be a system of representatives of the double quotient $\mathbf{G}(\mathbb{Q}) \setminus \mathbf{G}(\mathbb{A}_f)/\mathrm{K}$. For every $i \in I$, let $\Gamma_i = g_i \mathrm{K} g_i^{-1} \cap \mathbf{G}(\mathbb{Q})$. Then the Γ_i are neat arithmetic subgroups of $\mathbf{G}(\mathbb{Q})$,

$$M^{\mathrm{K}}(\mathbf{G}, \mathcal{X})(\mathbb{C}) = \coprod_{i \in I} \Gamma_i \setminus \mathcal{X},$$

and, for every $V \in \operatorname{Ob} \mathrm{Rep}_{\mathbf{G}}$,

$$\mathcal{F}^{\mathrm{K}} V = \bigoplus_{i \in I} \mathcal{F}^{\Gamma_i} V.$$

1.3 INTEGRAL MODELS

Notation is as in section 1.1. Let $(\mathbf{G}, \mathcal{X})$ be a Shimura datum such that $\mathbf{G}^{\mathrm{ad}}$ is simple and that the maximal parabolic subgroups of $\mathbf{G}$ satisfy the condition of section 1.1.

The goal here is to show that there exist integral models (i.e., models over a suitable localization of $\mathcal{O}_F$) of the varieties and sheaves of sections 1.1 and 1.2 such that Pink's theorem is still true. The exact conditions that we want these models to satisfy are given more precisely below (conditions (1)–(7)).

Fix a minimal parabolic subgroup $\mathbf{P}_0$ of $\mathbf{G}$, and let $(\mathbf{G}_1, \mathcal{X}_1), \dots, (\mathbf{G}_n, \mathcal{X}_n)$ be the Shimura data associated to the standard maximal parabolic subgroups of $\mathbf{G}$. We will also write $(\mathbf{G}_0, \mathcal{X}_0) = (\mathbf{G}, \mathcal{X})$. Note that, for every $i \in \{0, \dots, n\}$, $\mathbf{P}_0$ determines a minimal parabolic subgroup of $\mathbf{G}_i$. It is clear that, for every $i \in \{0, \dots, n\}$, the Shimura data associated to the standard maximal parabolic subgroups of $\mathbf{G}_i$ are the $(\mathbf{G}_j, \mathcal{X}_j)$, with $i + 1 \leq j \leq n$.

Remember that F is the reflex field of $(\mathbf{G}, \mathcal{X})$. It is also the reflex field of all the $(\mathbf{G}_i, \mathcal{X}_i)$ ([P1] 12.1 and 11.2(c)). Let $\overline{\mathbb{Q}}$ be the algebraic closure of $\mathbb{Q}$ in $\mathbb{C}$; as F is by definition a subfield of $\mathbb{C}$, it is included in $\overline{\mathbb{Q}}$. For every prime number p, fix an algebraic closure $\overline{\mathbb{Q}}_p$ of $\mathbb{Q}_p$ and an injection $F \subset \overline{\mathbb{Q}}_p$.

Fix a point x_0 of $\mathcal{X}$, and let $h_0 : \mathbb{S} \longrightarrow \mathbf{G}_\mathbb{R}$ be the morphism corresponding to x_0. Let w be the composition of h_0 and of the injection $\mathbb{G}_{m,\mathbb{R}} \subset \mathbb{S}$. Then w is independent of the choice of h_0, and it is defined over $\mathbb{Q}$ (cf. [P2] 5.4). An algebraic representation $\rho : \mathbf{G} \longrightarrow \mathbf{GL}(V)$ of $\mathbf{G}$ is said to be *pure of weight m* if $\rho \circ w$ is the multiplication by the character $\lambda \longmapsto \lambda^m$ of $\mathbb{G}_m$ (note that the sign convention here is not the same as in [P2] 5.4).

Consider the following data:

- for every $i \in \{0, \dots, n\}$, a set $\mathcal{K}_i$ of neat open compact subgroups of $\mathbf{G}_i(\mathbb{A}_f)$, stable by $\mathbf{G}(\mathbb{A}_f)$-conjugacy;
- for every $i \in \{0, \dots, n\}$, a subset A_i of $\mathbf{G}_i(\mathbb{A}_f)$ such that $1 \in A_i$;
- for every $i \in \{0, \dots, n\}$, a full abelian subcategory $\mathcal{R}_i$ of $\mathrm{Rep}_{\mathbf{G}_i}$, stable by taking direct factors.

These data should satisfy the following conditions. Let $i, j \in \{0, \dots, n\}$ be such that $j > i$ and $\mathrm{K} \in \mathcal{K}_i$. Let $\mathbf{P}$ be the standard maximal parabolic subgroup of $\mathbf{G}_i$ associated to $(\mathbf{G}_j, \mathcal{X}_j)$.

Then

(a) For every $g \in \mathbf{G}_i(\mathbb{A}_f)$,

$$(g\mathrm{K}g^{-1} \cap \mathbf{Q}_P(\mathbb{A}_f))/(g\mathrm{K}g^{-1} \cap \mathbf{N}_P(\mathbb{A}_f)) \in \mathcal{K}_j,$$

and, for every $g \in \mathbf{G}_i(\mathbb{A}_f)$ and every standard parabolic subgroup $\mathbf{P}'$ of $\mathbf{G}_i$ such that $\mathbf{Q}_P \subset \mathbf{P}' \subset \mathbf{P}$,

$$(g\mathrm{K}g^{-1} \cap \mathbf{P}'(\mathbb{Q})\mathbf{N}_{P'}(\mathbb{A}_f)\mathbf{Q}_P(\mathbb{A}_f))/(g\mathrm{K}g^{-1} \cap \mathbf{L}_{P'}(\mathbb{Q})\mathbf{N}_{P'}(\mathbb{A}_f)) \in \mathcal{K}_j,$$

$$(g\mathrm{K}g^{-1} \cap \mathbf{P}'(\mathbb{A}_f))/(g\mathrm{K}g^{-1} \cap \mathbf{L}_{P'}(\mathbb{A}_f)\mathbf{N}_{P'}(\mathbb{A}_f)) \in \mathcal{K}_j.$$

(b) Let $g \in A_i$ and $\mathrm{K}' \in \mathcal{K}_i$ be such that $\mathrm{K}' \subset g\mathrm{K}g^{-1}$. Let $h \in \mathbf{P}(\mathbb{Q})\mathbf{Q}_P(\mathbb{A}_f) \setminus \mathbf{G}(\mathbb{A}_f)/\mathrm{K}$ and $h' \in \mathbf{P}(\mathbb{Q})\mathbf{Q}_P(\mathbb{A}_f) \setminus \mathbf{G}(\mathbb{A}_f)/\mathrm{K}'$ be such that $\mathbf{P}(\mathbb{Q})\mathbf{Q}_P(\mathbb{A}_f)h\mathrm{K} = \mathbf{P}(\mathbb{Q})\mathbf{Q}_P(\mathbb{A}_f)h'g\mathrm{K}$. Then there exist $p \in \mathbf{L}_P(\mathbb{Q})$ and $q \in \mathbf{Q}_P(\mathbb{A}_f)$ such that $pqh\mathrm{K} = h'g\mathrm{K}$ and the image of q in $\mathbf{G}_j(\mathbb{A}_f) = \mathbf{Q}_P(\mathbb{A}_f)/\mathbf{N}_P(\mathbb{A}_f)$ is in A_j.

(c) For every $g \in \mathbf{G}_i(\mathbb{A}_f)$ and $V \in \mathrm{Ob}\,\mathcal{R}_i$,

$$R\Gamma(\Gamma_L, R\Gamma(\mathrm{Lie}(\mathbf{N}_P), V)) \in \mathrm{Ob}\, D^b(\mathcal{R}_j),$$

where

$$\Gamma_L = (g\mathrm{K}g^{-1} \cap \mathbf{P}(\mathbb{Q})\mathbf{Q}_P(\mathbb{A}_f))/(g\mathrm{K}g^{-1} \cap \mathbf{Q}_P(\mathbb{A}_f)).$$

Let Σ be a finite set of prime numbers such that the groups $\mathbf{G}_0, \ldots, \mathbf{G}_n$ are unramified outside Σ. For every $p \notin \Sigma$, fix $\mathbb{Z}_p$-models of these groups such that the group of $\mathbb{Z}_p$-points is hyperspecial. Let

$$\mathbb{A}_\Sigma = \prod_{p \in \Sigma} \mathbb{Q}_p.$$

Fix $\ell \in \Sigma$ and a place λ of K above ℓ, and consider the following conditions on Σ:

(1) For every $i \in \{0, \ldots, n\}$, $A_i \subset \mathbf{G}_i(\mathbb{A}_\Sigma)$ and every $\mathbf{G}(\mathbb{A}_f)$-conjugacy class in $\mathcal{K}_i$ has a representative of the form $\mathrm{K}_\Sigma \mathrm{K}^\Sigma$, with $\mathrm{K}_\Sigma \subset \mathbf{G}_i(\mathbb{A}_\Sigma)$ and $\mathrm{K}^\Sigma = \prod_{p \notin \Sigma} \mathbf{G}_i(\mathbb{Z}_p)$.
(2) For every $i \in \{0, \ldots, n\}$ and $\mathrm{K} \in \mathcal{K}_i$, there exists a smooth quasi-projective scheme $\mathcal{M}^{\mathrm{K}}(\mathbf{G}_i, \mathcal{X}_i)$ over $\mathrm{Spec}(\mathcal{O}_F[1/\Sigma])$ whose generic fiber is $M^{\mathrm{K}}(\mathbf{G}_i, \mathcal{X}_i)$.
(3) For every $i \in \{0, \ldots, n\}$ and $\mathrm{K} \in \mathcal{K}_i$, there exists a normal scheme $\mathcal{M}^{\mathrm{K}}(\mathbf{G}_i, \mathcal{X}_i)^*$, projective over $\mathrm{Spec}(\mathcal{O}_F[1/\Sigma])$, containing $\mathcal{M}^{\mathrm{K}}(\mathbf{G}_i, \mathcal{X}_i)$ as a dense open subscheme and with generic fiber $M^{\mathrm{K}}(\mathbf{G}_i, \mathcal{X}_i)^*$. Moreover, the morphisms $i_{P,g}$ (resp., $\overline{i}_{P,g}$) of section 1.1 extend to locally closed immersions (resp., finite morphisms) between the models over $\mathrm{Spec}(\mathcal{O}_F[1/\Sigma])$, and the boundary of $\mathcal{M}^{\mathrm{K}}(\mathbf{G}_i, \mathcal{X}_i)^* - \mathcal{M}^{\mathrm{K}}(\mathbf{G}_i, \mathcal{X}_i)$ is still the disjoint union of the images of the immersions $i_{P,g}$.
(4) For every $i \in \{0, \ldots, n\}$, $g \in A_i$ and $\mathrm{K}, \mathrm{K}' \in \mathcal{K}_i$ such that $\mathrm{K}' \subset g\mathrm{K}g^{-1}$, the morphism $\overline{T}_g : M^{\mathrm{K}'}(\mathbf{G}_i, \mathcal{X}_i)^* \longrightarrow M^{\mathrm{K}}(\mathbf{G}_i, \mathcal{X}_i)^*$ extends to a finite morphism $\mathcal{M}^{\mathrm{K}'}(\mathbf{G}_i, \mathcal{X}_i)^* \longrightarrow \mathcal{M}^{\mathrm{K}}(\mathbf{G}_i, \mathcal{X}_i)^*$, which will still be denoted by $\overline{T}_g$, whose restriction to the strata of $\mathcal{M}^{\mathrm{K}'}(\mathbf{G}_i, \mathcal{X}_i)^*$ (including the open stratum $\mathcal{M}^{\mathrm{K}'}(\mathbf{G}_i, \mathcal{X}_i)$) is étale.
(5) For every $i \in \{0, \ldots, n\}$ and $\mathrm{K} \in \mathcal{K}_i$, there exists a functor $\mathcal{F}^{\mathrm{K}}$ from $\mathcal{R}_i$ to the category of lisse λ-adic sheaves on $\mathcal{M}^{\mathrm{K}}(\mathbf{G}_i, \mathcal{X}_i)$ that, after passing to the special fiber, is isomorphic to the functor $\mathcal{F}^{\mathrm{K}}$ of section 1.2.
(6) For every $i \in \{0, \ldots, n\}$, $\mathrm{K} \in \mathcal{K}_i$, and $V \in \mathrm{Ob}\,\mathcal{R}_i$, the isomorphisms of Pink's theorem (1.2.3) extend to isomorphisms between λ-adic complexes on the $\mathrm{Spec}(\mathcal{O}_F[1/\Sigma])$-models.
(7) For every $i \in \{0, \ldots, n\}$, $\mathrm{K} \in \mathcal{K}_i$ and $V \in \mathrm{Ob}\,\mathcal{R}_i$, the sheaf $\mathcal{F}^{\mathrm{K}}V$ on $\mathcal{M}^{\mathrm{K}}(\mathbf{G}_i, \mathcal{X}_i)$ is mixed ([D2] 1.2.2). If moreover V is pure of weight m, then $\mathcal{F}^{\mathrm{K}}V$ is pure of weight $-m$.

The fact that suitable integral models exist for PEL Shimura varieties has been proved by Kai-Wen Lan, who constructed the toroidal and minimal compactifications of the integral models.

Proposition 1.3.1 *Suppose that the Shimura datum* $(\mathbf{G}, \mathcal{X})$ *is of the type considered in [K11] §5; more precisely, we suppose fixed data as in [Lan] 1.2. Let* Σ *be a finite set of prime numbers that contains all bad primes (in the sense of [Lan] 1.4.1.1). For every* $i \in \{0, \dots, n\}$*, let* $A_i = \mathbf{G}_i(\mathbb{A}_\Sigma)$*, let* $\mathcal{K}_i$ *be the union of the* $\mathbf{G}_i(\mathbb{A}_f)$*-conjugacy classes of neat open compact subgroups of the form* $\mathrm{K}_\Sigma \mathrm{K}^\Sigma$ *with* $\mathrm{K}_\Sigma \subset \prod_{p\in\Sigma} \mathbf{G}_i(\mathbb{Z}_p)$ *and* $\mathrm{K}^\Sigma = \prod_{p\notin\Sigma} \mathbf{G}_i(\mathbb{Z}_p)$*, and let* $\mathcal{R}_i = Rep_{\mathbf{G}_i}$*.*

Then the set Σ *satisfies conditions (1)–(7), and moreover the schemes* $\mathcal{M}^{\mathrm{K}}(\mathbf{G}_i, \mathcal{X}_i)$ *of (2) are the schemes representing the moduli problem of [Lan] 1.4.*

Proof. This is just putting together Lan's and Pink's results. Condition (1) is automatic. Condition (2) (in the more precise form given in the proposition) is a consequence of theorem 1.4.1.12 of [Lan]. Conditions (3) and (4) are implied by theorem 7.2.4.1 and proposition 7.2.5.1 of [Lan]. The construction of the sheaves in condition (5) is the same as in [P2] 5.1, once the integral models of condition (2) are known to exist. In [P2] 4.9, Pink observed that the proof of his theorem extends to integral models if toroidal compactifications and a minimal compactification of the integral model satisfying the properties of section 3 of [P2] have been constructed. This has been done by Lan (see, in addition to the results cited above, theorem 6.4.1.1 and propositions 6.4.2.3, 6.4.2.9 and 6.4.3.4 of [Lan]), so condition (6) is also satisfied. In the PEL case, $\mathbf{G}^{\mathrm{ad}}$ is automatically of abelian type in the sense of [P2] 5.6.2 (cf. [K11] §5). So $\mathbf{G}_i^{\mathrm{ad}}$ is of abelian type for all i, and condition (7) is implied by proposition 5.6.2 in [P2]. □

Remark 1.3.2 Let $(\mathbf{G}, \mathcal{X})$ be one of the Shimura data defined in 2.1, and let K be a neat open compact subgroup of $\mathbf{G}(\mathbb{A}_f)$. Then there exists a finite set S of primes such that $\mathrm{K} = \mathrm{K}_S \prod_{p\notin S} \mathbf{G}(\mathbb{Z}_p)$, with $\mathrm{K}_S \subset \prod_{v\in S} \mathbf{G}(\mathbb{Q}_v)$ (and with the $\mathbb{Z}$-structure on $\mathbf{G}$ defined in remark 2.1.1). Let Σ be the union of S and of all prime numbers that are ramified in E. Then Σ contains all bad primes, so proposition 1.3.1 above applies to Σ.

Remark 1.3.3 The convention we use here for the action of the Galois group on the canonical model is that of Pink ([P2] 5.5), which is different from the convention of Deligne (in [D1]) and hence also from the convention of Kottwitz (in [K11]); so what Kottwitz calls canonical model of the Shimura variety associated to the Shimura datum $(\mathbf{G}, \mathcal{X}, h^{-1})$ is here the canonical model of the Shimura variety associated to the Shimura datum $(\mathbf{G}, \mathcal{X}, h)$.

Let us indicate another way to find integral models when the Shimura datum is not necessarily PEL. The problem with this approach is that the set Σ of "bad" primes is unknown.

Proposition 1.3.4 *Let* $\mathcal{K}_i$ *and* A_i *be as above (and satisfying conditions (a) and (b)). Suppose that, for every* $i \in \{0, \dots, n\}$*,* $\mathcal{K}_i$ *is finite modulo* $\mathbf{G}_i(\mathbb{A}_f)$*-conjugacy and* A_i *is finite. If* $\mathbf{G}^{\mathrm{ad}}$ *is of abelian type (in the sense of [P2] 5.6.2), then there exists a finite set* Σ *of prime numbers satisfying conditions (1)–(7), with* $\mathcal{R}_i = Rep_{\mathbf{G}_i}$ *for every* $i \in \{0, \dots, n\}$*.*

In general, there exists a finite set Σ of prime numbers satisfying conditions (1)–(6), with $\mathcal{R}_i = \mathrm{Rep}_{\mathbf{G}_i}$ for every $i \in \{0, \ldots, n\}$. Let $\mathcal{R}'_i$, $0 \leq i \leq n$, be full subcategories of $\mathrm{Rep}_{\mathbf{G}_i}$, stable by taking direct factors and by isomorphism, containing the trivial representation, satisfying condition (c) and minimal for all these properties (this determines the $\mathcal{R}'_i$). Then there exists $\Sigma' \supset \Sigma$ finite such that Σ' and the $\mathcal{R}'_i$ satisfy condition (7).

This proposition will typically be applied to the following situation: $g \in \mathbf{G}(\mathbb{A}_f)$ and K, K$'$ are neat open compact subgroups of $\mathbf{G}(\mathbb{A}_f)$ such that $\mathrm{K}' \subset \mathrm{K} \cap g\mathrm{K}g^{-1}$, and we want to study the Hecke correspondence $(T_g, T_1) : M^{\mathrm{K}'}(\mathbf{G}, \mathcal{X})^* \longrightarrow (M^{\mathrm{K}}(\mathbf{G}, \mathcal{X})^*)^2$. In order to reduce this situation modulo p, choose sets $\mathcal{K}_i$ such that $\mathrm{K}, \mathrm{K}' \in \mathcal{K}_0$ and condition (a) is satisfied, and minimal for these properties, sets A_i such that $1, g \in A_0$ and condition (b) is satisfied, and minimal for these properties; take $\mathcal{R}_i = \mathrm{Rep}_{\mathbf{G}_i}$ if $\mathbf{G}^{\mathrm{ad}}$ is of abelian type and $\mathcal{R}_i$ equal to the $\mathcal{R}'_i$ defined in the proposition in the other cases; fix Σ such that conditions (1)–(7) are satisfied, and reduce modulo $p \notin \Sigma$.

Proof. First we show that, in the general case, there is a finite set Σ of prime numbers satisfying conditions (1)–(6), with $\mathcal{R}_i = \mathrm{Rep}_{\mathbf{G}_i}$. It is obviously possible to find Σ satisfying conditions (1)–(4). Proposition 3.6 of [W] implies that we can find Σ satisfying conditions (1)–(5). To show that there exists Σ satisfying conditions (1)–(6), reason as in the proof of proposition 3.7 of [W], using the generic base change theorem of Deligne (cf. SGA 4 1/2 [Th. finitude] théorème 1.9). As in the proof of proposition 1.3.1, if $\mathbf{G}^{\mathrm{ad}}$ is of abelian type, then condition (7) is true by proposition 5.6.2 of [P2]. In the general case, let $\mathcal{R}'_i$ be defined as in the statement of the proposition. Condition (7) for these subcategories is a consequence of proposition 5.6.1 of [P2] (reason as in the second proof of [P2] 5.6.6). □

Remark 1.3.5 Note that it is clear from the proof that, after replacing Σ by a bigger finite set, we can choose the integral models $\mathcal{M}^{\mathrm{K}}(\mathbf{G}_i, \mathcal{X}_i)$ to be any integral models specified before (as long as they satisfy the conditions of (2)).

When we later talk about reducing Shimura varieties modulo p, we will always implicitly fix Σ as in proposition 1.3.1 (or proposition 1.3.4) and take $p \notin \Sigma$. The prime number ℓ will be chosen among elements of Σ (or added to Σ).

1.4 WEIGHTED COHOMOLOGY COMPLEXES AND INTERSECTION COMPLEX

Let $(\mathbf{G}, \mathcal{X})$ be a Shimura datum and K be a neat open compact subgroup of $\mathbf{G}(\mathbb{A}_f)$. Assume that $\mathbf{G}$ satisfies the conditions of section 1.1 and that $\mathbf{G}^{\mathrm{ad}}$ is simple. Fix a minimal parabolic subgroup $\mathbf{P}_0$ of $\mathbf{G}$ and maximal standard parabolic subgroups $\mathbf{P}_1, \ldots, \mathbf{P}_n$ as before proposition 1.1.3. Fix prime numbers p and ℓ as at the end of section 1.3, and a place λ of K above ℓ. In this section, we will write $\mathbf{M}^{\mathrm{K}}(\mathbf{G}, \mathcal{X})$, etc. for the reduction modulo p of the varieties of 1.1.

Write $M_0 = M^{\mathrm{K}}(\mathbf{G}, \mathcal{X})$ and $d = \dim M_0$, and, for every $r \in \{1, \ldots, n\}$, denote by M_r the union of the boundary strata of $M^{\mathrm{K}}(\mathbf{G}, \mathcal{X})^*$ associated to $\mathbf{P}_r$, by d_r the dimension of M_r and by i_r the inclusion of M_r in $M^{\mathrm{K}}(\mathbf{G},\mathcal{X})^*$. Then $(M_0,\dots,M_n)$ is a stratification of $M^{\mathrm{K}}(\mathbf{G},\mathcal{X})^*$ in the sense of [M2] 3.3.1. Hence, for every $\underline{a}=(a_0,\dots,a_n)\in(\mathbb{Z}\cup\{\pm\infty\})^{n+1}$, the functors $w_{\leq\underline{a}}$ and $w_{>\underline{a}}$ of [M2] 3.3.2 are defined (on the category $D^b_m(M^{\mathrm{K}}(\mathbf{G},\mathcal{X})^*,K_\lambda)$ of mixed λ-adic complexes on $M^{\mathrm{K}}(\mathbf{G},\mathcal{X})^*$). We will recall the definition of the intersection complex and of the weighted cohomology complexes. Remember that j is the open immersion $M^{\mathrm{K}}(\mathbf{G},\mathcal{X})\longrightarrow M^{\mathrm{K}}(\mathbf{G},\mathcal{X})^*$.

Remark 1.4.1 We will need to use the fact that the sheaves $\mathcal{F}^{\mathrm{K}}V$ are mixed with known weights. So we fix categories $\mathcal{R}_0,\dots,\mathcal{R}_n$ as in section 1.3, satisfying conditions (c) and (7) of 1.3. If $\mathbf{G}^{\mathrm{ad}}$ is of abelian type, we can simply take $\mathcal{R}_0=\mathrm{Rep}_{\mathbf{G}}$.

Definition 1.4.2 (i) Let $V\in\mathrm{Ob}\,\mathrm{Rep}_{\mathbf{G}}$. The intersection complex on $M^{\mathrm{K}}(\mathbf{G},\mathcal{X})^*$ with coefficients in V is the complex

$$IC^{\mathrm{K}}V=(j_{!*}(\mathcal{F}^{\mathrm{K}}V[d]))[-d].$$

(ii) (cf. [M2] 4.1.3) Let $t_1,\dots,t_n\in\mathbb{Z}\cup\{\pm\infty\}$. For every $r\in\{1,\dots,n\}$, write $a_r=-t_r+d_r$. Define an additive triangulated functor

$$W^{\geq t_1,\dots,\geq t_n}:D^b(\mathcal{R}_0)\longrightarrow D^b_m(M^{\mathrm{K}}(\mathbf{G},\mathcal{X})^*,K_\lambda)$$

in the following way: for every $m\in\mathbb{Z}$, if $V\in\mathrm{Ob}\,D^b(\mathcal{R}_0)$ is such that all $\mathrm{H}^i V$, $i\in\mathbb{Z}$, are pure of weight m, then

$$W^{\geq t_1,\dots,\geq t_n}V=w_{\leq(-m+d,-m+a_1,\dots,-m+a_n)}Rj_*\mathcal{F}^{\mathrm{K}}V.$$

The definition of the weighted cohomology complex in (ii) was inspired by the work of Goresky, Harder and MacPherson ([GHM]). Proposition 4.1.5 of [M2] admits the following obvious generalization.

Proposition 1.4.3 *Let $t_1,\dots,t_n\in\mathbb{Z}$ be such that, for every $r\in\{1,\dots,n\}$, $d_r-d\leq t_r\leq 1+d_r-d$. Then, for every $V\in\mathrm{Ob}\,\mathcal{R}_0$, there is a canonical isomorphism*

$$IC^{\mathrm{K}}V\simeq W^{\geq t_1,\dots,\geq t_n}V.$$

We now want to calculate the restriction to boundary strata of the weighted cohomology complexes. The following theorem is a consequence of propositions 3.3.4 and 3.4.2 of [M2].

Theorem 1.4.4 *Let $\underline{a}=(a_0,\dots,a_n)\in(\mathbb{Z}\cup\{\pm\infty\})^{n+1}$. Then, for every $L\in\mathrm{Ob}\,D^b_m(M^{\mathrm{K}}(\mathbf{G},\mathcal{X}),K_\lambda)$ such that all perverse cohomology sheaves of L are pure of weight a_0, there is an equality of classes in the Grothendieck group of $D^b_m(M^{\mathrm{K}}(\mathbf{G},\mathcal{X})^*,K_\lambda)$:*

$$[w_{\leq\underline{a}}Rj_*L]=\sum_{1\leq n_1<\dots<n_r\leq n}(-1)^r[i_{n_r!}w_{\leq a_{n_r}}i^!_{n_r}\dots i_{n_1!}w_{\leq a_{n_1}}i^!_{n_1}j_!L].$$

Therefore it is enough to calculate the restriction to boundary strata of the complexes $i_{n_r!}w_{\leq a_{n_r}}i^!_{n_r}\dots i_{n_1!}w_{\leq a_{n_1}}i^!_{n_1}j_!\mathcal{F}^{\mathrm{K}}V$, $1\leq n_1<\dots<n_r\leq n$. The following proposition generalizes proposition 4.2.3 of [M2] and proposition 5.2.3 of [M1].

Proposition 1.4.5 *Let $n_1, \dots, n_r \in \{1, \dots, n\}$ be such that $n_1 < \cdots < n_r$, $a_1, \dots, a_r \in \mathbb{Z} \cup \{\pm\infty\}$, $V \in \operatorname{Ob} D^b(\mathcal{R}_0)$ and $g \in \mathbf{G}(\mathbb{A}_f)$. Write $\mathbf{P} = \mathbf{P}_{n_1} \cap \dots \cap \mathbf{P}_{n_r}$; remember that, in section 1.1, before proposition 1.1.3, we constructed a set $\mathcal{C}_P \simeq \mathbf{P}(\mathbb{Q})\mathbf{Q}_{n_r}(\mathbb{A}_f) \setminus \mathbf{G}(\mathbb{A}_f)/\mathrm{K}$ and a map from this set to the set of boundary strata of $M^{\mathrm{K}}(\mathbf{G}, \mathcal{X})^*$ associated to $\mathbf{P}_{n_r}$. For every $i \in \{1, \dots, r\}$, let $w_i : \mathbb{G}_m \longrightarrow \mathbf{G}_{n_i}$ be the cocharacter associated to the Shimura datum $(\mathbf{G}_{n_i}, \mathcal{X}_{n_i})$ as in section 1.3; the image of w_i is contained in the center of $\mathbf{G}_{n_i}$, and w_i can be seen as a cocharacter of $\mathbf{M}_P$. For every $i \in \{1, \dots, r\}$, write $t_i = -a_i + d_{n_i}$. Let*

$$L = i^*_{n_r,g} R i_{n_r *} w_{>a_r} i^*_{n_r} \dots R i_{n_1 *} w_{>a_1} i^*_{n_1} R j_* \mathcal{F}^{\mathrm{K}} V.$$

Then there is a canonical isomorphism

$$L \simeq \bigoplus_C T_{C*} L_C,$$

where the direct sum is over the set of $C = (X_1, \dots, X_r) \in \mathcal{C}_P$ that are sent to the stratum $\operatorname{Im}(i_{n_r,g})$, T_C is the obvious morphism $X_r \longrightarrow \operatorname{Im}(i_{n_r,g})$ (a finite étale morphism), and L_C is an λ-adic complex on X_r such that, if $h \in \mathbf{G}(\mathbb{A}_f)$ is a representative of C, there is an isomorphism

$$L_C \simeq \mathcal{F}^{\mathrm{H}/\mathrm{H}_L} R\Gamma(\mathrm{H}_L/\mathrm{K}_N, R\Gamma(\operatorname{Lie}(\mathbf{N}_P), V)_{<t_1,\dots,<t_r}),$$

where $\mathrm{H} = h\mathrm{K}h^{-1} \cap \mathbf{P}(\mathbb{Q})\mathbf{Q}_{n_r}(\mathbb{A}_f)$, $\mathrm{H}_L = h\mathrm{K}h^{-1} \cap \mathbf{P}(\mathbb{Q})\mathbf{N}_{n_r}(\mathbb{A}_f) \cap \mathbf{L}_{n_r}(\mathbb{Q})\mathbf{N}_{n_r}(\mathbb{A}_f)$, $\mathrm{K}_N = h\mathrm{K}h^{-1} \cap \mathbf{N}_P(\mathbb{Q})\mathbf{N}_{n_r}(\mathbb{A}_f)$ and, for every $i \in \{1, \dots, r\}$, the subscript "$< t_i$" means that the complex $R\Gamma(\operatorname{Lie}(\mathbf{N}_P), V)$ of representations of $\mathbf{M}_P$ is truncated by the weights of $w_i(\mathbb{G}_m)$ (cf. [M2] 4.1.1).

Remember that the Levi quotient $\mathbf{M}_P$ is the direct product of $\mathbf{G}_{n_r}$ and a Levi subgroup $\mathbf{L}_P$ of $\mathbf{L}_{n_r}$. Write $\Gamma_L = \mathrm{H}_L/\mathrm{K}_N$ and $X_L = \mathbf{L}_P(\mathbb{R})/\mathrm{K}_{L,\infty}\mathbf{A}_{L_P}(\mathbb{R})^0$, where $\mathrm{K}_{L,\infty}$ is a maximal compact subgroup of $\mathbf{L}_P(\mathbb{R})$ and $\mathbf{A}_{L_P}$ is, as in section 1.2, the maximal split subtorus of the center of $\mathbf{L}_P$; also remember that $q(\mathbf{L}_P) = \dim(X_L)/2$. Then Γ_L is a neat arithmetic subgroup of $\mathbf{L}_P(\mathbb{Q})$, and, for every $W \in \operatorname{Ob} D^b(\operatorname{Rep}_{\mathbf{L}_P})$,

$$R\Gamma(\Gamma_L, W) = R\Gamma(\Gamma_L \setminus X_L, \mathcal{F}^{\Gamma_L} W).$$

Write

$$R\Gamma_c(\Gamma_L, W) = R\Gamma_c(\Gamma_L \setminus X_L, \mathcal{F}^{\Gamma_L} W).$$

If $W \in \operatorname{Ob} D^b(\operatorname{Rep}_{\mathbf{M}_P})$, then this complex can be seen as an object of $D^b(\operatorname{Rep}_{\mathbf{G}_{n_r}})$, because it is the dual of $R\Gamma(\Gamma_L, W^*)[\dim(X_L)]$ (where W^* is the contragredient of W). Define in the same way a complex $R\Gamma_c(\mathrm{K}_L, W)$ for K_L a neat open compact subgroup of $\mathbf{L}_P(\mathbb{A}_f)$ and $W \in \operatorname{Ob} D^b(\operatorname{Rep}_{\mathbf{L}_P})$.

Corollary 1.4.6 *Write*

$$M = i^*_{n_r,g} i_{n_r!} w_{\le a_r} i^!_{n_r} \dots i_{n_1!} w_{\le a_1} i^!_{n_1} j_! \mathcal{F}^{\mathrm{K}} V.$$

Then there is a canonical isomorphism

$$M \simeq \bigoplus_C T_{C*} M_C,$$

where the sum is as in the proposition above and, for every $C = (X_1, \dots, X_r) \in \mathcal{C}_P$ *that is sent to the stratum* $\mathrm{Im}(i_{n_r,g})$, M_C *is an* λ*-adic complex on* X_r *such that, if* h *is a representative of* C, *then there is an isomorphism (with the notation of the proposition)*

$$M_C \simeq \mathcal{F}^{\mathrm{H}/\mathrm{H}_L} R\Gamma_c(\mathrm{H}_L/\mathrm{K}_N, R\Gamma(\mathrm{Lie}(\mathbf{N}_P), V)_{\geq t_1,\dots,\geq t_r})[-\dim(\mathbf{A}_{M_P}/\mathbf{A}_G)].$$

Proof. Let V^* be the contragredient of V. The complex dual to M is

$$\begin{aligned} D(M) &= i^*_{n_r,g} R i_{n_r*} w_{\geq -a_r} i^*_{n_r} \dots R i_{n_1*} w_{\geq -a_1} i^*_{n_1} R j_* D(\mathcal{F}^{\mathrm{K}} V) \\ &= i^*_{n_r,g} R i_{n_r*} w_{\geq -a_r} i^*_{n_r} \dots R i_{n_1*} w_{\geq -a_1} i^*_{n_1} R j_* (\mathcal{F}^{\mathrm{K}} V^*[2d](d)) \\ &= (i^*_{n_r,g} R i_{n_r*} w_{\geq 2d-a_r} i^*_{n_r} \dots R i_{n_1*} w_{\geq 2d-a_1} i^*_{n_1} R j_* \mathcal{F}^{\mathrm{K}} V^*)[2d](d). \end{aligned}$$

For every $i \in \{1, \dots, r\}$, let $s_i = -(2d - a_i - 1) + d_{n_i} = 1 - t_i - 2(d - d_{n_i})$. By proposition 1.4.5,

$$D(M) \simeq \bigoplus_C T_{C*} M'_C,$$

with

$$M'_C \simeq \mathcal{F}^{\mathrm{H}/\mathrm{H}_L} R\Gamma(\mathrm{H}_L/\mathrm{K}_N, R\Gamma(\mathrm{Lie}(\mathbf{N}_P), V^*)_{<s_1,\dots,<s_r})[2d](d).$$

Take $M_C = D(M'_C)$. It remains to prove the formula for M_C.

Let $m = \dim(\mathbf{N}_P)$. By lemma (10.9) of [GHM],

$$\begin{aligned} &R\Gamma(\mathrm{Lie}(\mathbf{N}_P), V)_{\geq t_1,\dots,\geq t_r} \\ &\quad \simeq R\,\mathrm{Hom}(R\Gamma(\mathrm{Lie}(\mathbf{N}_P), V^*)_{<s_1,\dots,<s_r}, \mathrm{H}^m(\mathrm{Lie}(\mathbf{N}_P), \mathbb{Q}))[-m], \end{aligned}$$

and $\mathrm{H}^m(\mathrm{Lie}(\mathbf{N}_P), \mathbb{Q})$ is the character $\gamma \longmapsto \det(\mathrm{Ad}(\gamma), \mathrm{Lie}(\mathbf{N}_P))^{-1}$ of $\mathbf{M}_P$ (only the case of groups $\mathbf{G}$ with anisotropic center is treated in [GHM], but the general case is similar). In particular, $\mathrm{H}_L/\mathrm{K}_N$ acts trivially on $\mathrm{H}^m(\mathrm{Lie}(\mathbf{N}_P), \mathbb{Q})$, and the group $w_r(\mathbb{G}_m)$ acts by the character $\lambda \longmapsto \lambda^{2(d-d_{n_r})}$ (w_r is defined as in proposition 1.4.5). Hence

$$M_C \simeq \mathcal{F}^{\mathrm{H}/\mathrm{H}_L} R\Gamma_c(\mathrm{H}_L/\mathrm{K}_N, R\Gamma(\mathrm{Lie}(\mathbf{N}_P), V)_{\geq t_1,\dots,\geq t_r})[a],$$

with

$$\begin{aligned} a &= 2d_{n_r} + m + 2q(\mathbf{L}_P) - 2d = 2q(\mathbf{G}_{n_r}) + 2q(\mathbf{L}_P) + \dim(\mathbf{N}_P) - 2q(\mathbf{G}) \\ &= -\dim(\mathbf{A}_{M_P}/\mathbf{A}_G). \end{aligned}$$

□

Proof of proposition 1.4.5. Let $C = (X_1, \dots, X_r) \in \mathcal{C}_P$. Let I_1 be the locally closed immersion $X_1 \longrightarrow M^{\mathrm{K}}(\mathbf{G}, \mathcal{X})$ and, for every $m \in \{1, \dots, r-1\}$, denote by j_m the open immersion $X_m \longrightarrow X^*_m$ and by I_{m+1} the locally closed immersion $X_{m+1} \longrightarrow X^*_m$ (where X^*_m is the Baily-Borel compactification of X_m). Define a complex L_C on X_r by

$$L_C = w_{>a_r} I^*_r R j_{r-1*} w_{>a_{r-1}} I^*_{r-1} \dots w_{>a_1} I^*_1 R j_* \mathcal{F}^{\mathrm{K}} V.$$

Let us show by induction on r that L is isomorphic to the direct sum of the $T_{C*} L_C$, for $C \in \mathcal{C}_P$ that is sent to the stratum $Y := \mathrm{Im}(i_{n_r,g})$. The statement is obvious if $r = 1$. Suppose that $r \geq 2$ and that the statement is true for $r-1$. Let $Y_1, \dots, Y_m$ be the boundary strata of $M^{\mathrm{K}}(\mathbf{G}, \mathcal{X})^*$ associated to $\mathbf{P}_{n_{r-1}}$ whose adherence contains Y. For every $i \in \{1, \dots, m\}$, let $u_i : Y_i \longrightarrow M^{\mathrm{K}}(\mathbf{G}, \mathcal{X})^*$ be the inclusion, and let

$$L_i = u_i^* Ri_{n_{r-1}*} w_{>a_{r-1}} i_{n_{r-1}}^* \dots Ri_{n_1 *} w_{>a_1} i_{n_1}^* Rj_* \mathcal{F}^{\mathrm{K}} V.$$

It is obvious that

$$L = \bigoplus_{i=1}^{m} i_{n_r,g}^* Ru_{i*} w_{>a_r} L_i.$$

Write $\mathbf{P}' = \mathbf{P}_{n_1} \cap \dots \cap \mathbf{P}_{n_{r-1}}$. Let $i \in \{1, \dots, m\}$. By the induction hypothesis, L_i is isomorphic to the direct sum of the $T_{C'*} L_{C'}$ over the set of $C' \in \mathcal{C}_{P'}$ that are sent to Y_i, where $L_{C'}$ is defined in the same way as L_C. Fix $C' = (X_1, \dots, X_{r-1})$ that is sent to Y_i; let us calculate $i_{n_r,g}^* Ru_{i*} w_{>a_r} T_{C'*} L_{C'}$. There is a commutative diagram, with squares cartesian up to nilpotent elements:

$$\begin{array}{ccccc} Y' & \xrightarrow{I'} & X_{r-1}^* & \xleftarrow{j_{r-1}} & X_{r-1} \\ {\scriptstyle T}\downarrow & & \downarrow{\scriptstyle \overline{T}_{C'}} & & \downarrow{\scriptstyle T_{C'}} \\ Y & \longrightarrow & \overline{Y}_i & \longleftarrow & Y_i \end{array},$$

where Y' is a disjoint union of boundary strata of X_{r-1}^* associated to the parabolic subgroup $(\mathbf{P}_{n_r} \cap \mathbf{Q}_{n_{r-1}})/\mathbf{N}_{r_{n-1}}$. Moreover, the vertical arrows are finite maps, and the maps T and $T_{C'}$ are étale. By the proper base change isomorphism and the fact that the functors $w_{>a}$ commute with taking the direct image by a finite étale morphism, there is an isomorphism

$$i_{n_r,g}^* Ru_{i*} w_{>a_r} T_{C'*} L_{C'} = T_* w_{>a_r} I'^* Rj_{r-1*} L_{C'}.$$

The right-hand side is the direct sum of the complexes

$$(T \circ I_r)_* w_{>a_r} I_r^* Rj_{r-1*} L_{C'} = T_{C*} L_C,$$

for $I_r : X_r \longrightarrow X_{r-1}^*$ in the set of boundary strata of X_{r-1}^* included in Y' and $C = (X_1, \dots, X_r)$. These calculations clearly imply the statement that we were trying to prove.

It remains to prove the formula for L_C given in the proposition. Again, use induction on r. If $r = 1$, the formula for L_C is a direct consequence of Pink's theorem (1.2.3) and of lemma 4.1.2 of [M2]. Supppose that $r \geq 2$ and that the result is known for $r-1$. Let $C = (X_1, \dots, X_r) \in \mathcal{C}_P$, and let $h \in \mathbf{G}(\mathbb{A}_f)$ be a representative of C. Write $\mathbf{P}' = \mathbf{P}_{r_1} \cap \dots \cap \mathbf{P}_{r_{n-1}}$, $C' = (X_1, \dots, X_{r-1})$,

$$\mathrm{H} = h\mathrm{K}h^{-1} \cap \mathbf{P}(\mathbb{Q})\mathbf{Q}_{n_r}(\mathbb{A}_f),$$

$$\mathrm{H}_L = h\mathrm{K}h^{-1} \cap \mathbf{P}(\mathbb{Q})\mathbf{N}_{n_r}(\mathbb{A}_f) \cap \mathbf{L}_{n_r}(\mathbb{Q})\mathbf{N}_{n_r}(\mathbb{A}_f) = \mathrm{H} \cap \mathbf{L}_{n_r}(\mathbb{Q})\mathbf{N}_{n_r}(\mathbb{A}_f),$$

$$\mathrm{K}_N = h\mathrm{K}h^{-1} \cap \mathbf{N}_P(\mathbb{Q})\mathbf{N}_{n_r}(\mathbb{A}_f),$$

$$\mathrm{H}' = h\mathrm{K}h^{-1} \cap \mathbf{P}'(\mathbb{Q})\mathbf{Q}_{n_{r-1}}(\mathbb{A}_f),$$

$$\mathrm{H}'_L = h\mathrm{K}h^{-1} \cap \mathbf{P}'(\mathbb{Q})\mathbf{N}_{n_{r-1}}(\mathbb{A}_f) \cap \mathbf{L}_{n_{r-1}}(\mathbb{Q})\mathbf{N}_{n_{r-1}}(\mathbb{A}_f) = \mathrm{H}' \cap \mathbf{L}_{n_{r-1}}(\mathbb{Q})\mathbf{N}_{n_{r-1}}(\mathbb{A}_f),$$

$$\mathrm{K}'_N = h\mathrm{K}h^{-1} \cap \mathbf{N}_{P'}(\mathbb{Q})\mathbf{N}_{n_{r-1}}(\mathbb{A}_f).$$

By the induction hypothesis, there is a canonical isomorphism

$$L_{C'} \simeq \mathcal{F}^{\mathrm{H}'/\mathrm{H}'_L} R\Gamma(\mathrm{H}'_L/\mathrm{K}'_N, R\Gamma(\mathrm{Lie}(\mathbf{N}_{P'}), V)_{<t_1,\dots,<t_{r-1}}).$$

Applying Pink's theorem, we get a canonical isomorphism

$$L_C \simeq w_{>a_r}\mathcal{F}^{\mathrm{H}/\mathrm{H}_L} R\Gamma(\mathrm{H}_L/\mathrm{H}'_L, R\Gamma(\mathrm{H}'_L/\mathrm{K}'_N, R\Gamma(\mathrm{Lie}(\mathbf{N}_{n_{r-1}}), V)_{<t_1,\dots,<t_{r-1}})).$$

There are canonical isomorphisms

$$\begin{aligned} R\Gamma(\mathrm{H}_L/\mathrm{H}'_L, R\Gamma(\mathrm{H}'_L/\mathrm{K}'_N, -)) &\simeq R\Gamma(\mathrm{H}_L/\mathrm{K}_N, R\Gamma(\mathrm{K}_N/\mathrm{K}'_N, -)) \\ &\simeq R\Gamma(\mathrm{H}_L/\mathrm{K}_N, R\Gamma(\mathrm{Lie}(\mathbf{N}_{n_r}/\mathbf{N}_{n_{r-1}}), -)) \end{aligned}$$

(the last isomorphism comes from van Est's theorem, cf. [GKM] §24). On the other hand, for every $i \in \{1, \dots, r-1\}$, the image of the cocharacter $w_i : \mathbb{G}_m \longrightarrow \mathbf{G}_{n_i}$ is contained in the center of $\mathbf{G}_{n_i}$; hence it commutes with $\mathbf{G}_{n_{r-1}}$. This implies that

$$\begin{aligned} &R\Gamma(\mathrm{Lie}(\mathbf{N}_{n_r}/\mathbf{N}_{n_{r-1}}), R\Gamma(\mathrm{Lie}(\mathbf{N}_{n_{r-1}}), V)_{<t_1,\dots,t_{r-1}}) \\ &\quad = R\Gamma(\mathrm{Lie}(\mathbf{N}_{n_r}), V)_{<t_1,\dots,<t_{r-1}}, \end{aligned}$$

so that

$$L_C \simeq w_{>a_r}\mathcal{F}^{\mathrm{H}/\mathrm{H}_L} R\Gamma(\mathrm{H}_L/\mathrm{K}_N, R\Gamma(\mathrm{Lie}(\mathbf{N}_{n_r}), V)_{<t_1,\dots,<t_{r-1}}).$$

To finish the proof, it suffices to apply lemma 4.1.2 of [M2] and to notice that the image of $w_r : \mathbb{G}_m \longrightarrow \mathbf{G}_{n_r}$ commutes with $\mathbf{L}_{n_r}(\mathbb{Q})$, hence also with its subgroup $\mathrm{H}_L/\mathrm{K}_N$. □

1.5 COHOMOLOGICAL CORRESPONDENCES

Notation 1.5.1 Let $(T_1, T_2) : X' \longrightarrow X_1 \times X_2$ be a correspondence of separated schemes of finite type over a finite field, and let $c : T_1^*L_1 \longrightarrow T_2^!L_2$ be a cohomological correspondence with support in (T_1, T_2). Denote by Φ the absolute Frobenius morphism of X_1. For every $j \in \mathbb{N}$, we write $\Phi^j c$ for the cohomological correspondence with support in $(\Phi^j \circ T_1, T_2)$ defined as the following composition of maps:

$$(\Phi^j \circ T_1)^*L_1 = T_1^*\Phi^{j*}L_1 \simeq T_1^*L_1 \xrightarrow{c} T_2^!L_2.$$

First we will define Hecke correspondences on the complexes of 1.2. Fix $\mathbf{M}$, $\mathbf{L}$ and $(\mathbf{G}, \mathcal{X})$ as in 1.2. Let $m_1, m_2 \in \mathbf{L}(\mathbb{Q})\mathbf{G}(\mathbb{A}_f)$ and $\mathrm{K}'_M, \mathrm{K}_M^{(1)}, \mathrm{K}_M^{(2)}$ be neat open compact subgroups of $\mathbf{M}(\mathbb{A}_f)$ such that $\mathrm{H}' \subset m_1\mathrm{H}^{(1)}m_1^{-1} \cap m_2\mathrm{H}^{(2)}m_2^{-1}$, where $\mathrm{H}' = \mathrm{K}'_M \cap \mathbf{L}(\mathbb{Q})\mathbf{G}(\mathbb{A}_f)$ and $\mathrm{H}^{(i)} = \mathrm{K}_M^{(i)} \cap \mathbf{L}(\mathbb{Q})\mathbf{G}(\mathbb{A}_f)$. This gives two finite étale morphisms $T_{m_i} : M(\mathbf{G}, \mathcal{X})/\mathrm{H}' \longrightarrow M(\mathbf{G}, \mathcal{X})/\mathrm{H}^{(i)}$, $i = 1, 2$. Write $\mathrm{H}_L^{(i)} = \mathrm{H}^{(i)} \cap \mathbf{L}(\mathbb{Q})$ and $\mathrm{H}'_L = \mathrm{H}' \cap \mathbf{L}(\mathbb{Q})$. Let $V \in \mathrm{Ob}\,\mathrm{Rep}_{\mathbf{M}}$. For $i = 1, 2$, write

$$L_i = \mathcal{F}^{\mathrm{H}^{(i)}/\mathrm{H}_L^{(i)}} R\Gamma(\mathrm{H}_L^{(i)}, V).$$

By [P2] 1.11.5, there are canonical isomorphisms

$$T_{m_i}^* L_i \simeq \mathcal{F}^{\mathrm{H}'/\mathrm{H}'_L} \theta_i^* R\Gamma(\mathrm{H}_L^{(i)}, V),$$

where $\theta_i^* R\Gamma(\mathrm{H}_L^{(i)}, V)$ is the inverse image by the morphism $\theta_i : \mathrm{H}'/\mathrm{H}'_L \longrightarrow \mathrm{H}^{(i)}/\mathrm{H}_L^{(i)}$, $h \longmapsto m_i^{-1} h m_i$, of the complex of $\mathrm{H}^{(i)}/\mathrm{H}_L^{(i)}$-modules $R\Gamma(\mathrm{H}_L^{(i)}, V)$. Using the injections $\mathrm{H}'_L \longrightarrow \mathrm{H}_L^{(i)}$, $h \longmapsto m_i^{-1} h m_i$, we get an adjunction morphism $\theta_1^* R\Gamma(\mathrm{H}_L^{(1)}, V) \xrightarrow{\mathrm{adj}} R\Gamma(\mathrm{H}'_L, V)$ and a trace morphism $R\Gamma(\mathrm{H}'_L, V) \xrightarrow{\mathrm{Tr}} \theta_2^* R\Gamma(\mathrm{H}_L^{(2)}, V)$ (this last morphism exists because the index of H'_L in $\mathrm{H}_L^{(2)}$ is finite); these morphisms are $\mathrm{H}'/\mathrm{H}'_L$-equivariant. The Hecke correspondence

$$c_{m_1,m_2} : T_{m_1}^* L_1 \longrightarrow T_{m_2}^! L_2 = T_{m_2}^* L_2$$

is the map

$$\begin{aligned} T_{m_1}^* L_1 &\simeq \mathcal{F}^{\mathrm{H}'/\mathrm{H}'_L} \theta_1^* R\Gamma(\mathrm{H}_L^{(1)}, V) \\ &\xrightarrow{\mathrm{adj}} \mathcal{F}^{\mathrm{H}'/\mathrm{H}'_L} R\Gamma(\mathrm{H}'_L, V) \xrightarrow{\mathrm{Tr}} \mathcal{F}^{\mathrm{H}'/\mathrm{H}'_L} \theta_2^* R\Gamma(\mathrm{H}_L^{(2)}, V) \simeq T_{m_2}^* L_2. \end{aligned}$$

Note that, if $\mathbf{L} = \{1\}$, then this correspondence is an isomorphism.

Remarks 1.5.2 (1) Assume that $\mathrm{K}'_M \subset m_1 \mathrm{K}_M^{(1)} m_1^{-1} \cap m_2 \mathrm{K}_M^{(2)} m_2^{-1}$, and write $\mathrm{K}'_L = \mathrm{K}'_M \cap \mathbf{L}(\mathbb{A}_f)$ and $\mathrm{K}_L^{(i)} = \mathrm{K}_M^{(i)} \cap \mathbf{L}(\mathbb{A}_f)$. Using the methods of [M1] 2.1.4 (and the fact that, for every open compact subgroup K_L of $\mathbf{L}(\mathbb{A}_f)$, $R\Gamma(\mathrm{K}_L, V) = \bigoplus_{i \in I} R\Gamma(g_i \mathrm{K}_L g_i^{-1} \cap \mathbf{L}(\mathbb{Q}), V)$, where $(g_i)_{i \in I}$ is a system of representatives of $\mathbf{L}(\mathbb{Q}) \setminus \mathbf{L}(\mathbb{A}_f)/\mathrm{K}_L$), it is possible to construct complexes $M_i = \mathcal{F}^{\mathrm{K}_M^{(i)}/\mathrm{K}_L^{(i)}} R\Gamma(\mathrm{K}_L^{(i)}, V)$ and $\mathcal{F}^{\mathrm{K}'_M/\mathrm{K}'_L} R\Gamma(\mathrm{K}'_L, V)$. There are a correspondence

$$(T_{m_1}, T_{m_2}) : M^{\mathrm{K}'_M/\mathrm{K}'_L}(\mathbf{G}, \mathcal{X}) \longrightarrow M^{\mathrm{K}_M^{(1)}/\mathrm{K}_L^{(1)}}(\mathbf{G}, \mathcal{X}) \times M^{\mathrm{K}_M^{(2)}/\mathrm{K}_L^{(2)}}(\mathbf{G}, \mathcal{X}),$$

and a cohomological correspondence, constructed as above,

$$c_{m_1,m_2} : T_{m_1}^* M_1 \longrightarrow T_{m_2}^! M_2.$$

(2) There are analogous correspondences, constructed by replacing $R\Gamma(\mathrm{H}_L^{(i)}, V)$ and $R\Gamma(\mathrm{H}'_L, V)$ (resp. $R\Gamma(\mathrm{K}_L^{(i)}, V)$ and $R\Gamma(\mathrm{K}'_L, V)$) with $R\Gamma_c(\mathrm{H}_L^{(i)}, V)$ and $R\Gamma_c(\mathrm{H}'_L, V)$ (resp. $R\Gamma_c(\mathrm{K}_L^{(i)}, V)$ and $R\Gamma_c(\mathrm{K}'_L, V)$). We will still use the notation c_{m_1,m_2} for these correspondences.

Use the notation of section 1.4, and fix $g \in \mathbf{G}(\mathbb{A}_f)$ and a second open compact subgroup K' of $\mathbf{G}(\mathbb{A}_f)$, such that $\mathrm{K}' \subset \mathrm{K} \cap g\mathrm{K}g^{-1}$. Fix prime numbers p and ℓ as at the end of 1.3. In particular, it is assumed that $g \in \mathbf{G}(\mathbb{A}_f^p)$ and that K (resp., K$'$) is of the form $\mathrm{K}^p\mathbf{G}(\mathbb{Z}_p)$ (resp., $\mathrm{K}'^p\mathbf{G}(\mathbb{Z}_p)$), with $\mathrm{K}^p \subset \mathbf{G}(\mathbb{A}_f^p)$ (resp., $\mathrm{K}'^p \subset \mathbf{G}(\mathbb{A}_f^p)$) and $\mathbf{G}(\mathbb{Z}_p)$ a hyperspecial maximal compact subgroup of $\mathbf{G}(\mathbb{Q}_p)$. As in section 1.4, we will use the notations $M^{\mathrm{K}}(\mathbf{G}, \mathcal{X})$, etc, for the reductions modulo p of the varieties of section 1.1.

Let Φ be the absolute Frobenius morphism of $M^{\mathrm{K}}(\mathbf{G}, \mathcal{X})^*$. For every $V \in \mathrm{Ob}\, D^b(\mathrm{Rep}_{\mathbf{G}})$ and $j \in \mathbb{Z}$, let $u_j : (\Phi^j T_g)^* \mathcal{F}^{\mathrm{K}} V \longrightarrow T_1^! \mathcal{F}^{\mathrm{K}} V$ be the cohomological correspondence $\Phi^j c_{g,1}$ on $\mathcal{F}^{\mathrm{K}} V$ (with support in $(\Phi^j T_g, T_1)$).

Let $V \in \mathrm{Ob}\, D^b(\mathcal{R}_0)$. By [M2] 5.1.2 and 5.1.3:

- for every $t_1, \dots, t_n \in \mathbb{Z} \cup \{\pm\infty\}$, the correspondence u_j extends in a unique way to a correspondence
$$\overline{u}_j : (\Phi^j \overline{T}_g)^* W^{\geq t_1, \dots, \geq t_n} V \longrightarrow \overline{T}_1^! W^{\geq t_1, \dots, \geq t_n} V;$$
- for every $n_1, \dots, n_r \in \{1, \dots, n\}$ such that $n_1 < \dots < n_r$ and every $a_1, \dots, a_r \in \mathbb{Z} \cup \{\pm\infty\}$, the correspondence u_j gives in a natural way a cohomological correspondence on $i_{n_r!} w_{\leq a_r} i_{n_r}^! \dots i_{n_1!} w_{\leq a_1} i_{n_1}^! j_! \mathcal{F}^{\mathrm{K}} V$ with support in $(\Phi^j \overline{T}_g, \overline{T}_1)$; write $i_{n_r!} w_{\leq a_r} i_{n_r}^! \dots i_{n_1!} w_{\leq a_1} i_{n_1}^! j_! u_j$ for this correspondence.

Moreover, there is an analog of theorem 1.4.4 for cohomological correspondences (cf. [M2] 5.1.5). The goal of this section is to calculate the correspondences $i_{n_r!} w_{\leq a_r} i_{n_r}^! \dots i_{n_1!} w_{\leq a_1} i_{n_1}^! j_! u_j$.

Fix $n_1, \dots, n_r \in \{1, \dots, n\}$ such that $n_1 < \dots < n_r$ and $a_1, \dots, a_r \in \mathbb{Z} \cup \{\pm\infty\}$, and write
$$L = i_{n_r!} w_{\leq a_r} i_{n_r}^! \dots i_{n_1!} w_{\leq a_1} i_{n_1}^! j_! \mathcal{F}^{\mathrm{K}} V,$$
$$u = i_{n_r!} w_{\leq a_r} i_{n_r}^! \dots i_{n_1!} w_{\leq a_1} i_{n_1}^! j_! u_j.$$

Use the notation of corollary 1.4.6. By this corollary, there is an isomorphism
$$L \simeq \bigoplus_{C \in \mathcal{C}_P} (i_C T_C)_! L_C,$$
where, for every $C = (X_1, \dots, X_r) \in \mathcal{C}_P$, i_C is the inclusion in $M^{\mathrm{K}}(\mathbf{G}, \mathcal{X})^*$ of the boundary stratum image of X_r (i.e., of the stratum $\mathrm{Im}(i_{n_r,h})$, if $h \in \mathbf{G}(\mathbb{A}_f)$ is a representative of C). Hence the correspondence u can be seen as a matrix $(u_{C_1,C_2})_{C_1,C_2 \in \mathcal{C}_P}$, and we want to calculate the entries of this matrix.

Let $\mathcal{C}'_P$ be the analog of the set $\mathcal{C}_P$ obtained when K is replaced with K′. The morphisms $\overline{T}_g, \overline{T}_1$ define maps $T_g, T_1 : \mathcal{C}'_P \longrightarrow \mathcal{C}_P$, and these maps correspond via the bijections $\mathcal{C}_P \simeq \mathbf{P}(\mathbb{Q})\mathbf{Q}_{n_r}(\mathbb{A}_f) \backslash \mathbf{G}(\mathbb{A}_f)/\mathrm{K}$ and $\mathcal{C}'_P \simeq \mathbf{P}(\mathbb{Q})\mathbf{Q}_{n_r}(\mathbb{A}_f) \backslash \mathbf{G}(\mathbb{A}_f)/\mathrm{K}'$ of proposition 1.1.3 to the maps induced by $h \longmapsto hg$ and $h \longmapsto h$.

Let $C_1 = (X_1^{(1)}, \dots, X_r^{(1)})$, $C_2 = (X_1^{(2)}, \dots, X_r^{(2)}) \in \mathcal{C}_P$, and choose representatives $h_1, h_2 \in \mathbf{G}(\mathbb{A}_f)$ of C_1 and C_2. Let $C' = (X'_1, \dots, X'_r) \in \mathcal{C}'_P$ be such that $T_g(C') = C_1$ and $T_1(C') = C_2$. Fix a representative $h' \in \mathbf{G}(\mathbb{A}_f)$ of C'. There exist $q_1, q_2 \in \mathbf{P}(\mathbb{Q})\mathbf{Q}_{n_r}(\mathbb{A}_f)$ such that $q_1 h' \in h_1 g\mathrm{K}$ and $q_2 h' \in h_2 \mathrm{K}$. Let $\overline{q}_1, \overline{q}_2$ be the images of q_1, q_2 in $\mathbf{L}_{n_r}(\mathbb{Q})\mathbf{G}_{n_r}(\mathbb{A}_f)$. The following diagrams are commutative:

$$\begin{array}{ccc} X'_r & \xrightarrow{\;i_{C'} T_{C'}\;} & M^{\mathrm{K}'}(\mathbf{G}, \mathcal{X})^* \\ {\scriptstyle T_{\overline{q}_1}} \downarrow & & \downarrow {\scriptstyle \overline{T}_g} \\ X_r^{(1)} & \xrightarrow{\;i_{C_1} T_{C^{(1)}}\;} & M^{\mathrm{K}}(\mathbf{G}, \mathcal{X})^* \end{array} \qquad \begin{array}{ccc} X'_r & \xrightarrow{\;i_{C'} T_{C'}\;} & M^{\mathrm{K}'}(\mathbf{G}, \mathcal{X})^* \\ {\scriptstyle T_{\overline{q}_2}} \downarrow & & \downarrow {\scriptstyle \overline{T}_1} \\ X_r^{(2)} & \xrightarrow{\;i_{C_2} T_{C^{(2)}}\;} & M^{\mathrm{K}}(\mathbf{G}, \mathcal{X})^* \end{array}$$

By corollary 1.4.6, there are isomorphisms
$$L_{C_1} \simeq \mathcal{F}^{\mathrm{H}^{(1)}/\mathrm{H}_L^{(1)}} R\Gamma_c(\mathrm{H}_L^{(1)}/\mathrm{K}_N^{(1)}, R\Gamma(\mathrm{Lie}(\mathbf{N}_{n_r}), V)_{\geq t_1, \dots, \geq t_r})[a]$$

and

$$L_{C_2} \simeq \mathcal{F}^{\mathrm{H}^{(2)}/\mathrm{H}_L^{(2)}} R\Gamma_c(\mathrm{H}_L^{(2)}/\mathrm{K}_N^{(2)}, R\Gamma(\mathrm{Lie}(\mathbf{N}_{n_r}), V)_{\geq t_1,\dots,\geq t_r})[a],$$

where $t_1, \dots, t_r$ are defined as in proposition 1.4.5, $a = -\dim(\mathbf{A}_{M_P}/\mathbf{A}_G)$, $\mathrm{H}^{(i)} = h_i\mathrm{K}h_i^{-1} \cap \mathbf{P}(\mathbb{Q})\mathbf{Q}_{n_r}(\mathbb{A}_f)$, $\mathrm{H}_L^{(i)} = \mathrm{H}^{(i)} \cap \mathbf{L}_{n_r}(\mathbb{Q})\mathbf{N}_{n_r}(\mathbb{A}_f)$ and $\mathrm{K}_N^{(i)} = \mathrm{H}^{(i)} \cap \mathbf{N}_{n_r}(\mathbb{A}_f)$. We get a cohomological correspondence

$$\Phi^j c_{\overline{q}_1,\overline{q}_2} : (\Phi^j T_{\overline{q}_1})^* L_{C_1} \longrightarrow T_{\overline{q}_2}^! L_{C_2}.$$

Define a cohomological correspondence

$$u_{C'} : (\Phi^j \overline{T}_g)^*(i_{C_1} T_{C_1})_! L_{C_1} \longrightarrow \overline{T}_1^!(i_{C_2} T_{C_2})_! L_{C_2}$$

by taking the direct image with compact support of the previous correspondence by $(i_{C_1} T_{C_1}, i_{C_2} T_{C_2})$ (the direct image of a correspondence by a proper morphism is defined in [SGA 5] III 3.3; the direct image by a locally closed immersion is defined in [M2] 5.1.1 (following [F] 1.3.1), and the direct image with compact support is defined by duality). Finally, write

$$N_{C'} = [\mathrm{K}_N^{(2)} : h_2\mathrm{K}'h_2^{-1} \cap \mathbf{N}_{n_r}(\mathbb{A}_f)].$$

Proposition 1.5.3 *The coefficient u_{C_1,C_2} in the above matrix is equal to*

$$\sum_{C'} N_{C'} u_{C'},$$

where the sum is taken over the set of $C' \in \mathcal{C}'_P$ such that $T_g(C') = C_1$ and $T_1(C') = C_2$.

This proposition generalizes (the dual version of) theorem 5.2.2 of [M2] and can be proved in exactly the same way (by induction on r, as in the proof of proposition 1.4.5). The proof of theorem 5.2.2 of [M2] uses proposition 2.2.3 of [M2] (via the proof of corollary 5.2.4), but this proposition is simply a reformulation of proposition 4.8.5 of [P2], and it is true as well for the Shimura varieties considered here.

1.6 THE FIXED POINT FORMULAS OF KOTTWITZ AND GORESKY-KOTTWITZ-MACPHERSON

In this section, we recall two results about the fixed points of Hecke correspondences, which will be used in 1.7.

Theorem 1.6.1 *([K11] 19.6) Notation is as in 1.5. Assume that the Shimura datum $(\mathbf{G}, \mathcal{X})$ is of the type considered in [K11] §5, and that we are not in case (D) of that article (i.e., that $\mathbf{G}$ is not an orthogonal group). Fix an algebraic closure $\mathbb{F}$ of $\mathbb{F}_p$. Let $V \in \mathrm{Ob}\,\mathrm{Rep}_{\mathbf{G}}$. For every $j \geq 1$, denote by $T(j,g)$ the sum over the set of fixed points in $M^{\mathrm{K}'}(\mathbf{G},\mathcal{X})(\mathbb{F})$ of the correspondence $(\Phi^j \circ T_g, T_1)$ of the naive local terms (cf. [P3] 1.5) of the cohomological correspondence u_j on $\mathcal{F}^{\mathrm{K}}V$ defined in section 1.5. Then*

$$T(j,g) = \sum_{(\gamma_0;\gamma,\delta)\in C_{\mathbf{G},j}} c(\gamma_0;\gamma,\delta) O_\gamma(f^p) TO_\delta(\phi_j^{\mathbf{G}}) \mathrm{Tr}(\gamma_0, V).$$

Let us explain briefly the notation (see [K9] §§2 and 3 for more detailed explanations).

The function $f^p \in C_c^\infty(\mathbf{G}(\mathbb{A}_f^p))$ is defined by the formula

$$f^p = \frac{\mathbb{1}_{g\mathrm{K}^p}}{\mathrm{vol}(\mathrm{K}'^p)}.$$

For every $\gamma \in \mathbf{G}(\mathbb{A}_f^p)$, write

$$O_\gamma(f^p) = \int_{\mathbf{G}(\mathbb{A}_f^p)_\gamma \backslash \mathbf{G}(\mathbb{A}_f^p)} f^p(x^{-1}\gamma x) d\overline{x},$$

where $\mathbf{G}(\mathbb{A}_f^p)_\gamma$ is the centralizer of γ in $\mathbf{G}(\mathbb{A}_f^p)$.

Remember that we fixed an injection $F \subset \overline{\mathbb{Q}}_p$; this determines a place $\wp$ of F over p. Let $\mathbb{Q}_p^{nr}$ be the maximal unramified extension of $\mathbb{Q}_p$ in $\overline{\mathbb{Q}}_p$, L be the unramified extension of degree j of $F_\wp$ in $\overline{\mathbb{Q}}_p$, $r = [L : \mathbb{Q}_p]$, ϖ_L be a uniformizer of L and $\sigma \in \mathrm{Gal}(\mathbb{Q}_p^{nr}/\mathbb{Q}_p)$ be the element lifting the arithmetic Frobenius morphism of $\mathrm{Gal}(\mathbb{F}/\mathbb{F}_p)$. Let $\delta \in \mathbf{G}(L)$. Define the norm $N\delta$ of δ by

$$N\delta = \delta\sigma(\delta)\ldots\sigma^{r-1}(\delta) \in \mathbf{G}(L).$$

The σ-centralizer of δ in $\mathbf{G}(L)$ is by definition

$$\mathbf{G}(L)_\delta^\sigma = \{x \in \mathbf{G}(L) | x\delta = \delta\sigma(x)\}.$$

We say that $\delta' \in \mathbf{G}(L)$ is σ-conjugate to δ in $\mathbf{G}(L)$ if there exists $x \in \mathbf{G}(L)$ such that $\delta' = x^{-1}\delta\sigma(x)$.

By definition of the reflex field F, the conjugacy class of cocharacters $h_x \circ \mu_0 : \mathbb{G}_{m,\mathbb{C}} \longrightarrow \mathbf{G}_\mathbb{C}$, $x \in \mathcal{X}$, of section 1.1 is defined over F. Choose an element μ in this conjugacy class that factors through a maximal split torus of $\mathbf{G}$ over $\mathcal{O}_L$ (cf. [K9] §3 p. 173), and write

$$\phi_j^\mathbf{G} = \mathbb{1}_{\mathbf{G}(\mathcal{O}_L)\mu(\varpi_L^{-1})\mathbf{G}(\mathcal{O}_L)} \in \mathcal{H}(\mathbf{G}(L), \mathbf{G}(\mathcal{O}_L)).$$

($\mathcal{H}(\mathbf{G}(L), \mathbf{G}(\mathcal{O}_L))$ is the Hecke algebra of functions with compact support on $\mathbf{G}(L)$ that are bi-invariant by $\mathbf{G}(\mathcal{O}_L)$.) For every $\delta \in \mathbf{G}(L)$ and $\phi \in C_c^\infty(\mathbf{G}(L))$, write

$$TO_\delta(\phi) = \int_{\mathbf{G}(L)_\delta^\sigma \backslash \mathbf{G}(L)} \phi(y^{-1}\delta\sigma(y)) d\overline{y}.$$

Let $\widehat{\mathbf{T}}$ be a maximal torus of $\widehat{\mathbf{G}}$. The conjugacy class of cocharacters $h_x \circ \mu_0$, $x \in \mathcal{X}$, corresponds to a Weyl group orbit of characters of $\widehat{\mathbf{T}}$; denote by μ_1 the restriction to $Z(\widehat{\mathbf{G}})$ of any of these characters (this does not depend on the choices).

It remains to define the set $C_{\mathbf{G},j}$ indexing the sum of the theorem and the coefficients $c(\gamma_0; \gamma, \delta)$. Consider the set of triples $(\gamma_0; \gamma, \delta) \in \mathbf{G}(\mathbb{Q}) \times \mathbf{G}(\mathbb{A}_f^p) \times \mathbf{G}(L)$ satisfying the following conditions (we will later write (C) for the list of these conditions):

- γ_0 is semisimple and elliptic in $\mathbf{G}(\mathbb{R})$ (i.e., there exists an elliptic maximal torus $\mathbf{T}$ of $\mathbf{G}_\mathbb{R}$ such that $\gamma_0 \in \mathbf{T}(\mathbb{R})$).
- For every place $v \neq p, \infty$ of $\mathbb{Q}$, γ_v (the local component of γ at v) is $\mathbf{G}(\overline{\mathbb{Q}}_v)$-conjugate to γ_0.

- $N\delta$ and γ_0 are $\mathbf{G}(\overline{\mathbb{Q}}_p)$-conjugate.
- The image of the σ-conjugacy class of δ by the map $B(\mathbf{G}_{\mathbb{Q}_p}) \longrightarrow X^*(Z(\widehat{\mathbf{G}})^{\mathrm{Gal}(\overline{\mathbb{Q}}_p/\mathbb{Q}_p)})$ of [K9] 6.1 is the restriction of $-\mu_1$ to $Z(\widehat{\mathbf{G}})^{\mathrm{Gal}(\overline{\mathbb{Q}}_p/\mathbb{Q}_p)}$.

Two triples $(\gamma_0; \gamma, \delta)$ and $(\gamma_0'; \gamma', \delta')$ are called equivalent if γ_0 and γ_0' are $\mathbf{G}(\overline{\mathbb{Q}})$-conjugate, γ and γ' are $\mathbf{G}(\mathbb{A}_f^p)$-conjugate, and δ and δ' are σ-conjugate in $\mathbf{G}(L)$.

Let $(\gamma_0; \gamma, \delta)$ be a triple satisfying conditions (C). Let I_0 be the centralizer of γ_0 in $\mathbf{G}$. There is a canonical morphism $Z(\widehat{\mathbf{G}}) \longrightarrow Z(\widehat{I_0})$, and the exact sequence

$$1 \longrightarrow Z(\widehat{\mathbf{G}}) \longrightarrow Z(\widehat{I_0}) \longrightarrow Z(\widehat{I_0})/Z(\widehat{\mathbf{G}}) \longrightarrow 1$$

induces a morphism

$$\pi_0((Z(\widehat{I_0})/Z(\widehat{\mathbf{G}}))^{\mathrm{Gal}(\overline{\mathbb{Q}}/\mathbb{Q})}) \longrightarrow \mathrm{H}^1(\mathbb{Q}, Z(\widehat{\mathbf{G}})).$$

Denote by $\mathfrak{K}(I_0/\mathbb{Q})$ the inverse image by this morphism of the subgroup

$$\mathrm{Ker}^1(\mathbb{Q}, Z(\widehat{\mathbf{G}})) := \mathrm{Ker}\left(\mathrm{H}^1(\mathbb{Q}, Z(\widehat{\mathbf{G}})) \longrightarrow \prod_{v \text{ place of } \mathbb{Q}} \mathrm{H}^1(\mathbb{Q}_v, Z(\widehat{\mathbf{G}}))\right).$$

In [K9] §2, Kottwitz defines an element $\alpha(\gamma_0; \gamma, \delta) \in \mathfrak{K}(I_0/\mathbb{Q})^D$ (where, for every group A, $A^D = \mathrm{Hom}(A, \mathbb{C}^\times)$); this element depends only on the equivalence class of $(\gamma_0; \gamma, \delta)$. For every place $v \neq p, \infty$ of $\mathbb{Q}$, denote by $I(v)$ the centralizer of γ_v in $\mathbf{G}_{\mathbb{Q}_v}$; as γ_0 and γ_v are $\mathbf{G}(\overline{\mathbb{Q}}_v)$-conjugate, the group $I(v)$ is an inner form of I_0 over $\mathbb{Q}_v$. On the other hand, there exists a $\mathbb{Q}_p$-group $I(p)$ such that $I(p)(\mathbb{Q}_p) = \mathbf{G}(L)^\sigma_\delta$, and this group is an inner form of I_0 over $\mathbb{Q}_p$. There is a similar object for the infinite place: in the beginning of [K9] §3, Kottwitz defines an inner form $I(\infty)$ of I_0; $I(\infty)$ is an algebraic group over $\mathbb{R}$, anisotropic modulo $\mathbf{A}_G$. Kottwitz shows that, if $\alpha(\gamma_0; \gamma, \delta) = 1$, then there exists an inner form I of I_0 over $\mathbb{Q}$ such that, for every place v of $\mathbb{Q}$, $I_{\mathbb{Q}_v}$ and $I(v)$ are isomorphic (Kottwitz's statement is more precise, cf. [K9] pp. 171–172).

The set $C_{\mathbf{G},j}$ indexing the sum of the theorem is the set of equivalence classes of triples $(\gamma_0; \gamma, \delta)$ satisfying conditions (C) and such that $\alpha(\gamma_0; \gamma, \delta) = 1$. For every $(\gamma_0; \gamma, \delta)$ in $C_{\mathbf{G},j}$, let

$$c(\gamma_0; \gamma, \delta) = \mathrm{vol}(I(\mathbb{Q}) \setminus I(\mathbb{A}_f)) |\mathrm{Ker}(\mathrm{Ker}^1(\mathbb{Q}, I_0) \longrightarrow \mathrm{Ker}^1(\mathbb{Q}, \mathbf{G}))|.$$

Finally, the Haar measures are normalized as in [K9] §3. Take on $\mathbf{G}(\mathbb{A}_f^p)$ (resp., $\mathbf{G}(\mathbb{Q}_p)$, resp., $\mathbf{G}(L)$) the Haar measure such that the volume of K^p (resp., $\mathbf{G}(\mathbb{Z}_p)$, resp., $\mathbf{G}(\mathcal{O}_L)$) is equal to 1. Take on $I(\mathbb{A}_f^p)$ (resp., $I(\mathbb{Q}_p)$) a Haar measure such that the volume of every open compact subgroup is a rational number, and use inner twistings to transport these measures to $\mathbf{G}(\mathbb{A}_f^p)_\gamma$ and $\mathbf{G}(L)^\sigma_\delta$.

Remark 1.6.2 If $\mathrm{K}' = \mathrm{K} \cap g\mathrm{K}g^{-1}$, we may replace f^p with the function

$$\frac{\mathbb{1}_{\mathrm{K}^p g \mathrm{K}^p}}{\mathrm{vol}(\mathrm{K}^p)} \in \mathcal{H}(\mathbf{G}(\mathbb{A}_f^p), \mathrm{K}^p) := C_c^\infty(\mathrm{K}^p \setminus \mathbf{G}(\mathbb{A}_f^p)/\mathrm{K}^p)$$

(cf. [K11] §16 p. 432).

Remark 1.6.3 There are two differences between the formula given here and formula (19.6) of [K11]:

(1) Kottwitz considers the correspondence $(T_g, \Phi^j \circ T_1)$ (and not $(\Phi^j \circ T_g, T_1)$) and does not define the naive local term in the same way as Pink (cf. [K11] §16 p. 433). But is is easy to see (by comparing the definitions of the naive local terms and composing Kottwitz's correspondence by $T_{g^{-1}}$) that the number $T(j, f)$ of [K11] (19.6) is equal to $T(j, g^{-1})$. This explains that the function of $C_c^\infty(\mathbf{G}(\mathbb{A}_f^p))$ appearing in theorem 1.6.1 is $\mathrm{vol}(\mathrm{K}'^p)^{-1}\mathbb{1}_{g\mathrm{K}^p}$, instead of the function $\widetilde{f}^p = \mathrm{vol}(\mathrm{K}'^p)^{-1}\mathbb{1}_{\mathrm{K}^p g^{-1}}$ of [K11] §16 p. 432. (Kottwitz also takes systematically $\mathrm{K}' = \mathrm{K} \cap g\mathrm{K}g^{-1}$, but his result generalizes immediately to the case where K' is of finite index in $\mathrm{K} \cap g\mathrm{K}g^{-1}$.)

(2) Below formula (19.6) of [K11], Kottwitz notes that this formula is true for the canonical model of a Shimura variety associated to the datum $(\mathbf{G}, \mathcal{X}, h^{-1})$ (and not $(\mathbf{G}, \mathcal{X}, h)$). The normalization of the global class field isomorphism used in [K9], [K11], and here are the same (it is also the normalization of [D1] 0.8 and [P2] 5.5). However, the convention for the action of the Galois group on the special points of the canonical model that is used here is the convention of [P2] 5.5, and it differs (by a sign) from the convention of [D1] 2.2.4 (because the reciprocity morphism of [P2] 5.5 is the inverse of the reciprocity morphism of [D1] 2.2.3). As Kottwitz uses Deligne's conventions, what he calls canonical model of a Shimura variety associated to the datum $(\mathbf{G}, \mathcal{X}, h^{-1})$ is what is called here canonical model of a Shimura variety associated to the datum $(\mathbf{G}, \mathcal{X}, h)$.

Remark 1.6.4 Actually, Kottwitz proves a stronger result in [K11] §19: For every $\gamma \in \mathbf{G}(\mathbb{A}_f^p)$, let $N(\gamma)$ be the number of fixed points x' in $M^{\mathrm{K}'}(\mathbf{G}, \mathcal{X})(\mathbb{F})$ that can be represented by an element $\widetilde{x}$ of $M(\mathbf{G}, \mathcal{X})(\mathbb{F})$ such that there exist $k \in \mathrm{K}$ and $g \in \mathbf{G}(\mathbb{A}_f)$ with $\Phi^j(\widetilde{x})g = \widetilde{x}k$ and gk^{-1} $\mathbf{G}(\mathbb{A}_f^p)$-conjugate to γ (this condition depends only on x', and not on the choice of $\widetilde{x}$). Then

$$N(\gamma) = \sum_{\delta} c(\gamma_0; \gamma, \delta) O_\gamma(f^p) T O_\delta(\phi_j^{\mathbf{G}}),$$

where the sum is taken over the set of σ-conjugacy classes of $\delta \in \mathbf{G}(L)$ such that there exists $\gamma_0 \in \mathbf{G}(\mathbb{Q})$ such that the triple $(\gamma_0; \gamma, \delta)$ is in $C_{\mathbf{G},j}$ (if such a γ_0 exists, it is unique up to $\mathbf{G}(\overline{\mathbb{Q}})$-conjugacy, because, for every place $v \neq p, \infty$ of $\mathbb{Q}$, it is conjugate under $\mathbf{G}(\overline{\mathbb{Q}}_v)$ to the component at v of γ). Moreover, if x' is a fixed point contributing to $N(\gamma)$, then the naive local term at x' is $\mathrm{Tr}(\gamma_\ell, V)$ (where γ_ℓ is the ℓ-adic component of γ).

Remark 1.6.5 Some of the Shimura varieties that will be used later are not of the type considered in [K11] §5, so we will need another generalization of Kottwitz's result, in a very particular (and easy) case. Let $(\mathbf{G}, \mathcal{X}, h)$ be a Shimura datum (in the sense of section 1.1) such that $\mathbf{G}$ is a torus. Let $\mathcal{Y}$ be the image of $\mathcal{X}$ by the morphism $h : \mathcal{X} \longrightarrow \mathrm{Hom}(\mathbb{S}, \mathbf{G})$ ($\mathcal{Y}$ is a point because $\mathbf{G}$ is commutative, but the cardinality of $\mathcal{X}$ can be greater than 1 in general; remember that the morphism

h is assumed to have finite fibers, but that it is not assumed to be injective). Let $\mathbf{G}(\mathbb{R})^+$ be the subgroup of $\mathbf{G}(\mathbb{R})$ stabilizing a connected component of $\mathcal{X}$ (this group does not depend on the choice of the connected component) and $\mathbf{G}(\mathbb{Q})^+ = \mathbf{G}(\mathbb{Q}) \cap \mathbf{G}(\mathbb{R})^+$. The results of theorem 1.6.1 and of remark 1.6.4 are true for the Shimura datum $(\mathbf{G}, \mathcal{Y})$ (in this case, they are a consequence of the description of the action of the Galois group on the special points of the canonical model, cf. [P2] 5.5). For the Shimura datum $(\mathbf{G}, \mathcal{X})$, these results are also true if the following changes are made:

- multiply the formula giving the trace in theorem 1.6.1 and the formula giving the number of fixed points in remark 1.6.4 by $|\mathcal{X}|$;
- replace $C_{\mathbf{G},j}$ with the subset of triples $(\gamma_0; \gamma, \delta) \in C_{\mathbf{G},j}$ such that $\gamma_0 \in \mathbf{G}(\mathbb{Q})^+$.

This fact is also an easy consequence of [P2] 5.5.

The fixed point formula of Goresky, Kottwitz and MacPherson applies to a different situation, that of the end of 1.2. Use the notation introduced there. Let $V \in \operatorname{Ob}\operatorname{Rep}_{\mathbf{G}}$, $g \in \mathbf{G}(\mathbb{A}_f)$, and let K, K' be neat open compact subgroups of $\mathbf{G}(\mathbb{A}_f)$ such that $\mathrm{K}' \subset \mathrm{K} \cap g\mathrm{K}g^{-1}$. This gives two finite étale morphisms T_g, $T_1 : M^{\mathrm{K}'}(\mathbf{G}, \mathcal{X})(\mathbb{C}) \longrightarrow M^{\mathrm{K}}(\mathbf{G}, \mathcal{X})(\mathbb{C})$. Define a cohomological correspondence

$$u_g : T_g^* \mathcal{F}^{\mathrm{K}} V \xrightarrow{\sim} T_1^! \mathcal{F}^{\mathrm{K}} V$$

as at the beginning of section 1.5. The following theorem is a particular case of theorem 7.14.B of [GKM] (cf. [GKM] (7.17)).

Theorem 1.6.6 *The trace of the correspondence u_g on the cohomology with compact support $R\Gamma_c(M^{\mathrm{K}}(\mathbf{G}, \mathcal{X})(\mathbb{C}), \mathcal{F}^{\mathrm{K}} V)$ is equal to*

$$\sum_{\mathbf{M}} (-1)^{\dim(\mathbf{A}_M/\mathbf{A}_G)} (n_M^G)^{-1} \sum_{\gamma} \iota^M(\gamma)^{-1} \chi(\mathbf{M}_\gamma) O_\gamma(f_{\mathbf{M}}^\infty) |D_M^G(\gamma)|^{1/2} \mathrm{Tr}(\gamma, V),$$

where the first sum is taken over the set of $\mathbf{G}(\mathbb{Q})$-conjugacy classes of cuspidal Levi subgroups $\mathbf{M}$ of $\mathbf{G}$ and, for every $\mathbf{M}$, the second sum is taken over the set γ of semisimple $\mathbf{M}(\mathbb{Q})$-conjugacy classes that are elliptic in $\mathbf{M}(\mathbb{R})$.

Let us explain the notation.

- $f^\infty = \dfrac{\mathbb{1}_{g\mathrm{K}}}{\mathrm{vol}(\mathrm{K}')} \in C_c^\infty(\mathbf{G}(\mathbb{A}_f))$, and $f_{\mathbf{M}}^\infty$ is the constant term of f^∞ at $\mathbf{M}$ (cf. [GKM] (7.13.2)).
- Let $\mathbf{M}$ be a Levi subgroup of $\mathbf{G}$. Let $\mathbf{A}_M$ be the maximal ($\mathbb{Q}$-)split subtorus of the center of $\mathbf{M}$ and
$$n_M^G = |\operatorname{Nor}_{\mathbf{G}}(\mathbf{M})(\mathbb{Q})/\mathbf{M}(\mathbb{Q})|.$$
$\mathbf{M}$ is called *cuspidal* if the group $\mathbf{M}_{\mathbb{R}}$ has a maximal ($\mathbb{R}$-)torus $\mathbf{T}$ such that $\mathbf{T}/\mathbf{A}_{M,\mathbb{R}}$ is anisotropic.
- Let $\mathbf{M}$ be a Levi subgroup of $\mathbf{G}$ and $\gamma \in \mathbf{M}(\mathbb{Q})$. Let $\mathbf{M}^\gamma$ be the centralizer of γ in $\mathbf{M}$, $\mathbf{M}_\gamma = (\mathbf{M}^\gamma)^0$,
$$\iota^M(\gamma) = |\mathbf{M}^\gamma(\mathbb{Q})/\mathbf{M}_\gamma(\mathbb{Q})|$$

and

$$D_M^G(\gamma) = \det(1 - \mathrm{Ad}(\gamma), \mathrm{Lie}(\mathbf{G})/\mathrm{Lie}(\mathbf{M})).$$

- $\chi(\mathbf{M}_\gamma)$ is the Euler characteristic of $\mathbf{M}_\gamma$, cf. [GKM] (7.10).

Remark 1.6.7 According to [GKM] 7.14.B, the formula of the theorem should use $\mathrm{Tr}(\gamma, V^*)$ (or $\mathrm{Tr}(\gamma^{-1}, V)$) and not $\mathrm{Tr}(\gamma, V)$. The difference between the formula given here and that of [GKM] comes from the fact that [GKM] uses a different convention to define the trace of u_g (cf. [GKM] (7.7)); the convention used here is that of [SGA 5] III and of [P1].

1.7 THE FIXED POINT FORMULA

Use the notation introduced before proposition 1.5.3 and in section 1.6. Assume that the Shimura data $(\mathbf{G}, \mathcal{X})$ and $(\mathbf{G}_i, \mathcal{X}_i)$, $1 \leq i \leq n-1$, are of the type considered [K11] §5, with case (D) excluded. (In particular, $\mathbf{G}^{\mathrm{ad}}$ is of abelian type, so we can take $\mathcal{R}_0 = \mathrm{Rep}_{\mathbf{G}}$, i.e., choose any $V \in \mathrm{Ob}\, D^b(\mathrm{Rep}_{\mathbf{G}})$.) Assume moreover that $(\mathbf{G}_n, \mathcal{X}_n)$ is of the type considered in [K11] §5 (case (D) excluded) or that $\mathbf{G}_n$ is a torus.

We want to calculate the trace of the cohomological correspondence

$$\overline{u}_j : (\Phi^j \overline{T}_g)^* W^{\geq t_1, \dots, \geq t_n} V \longrightarrow \overline{T}_1^! W^{\geq t_1, \dots, \geq t_n} V.$$

Assume that $w(\mathbb{G}_m)$ acts on the $\mathrm{H}^i V$, $i \in \mathbb{Z}$, by $t \longmapsto t^m$, for a certain $m \in \mathbb{Z}$ (where $w : \mathbb{G}_m \longrightarrow \mathbf{G}$ is the cocharacter of 1.3).

Let

$$f^{\infty,p} = \mathrm{vol}(\mathrm{K}'^p)^{-1} \mathbb{1}_{g\mathrm{K}^p}.$$

Let $\mathbf{P}$ be a standard parabolic subgroup of $\mathbf{G}$. Write $\mathbf{P} = \mathbf{P}_{n_1} \cap \dots \cap \mathbf{P}_{n_r}$, with $n_1 < \dots < n_r$. Let

$$T_P = m_P \sum_{\mathbf{L}} (-1)^{\dim(\mathbf{A}_L/\mathbf{A}_{L_P})} (n_L^{L_P})^{-1} \sum_{\gamma_L} \iota^L(\gamma_L)^{-1} \chi(\mathbf{L}_{\gamma_L}) |D_L^{L_P}(\gamma_L)|^{1/2}$$

$$\sum_{(\gamma_0;\gamma,\delta) \in C'_{\mathbf{G}_{n_r},j}} c(\gamma_0;\gamma,\delta) O_{\gamma_L\gamma}(f^{\infty,p}_{\mathbf{LG}_{n_r}}) O_{\gamma_L}(\mathbb{1}_{\mathbf{L}(\mathbb{Z}_p)}) \delta^{1/2}_{P(\mathbb{Q}_p)}(\gamma_0) T O_\delta(\phi_j^{\mathbf{G}_{n_r}})$$

$$\times \delta^{1/2}_{P(\mathbb{R})}(\gamma_L\gamma_0) \, \mathrm{Tr}(\gamma_L\gamma_0, R\Gamma(\mathrm{Lie}(\mathbf{N}_P), V)_{\geq t_{n_1}+m, \dots, \geq t_{n_r}+m}),$$

where the first sum is taken over the set of $\mathbf{L}_P(\mathbb{Q})$-conjugacy classes of cuspidal Levi subgroups $\mathbf{L}$ of $\mathbf{L}_P$, the second sum is taken over the set of semisimple conjugacy classes $\gamma_L \in \mathbf{L}(\mathbb{Q})$ that are elliptic in $\mathbf{L}(\mathbb{R})$, and

- $\mathbf{L}(\mathbb{Z}_p)$ is a hyperspecial maximal compact subgroup of $\mathbf{L}(\mathbb{Q}_p)$;
- $m_P = 1$ if $n_r < n$ or if $(\mathbf{G}_n, \mathcal{X}_n)$ is of the type considered in [K11] §5, and $m_P = |\mathcal{X}_{n_r}|$ if $n_r = n$ and $\mathbf{G}_{n_r}$ is a torus;
- $C'_{\mathbf{G}_{n_r},j} = C_{\mathbf{G}_{n_r},j}$ if $n_r < n$ or if $(\mathbf{G}_n, \mathcal{X}_n)$ is of the type considered in [K11] §5, and, if $\mathbf{G}_n$ is a torus, $C'_{\mathbf{G}_n,j}$ is the subset of $C_{\mathbf{G}_n,j}$ defined in remark 1.6.5.

Write also

$$T_G = \sum_{(\gamma_0;\gamma,\delta)\in C_{\mathbf{G},j}} c(\gamma_0;\gamma,\delta) O_\gamma(f^{\infty,p}) T O_\delta(\phi_j^{\mathbf{G}}) \operatorname{Tr}(\gamma_0, V).$$

Theorem 1.7.1 *If j is positive and big enough, then*

$$\operatorname{Tr}(\overline{u}_j, R\Gamma(M^{\mathrm{K}}(\mathbf{G},\mathcal{X})^*_{\mathbb{F}}, (W^{\geq t_1,\dots,\geq t_n}V)_{\mathbb{F}})) = T_G + \sum_{\mathbf{P}} T_P,$$

where the sum is taken over the set of standard parabolic subgroups of $\mathbf{G}$. *Moreover, if $g = 1$ and $\mathrm{K} = \mathrm{K}'$, then this formula is true for every $j \in \mathbb{N}^*$.*

Proof. For every $i \in \{1,\dots,n\}$, let $a_i = -t_i - m + \dim(M_i)$. For every standard parabolic subgroup $\mathbf{P} = \mathbf{P}_{n_1} \cap \cdots \cap \mathbf{P}_{n_r}$, with $n_1 < \cdots < n_r$, let

$$T'_P = (-1)^r \operatorname{Tr}(i_{n_r !} w_{\leq a_{n_r}} i^!_{n_r} \dots i_{n_1 !} w_{\leq a_{n_1}} i^!_{n_1} \overline{u}_j).$$

Let

$$T'_G = \operatorname{Tr}(\overline{u}_j, R\Gamma(M^{\mathrm{K}}(\mathbf{G},\mathcal{X})^*_{\mathbb{F}}, (j_! \mathcal{F}^{\mathrm{K}} V)_{\mathbb{F}})).$$

Then, by the dual of proposition 5.1.5 of [M2] and the definition of $W^{\geq t_1,\dots,\geq t_n}V$,

$$\operatorname{Tr}(\overline{u}_j, R\Gamma(M^{\mathrm{K}}(\mathbf{G},\mathcal{X})^*_{\mathbb{F}}, (W^{\geq t_1,\dots,\geq t_n}V)_{\mathbb{F}})) = T'_G + \sum_{\mathbf{P}} T'_P,$$

where the sum is taken over the set of standard parabolic subgroups of $\mathbf{G}$. So we want to show that $T'_G = T_G$ and $T'_P = T_P$. Fix $\mathbf{P} \neq \mathbf{G}$ (and $n_1,\dots,n_r$). It is easy to see that

$$\dim(\mathbf{A}_{M_P}/\mathbf{A}_G) = r.$$

Let $h \in \mathbf{G}(\mathbb{A}_f^p)$. Write

$$\mathrm{K}_{N,h} = h\mathrm{K}h^{-1} \cap \mathbf{N}(\mathbb{A}_f),$$

$$\mathrm{K}_{P,h} = h\mathrm{K}h^{-1} \cap \mathbf{P}(\mathbb{A}_f),$$

$$\mathrm{K}_{M,h} = \mathrm{K}_{P,h}/\mathrm{K}_{N,h},$$

$$\mathrm{K}_{L,h} = (h\mathrm{K}h^{-1} \cap \mathbf{L}_P(\mathbb{A}_f)\mathbf{N}_P(\mathbb{A}_f))/\mathrm{K}_{N,h},$$

$$\mathrm{H}_h = h\mathrm{K}h^{-1} \cap \mathbf{P}(\mathbb{Q})\mathbf{Q}_{n_r}(\mathbb{A}_f),$$

$$\mathrm{H}_{L,h} = h\mathrm{K}h^{-1} \cap \mathbf{L}_P(\mathbb{Q})\mathbf{N}_P(\mathbb{A}_f).$$

Define in the same way groups $\mathrm{K}'_{N,h}$, etc., by replacing K with K′. If there exists $q \in \mathbf{P}(\mathbb{Q})\mathbf{Q}_{n_r}(\mathbb{A}_f)$ such that $qh\mathrm{K} = hg\mathrm{K}$, let $\overline{q}$ be the image of q in $\mathbf{M}_P(\mathbb{A}_f)$, and let u_h be the cohomological correspondence on $\mathcal{F}^{\mathrm{H}_h/\mathrm{H}_{L,h}} R\Gamma_c(\mathrm{H}_{L,h}, R\Gamma(\mathrm{Lie}(\mathbf{N}_{n_r}), V)_{\geq t_{n_1},\dots,\geq t_{n_r}})[a]$ with support in $(\Phi^j T_{\overline{q}}, T_1)$ equal to $\Phi^j c_{\overline{q},1}$ (we may assume that $q \in \mathbf{P}(\mathbb{A}_f^p)$, hence that $\overline{q} \in \mathbf{M}_P(\mathbb{A}_f^p)$). This correspondence is called $u_{C'}$ in section 1.5, where C' is the image of h in C'_P. If there is no such $q \in \mathbf{P}(\mathbb{Q})\mathbf{Q}_{n_r}(\mathbb{A}_f)$, take $u_h = 0$. Similarly, if there exists $q \in \mathbf{P}(\mathbb{A}_f)$ such that

$qh\mathrm{K} = hg\mathrm{K}$, let $\overline{q}$ be the image of q in $\mathbf{M}_P(\mathbb{A}_f)$, and let v_h be the cohomological correspondence on $\mathcal{F}^{\mathrm{K}_{M,h}/\mathrm{K}_{L,h}} R\Gamma_c(\mathrm{K}_{L,h}, R\Gamma(\mathrm{Lie}(\mathbf{N}_P), V)_{\geq t_{n_1},\dots,\geq t_{n_r}})[a]$ with support in $(\Phi^j T_{\overline{q}}, T_1)$ equal to $\Phi^j c_{\overline{q},1}$ (we may assume that $q \in \mathbf{P}(\mathbb{A}_f^p)$). If there is no such $q \in \mathbf{P}(\mathbb{A}_f)$, take $v_h = 0$. Finally, let $N_h = [\mathrm{K}_{N,h} : \mathrm{K}'_{N,h}]$.

Let $h \in \mathbf{G}(\mathbb{A}_f^p)$ be such that there exists $q \in \mathbf{P}(\mathbb{A}_f)$ with $qh\mathrm{K} = hg\mathrm{K}$. By proposition 1.7.2 below,

$$\mathrm{Tr}(v_h) = \sum_{h'} \mathrm{Tr}(u_{h'}),$$

where the sum is taken over a system of representatives $h' \in \mathbf{G}(\mathbb{A}_f^p)$ of the double classes in $\mathbf{P}(\mathbb{Q})\mathbf{Q}_{n_r}(\mathbb{A}_f) \setminus \mathbf{G}(\mathbb{A}_f)/\mathrm{K}'$ that are sent to the class of h in $\mathbf{P}(\mathbb{A}_f) \setminus \mathbf{G}(\mathbb{A}_f)/K'$ (apply proposition 1.7.2 with $\mathbf{M} = \mathbf{M}_P$, $\mathrm{K}_M = \mathrm{K}_{M,h}$, m equal to the image of q in $\mathbf{M}_P(\mathbb{A}_f)$). On the other hand, by proposition 1.5.3,

$$T'_P = (-1)^r \sum_h N_h \,\mathrm{Tr}(u_h),$$

where the sum is taken over a system of representatives $h \in \mathbf{G}(\mathbb{A}_f^p)$ of the double classes in $\mathbf{P}(\mathbb{Q})\mathbf{Q}_{n_r}(\mathbb{A}_f) \setminus \mathbf{G}(\mathbb{A}_f)/\mathrm{K}'$. Hence

$$T'_P = (-1)^r \sum_h N_h \,\mathrm{Tr}(v_h),$$

where the sum is taken over a system of representatives $h \in \mathbf{G}(\mathbb{A}_f^p)$ of the double classes in $\mathbf{P}(\mathbb{A}_f) \setminus \mathbf{G}(\mathbb{A}_f)/\mathrm{K}'$.

Let $h \in \mathbf{G}(\mathbb{A}_f^p)$. Assume that there exists $q \in \mathbf{P}(\mathbb{A}_f^p)$ such that $qh\mathrm{K} = hg\mathrm{K}$. Let $\overline{q}$ be the image of q in $\mathbf{M}_P(\mathbb{A}_f^p)$. Write $\overline{q} = q_L q_H$, with $q_L \in \mathbf{L}_P(\mathbb{A}_f^p)$ and $q_H \in \mathbf{G}_{n_r}(\mathbb{A}_f^p)$. Let

$$f_{G,h}^{\infty,p} = \mathrm{vol}(\mathrm{K}'_{M,h}/\mathrm{K}'_{L,h})^{-1} \mathbb{1}_{q_H(\mathrm{K}_{M,h}/\mathrm{K}_{L,h})}.$$

Notice that $\mathrm{K}'_{L,h} \subset \mathrm{K}_{L,h} \cap q_L \mathrm{K}_{L,h} q_L^{-1}$. Let u_{q_L} be the endomorphism of $R\Gamma_c(\mathrm{K}_{L,h}, R\Gamma(\mathrm{Lie}(\mathbf{N}_P), V)_{\geq t_{n_1},\dots,\geq t_{n_r}})$ induced by the cohomological correspondence $c_{q_L,1}$.

To calculate the trace of v_h, we will use Deligne's conjecture, which has been proved by Pink (cf. [P3]) assuming some hypotheses (that are satisfied here), and in general by Fujiwara ([F]) and Varshavsky ([V]). This conjecture (which should now be called a theorem) says that, if j is big enough, then the fixed points of the correspondence between schemes underlying v_h are all isolated, and that the trace of v_h is the sum over these fixed points of the naive local terms. By theorem 1.6.1 and remarks 1.6.4 and 1.6.5, if j is big enough, then

$$\mathrm{Tr}(v_h) = (-1)^r m_P \sum_{(\gamma_0;\gamma,\delta)\in C'_{\mathbf{G}_{n_r},j}} c(\gamma_0;\gamma,\delta) O_\gamma(f_{G,h}^{\infty,p}) T O_\delta(\phi_j^{\mathbf{G}_{n_r}}) \,\mathrm{Tr}(u_{q_L}\gamma_0, R\Gamma_c(\mathrm{K}_{L,h}, R\Gamma(\mathrm{Lie}(\mathbf{N}_P), V)_{\geq t_{n_1},\dots,\geq t_{n_r}})).$$

Let

$$f_{L_P,h}^{\infty} = \mathrm{vol}(\mathrm{K}'_{L,h})^{-1} \mathbb{1}_{q_L \mathrm{K}_{L,h}}.$$

Then

$$f^{\infty}_{L_P,h} = \mathbb{1}_{\mathbf{L}_P(\mathbb{Z}_p)} f^{\infty,p}_{L_P,h},$$

with $f^{\infty,p}_{L_P,h} \in C_c^{\infty}(\mathbf{L}_P(\mathbb{A}_f^p))$. By theorem 1.6.6, for every $\gamma_0 \in \mathbf{G}_{n_r}(\mathbb{Q})$,

$$\begin{aligned}
&\mathrm{Tr}(u_{q_L}\gamma_0, R\Gamma_c(\mathrm{K}_{L,h}, R\Gamma(\mathrm{Lie}(\mathbf{N}_P), V)_{\geq t_{n_1},\dots,\geq t_{n_r}})) \\
&= \sum_{\mathbf{L}} (-1)^{\dim(\mathbf{A}_L/\mathbf{A}_{L_P})}(n_L^{L_P})^{-1} \\
&\sum_{\gamma_L} \iota^L(\gamma_L)^{-1}\chi(\mathbf{L}_{\gamma_L})|D_L^{L_P}(\gamma_L)|^{1/2} O_{\gamma_L}((f^{\infty}_{L_P,h})_L) \\
&\quad \times \mathrm{Tr}(\gamma_L\gamma_0, R\Gamma(\mathrm{Lie}(\mathbf{N}_P), V)_{\geq t_{n_1},\dots,\geq t_{n_r}}),
\end{aligned}$$

where the first sum is taken over the set of conjugacy classes of cuspidal Levi subgroups $\mathbf{L}$ of $\mathbf{L}_P$ and the second sum is taken over the set of semisimple conjugacy classes γ_L of $\mathbf{L}(\mathbb{Q})$ that are elliptic in $\mathbf{L}(\mathbb{R})$. To show that $T'_P = T_P$, it is enough to show that, for every Levi subgroup $\mathbf{L}$ of $\mathbf{L}_P$, for every $\gamma_L \in \mathbf{L}(\mathbb{Q})$ and every $(\gamma_0; \gamma, \delta) \in C_{\mathbf{G}_{n_r},j}$,

$$\sum_h N_h O_{\gamma_L}((f^{\infty}_{L_P,h})_L) O_\gamma(f^{\infty,p}_{G,h}) = O_{\gamma_L\gamma}(f^{\infty,p}_{\mathbf{LG}_{n_r}})\delta^{1/2}_{P(\mathbb{Q}_p)}(\gamma_L\gamma_0) O_{\gamma_L}(\mathbb{1}_{\mathbf{L}(\mathbb{Z}_p)})\delta^{1/2}_{P(\mathbb{R})}(\gamma_0),$$

where the sum is taken over a system of representatives $h \in \mathbf{G}(\mathbb{A}_f^p)$ of the double classes in $\mathbf{P}(\mathbb{A}_f) \setminus \mathbf{G}(\mathbb{A}_f)/\mathrm{K}'$ (with $f^{\infty}_{L_P,h} = 0$ and $f^{\infty,p}_{G,h} = 0$ if there is no $q \in \mathbf{P}(\mathbb{A}_f)$ such that $qh\mathrm{K} = hg\mathrm{K}$).

Fix a parabolic subgroup $\mathbf{R}$ of $\mathbf{L}_P$ with Levi subgroup $\mathbf{L}$, and let $\mathbf{P}' = \mathbf{R}\mathbf{G}_{n_r}\mathbf{N}_P$ (a parabolic subgroup of $\mathbf{G}$ with Levi subgroup $\mathbf{LG}_{n_r}$). Fix a system of representatives $(h_i)_{i\in I}$ in $\mathbf{G}(\mathbb{A}_f^p)$ of $\mathbf{P}(\mathbb{A}_f) \setminus \mathbf{G}(\mathbb{A}_f)/\mathrm{K}'$. For every $i \in I$, fix a system of representatives $(m_{ij})_{j\in J_i}$ in $\mathbf{L}_P(\mathbb{A}_f^p)$ of $\mathbf{R}(\mathbb{A}_f) \setminus \mathbf{L}_P(\mathbb{A}_f)/\mathrm{K}'_{L,h_i}$. Then $(m_{ij}h_i)_{i,j}$ is a system of representatives of $\mathbf{P}'(\mathbb{A}_f) \setminus \mathbf{G}(\mathbb{A}_f)/\mathrm{K}'$. By lemma 1.7.4 below,

$$O_{\gamma_L\gamma}(f^{\infty,p}_{\mathbf{LG}_{n_r}}) = \delta^{1/2}_{P'(\mathbb{A}_f^p)}(\gamma_L\gamma)\sum_{i,j} r(m_{ij}h_i) O_{\gamma_L\gamma}(f_{P',m_{ij}h_i}),$$

where

$$r(m_{ij}h_i) = [(m_{ij}h_i)\mathrm{K}(m_{ij}h_i)^{-1} \cap \mathbf{N}_{P'}(\mathbb{A}_f) : (m_{ij}h_i)\mathrm{K}'(m_{ij}h_i)^{-1} \cap \mathbf{N}_{P'}(\mathbb{A}_f)]$$

and $f_{P',m_{ij}h_i}$ is equal to the product of

$$\mathrm{vol}\left(((m_{ij}h_i)\mathrm{K}'(m_{ij}h_i)^{-1} \cap \mathbf{P}'(\mathbb{A}_f))/((m_{ij}h_i)\mathrm{K}'(m_{ij}h_i)^{-1} \cap \mathbf{N}_{P'}(\mathbb{A}_f))\right)^{-1}$$

and of the characteristic function of the image in $(\mathbf{LG}_{n_r})(\mathbb{A}_f^p) = \mathbf{M}_{P'}(\mathbb{A}_f^p)$ of $(m_{ij}h_i)g\mathrm{K}(m_{ij}h_i)^{-1} \cap \mathbf{P}'(\mathbb{A}_f^p)$. Note that

$$r(m_{ij}h_i) = N_{h_i} r'(m_{ij}),$$

where

$$r'(m_{ij}) = [m_{ij}\mathrm{K}_{L,h_i}m_{ij}^{-1} \cap \mathbf{N}_R(\mathbb{A}_f) : m_{ij}\mathrm{K}'_{L,h_i}m_{ij}^{-1} \cap \mathbf{N}_R(\mathbb{A}_f)],$$

that

$$\delta_{P'(\mathbb{A}_f^p)}(\gamma_L\gamma) = \delta_{R(\mathbb{A}_f^p)}(\gamma_L)\delta_{P(\mathbb{A}_f^p)}(\gamma_L\gamma),$$

and that

$$f_{P',m_{ij}h_i} = f_{R,m_{ij}} f_{G,h_i}^{\infty,p},$$

where $f_{R,m_{ij}}$ is the product of

$$\mathrm{vol}\left((m_{ij}\mathrm{K}'_{L,h_i}m_{ij}^{-1} \cap \mathbf{R}(\mathbb{A}_f))/(m_{ij}\mathrm{K}'_{L,h_i}m_{ij}^{-1} \cap \mathbf{N}_R(\mathbb{A}_f))\right)^{-1}$$

and of the characteristic function of the image in $\mathbf{L}(\mathbb{A}_f^p) = \mathbf{M}_R(\mathbb{A}_f^p)$ of $(m_{ij}h_i)g\mathrm{K}(m_{ij}h_i)^{-1} \cap \mathbf{R}(\mathbb{A}_f)\mathbf{N}_P(\mathbb{A}_f)$. By applying lemma 1.7.4 again, we find, for every $i \in I$,

$$\sum_{j\in J_i} r'(m_{ij})O_{\gamma_L}(f_{R,m_{ij}}) = \delta_{R(\mathbb{A}_f^p)}^{-1/2}(\gamma_L)O_{\gamma_L}((f_{L_P,h_i}^{\infty,p})_{\mathbf{L}}).$$

Finally,

$$\begin{aligned}
&\sum_{i\in I} N_{h_i}O_{\gamma_L}((f_{L_P,h_i}^{\infty})_{\mathbf{L}})O_\gamma(f_{G,h_i}^{\infty,p})\\
&= O_{\gamma_L}((\mathbb{1}_{\mathbf{L}_P(\mathbb{Z}_p)})_{\mathbf{L}})\sum_{i\in I} N_{h_i}O_{\gamma_L}((f_{L_P,h_i}^{\infty,p})_{\mathbf{L}})O_\gamma(f_{G,h_i}^{\infty,p})\\
&= O_{\gamma_L}((\mathbb{1}_{\mathbf{L}_P(\mathbb{Z}_p)})_{\mathbf{L}})\sum_{i\in I} N_{h_i}O_\gamma(f_{G,h_i}^{\infty,p})\delta_{R(\mathbb{A}_f^p)}^{1/2}(\gamma_L)\sum_{j\in J_i} r'(m_{ij})O_{\gamma_L}(f_{R,m_{ij}})\\
&= O_{\gamma_L}((\mathbb{1}_{\mathbf{L}_P(\mathbb{Z}_p)})_{\mathbf{L}})\delta_{R(\mathbb{A}_f^p)}^{1/2}(\gamma_L)\sum_{i\in I}\sum_{j\in J_i} r(m_{ij}h_i)O_{\gamma_L\gamma}(f_{P',m_{ij}h_i})\\
&= O_{\gamma_L}((\mathbb{1}_{\mathbf{L}_P(\mathbb{Z}_p)})_{\mathbf{L}})\delta_{R(\mathbb{A}_f^p)}^{1/2}(\gamma_L)\delta_{P'(\mathbb{A}_f^p)}^{-1/2}(\gamma_L\gamma)O_{\gamma_L\gamma}(f_{\mathbf{LG}_{nr}}^{\infty,p})\\
&= O_{\gamma_L}((\mathbb{1}_{\mathbf{L}_P(\mathbb{Z}_p)})_{\mathbf{L}})\delta_{P(\mathbb{A}_f^p)}^{-1/2}(\gamma_L\gamma)O_{\gamma_L\gamma}(f_{\mathbf{LG}_{nr}}^{\infty,p}).
\end{aligned}$$

To finish the proof, it suffices to notice that $(\mathbb{1}_{\mathbf{L}_P(\mathbb{Z}_p)})_{\mathbf{L}} = \mathbb{1}_{\mathbf{L}(\mathbb{Z}_p)}$, that $\delta_{P(\mathbb{A}_f^p)}^{-1/2}(\gamma_L\gamma) = \delta_{P(\mathbb{A}_f^p)}^{-1/2}(\gamma_L\gamma_0)$, that, as $\gamma_L\gamma_0 \in \mathbf{M}_P(\mathbb{Q})$, the product formula gives

$$\delta_{P(\mathbb{A}_f^p)}^{-1/2}(\gamma_L\gamma_0) = \delta_{P(\mathbb{Q}_p)}^{1/2}(\gamma_L\gamma_0)\delta_{P(\mathbb{R})}^{1/2}(\gamma_L\gamma_0)$$

and that

$$\delta_{P(\mathbb{Q}_p)}(\gamma_L\gamma_0) = \delta_{P(\mathbb{Q}_p)}(\gamma_L)\delta_{P(\mathbb{Q}_p)}(\gamma_0) = \delta_{P(\mathbb{Q}_p)}(\gamma_0)$$

if $O_{\gamma_L}(\mathbb{1}_{\mathbf{L}(\mathbb{Z}_p)}) \neq 0$ (because this implies that γ_L is conjugate in $\mathbf{L}(\mathbb{Q}_p)$ to an element of $\mathbf{L}(\mathbb{Z}_p)$).

If j is big enough, we can calculate T'_G using theorem 1.6.1 and Deligne's conjecture. It is obvious $T'_G = T_G$.

If $g = 1$ and $\mathrm{K} = \mathrm{K}'$, then $\overline{u}_j$ is simply the cohomological correspondence induced by Φ^j. In this case, we can calculate the trace of $\overline{u}_j$, for every $j \in \mathbb{N}^*$, using Grothendieck's trace formula (cf. [SGA 4 1/2] Rapport 3.2). □

Proposition 1.7.2 *Let* $\mathbf{M}$, $\mathbf{L}$ *and* $(\mathbf{G}, \mathcal{X})$ *be as in section 1.2. Let* $m \in \mathbf{M}(\mathbb{A}_f)$, *and let* K'_M, K_M *be neat open compact subgroups of* $\mathbf{M}(\mathbb{A}_f)$ *such that* $\mathrm{K}'_M \subset \mathrm{K}_M \cap m\mathrm{K}_M m^{-1}$. *Let* $\mathrm{K}_L = \mathrm{K}_M \cap \mathbf{L}(\mathbb{A}_f)$ *and* $\mathrm{K} = \mathrm{K}_M/\mathrm{K}_L$. *Consider a system of* *representatives* $(m_i)_{i\in I}$ *of the set of double classes* $c \in \mathbf{L}(\mathbb{Q})\mathbf{G}(\mathbb{A}_f) \setminus \mathbf{M}(\mathbb{A}_f)/\mathrm{K}'_M$ *such that* $cm\mathrm{K}_M = c\mathrm{K}_M$. *For every* $i \in I$, *fix* $l_i \in \mathbf{L}(\mathbb{Q})$ *and* $g_i \in \mathbf{G}(\mathbb{A}_f)$ *such that* $l_i g_i m_i \in m_i m \mathrm{K}_M$. *Assume that the Shimura varieties and the morphisms that we get from the above data have good reduction modulo* p *as in section 1.3 (in particular,* K_M *and* K'_M *are hyperspecial at* p, *and* $m, m_i \in \mathbf{M}(\mathbb{A}_f^p)$, $g_i \in \mathbf{G}(\mathbb{A}_f^p)$*). Let* $\mathbb{F}_q$ *be the field of definition of these varieties and* $\mathbb{F}$ *be an algebraic closure of* $\mathbb{F}_q$.

For every $i \in I$, *let* $\mathrm{H}_i = m_i\mathrm{K}_M m_i^{-1} \cap \mathbf{L}(\mathbb{Q})\mathbf{G}(\mathbb{A}_f)$, $\mathrm{H}_{i,L} = \mathrm{H}_i \cap \mathbf{L}(\mathbb{Q})$ *and* $\mathrm{K}_i = \mathrm{H}_i/\mathrm{H}_{i,L}$. *Fix* $V \in \mathrm{Ob}\,\mathrm{Rep}_{\mathbf{G}}$. *Let*

$$L = \mathcal{F}^{\mathrm{K}} R\Gamma(\mathrm{K}_L, V),$$

$$L_i = \mathcal{F}^{\mathrm{K}_i} R\Gamma(\mathrm{H}_{i,L}, V),$$

$$M = \mathcal{F}^{\mathrm{K}} R\Gamma_c(\mathrm{K}_L, V),$$

$$M_i = \mathcal{F}^{\mathrm{K}} R\Gamma_c(\mathrm{H}_{i,L}, V).$$

Then, for every $\sigma \in \mathrm{Gal}(\mathbb{F}/\mathbb{F}_q)$,

(1) $\sum_{i\in I} \mathrm{Tr}(\sigma c_{l_i g_i,1}, R\Gamma(M^{\mathrm{K}_i}(\mathbf{G},\mathcal{X})_{\mathbb{F}}, L_{i,\mathbb{F}})) = \mathrm{Tr}(\sigma c_{m,1}, R\Gamma(M^{\mathrm{K}}(\mathbf{G},\mathcal{X})_{\mathbb{F}}, L_{\mathbb{F}}))$.

(2) $\sum_{i\in I} \mathrm{Tr}(\sigma c_{l_i g_i,1}, R\Gamma_c(M^{\mathrm{K}_i}(\mathbf{G},\mathcal{X})_{\mathbb{F}}, M_{i,\mathbb{F}})) = \mathrm{Tr}(\sigma c_{m,1}, R\Gamma_c(M^{\mathrm{K}}(\mathbf{G},\mathcal{X})_{\mathbb{F}}, M_{\mathbb{F}}))$.

Proof. Write $m = lg$, with $l \in \mathbf{L}(\mathbb{A}_f)$ and $g \in \mathbf{G}(\mathbb{A}_f)$. We may assume that $m_i \in \mathbf{L}(\mathbb{A}_f)$, hence $g_i = g$, for every $i \in I$. Let $\mathrm{K}^0 = \mathrm{H}_i \cap \mathbf{G}(\mathbb{A}_f) = m\mathrm{K}_M m^{-1} \cap \mathbf{G}(\mathbb{A}_f)$.

Point (1) implies point (2) by duality.

Let us prove (1). Let c_m be the endomorphism of $R\Gamma(\mathrm{K}_M, V)$ equal to

$$R\Gamma(\mathrm{K}_M, V) \longrightarrow R\Gamma(\mathrm{K}'_M, V) \xrightarrow{\mathrm{Tr}} R\Gamma(\mathrm{K}_M, V),$$

where the first map is induced by the injection $\mathrm{K}'_M \longrightarrow \mathrm{K}_M$, $k \longmapsto m^{-1}km$, and the second map is the trace morphism associated to the injection $\mathrm{K}'_M \subset \mathrm{K}_M$. Define in the same way, for every $i \in I$, an endomorphism $c_{l_i g_i}$ of $R\Gamma(\mathrm{H}_i, V)$. Then

$$R\Gamma(\mathrm{K}_M, V) \simeq \bigoplus_{i\in I} R\Gamma(\mathrm{H}_i, V)$$

and $c_m = \bigoplus_{i\in I} c_{l_i g_i}$, so it is enough to show that this decomposition is $\mathrm{Gal}(\mathbb{F}/\mathbb{F}_q)$-equivariant. Let $\sigma \in \mathrm{Gal}(\mathbb{F}/\mathbb{F}_q)$. Then σ induces an endomorphism of $R\Gamma(\mathrm{K}^0, V) = R\Gamma(M^{\mathrm{K}^0}(\mathbf{G},\mathcal{X})_{\mathbb{F}}, \mathcal{F}^{\mathrm{K}^0} V_{\mathbb{F}})$, that will still be denoted by σ, and, by the lemma below, the endomorphism of $R\Gamma(\mathrm{K}_M, V)$ (resp., $R\Gamma(\mathrm{H}_i, V)$) induced by σ is

$$R\Gamma(\mathrm{K}_M/(\mathrm{K}_M \cap \mathbf{L}(\mathbb{A}_f)), \sigma)$$

$$(\text{resp.,} \quad R\Gamma(\mathrm{H}_i/(\mathrm{H}_i \cap \mathbf{L}(\mathbb{Q})), \sigma)).$$

This finishes the proof. □

Lemma 1.7.3 *Let* $\mathbf{M}$, $\mathbf{L}$ *and* $(\mathbf{G}, \mathcal{X})$ *be as in the proposition above. Let* K_M *be a neat open compact subgroup of* $\mathbf{M}(\mathbb{A}_f)$. *Let* $\mathrm{K}_L = \mathrm{K}_M \cap \mathbf{M}(\mathbb{A}_f)$, $\mathrm{K}_G = \mathrm{K}_M \cap \mathbf{G}(\mathbb{A}_f)$, $\mathrm{H} = \mathrm{K}_M \cap \mathbf{L}(\mathbb{Q})\mathbf{G}(\mathbb{A}_f)$, $\mathrm{H}_L = \mathrm{K}_M \cap \mathbf{L}(\mathbb{Q})$, $\mathrm{K} = \mathrm{K}_M/\mathrm{K}_L$ *and* $\mathrm{K}' = \mathrm{H}/\mathrm{H}_L$. *Let* $V \in \operatorname{Ob}\operatorname{Rep}_{\mathbf{M}}$ *and* $\sigma \in \operatorname{Gal}(\mathbb{F}/\mathbb{F}_q)$. *The element* σ *induces an endomorphism of* $R\Gamma(\mathrm{K}_G, V) = R\Gamma(M^{\mathrm{K}_G}(\mathbf{G}, \mathcal{X})_{\mathbb{F}}, \mathcal{F}^{\mathrm{K}_G} V_{\mathbb{F}})$ *(resp.,* $R\Gamma(\mathrm{K}_M, V) = R\Gamma(M^{\mathrm{K}}(\mathbf{G}, \mathcal{X})_{\mathbb{F}}, \mathcal{F}^{\mathrm{K}} R\Gamma(\mathrm{K}_L, V)_{\mathbb{F}})$, *resp.,* $R\Gamma(\mathrm{H}, V) = R\Gamma(M^{\mathrm{K}'}(\mathbf{G}, \mathcal{X})_{\mathbb{F}}, \mathcal{F}^{\mathrm{K}'} R\Gamma(\mathrm{H}_L, V)_{\mathbb{F}})$*), that will be denoted by* φ_0 *(resp.* φ*, resp.* φ'*). Then*

$$\varphi = R\Gamma(\mathrm{K}_M/\mathrm{K}_G, \varphi_0)$$

and

$$\varphi' = R\Gamma(\mathrm{H}/\mathrm{K}_G, \varphi_0).$$

Proof. The two equalities are proved in the same way. Let us prove the first one. Let $Y = M^{\mathrm{K}_G}(\mathbf{G}, \mathcal{X})$, $X = M^{\mathrm{K}}(\mathbf{G}, \mathcal{X})$, let $f : Y \longrightarrow X$ be the (finite étale) morphism T_1 and $L = \mathcal{F}^{\mathrm{K}} R\Gamma(\mathrm{K}_L, V)$. Then, $f^*L = \mathcal{F}^{\mathrm{K}_G} R\Gamma(\mathrm{K}_L, V)$ by [P1] (1.11.5), and L is canonically a direct factor of $f_* f^* L$ because f is finite étale, so it is enough to show that the endomorphism of

$$R\Gamma(Y_{\mathbb{F}}, f^*L) = R\Gamma(\mathrm{K}_G, R\Gamma(\mathrm{K}_L, V)) = R\Gamma(\mathrm{K}_L, R\Gamma(\mathrm{K}_G, V))$$

induced by σ is equal to $R\Gamma(\mathrm{K}_L, \varphi_0)$. The complex $M = \mathcal{F}^{\mathrm{K}_G} V$ on Y is a complex of K_L-sheaves in the sense of [P2] (1.2), and $R\Gamma(\mathrm{K}_L, M) = f^*L$ by [P2] (1.9.3). To conclude, apply [P2] (1.6.4). □

The following lemma of [GKM] is used in the proof of theorem 1.7.1. Let $\mathbf{G}$ be a connected reductive group over $\mathbb{Q}$, $\mathbf{M}$ a Levi subgroup of $\mathbf{G}$ and $\mathbf{P}$ a parabolic subgroup of $\mathbf{G}$ with Levi subgroup $\mathbf{M}$. Let $\mathbf{N}$ be the unipotent radical of $\mathbf{P}$. If $f \in C_c^\infty(\mathbf{G}(\mathbb{A}_f))$, the constant term $f_M \in C_c^\infty(\mathbf{M}(\mathbb{A}_f))$ of f at $\mathbf{M}$ is defined in [GKM] (7.13) (the function f_M depends on the choice of $\mathbf{P}$, but its orbital integrals do not depend on that choice). For every $g \in \mathbf{M}(\mathbb{A}_f)$, let

$$\delta_{P(\mathbb{A}_f)}(g) = |\det(\mathrm{Ad}(g), \mathrm{Lie}(\mathbf{N}) \otimes \mathbb{A}_f)|_{\mathbb{A}_f}.$$

Let $g \in \mathbf{G}(\mathbb{A}_f)$ and let K', K be open compact subgroups of $\mathbf{G}(\mathbb{A}_f)$ such that $\mathrm{K}' \subset g\mathrm{K}g^{-1}$. For every $h \in \mathbf{G}(\mathbb{A}_f)$, let $\mathrm{K}_M(h)$ be the image in $\mathbf{M}(\mathbb{A}_f)$ of $hg\mathrm{K}h^{-1} \cap \mathbf{P}(\mathbb{A}_f)$,

$$f_{P,h} = \mathrm{vol}((h\mathrm{K}'h^{-1} \cap \mathbf{P}(\mathbb{A}_f))/(h\mathrm{K}'h^{-1} \cap \mathbf{N}(\mathbb{A}_f)))^{-1} \mathbb{1}_{\mathrm{K}_M(h)} \in C_c^\infty(\mathbf{M}(\mathbb{A}_f)),$$

and

$$r(h) = [h\mathrm{K}h^{-1} \cap \mathbf{N}(\mathbb{A}_f) : h\mathrm{K}'h^{-1} \cap \mathbf{N}(\mathbb{A}_f)].$$

(Note that, if there is no element $q \in \mathbf{P}(\mathbb{A}_f)$ such that $qh\mathrm{K} = hg\mathrm{K}$, then $\mathrm{K}_M(h)$ is empty, hence $f_{P,h} = 0$.) Let

$$f = \mathrm{vol}(\mathrm{K}')^{-1} \mathbb{1}_{g\mathrm{K}}$$

and

$$f_P = \sum_h r(h) f_{P,h},$$

where the sum is taken over a system of representatives of the double quotient $\mathbf{P}(\mathbb{A}_f) \setminus \mathbf{G}(\mathbb{A}_f)/\mathrm{K}'$.

Lemma 1.7.4 *([GKM] 7.13.A) The functions f_M and $\delta_{P(\mathbb{A}_f)}^{1/2} f_P$ have the same orbital integrals.*

In [GKM], the g is on the right of the K (and not on the left), and $\delta_{P(\mathbb{A}_f)}^{-1/2}$ appears in the formula instead of $\delta_{P(\mathbb{A}_f)}^{1/2}$, but it is easy to see that their proof adapts to the case considered here. There are obvious variants of this lemma obtained by replacing $\mathbb{A}_f$ with $\mathbb{A}_f^p$ or $\mathbb{Q}_p$, where p is a prime number.

Remark 1.7.5 The above lemma implies in particular that the function $\gamma \longmapsto O_\gamma(f_{\mathbf{M}})$ on $\mathbf{M}(\mathbb{A}_f)$ has its support contained in a set of the form $\bigcup_{m \in \mathbf{M}(\mathbb{A}_f)} mXm^{-1}$, where X is a compact subset of $\mathbf{M}(\mathbb{A}_f)$, because the support of $\gamma \longmapsto O_\gamma(f_{\mathbf{M}})$ is contained in the union of the conjugates of $\mathrm{K}_M(h)$, for h in a system of representatives of the finite set $\mathbf{P}(\mathbb{A}_f) \setminus \mathbf{G}(\mathbb{A}_f)/\mathrm{K}'$. Moreover, if $g = 1$, then we may assume that X is a finite union of compact subgroups of $\mathbf{M}(\mathbb{A}_f)$, that are neat if K is neat (because the $\mathrm{K}_M(h)$ are subgroups of $\mathbf{M}(\mathbb{A}_f)$ in that case).

Chapter Two

The groups

In the next chapters, we will apply the fixed point formula to certain unitary groups over $\mathbb{Q}$. The goal of this chapter is to define these unitary groups and their Shimura data, and to recall the description of their parabolic subgroups and of their endoscopic groups.

2.1 DEFINITION OF THE GROUPS AND OF THE SHIMURA DATA

For $n \in \mathbb{N}^*$, write

$$I = I_n = \begin{pmatrix} 1 & & 0 \\ & \ddots & \\ 0 & & 1 \end{pmatrix} \in \mathbf{GL}_n(\mathbb{Z})$$

and

$$A_n = \begin{pmatrix} 0 & & 1 \\ & \cdot^{\cdot^{\cdot}} & \\ 1 & & 0 \end{pmatrix} \in \mathbf{GL}_n(\mathbb{Z}).$$

Let $E = \mathbb{Q}[\sqrt{-b}]$ ($b \in \mathbb{N}^*$ square-free) be an imaginary quadratic extension of $\mathbb{Q}$. The nontrivial automorphism of E will be denoted by $\overline{}$. Fix once and for all an injection $E \subset \overline{\mathbb{Q}} \subset \mathbb{C}$, and an injection $\overline{\mathbb{Q}} \subset \overline{\mathbb{Q}}_p$ for every prime number p.

Let $n \in \mathbb{N}^*$ and let $J \in \mathbf{GL}_n(\mathbb{Q})$ be a symmetric matrix. Define an algebraic group $\mathbf{GU}(J)$ over $\mathbb{Q}$ by

$$\mathbf{GU}(J)(A) = \{g \in \mathbf{GL}_n(E \otimes_{\mathbb{Q}} A) | g^* J g = c(g) J, c(g) \in A^\times\},$$

for every $\mathbb{Q}$-algebra A (for $g \in \mathbf{GL}_n(E \otimes_{\mathbb{Q}} A)$, we write $g^* = {}^t\overline{g}$). The group $\mathbf{GU}(J)$ comes with two morphisms of algebraic groups over $\mathbb{Q}$:

$$c : \mathbf{GU}(J) \longrightarrow \mathbb{G}_m \text{ and } \det : \mathbf{GU}(J) \longrightarrow R_{E/\mathbb{Q}}\mathbb{G}_m.$$

Let $\mathbf{U}(J) = \mathrm{Ker}(c)$ and $\mathbf{SU}(J) = \mathrm{Ker}(c) \cap \mathrm{Ker}(\det)$.

The group $\mathbf{SU}(J)$ is the derived group of $\mathbf{GU}(J)$ and $\mathbf{U}(J)$. The groups $\mathbf{GU}(J)$ and $\mathbf{U}(J)$ are connected reductive, and the group $\mathbf{SU}(J)$ is semisimple and simply connected.

Let $p, q \in \mathbb{N}$ be such that $n := p + q \geq 1$. Let

$$J = J_{p,q} := \begin{pmatrix} I_p & 0 \\ 0 & -I_q \end{pmatrix},$$

and set $\mathbf{GU}(p,q) = \mathbf{GU}(J)$, $\mathbf{U}(p,q) = \mathbf{U}(J)$ and $\mathbf{SU}(p,q) = \mathbf{SU}(J)$. If $q = 0$, we also write $\mathbf{GU}(p) = \mathbf{GU}(p,q)$, etc. These groups are quasi-split over $\mathbb{Q}$ if and only if $|p-q| \leq 1$. The semisimple $\mathbb{Q}$-rank and the semisimple $\mathbb{R}$-rank of $\mathbf{GU}(p,q)$ are both equal to $\min(p,q)$.

Let $n \in \mathbb{N}^*$. Let

$$\mathbf{GU}^*(n) = \begin{cases} \mathbf{GU}(n/2, n/2) & \text{if } n \text{ is even,} \\ \mathbf{GU}((n+1)/2, (n-1)/2) & \text{if } n \text{ is odd.} \end{cases}$$

The group $\mathbf{GU}^*(n)$ is the quasi-split inner form of $\mathbf{GU}(J)$, for every symmetric $J \in \mathbf{GL}_n(\mathbb{Q})$. Write $\mathbf{U}^*(n) = \mathrm{Ker}(c : \mathbf{GU}^*(n) \longrightarrow \mathbb{G}_m)$ and $\mathbf{SU}^*(n) = \mathrm{Ker}(\det : \mathbf{U}^*(n) \longrightarrow R_{E/\mathbb{Q}}\mathbb{G}_m)$.

Finally, let $\mathbf{GU}^*(0) = \mathbf{GU}(0,0) = \mathbb{G}_m$, and $(c : \mathbf{GU}^*(0) \longrightarrow \mathbb{G}_m) = \mathrm{id}$.

Let $n_1, \dots, n_r \in \mathbb{N}$ and let $J_1 \in \mathbf{GL}_{n_1}(\mathbb{Q}), \dots, J_r \in \mathbf{GL}_{n_r}(\mathbb{Q})$ be symmetric matrices. Write

$$\begin{aligned} &\mathbf{G}(\mathbf{U}(J_1) \times \cdots \times \mathbf{U}(J_r)) \\ &= \{(g_1, \dots, g_r) \in \mathbf{GU}(J_1) \times \cdots \times \mathbf{GU}(J_r) | c(g_1) = \cdots = c(g_r)\}. \end{aligned}$$

Similarly, write

$$\begin{aligned} &\mathbf{G}(\mathbf{U}^*(n_1) \times \cdots \times \mathbf{U}^*(n_r)) \\ &= \{(g_1, \dots, g_r) \in \mathbf{GU}^*(n_1) \times \cdots \times \mathbf{GU}^*(n_r) | c(g_1) = \cdots = c(g_r)\}. \end{aligned}$$

Remark 2.1.1 If the matrix J is in $\mathbf{GL}_n(\mathbb{Z})$, then there is an obvious way to extend $\mathbf{GU}(J)$ to a group scheme $\mathcal{G}$ over $\mathbb{Z}$: for every $\mathbb{Z}$-algebra A, set

$$\mathcal{G}(A) = \{g \in \mathbf{GL}_n(A \otimes_{\mathbb{Z}} \mathcal{O}_E) | g^* J g = c(g) J, c(g) \in A^\times\}.$$

If ℓ is a prime number unramified in E, then $\mathbf{G}_{\mathbb{Q}_\ell}$ is unramified and $\mathcal{G}_{\mathbb{F}_\ell}$ is a connected reductive algebraic group over $\mathbb{F}_\ell$.

In particular, this construction applies to the groups $\mathbf{GU}(p,q)$ and $\mathbf{GU}^*(n)$.

We now define the Shimura data. Let as before $\mathbb{S} = R_{\mathbb{C}/\mathbb{R}}\mathbb{G}_m$.

Let $p, q \in \mathbb{N}$ be such that $n := p + q \geq 1$, and let $\mathbf{G} = \mathbf{GU}(p,q)$. If $p \neq q$ (resp., $p = q$), let $\mathcal{X}$ be the set of q-dimensional subspaces of $\mathbb{C}^n$ on which the hermitian form $(v, w) \longmapsto {}^t\overline{v} J_{p,q} w$ is negative definite (resp., positive or negative definite). Let $x_0 \in \mathcal{X}$ be the subspace of $\mathbb{C}^n$ generated by the q vectors $e_{n+1-q}, \dots, e_n$, where $(e_1, \dots, e_n)$ is the canonical basis of $\mathbb{C}^n$.

The group $\mathbf{G}(\mathbb{R})$ acts on $\mathcal{X}$ via the injection $\mathbf{G}(\mathbb{R}) \subset \mathbf{GL}_n(\mathbb{R} \otimes_{\mathbb{Q}} E) \simeq \mathbf{GL}_n(\mathbb{C})$, and this action is transitive. Define a $\mathbf{G}(\mathbb{R})$-equivariant morphism $h : \mathcal{X} \longrightarrow \mathrm{Hom}(\mathbb{S}, \mathbf{G}_{\mathbb{R}})$ by

$$h_0 = h(x_0) = \begin{cases} \mathbb{S} & \longrightarrow \quad \mathbf{G}_{\mathbb{R}}, \\ z & \longmapsto \quad \begin{pmatrix} z I_p & 0 \\ 0 & \overline{z} I_q \end{pmatrix} \end{cases}$$

Then $(\mathbf{G}, \mathcal{X}, h)$ is a Shimura datum in the sense of [P1] 2.1.

The group $\mathbb{S}(\mathbb{C}) = (\mathbb{C} \otimes_{\mathbb{R}} \mathbb{C})^{\times}$ is isomorphic to $\mathbb{C}^{\times} \times \mathbb{C}^{\times}$ by the morphism $a \otimes 1 + b \otimes i \longmapsto (a + ib, a - ib)$. Let $r : \mathbb{G}_{m,\mathbb{C}} \longrightarrow \mathbb{S}_{\mathbb{C}}$ be the morphism $z \longmapsto (z, 1)$, and let $\mu = h_0 \circ r : \mathbb{G}_{m,\mathbb{C}} \longrightarrow \mathbf{G}_{\mathbb{C}}$.

Identify $\mathbf{G}_E$ with a subgroup of $\mathbf{GL}_{n,E} \times \mathbf{GL}_{n,E}$ by the isomorphism $(R_{E/\mathbb{Q}}\mathbf{GL}_{n,\mathbb{Q}})_E \simeq \mathbf{GL}_{n,E} \times \mathbf{GL}_{n,E}$ that sends $X \otimes 1 + Y \otimes \sqrt{-b}$ to $(X + \sqrt{-b}Y, X - \sqrt{-b}Y)$. Then, for every $z \in (R_{E/\mathbb{Q}}\mathbb{G}_{m,\mathbb{Q}})$,

$$\mu(z) = \begin{pmatrix} (z,1)I_p & 0 \\ 0 & (1,z)I_q \end{pmatrix}.$$

Notation 2.1.2 Let $p' \in \{1, \ldots, n\}$. Define a cocharacter $\mu_{p'} : \mathbb{G}_{m,E} \longrightarrow \mathbf{G}_E$ by

$$\mu_{p'}(z) = \begin{pmatrix} (z,1)I_{p'} & \\ 0 & (1,z)I_{n-p'} \end{pmatrix}.$$

2.2 PARABOLIC SUBGROUPS

Let $\mathbf{G}$ be a connected reductive algebraic group over $\mathbb{Q}$. Fix a minimal parabolic subgroup $\mathbf{P}_0$ of $\mathbf{G}$. Remember that a parabolic subgroup of $\mathbf{G}$ is called *standard* if it contains $\mathbf{P}_0$. Fix a Levi subgroup $\mathbf{M}_0$ of $\mathbf{P}_0$. Then a Levi subgroup $\mathbf{M}$ of $\mathbf{G}$ will be called *standard* if $\mathbf{M}$ is a Levi subgroup of a standard parabolic subgroup and $\mathbf{M} \supset \mathbf{M}_0$. Any parabolic subgroup of $\mathbf{G}$ is $\mathbf{G}(\mathbb{Q})$-conjugate to a unique standard parabolic subgroup, so it is enough to describe the standard parabolic subgroups.

Let $p, q \in \mathbb{N}$ be such that $n := p + q \geq 1$. We are interested in the parabolic subgroups of $\mathbf{GU}(p, q)$. As $\mathbf{GU}(p, q) = \mathbf{GU}(q, p)$, we may assume that $p \geq q$. Then the matrix $J_{p,q}$ is $\mathbf{GL}_n(\mathbb{Q})$-conjugate to

$$A_{p,q} := \begin{pmatrix} 0 & 0 & A_q \\ 0 & I_{p-q} & 0 \\ A_q & 0 & 0 \end{pmatrix},$$

so $\mathbf{GU}(p, q)$ is isomorphic to the unitary group $\mathbf{G} := \mathbf{GU}(A_{p,q})$, and it is enough to describe the parabolic subgroups of $\mathbf{G}$. A maximal torus of $\mathbf{G}$ is the diagonal torus

$$\mathbf{T} = \left\{ \begin{pmatrix} \lambda_1 & & 0 \\ & \ddots & \\ 0 & & \lambda_n \end{pmatrix}, \lambda_1, \ldots, \lambda_n \in R_{E/\mathbb{Q}}\mathbb{G}_m, \right.$$

$$\left. \lambda_1\overline{\lambda}_n = \cdots = \lambda_q\overline{\lambda}_{p+1} = \lambda_{q+1}\overline{\lambda}_{q+1} = \cdots = \lambda_p\overline{\lambda}_p \in \mathbb{G}_m \right\}.$$

The maximal split subtorus of $\mathbf{T}$ is

$$\mathbf{S} = \left\{ \lambda \begin{pmatrix} \lambda_1 & & & & & & 0 \\ & \ddots & & & & & \\ & & \lambda_q & & & & \\ & & & I_{p-q} & & & \\ & & & & \lambda_q^{-1} & & \\ & & & & & \ddots & \\ 0 & & & & & & \lambda_1^{-1} \end{pmatrix}, \lambda, \lambda_1, \dots, \lambda_q \in \mathbb{G}_m \right\}$$

if $p > q$, and

$$\mathbf{S} = \left\{ \begin{pmatrix} \lambda\lambda_1 & & & & & 0 \\ & \ddots & & & & \\ & & \lambda\lambda_q & & & \\ & & & \lambda_q^{-1} & & \\ & & & & \ddots & \\ 0 & & & & & \lambda_1^{-1} \end{pmatrix}, \lambda, \lambda_1, \dots, \lambda_q \in \mathbb{G}_m \right\}$$

if $p = q$.

A minimal parabolic subgroup of $\mathbf{G}$ containing $\mathbf{S}$ is

$$\mathbf{P}_0 = \left\{ \begin{pmatrix} A & & * \\ & B & \\ 0 & & C \end{pmatrix}, A, C \in R_{E/\mathbb{Q}}\mathbf{B}_q, B \in R_{E/\mathbb{Q}}\mathbf{GL}_{p-q} \right\} \cap \mathbf{G},$$

where $\mathbf{B}_q \subset \mathbf{GL}_q$ is the subgroup of upper triangular matrices.

The standard parabolic subgroups of $\mathbf{G}$ are indexed by the subsets of $\{1, \dots, q\}$ in the following way.

Let $S \subset \{1, \dots, q\}$. Write $S = \{r_1, r_1 + r_2, \dots, r_1 + \dots + r_m\}$ with $r_1, \dots, r_m \in \mathbb{N}^*$, and let $r = r_1 + \dots + r_m$. The standard parabolic subgroup $\mathbf{P}_S$ corresponding to S is the intersection of $\mathbf{G}$ and of the group

$$\begin{pmatrix} R_{E/\mathbb{Q}}\mathbf{GL}_{r_1} & & & & & & * \\ & \ddots & & & & & \\ & & R_{E/\mathbb{Q}}\mathbf{GL}_{r_m} & & & & \\ & & & \mathbf{GU}(A_{p-r,q-r}) & & & \\ & & & & R_{E/\mathbb{Q}}\mathbf{GL}_{r_m} & & \\ & & & & & \ddots & \\ 0 & & & & & & R_{E/\mathbb{Q}}\mathbf{GL}_{r_1} \end{pmatrix}.$$

In particular, the standard maximal parabolic subgroups of $\mathbf{G}$ are the

$$\mathbf{P}_r := \mathbf{P}_{\{r\}} = \begin{pmatrix} R_{E/\mathbb{Q}}\mathbf{GL}_r & & * \\ & \mathbf{GU}(A_{p-r,q-r}) & \\ 0 & & R_{E/\mathbb{Q}}\mathbf{GL}_r \end{pmatrix} \cap \mathbf{G}$$

for $r \in \{1, \dots, q\}$, and $\mathbf{P}_S = \bigcap_{r \in S} \mathbf{P}_r$. Note that $\mathbf{P}_0 = \mathbf{P}_{\{1,\dots,q\}}$.

Let $\mathbf{N}_S$ (or $\mathbf{N}_{P_S}$) be the unipotent radical of $\mathbf{P}_S$, $\mathbf{M}_S$ (or $\mathbf{M}_{P_S}$) the obvious Levi subgroup (of block diagonal matrices) and $\mathbf{A}_{M_S}$ the maximal split subtorus of the center of $\mathbf{M}_S$. Write as before $S = \{r_1, \dots, r_1 + \dots + r_m\}$ and $r = r_1 + \dots + r_m$. Then there is an isomorphism

$$\mathbf{M}_S \xrightarrow{\sim} R_{E/\mathbb{Q}}\mathbf{GL}_{r_1} \times \dots \times R_{E/\mathbb{Q}}\mathbf{GL}_{r_m} \times \mathbf{GU}(p-r, q-r)$$

that sends $\operatorname{diag}(g_1, \dots, g_m, g, h_m, \dots, h_1)$ to $(c(g)^{-1}g_1, \dots, c(g)^{-1}g_m, g)$.

The inverse image by this isomorphism of $R_{E/\mathbb{Q}}\mathbf{GL}_{r_1} \times \dots \times R_{E/\mathbb{Q}}\mathbf{GL}_{r_m}$ is called the *linear part* of $\mathbf{M}_S$ and denoted by $\mathbf{L}_S$ (or $\mathbf{L}_{P_S}$). The inverse image of $\mathbf{GU}(p-r, q-r)$ is called the *hermitian part* of $\mathbf{M}_S$ and denoted by $\mathbf{G}_r$ (or $\mathbf{G}_{P_S}$). Note that the maximal parabolic subgroups of $\mathbf{G}$ satisfy the condition of section 1.1.

2.3 ENDOSCOPIC GROUPS

In this section, we want to study the elliptic endoscopic triples for the groups $\mathbf{G}$ defined in section 2.1. It is enough to consider the quasi-split forms. We will use the definition of elliptic endoscopic triples and isomorphisms of endoscopic triples given in [K4] 7.4, 7.5.

Let $n_1, \dots, n_r \in \mathbb{N}^*$ and $\mathbf{G} = \mathbf{G}(\mathbf{U}^*(n_1) \times \dots \times \mathbf{U}^*(n_r))$; here we use the hermitian forms $A_{p,q}$ of section 2.2 to define $\mathbf{G}$. We first calculate the dual group $\widehat{\mathbf{G}}$ of $\mathbf{G}$. As $\mathbf{G}$ splits over E, the action of $\operatorname{Gal}(\overline{\mathbb{Q}}/\mathbb{Q})$ on $\widehat{\mathbf{G}}$ factors through $\operatorname{Gal}(E/\mathbb{Q})$. Let τ be the nontrivial element of $\operatorname{Gal}(E/\mathbb{Q})$.

Let φ be the isomorphism from $\mathbf{G}_E \subset \mathbf{GL}_{n_1, E\otimes E} \times \dots \times \mathbf{GL}_{n_r, E\otimes E}$ to $\mathbb{G}_{m,E} \times \mathbf{GL}_{n_1,E} \times \dots \times \mathbf{GL}_{n_r,E}$ that sends $g = (X_1 \otimes 1 + Y_1 \otimes \sqrt{-b}, \dots, X_r \otimes 1 + Y_r \otimes \sqrt{-b}) \in \mathbf{G}_E$ to $(c(g), X_1 + \sqrt{-b}Y_1, \dots, X_r + \sqrt{-b}Y_r)$. Let $\mathbf{T}$ be the diagonal torus of $\mathbf{G}$ (a maximal torus of $\mathbf{G}$) and $\mathbf{B}$ be the subgroup of upper triangular matrices in $\mathbf{G}$ (this is a Borel subgroup of $\mathbf{G}$ because of the choice of the hermitian form). There is a canonical isomorphism

$$\mathbf{T} = \{((\lambda_{1,1}, \dots, \lambda_{1,n_1}), \dots, (\lambda_{r,1}, \dots, \lambda_{r,n_r})) \in R_{E/\mathbb{Q}}\mathbb{G}_m^{n_1+\dots+n_r} \mid \exists \lambda \in \mathbb{G}_m, \forall i \in \{1, \dots, r\}, \forall j \in \{1, \dots, n_i\}, \lambda_{i,j}\overline{\lambda}_{i,n_i+1-j} = \lambda\}.$$

The restriction of φ to $\mathbf{T}_E$ induces an isomorphism

$$\mathbf{T}_E \xrightarrow{\sim} \mathbb{G}_{m,E} \times \mathbb{G}_{m,E}^{n_1} \times \dots \times \mathbb{G}_{m,E}^{n_r}.$$

For every $i \in \{1, \ldots, r\}$ and $j \in \{1, \ldots, n_i\}$, let $e_{i,j}$ be the character of $\mathbf{T}$ defined by

$$e_{i,j}(\varphi^{-1}((\lambda, (\lambda_{1,1}, \ldots, \lambda_{1,n_1}), \ldots, (\lambda_{r,1}, \ldots, \lambda_{r,n_r})))) = \lambda_{i,j}.$$

Then the group of characters of $\mathbf{T}$ is

$$X^*(\mathbf{T}) = \mathbb{Z}c \oplus \bigoplus_{i=1}^{r} \bigoplus_{j=1}^{n_i} \mathbb{Z}e_{i,j},$$

and $\mathrm{Gal}(E/\mathbb{Q})$ acts on $X^*(\mathbf{T})$ by

$$\tau(c) = c,$$

$$\tau(e_{i,j}) = c - e_{i,n_i+1-j}.$$

Hence the dual torus of $\mathbf{T}$ is

$$\widehat{\mathbf{T}} = \mathbb{C}^\times \times (\mathbb{C}^\times)^{n_1} \times \cdots \times (\mathbb{C}^\times)^{n_r},$$

with the action of $\mathrm{Gal}(E/\mathbb{Q})$ given by

$$\tau((\lambda, (\lambda_{i,j})_{1 \le i \le r, 1 \le j \le n_i})) = \left(\lambda \prod_{i,j} \lambda_{i,j}, (\lambda_{i,n_i+1-j}^{-1})_{1 \le i \le r, 1 \le j \le n_i}\right).$$

The set of roots of $\mathbf{T}$ in $\mathrm{Lie}(\mathbf{G})$ is

$$\Phi = \Phi(\mathbf{T}, \mathbf{G}) = \{e_{i,j} - e_{i,j'}, 1 \le i \le r, 1 \le j, j' \le n_i, j \ne j'\}.$$

The subset of simple roots determined by $\mathbf{B}$ is

$$\Delta = \{\alpha_{i,j} = e_{i,j+1} - e_{i,j}, 1 \le i \le r, 1 \le j \le n_i - 1\}.$$

The group $\mathrm{Gal}(E/\mathbb{Q})$ acts on Δ by

$$\tau(\alpha_{i,j}) = \alpha_{i,n_i-j}.$$

For every $n \in \mathbb{N}^*$, let $\Phi_n \in \mathbf{GL}_n(\mathbb{Z})$ be the matrix with entries

$$(\Phi_n)_{ij} = (-1)^{i+1}\delta_{i,n+1-j}.$$

The dual group of $\mathbf{G}$ is

$$\widehat{\mathbf{G}} = \mathbb{C}^\times \times \mathbf{GL}_{n_1}(\mathbb{C}) \times \cdots \times \mathbf{GL}_{n_r}(\mathbb{C}),$$

with $\widehat{\mathbf{T}}$ immersed diagonally. The action of $\mathrm{Gal}(E/\mathbb{Q})$ that respects the obvious splitting is

$$\tau((\lambda, g_1, \ldots, g_r)) = (\lambda \det(g_1) \ldots \det(g_r), \Phi_{n_1}^{-1}({}^t g_1)^{-1}\Phi_{n_1}, \ldots, \Phi_{n_r}^{-1}({}^t g_r)^{-1}\Phi_{n_r}).$$

Proposition 2.3.1 *For every $i \in \{1, \dots, r\}$, let $n_i^+, n_i^- \in \mathbb{N}$ be such that $n_i = n_i^+ + n_i^-$. Suppose that $n_1^- + \cdots + n_r^-$ is even. Set*

$$s = \left(1, \mathrm{diag}(\overbrace{1, \dots, 1}^{n_1^+}, \overbrace{-1, \dots, -1}^{n_1^-}), \dots, \mathrm{diag}(\overbrace{1, \dots, 1}^{n_r^+}, \overbrace{-1, \dots, -1}^{n_r^-})\right) \in \widehat{\mathbf{G}},$$

$$\mathbf{H} = \mathbf{G}(\mathbf{U}^*(n_1^+) \times \mathbf{U}^*(n_1^-) \times \cdots \times \mathbf{U}^*(n_r^+) \times \mathbf{U}^*(n_r^-)),$$

and define

$$\eta_0 : \widehat{\mathbf{H}} = \mathbb{C}^\times \times \mathbf{GL}_{n_1^+}(\mathbb{C}) \times \mathbf{GL}_{n_1^-}(\mathbb{C}) \times \cdots \times \mathbf{GL}_{n_r^+}(\mathbb{C}) \times \mathbf{GL}_{n_r^-}(\mathbb{C})$$

$$\longrightarrow \widehat{\mathbf{G}} = \mathbb{C}^\times \times \mathbf{GL}_{n_1}(\mathbb{C}) \times \cdots \times \mathbf{GL}_{n_r}(\mathbb{C})$$

by

$$\eta_0((\lambda, g_1^+, g_1^-, \dots, g_r^+, g_r^-)) = \left(\lambda, \begin{pmatrix} g_1^+ & 0 \\ 0 & g_1^- \end{pmatrix}, \dots, \begin{pmatrix} g_r^+ & 0 \\ 0 & g_r^- \end{pmatrix}\right).$$

Then $(\mathbf{H}, s, \eta_0)$ is an elliptic endoscopic triple for $\mathbf{G}$. The group $\Lambda(\mathbf{H}, s, \eta_0)$ of [K4] 7.5 is isomorphic to $(\mathbb{Z}/2\mathbb{Z})^I$, where $I = \{i \in \{1, \dots, r\} | n_i^+ = n_i^-\}$.

Moreover, the elliptic endoscopic triples for $\mathbf{G}$ determined by $((n_1^+, n_1^-), \dots, (n_r^+, n_r^-))$ and $((m_1^+, m_1^-), \dots, (m_r^+, m_r^-))$ are isomorphic if and only if, for every $i \in \{1, \dots, r\}$, $(n_i^+, n_i^-) = (m_i^+, m_i^-)$ or $(n_i^+, n_i^-) = (m_i^-, m_i^+)$.

Finally, every elliptic endoscopic triple for $\mathbf{G}$ is isomorphic to one of the triples defined above.

Note that an elliptic endoscopic triple $(\mathbf{H}, s, \eta_0)$ is uniquely determined by s and that, for every elliptic endoscopic triple $(\mathbf{H}, s, \eta_0)$, the group $\mathbf{H}_{\mathbb{R}}$ has an elliptic maximal torus.

Proof. Let $(\mathbf{H}, s, \eta_0)$ be determined by $((n_1^+, n_1^-), \dots, (n_r^+, n_r^-))$ as above. To show that $(\mathbf{H}, s, \eta_0)$ is an endoscopic triple for $\mathbf{G}$, we have to check conditions (7.4.1)–(7.4.3) of [K4]. Conditions (7.4.1) and (7.4.2) are obviously satisfied, and condition (7.4.3) is a consequence of the fact that $s \in Z(\widehat{\mathbf{H}})^{\mathrm{Gal}(E/\mathbb{Q})}$. (Note that the condition "$n_1^- + \cdots + n_r^-$ even" is necessary for $s \in Z(\widehat{\mathbf{H}})$ to be fixed by $\mathrm{Gal}(E/\mathbb{Q})$.)

We next show that $(\mathbf{H}, s, \eta_0)$ is elliptic. The center of $\widehat{\mathbf{H}}$ is

$$Z(\widehat{\mathbf{H}}) = \{(\lambda, \lambda_1^+ I_{n_1^+}, \lambda_1^- I_{n_1^-}, \dots, \lambda_r^+ I_{n_r^+}, \lambda_r^- I_{n_r^-}), \lambda, \lambda_1^+, \lambda_1^-, \dots, \lambda_r^+, \lambda_r^- \in \mathbb{C}^\times\},$$

with the action of $\mathrm{Gal}(E/\mathbb{Q})$ given by

$$\tau((\lambda, \lambda_1^+ I_{n_1^+}, \lambda_1^- I_{n_1^-}, \dots, \lambda_r^+ I_{n_r}^+, \lambda_r^- I_{n_r^-}))$$

$$= (\lambda(\lambda_1^+)^{n_1^+}(\lambda_1^-)^{n_1^-} \dots (\lambda_r^+)^{n_r^+}(\lambda_r^-)^{n_r^-},$$

$$(\lambda_1^+)^{-1} I_{n_1^+}, (\lambda_1^-)^{-1} I_{n_1^-}, \dots, (\lambda_r^+)^{-1} I_{n_r^+}, (\lambda_r^-)^{-1} I_{n_r^-}).$$

Hence $(Z(\widehat{\mathbf{H}})^{\mathrm{Gal}(E/\mathbb{Q})})^0 = \mathbb{C}^\times \times \{1\} \subset Z(\widehat{\mathbf{G}})$, and $(\mathbf{H}, s, \eta_0)$ is elliptic.

We want to calculate the group of outer automorphisms of $(\mathbf{H}, s, \eta_0)$. It is the same to calculate the group of outer automorphisms of the endoscopic data (s, ρ) associated to $(\mathbf{H}, s, \eta_0)$ (cf. [K4] 7.2 and 7.6). Let

$$I = \{i \in \{1, \dots, r\} | n_i^+ = n_i^-\}.$$

Let $g \in \widehat{\mathbf{G}}$ be such that $\mathrm{Int}(g)(\eta_0(\widehat{\mathbf{H}})) = \eta_0(\widehat{\mathbf{H}})$.

Let $a, b \in \mathbb{N}$ be such that $a + b = n > 0$, and

$$\mathbf{G}' = \begin{pmatrix} \mathbf{GL}_a & 0 \\ 0 & \mathbf{GL}_b \end{pmatrix} \subset \mathbf{GL}_n.$$

If $a \neq b$, then the normalizer of $\mathbf{G}'$ in $\mathbf{GL}_n$ is $\mathbf{G}'$. If $a = b$, then the normalizer of $\mathbf{G}'$ in $\mathbf{GL}_n$ is generated by $\mathbf{G}'$ and by

$$I_{a,b} := \begin{pmatrix} 0 & I_a \\ I_b & 0 \end{pmatrix}.$$

By applying this remark to $\eta_0(\widehat{\mathbf{H}}) \subset \widehat{\mathbf{G}}$, we find that g is in the subgroup of $\widehat{\mathbf{G}}$ generated by $\eta_0(\widehat{\mathbf{H}})$ and by the elements $(1, \dots, 1, I_{n_i^+, n_i^-}, 1, \dots, 1)$, $i \in I$. It is easy to see that all the elements of this group define automorphisms of (s, ρ). Hence

$$\Lambda(\mathbf{H}, s, \eta_0) = \Lambda(s, \rho) = \mathrm{Aut}(s, \rho)/\,\mathrm{Int}(\widehat{\mathbf{H}}) \simeq (\mathbb{Z}/2\mathbb{Z})^I.$$

The statement about isomorphisms between the endoscopic triples defined in the proposition is obvious.

Let $(\mathbf{H}, s, \eta_0)$ be an elliptic endoscopic triple for $\mathbf{G}$. We want to show that $(\mathbf{H}, s, \eta_0)$ is isomorphic to one of the triples defined above. We may assume (without changing the isomorphism class of $(\mathbf{H}, s, \eta_0)$) that $s \in \widehat{\mathbf{T}}$. We know that $\mathrm{Ker}^1(\mathbb{Q}, \mathbf{G}) = \{1\}$ by lemma 2.3.3 below, so condition (7.4.3) of [K4] implies that the image of s in $\pi_0((Z(\widehat{\mathbf{H}})/Z(\widehat{\mathbf{G}}))^{\mathrm{Gal}(\overline{\mathbb{Q}}/\mathbb{Q})})$ comes from an element of $Z(\widehat{\mathbf{H}})^{\mathrm{Gal}(\overline{\mathbb{Q}}/\mathbb{Q})}$. As $(\mathbf{H}, s, \eta_0)$ is elliptic,

$$\pi_0((Z(\widehat{\mathbf{H}})/Z(\widehat{\mathbf{G}}))^{\mathrm{Gal}(\overline{\mathbb{Q}}/\mathbb{Q})}) = (Z(\widehat{\mathbf{H}})/Z(\widehat{\mathbf{G}}))^{\mathrm{Gal}(\overline{\mathbb{Q}}/\mathbb{Q})},$$

so the image of s in $Z(\widehat{\mathbf{H}})/Z(\widehat{\mathbf{G}})$ comes from an element of $Z(\widehat{\mathbf{H}})^{\mathrm{Gal}(\overline{\mathbb{Q}}/\mathbb{Q})}$. We may assume that s is fixed by $\mathrm{Gal}(\overline{\mathbb{Q}}/\mathbb{Q})$ (because replacing s by a $Z(\widehat{\mathbf{G}})$-translate does not change the isomorphism class of $(\mathbf{H}, s, \eta_0)$).

Let us first suppose that $r = 1$. Write $n = n_1$. We may assume that

$$s = \left(1, \begin{pmatrix} \lambda_1 I_{m_1} & 0 & 0 \\ 0 & \ddots & 0 \\ 0 & 0 & \lambda_t I_{m_t} \end{pmatrix}\right),$$

with $\lambda_1, \dots, \lambda_t \in \mathbb{C}^\times$, $\lambda_i \neq \lambda_j$ if $i \neq j$ and $m_1, \dots, m_t \in \mathbb{N}^*$ such that $m_1 + \dots + m_t = n$. Then $\widehat{\mathbf{H}} = \mathrm{Cent}_{\widehat{\mathbf{G}}}(s) \simeq \mathbb{C}^\times \times \mathbf{GL}_{m_1}(\mathbb{C}) \times \dots \times \mathbf{GL}_{m_t}(\mathbb{C})$ and $Z(\widehat{\mathbf{H}}) \simeq \mathbb{C}^\times \times (\mathbb{C}^\times)^t$.

As $(\mathbf{H}, s, \eta_0)$ is elliptic, we must have $(Z(\widehat{\mathbf{H}})^{\mathrm{Gal}(\overline{\mathbb{Q}}/\mathbb{Q})})^0 \subset Z(\widehat{\mathbf{G}})^{\mathrm{Gal}(\overline{\mathbb{Q}}/\mathbb{Q})} \subset \mathbb{C}^\times \times \{\pm I_n\}$. The only way $Z(\widehat{\mathbf{H}})^{\mathrm{Gal}(\overline{\mathbb{Q}}/\mathbb{Q})}/Z(\widehat{\mathbf{G}})^{\mathrm{Gal}(\overline{\mathbb{Q}}/\mathbb{Q})}$ can be finite is if

$$Z(\widehat{\mathbf{H}})^{\mathrm{Gal}(\overline{\mathbb{Q}}/\mathbb{Q})} \subset \mathbb{C}^\times \times \{\pm 1\}^t.$$

But $s = (1, \lambda_1, \dots, \lambda_t) \in Z(\widehat{\mathbf{H}})^{\mathrm{Gal}(\overline{\mathbb{Q}}/\mathbb{Q})}$ and the λ_i are pairwise distinct, so $t \leq 2$. If $t = 1$, then $s \in Z(\widehat{\mathbf{G}})$ and $(\mathbf{H}, s, \eta_0)$ is isomorphic to the trivial endoscopic triple $(\mathbf{G}, 1, \mathrm{id})$.

Suppose that $t = 2$. We may assume that $\lambda_1 = 1$ and $\lambda_2 = -1$. By condition (7.1.1) of [K4],

$$\tau((\lambda, \lambda_1, \lambda_2)) = (\lambda\lambda_1^{m_1}\lambda_2^{m_2}, \lambda_{w(1)}^{-1}, \lambda_{w(2)}^{-1}),$$

for a permutation $w \in \mathfrak{S}_2$. In particular, $(-1)^{m_2} = 1$, so m_2 is even.

It remains to determine the morphism $\rho : \mathrm{Gal}(\overline{\mathbb{Q}}/\mathbb{Q}) \longrightarrow \mathrm{Out}(\widehat{\mathbf{H}})$ associated to $(\mathbf{H}, s, \eta_0)$ in [K4] 7.6. As the derived group of $\widehat{\mathbf{G}}$ is simply connected and $\mathbf{G}$ splits over E, $\mathbf{H}$ also splits over E (cf. definition 1.8.1 in [Ng]). So the action of $\mathrm{Gal}(\overline{\mathbb{Q}}/\mathbb{Q})$ on $\widehat{\mathbf{H}}$ factors through $\mathrm{Gal}(E/\mathbb{Q})$, and in particular ρ factors through $\mathrm{Gal}(E/\mathbb{Q})$. By condition (7.4.2) of [K4], there exists $g_\tau \in \widehat{\mathbf{G}}$ such that (g_τ, τ) normalizes $\widehat{\mathbf{H}}$ in $\widehat{\mathbf{G}} \rtimes \mathrm{Gal}(\overline{\mathbb{Q}}/\mathbb{Q})$ and that $\rho(\tau) = \mathrm{Int}((g_\tau, \tau))$ in $\mathrm{Out}(\widehat{\mathbf{H}})$. Hence $g_\tau = gw_0$, with $g \in \mathrm{Nor}_{\widehat{\mathbf{G}}}(\widehat{\mathbf{H}})$ and $w_0 = \begin{pmatrix} 0 & I_{m_1} \\ I_{m_2} & 0 \end{pmatrix}$.

Suppose first that $m_1 \neq m_2$. Then $\mathrm{Nor}_{\widehat{\mathbf{G}}}(\widehat{\mathbf{H}}) = \widehat{\mathbf{H}}$, so $\rho(\tau) = \mathrm{Int}((w_0, \tau))$. It is now clear that $(\mathbf{H}, s, \eta_0)$ is isomorphic to one of the triples defined above.

Suppose that $m_1 = m_2$. Then $\mathrm{Nor}_{\widehat{\mathbf{G}}}(\widehat{\mathbf{H}})$ is the subgroup of $\widehat{\mathbf{G}}$ generated by $\widehat{\mathbf{H}}$ and w_0. Hence $\rho(\tau) = \mathrm{Int}((1, \tau))$ or $\mathrm{Int}((w_0, \tau))$. If $\rho(\tau) = \mathrm{Int}((1, \tau))$, then

$$\begin{aligned} Z(\widehat{\mathbf{H}})^{\mathrm{Gal}(E/\mathbb{Q})} &\simeq \{(\lambda, \lambda_1, \lambda_2) \in (\mathbb{C}^\times)^3 | (\lambda_1\lambda_2)^{m_1} = 1 \text{ and } \lambda_1 = \lambda_2^{-1}\} \\ &= \{(\lambda, \lambda_1, \lambda_1^{-1}), \lambda, \lambda_1 \in \mathbb{C}^\times\}, \end{aligned}$$

and s is not in the image of $Z(\widehat{\mathbf{H}})^{\mathrm{Gal}(E/\mathbb{Q})}$. Hence $\rho(\tau) = \mathrm{Int}((w_0, \tau))$, and $(\mathbf{H}, s, \eta_0)$ is isomorphic to one of the triples defined above.

If $r > 1$, the reasoning is the same (but with more complicated notation). □

Fix $n_1^+, n_1^-, \dots, n_r^+, n_r^- \in \mathbb{N}$ such that $n_i^+ + n_i^- = n_i$ for every $i \in \{1, \dots, r\}$ and that $n_1^- + \dots + n_r^-$ is even. Let $(\mathbf{H}, s, \eta_0)$ be the elliptic endoscopic triple for $\mathbf{G}$ associated to these data as in proposition 2.3.1. The derived group of $\mathbf{G}$ is simply connected, so, by proposition 1 of [L2], there exists a L-morphism $\eta : {}^L\mathbf{H} := \widehat{\mathbf{H}} \rtimes W_\mathbb{Q} \longrightarrow {}^L\mathbf{G} := \widehat{\mathbf{G}} \rtimes W_\mathbb{Q}$ extending $\eta_0 : \widehat{\mathbf{H}} \longrightarrow \widehat{\mathbf{G}}$. We want to give an explicit formula for such a η.

For every place v of $\mathbb{Q}$, we fixed in section 2.1 an injection $\overline{\mathbb{Q}} \subset \overline{\mathbb{Q}}_v$; this gives a morphism $\mathrm{Gal}(\overline{\mathbb{Q}}_v/\mathbb{Q}_v) \longrightarrow \mathrm{Gal}(\overline{\mathbb{Q}}/\mathbb{Q})$, and we fix a morphism $W_{\mathbb{Q}_v} \longrightarrow W_\mathbb{Q}$ above this morphism of Galois groups.

Let $\omega_{E/\mathbb{Q}} : \mathbb{A}^\times/\mathbb{Q}^\times \longrightarrow \{\pm 1\}$ be the quadratic character of $E/\mathbb{Q}$. (Note that, for every prime number p unramified in E, the character $\omega_{E/\mathbb{Q}}$ is unramified at p.)

The following proposition is the adaptation to unitary similitude groups of [Ro2] 1.2 and is easy to prove.

Proposition 2.3.2 *Let $\mu : W_E \longrightarrow \mathbb{C}^\times$ be the character corresponding by the class field isomorphism $W_E^{ab} \simeq \mathbb{A}_E^\times/E^\times$ to a character extending $\omega_{E/\mathbb{Q}}$. We may, and will, assume that μ is unitary.*[1] *Let $c \in W_\mathbb{Q}$ be an element lifting the nontrivial element of* Gal$(E/\mathbb{Q})$. *Define a morphism $\varphi : W_\mathbb{Q} \longrightarrow {}^L\mathbf{G}$ in the following way:*

- $\varphi(c) = (A, c)$, *where*

$$A = \left(1, \left(\begin{pmatrix} \Phi_{n_1^+} & 0 \\ 0 & (-1)^{n_1^+}\Phi_{n_1^-} \end{pmatrix} \Phi_{n_1}^{-1}, \dots, \begin{pmatrix} \Phi_{n_r^+} & 0 \\ 0 & (-1)^{n_r^+}\Phi_{n_r^-} \end{pmatrix} \Phi_{n_r}^{-1}\right)\right);$$

- *on W_E, φ is given by*

$$\varphi_{|W_E} = \left(1, \left(\begin{pmatrix} \mu^{n_1^-} I_{n_1^+} & 0 \\ 0 & \mu^{-n_1^+} I_{n_1^-} \end{pmatrix}, \dots, \begin{pmatrix} \mu^{n_r^-} I_{n_r^+} & 0 \\ 0 & \mu^{-n_r^+} I_{n_r^-} \end{pmatrix}\right), \mathrm{id}\right).$$

Then φ is well-defined, and $\eta : {}^L\mathbf{H} \longrightarrow {}^L\mathbf{G}$, $(h, w) \longmapsto (\eta_0(h), 1)\varphi(w)$, is a L-morphism extending η_0.

For every place v of $\mathbb{Q}$, let φ_v be the composition of φ and of the morphism $W_{\mathbb{Q}_v} \longrightarrow W_\mathbb{Q}$. We have the following consequences of the properties of φ in the proposition;

Let p be a prime number unramified in E, and fix $\sigma \in W_{\mathbb{Q}_p}$ lifting the arithmetic Frobenius. Set $r = 1$ if p splits totally in E, and $r = 2$ if p is inert in E. Then

$$\varphi_p(\sigma^r) = (1, (I_{n_1}, \dots, I_{n_r}), \sigma^r).$$

On the other hand, there exists an odd integer $C \in \mathbb{Z}$ such that, for every $z \in \mathbb{C}^\times = W_\mathbb{C}$,

$$\varphi_\infty(z) = ((1, (B_1(z), \dots, B_r(z))), z),$$

with

$$B_i(z) = \begin{pmatrix} z^{Cn_i^-/2}\overline{z}^{-Cn_i^-/2} I_{n_i^+} & 0 \\ 0 & z^{-Cn_i^+/2}\overline{z}^{Cn_i^+/2} I_{n_i^-} \end{pmatrix}.$$

We finish this section by a calculation of Tamagawa numbers.

Lemma 2.3.3 *(i) Let $n_1, \dots, n_r \in \mathbb{N}^*$ and $\mathbf{G} = \mathbf{G}(\mathbf{U}(a_1, b_1) \times \dots \times \mathbf{U}(a_r, b_r))$, with $a_i + b_i = n_i$. Then* $\mathrm{Ker}^1(\mathbb{Q}, \mathbf{G}) = \{1\}$, *and* $Z(\widehat{\mathbf{G}})^{\mathrm{Gal}(E/\mathbb{Q})} \simeq \mathbb{C}^\times \times \{(\epsilon_1, \dots, \epsilon_r) \in \{\pm 1\}^r | \epsilon_1^{n_1} \dots \epsilon_r^{n_r} = 1\}$. *Hence the Tamagawa number of $\mathbf{G}$ is*

$$\tau(\mathbf{G}) = \begin{cases} 2^r & \text{if all the } n_i \text{ are even,} \\ 2^{r-1} & \text{otherwise.} \end{cases}$$

(ii) Let F be a finite extension of $\mathbb{Q}$ and $\mathbf{L} = R_{F/\mathbb{Q}}\mathbf{GL}_{n,F}$, with $n \in \mathbb{N}^*$. Then $\tau(\mathbf{L}) = 1$.

---

[1] In this case, we may even assume that μ is of finite order, but we will not need this fact.

Proof. Remember that, by [K4] 4.2.2 and 5.1.1, [K8], and [C], for every connected reductive algebraic group $\mathbf{G}$ over $\mathbb{Q}$,

$$\tau(\mathbf{G}) = |\pi_0(Z(\widehat{\mathbf{G}})^{\mathrm{Gal}(\overline{\mathbb{Q}}/\mathbb{Q})})|.|\operatorname{Ker}^1(\mathbb{Q}, \mathbf{G})|^{-1}.$$

(i) It is enough to prove the first two statements. By [K11] §7, the canonical morphism

$$\operatorname{Ker}^1(\mathbb{Q}, Z(\mathbf{G})) \longrightarrow \operatorname{Ker}^1(\mathbb{Q}, \mathbf{G})$$

is an isomorphism. The center of $\mathbf{G}$ is $\{(\lambda_1, \dots, \lambda_r) \in (R_{E/\mathbb{Q}}\mathbb{G}_m)^r | \lambda_1\overline{\lambda}_1 = \dots = \lambda_r\overline{\lambda}_r\}$, so it is isomorphic to $R_{E/\mathbb{Q}}\mathbb{G}_m \times \mathbf{U}(1)^{r-1}$ (by the map $(\lambda_1, \dots, \lambda_r) \longmapsto (\lambda_1, \lambda_2\lambda_1^{-1}, \dots, \lambda_r\lambda_1^{-1})$). As

$$\mathrm{H}^1(\mathbb{Q}, R_{E/\mathbb{Q}}\mathbb{G}_m) = \mathrm{H}^1(E, \mathbb{G}_m) = \{1\},$$

it remains to show that $\operatorname{Ker}^1(\mathbb{Q}, \mathbf{U}(1)) = \{1\}$. Let $c : \mathrm{Gal}(\overline{\mathbb{Q}}/\mathbb{Q}) \longrightarrow \mathbf{U}(1)(\overline{\mathbb{Q}})$ be a 1-cocyle representing an element of $\operatorname{Ker}^1(\mathbb{Q}, \mathbf{U}(1))$. Note that $\mathbf{U}(1)(\overline{\mathbb{Q}}) \simeq \overline{\mathbb{Q}}^\times$, that $\mathrm{Gal}(\overline{\mathbb{Q}}/\mathbb{Q})$ acts on $\mathbf{U}(1)(\overline{\mathbb{Q}})$ via its quotient $\mathrm{Gal}(E/\mathbb{Q})$, and that $\tau \in \mathrm{Gal}(E/\mathbb{Q})$ acts by $t \longmapsto t^{-1}$. In particular, the restriction of c to $\mathrm{Gal}(\overline{\mathbb{Q}}/E)$ is a group morphism $\mathrm{Gal}(\overline{\mathbb{Q}}/E) \longrightarrow \overline{\mathbb{Q}}^\times$. As this restriction is locally trivial, the Čebotarev density theorem implies that $c(\mathrm{Gal}(\overline{\mathbb{Q}}/E)) = 1$. So we can see c as a 1-cocycle $\mathrm{Gal}(E/\mathbb{Q}) \longrightarrow \mathbf{U}(1)(\overline{\mathbb{Q}})$. As $\mathrm{Gal}(E/\mathbb{Q}) \simeq \mathrm{Gal}(\mathbb{C}/\mathbb{R})$ and c is locally a coboundary, this implies that c is a coboundary.

By the description of $\widehat{\mathbf{G}}$ given above,

$$Z(\widehat{\mathbf{G}}) = \{(\lambda, \lambda_1 I_{n_1}, \dots, \lambda_r I_{n_r}), \lambda, \lambda_1, \dots, \lambda_r \in \mathbb{C}^\times\},$$

with the action of $\mathrm{Gal}(E/\mathbb{Q})$ given by

$$\tau((\lambda, \lambda_1 I_{n_1}, \dots, \lambda_r I_{n_r})) = (\lambda\lambda_1^{n_1} \dots, \lambda_r^{n_r}, \lambda_1^{-1} I_{n_1}, \dots, \lambda_r^{-1} I_{n_r}).$$

The second statement is now clear.

(ii) It suffices to show that $\operatorname{Ker}^1(\mathbb{Q}, \mathbf{L}) = \{1\}$ and that $Z(\widehat{\mathbf{L}})^{\mathrm{Gal}(F/\mathbb{Q})}$ is connected. The first equality comes from the fact that

$$\mathrm{H}^1(\mathbb{Q}, \mathbf{L}) = \mathrm{H}^1(F, \mathbf{GL}_n) = \{1\}.$$

On the other hand, $\widehat{\mathbf{L}} = \mathbf{GL}_n(\mathbb{C})^{[F:\mathbb{Q}]}$, with the obvious action of $\mathrm{Gal}(F/\mathbb{Q})$, so $Z(\widehat{\mathbf{L}})^{\mathrm{Gal}(F/\mathbb{Q})} \simeq \mathbb{C}^\times$ is connected. □

2.4 LEVI SUBGROUPS AND ENDOSCOPIC GROUPS

In this section, we recall some notions defined in section 7 of [K13]. Notation and definitions are as in section 7 of [K4].

Let $\mathbf{G}$ be a connected reductive group over a local or global field F. Let $\mathcal{E}(\mathbf{G})$ be the set of isomorphism classes of elliptic endoscopic triples for $\mathbf{G}$ (in the sense of [K4] 7.4) and $\mathcal{L}(\mathbf{G})$ be the set of $\mathbf{G}(F)$-conjugacy classes of Levi subgroups of $\mathbf{G}$. Let $\mathbf{M}$ be a Levi subgroup of $\mathbf{G}$. There is a canonical $\mathrm{Gal}(\overline{F}/F)$-equivariant embedding $Z(\widehat{\mathbf{G}}) \longrightarrow Z(\widehat{\mathbf{M}})$.

Definition 2.4.1 ([K13] 7.1) An endoscopic **G**-triple for **M** is an endoscopic triple $(\mathbf{M}', s_M, \eta_{M,0})$ for **M** such that

(i) the image of s_M in $Z(\widehat{\mathbf{M}'})/Z(\widehat{\mathbf{G}})$ is fixed by $\mathrm{Gal}(\overline{F}/F)$;

(ii) the image of s_M in $\mathrm{H}^1(F, Z(\widehat{\mathbf{G}}))$ (via the morphism $\pi_0((Z(\widehat{\mathbf{M}'})/Z(\widehat{\mathbf{G}}))^{\mathrm{Gal}(\overline{F}/F)}) \longrightarrow \mathrm{H}^1(F, Z(\widehat{\mathbf{G}}))$ of [K4] 7.1) is trivial if F is local, and in $\mathrm{Ker}^1(F, Z(\widehat{\mathbf{G}}))$ if F is global.

The **G**-triple $(\mathbf{M}', s_M, \eta_{M,0})$ is called elliptic if it is elliptic as an endoscopic triple for **M**.

Let $(\mathbf{M}'_1, s_1, \eta_{1,0})$ and $(\mathbf{M}'_2, s_2, \eta_{2,0})$ be endoscopic **G**-triples for **M**. An isomorphism of endoscopic **G**-triples from $(\mathbf{M}'_1, s_1, \eta_{1,0})$ to $(\mathbf{M}'_2, s_2, \eta_{2,0})$ is an isomorphism $\alpha : \mathbf{M}'_1 \longrightarrow \mathbf{M}'_2$ of endoscopic triples for **M** (in the sense of [K4] 7.5) such that the images of s_1 and $\widehat{\alpha}(s_2)$ in $Z(\widehat{\mathbf{M}'_1})/Z(\widehat{\mathbf{G}})$ are equal.

Let $(\mathbf{M}', s_M, \eta_{M,0})$ be an endoscopic **G**-triple for **M**. Then there is an isomorphism class of endoscopic triples for **G** associated to $(\mathbf{M}', s_M, \eta_{M,0})$ in the following way (cf. [K13] 3.7 and 7.4): There is a canonical $\widehat{\mathbf{G}}$-conjugacy class of embeddings ${}^L\mathbf{M} \longrightarrow {}^L\mathbf{G}$; fix an element in this class, and use it to see ${}^L\mathbf{M}$ as a subgroup of ${}^L\mathbf{G}$. Define a subgroup $\mathcal{M}$ of ${}^L\mathbf{M}$ as follows: an element $x \in {}^L\mathbf{M}$ is in $\mathcal{M}$ if and only if there exists $y \in {}^L\mathbf{M}'$ such that the images of x and y by the projections ${}^L\mathbf{M} \longrightarrow W_F$ and ${}^L\mathbf{M}' \longrightarrow W_F$ are the same and that $\mathrm{Int}(x) \circ \eta_0 = \eta_0 \circ \mathrm{Int}(y)$. Then the restriction to $\mathcal{M}$ of the projection ${}^L\mathbf{M} \longrightarrow W_F$ is surjective, and $\mathcal{M} \cap \widehat{\mathbf{M}} = \eta_0(\widehat{\mathbf{M}'})$. Moreover, $\mathcal{M}$ is a closed subgroup of ${}^L\mathbf{M}$.[2] Set $\widehat{\mathbf{H}} = \mathrm{Cent}_{\widehat{\mathbf{G}}}(s_M)^0$, and $\mathcal{H} = \mathcal{M}\widehat{\mathbf{H}}$. Then $\mathcal{H}$ is a closed subgroup of ${}^L\mathbf{G}$, the restriction to $\mathcal{H}$ of the projection ${}^L\mathbf{G} \longrightarrow W_F$ is surjective, and $\mathcal{H} \cap \widehat{\mathbf{G}} = \widehat{\mathbf{H}}$. Hence $\mathcal{H}$ induces a morphism $\rho : W_F \longrightarrow \mathrm{Out}(\widehat{\mathbf{H}})$. Moreover, there exist a finite extension K of F and a closed subgroup $\mathcal{H}_K$ of $\widehat{\mathbf{G}} \rtimes \mathrm{Gal}(K/F)$ such that $\mathcal{H}$ is the inverse image of $\mathcal{H}_K$.[3] Hence ρ factors through $W_F \longrightarrow \mathrm{Gal}(K/F)$, and ρ can be seen as a morphism $\mathrm{Gal}(\overline{F}/F) \longrightarrow \mathrm{Out}(\widehat{\mathbf{H}})$. It is easy to see that $(s_M \mod Z(\widehat{\mathbf{G}}), \rho)$ is an endoscopic datum for **G** (in the sense of [K4]), and that its isomorphism class depends only on the isomorphism class of $(\mathbf{M}', s_M, \eta_{M,0})$. We associate to $(\mathbf{M}', s_M, \eta_{M,0})$ the isomorphism class of endoscopic triples for **G** corresponding to $(s_M \mod Z(\widehat{\mathbf{G}}), \rho)$ (cf. [K4] 7.6).

Let $\mathcal{E}_{\mathbf{G}}(\mathbf{M})$ be the set of isomorphism classes of elliptic endoscopic **G**-triples $(\mathbf{M}', s_M, \eta_{M,0})$ for **M** such that the isomorphism class of endoscopic triples for **G** associated to $(\mathbf{M}', s_M, \eta_{M,0})$ is elliptic. There are obvious maps $\mathcal{E}_{\mathbf{G}}(\mathbf{M}) \longrightarrow \mathcal{E}(\mathbf{M})$ and $\mathcal{E}_{\mathbf{G}}(\mathbf{M}) \longrightarrow \mathcal{E}(\mathbf{G})$. For every endoscopic **G**-triple $(\mathbf{M}', s_M, \eta_{M,0})$ for **M**, let $\mathrm{Aut}_G(\mathbf{M}', s_M, \eta_{M,0})$ be the group of **G**-automorphisms of $(\mathbf{M}', s_M, \eta_{M,0})$ and

[2] [K13] 3.4: Let K be a finite extension of F over which $\mathbf{M}'$ and **M** split. Define a subgroup $\mathcal{M}_K$ of $\widehat{\mathbf{M}} \rtimes \mathrm{Gal}(K/F)$ in the same way as $\mathcal{M}$. This subgroup is obviously closed, and $\mathcal{M}$ is the inverse image of $\mathcal{M}_K$.

[3] [K13] 3.5: Let K' be a finite extension of F over which **G** splits. The group $\widehat{\mathbf{H}}$ is of finite index in its normalizer in $\widehat{\mathbf{G}}$, so the group $\mathcal{H}$ is of finite index in its normalizer $\mathcal{N}$ in ${}^L\mathbf{G}$. Hence the intersection of $\mathcal{H}$ with the subgroup $\widehat{\mathbf{H}} \times W_{K'}$ of $\mathcal{N}$ is a closed subgroup of finite index of $\widehat{\mathbf{H}} \times W_{K'}$; so it is also an open subgroup. Hence $\mathcal{H}$ contains an open subgroup of $\widehat{\mathbf{H}} \times W_{K'}$, i.e., it contains a subgroup $\widehat{\mathbf{H}} \times W_K$, with K a finite extension of K'. The sought-for group $\mathcal{H}_K$ is $\widehat{\mathbf{H}} \times W_K$.

$\Lambda_G(\mathbf{M}', s_M, \eta_{M,0}) = \mathrm{Aut}_G(\mathbf{M}', s_M, \eta_{M,0})/\mathbf{M}'_{ad}(F)$ be the group of outer $\mathbf{G}$-automorphisms; if $\mathbf{M} = \mathbf{G}$, we will omit the subscript $\mathbf{G}$.

Remember that we write $n_M^G = |\operatorname{Nor}_{\mathbf{G}}(\mathbf{M})(F)/\mathbf{M}(F)|$ (cf. 1.6).

Lemma 2.4.2 below is a particular case of lemma 7.2 of [K13]. As [K13] is (as yet) unpublished, we prove lemma 2.4.2 by a direct calculation. Assume that $\mathbf{G}$ is one of the unitary groups of 2.1 (and that $F = \mathbb{Q}$). If $(\mathbf{M}', s_M, \eta_{M,0}) \in \mathcal{E}_{\mathbf{G}}(\mathbf{M})$ and if $(\mathbf{H}, s, \eta_0)$ is its image in $\mathcal{E}(\mathbf{G})$, then it is easy to see that $\mathbf{M}'$ determines a $\mathbf{H}(\mathbb{Q})$-conjugacy class of Levi subgroups of $\mathbf{H}$.[4]

Lemma 2.4.2 *Assume that* $\mathbf{G}$ *is quasi-split. Let* $\varphi : \coprod_{(\mathbf{H},s,\eta_0)\in\mathcal{E}(\mathbf{G})} \mathcal{L}(\mathbf{H}) \longrightarrow \mathbb{C}$. *Then*

$$\sum_{(\mathbf{H},s,\eta_0)\in\mathcal{E}(\mathbf{G})} |\Lambda(\mathbf{H}, s, \eta_0)|^{-1} \sum_{\mathbf{M}_H\in\mathcal{L}(\mathbf{H})} (n_{M_H}^H)^{-1}\varphi(\mathbf{H}, \mathbf{M}_H)$$
$$= \sum_{\mathbf{M}\in\mathcal{L}(\mathbf{G})} (n_M^G)^{-1} \sum_{(\mathbf{M}',s_M,\eta_{M,0})\in\mathcal{E}_{\mathbf{G}}(\mathbf{M})} |\Lambda_G(\mathbf{M}', s_M, \eta_{M,0})|^{-1}\varphi(\mathbf{H}, \mathbf{M}_H),$$

where, in the second sum, $(\mathbf{H}, s, \eta_0)$ *is the image of* $(\mathbf{M}', s_M, \eta_{M,0})$ *in* $\mathcal{E}(\mathbf{G})$ *and* $\mathbf{M}_H$ *is the element of* $\mathcal{L}(\mathbf{H})$ *associated to* $(\mathbf{M}', s_M, \eta_{M,0})$.

(As $\mathbf{M}'$ and $\mathbf{M}_H$ are isomorphic, we will sometimes write $\mathbf{M}'$ instead of $\mathbf{M}_H$.)

We will use this lemma only for functions φ_H that vanish when their second argument is not a cuspidal Levi subgroup (see theorem 1.6.6 for the definition of a cuspidal Levi subgroup). In that case, the lemma is an easy consequence of lemma 2.4.3 below, which is proved in the same way as proposition 2.3.1.

In the next lemma, we consider only the case of the group $\mathbf{GU}^*(n)$ in order to simplify the notation. The case of $\mathbf{G}(\mathbf{U}^*(n_1) \times \cdots \times \mathbf{U}^*(n_r))$ is similar.

Lemma 2.4.3 *Let* $n \in \mathbb{N}^*$ *and* $\mathbf{G} = \mathbf{GU}^*(n)$. *Let* $\mathbf{T}$ *be the diagonal torus of* $\mathbf{G}$, *and identify* $\widehat{\mathbf{T}}$ *with* $\mathbb{C}^\times \times (\mathbb{C}^\times)^n$ *as in section 2.3. Let* $\mathbf{M}$ *be a cuspidal Levi subgroup of* $\mathbf{G}$. *Then* $\mathbf{M}$ *is isomorphic to* $(R_{E/\mathbb{Q}}\mathbb{G}_m)^r \times \mathbf{GU}^*(m)$, *with* $r, m \in \mathbb{N}$ *such that* $n = m + 2r$. *Let* $\mathbf{T}_M$ *be the diagonal torus of* $\mathbf{M}$. *The dual group* $\widehat{\mathbf{M}}$ *is isomorphic to the Levi subgroup*

$$\mathbb{C}^\times \times \begin{pmatrix} * & 0 & & & & \\ & \ddots & & 0 & & \\ 0 & & * & & & \\ & & & \mathbf{GL}_m(\mathbb{C}) & & \\ & & & & * & 0 \\ & 0 & & & & \ddots \\ & & & & 0 & * \end{pmatrix}$$

---

[4]In the case of unitary groups, this can be seen simply by writing explicit formulas for $\mathbf{H}$, $\mathbf{M}$ and $\mathbf{M}'$. Actually, this fact is true in greater generality and proved in [K13] 7.4 (but we will not need this here): with notation as above, the group $\mathcal{M}$ is a Levi subgroup of $\mathcal{H}$ (for a suitable definition of "Levi subgroup" in that context), and gives a conjugacy class of Levi subgroups of $\mathbf{H}$ because $\mathbf{H}$ is quasi-split.

(with blocks of size r, m, r) of $\widehat{\mathbf{G}}$. *Fix an isomorphism* $\widehat{\mathbf{T}}_M \simeq \widehat{\mathbf{T}}$ *compatible with this identification.*

Then an element $(\mathbf{M}', s_M, \eta_{M,0})$ *of* $\mathcal{E}_{\mathbf{G}}(\mathbf{M})$ *is uniquely determined by* s_M. *If we assume (as we may) that* $s_M \in \widehat{\mathbf{T}}_M \simeq \widehat{\mathbf{T}}$, *then* $s_M \in Z(\widehat{\mathbf{G}})(\{1\} \times \{\pm 1\}^n)^{\mathrm{Gal}(\overline{\mathbb{Q}}/\mathbb{Q})}$. *For every* $A \subset \{1, \dots, r\}$ *and* $m_1, m_2 \in \mathbb{N}$ *such that* $m = m_1 + m_2$ *and that* m_2 *is even, set*

$$s_{A,m_1,m_2} = \left(s_1, \dots, s_r, \overbrace{1, \dots, 1}^{m_1}, \overbrace{-1, \dots, -1}^{m_2}, s_r, \dots, s_1 \right),$$

with $s_i = -1$ *if* $i \in A$ *and* $s_i = 1$ *if* $i \notin A$. *If* $r < n/2$, *then the set of* $(1, s_{A,m_1,m_2})$ *is a system of representatives of the set of equivalence classes of possible* s_M. *If* $r = n/2$ *(so* $m = 0$*), then every possible* s_M *is equivalent to a* $(1, s_{A,0,0})$, *and* $(1, s_{A,0,0})$ *and* $(1, s_{A',0,0})$ *are equivalent if and only if* $\{1, \dots, r\} = A \sqcup A'$.

Let $s_M = (1, (s_1, \dots, s_n)) \in (\{1\} \times \{\pm 1\}^n)^{\mathrm{Gal}(\overline{\mathbb{Q}}/\mathbb{Q})}$. *Let* $(\mathbf{M}', s_M, \eta_{M,0})$ *be the element of* $\mathcal{E}_{\mathbf{G}}(\mathbf{M})$ *associated to* s_M, *and* $(\mathbf{H}, s, \eta_0)$ *be its image in* $\mathcal{E}(\mathbf{G})$. *Let* $n_1 = |\{i \in \{1, \dots, n\} | s_i = 1\}$, $n_2 = n - n_1$, $m_1 = |\{i \in \{r+1, \dots, r+m\} | s_i = 1\}|$, $m_2 = m - m_1$, $r_1 = (n_1 - m_1)/2$, $r_2 = (n_2 - m_2)/2$ *(*r_1 *and* r_2 *are integers by the condition on* s_M*). Then* $\mathbf{H} = \mathbf{G}(\mathbf{U}^*(n_1) \times \mathbf{U}^*(n_2))$, $\mathbf{M}' = (R_{E/\mathbb{Q}}\mathbb{G}_m)^r \times \mathbf{G}(\mathbf{U}^*(m_1) \times \mathbf{U}^*(m_2))$, *and* $n_{M'}^H = 2^r (r_1)!(r_2)!$. *Moreover,* $|\Lambda_G(\mathbf{M}', s_M, \eta_{M,0})|$ *is equal to* 1 *if* $\mathbf{M} \neq \mathbf{G}$.

We end this section by recalling a result of [K13] 7.3. Assume again that **G** is any connected reductive group on a local or global field F. Let **M** be a Levi subgroup of **G**.

Definition 2.4.4 Let $\gamma \in \mathbf{M}(F)$ be semisimple. An endoscopic **G**-quadruple for $(\mathbf{M}, \gamma)$ is a quadruple $(\mathbf{M}', s_M, \eta_{M,0}, \gamma')$, where $(\mathbf{M}', s_M, \eta_{M,0})$ is an endoscopic **G**-triple for **M** and $\gamma' \in \mathbf{M}'(F)$ is a semisimple $(\mathbf{M}, \mathbf{M}')$-regular element such that γ is an image of γ' (the unexplained expressions in this sentence are defined in [K7] 3). An isomorphism of endoscopic **G**-quadruples $\alpha : (\mathbf{M}'_1, s_{M,1}, \eta_{M,0,1}, \gamma'_1) \longrightarrow (\mathbf{M}'_2, s_{M,2}, \eta_{M,0,2}, \gamma'_2)$ is an isomorphism of endoscopic **G**-triples $\alpha : \mathbf{M}'_1 \longrightarrow \mathbf{M}'_2$ such that $\alpha(\gamma'_1)$ and γ'_2 are stably conjugate.

Let I be a connected reductive subgroup of **G** that contains a maximal torus of **G**. There is a canonical $\mathrm{Gal}(\overline{F}/F)$-equivariant inclusion $Z(\widehat{\mathbf{G}}) \subset Z(\widehat{I})$. Let $\mathfrak{K}_{\mathbf{G}}(I/F)$ be the set of elements in $(Z(\widehat{I})/Z(\widehat{\mathbf{G}}))^{\mathrm{Gal}(\overline{F}/F)}$ whose image by the morphism $(Z(\widehat{I})/Z(\widehat{\mathbf{G}}))^{\mathrm{Gal}(\overline{F}/F)} \longrightarrow \mathbf{H}^1(F, Z(\widehat{\mathbf{G}}))$ (coming from the exact sequence $1 \longrightarrow Z(\widehat{\mathbf{G}}) \longrightarrow Z(\widehat{I}) \longrightarrow Z(\widehat{I})/Z(\widehat{\mathbf{G}}) \longrightarrow 1$) is trivial if F is local and locally trivial if F is global.[5] If I is included in **M**, there is an obvious morphism $\mathfrak{K}_{\mathbf{G}}(I/F) \longrightarrow \mathfrak{K}_{\mathbf{M}}(I/F)$.

Fix $\gamma \in \mathbf{M}(F)$ semisimple, and let $I = \mathrm{Cent}_{\mathbf{M}}(\gamma)^0$. Let $(\mathbf{M}', s_M, \eta_{M,0}, \gamma')$ be an endoscopic **G**-quadruple for $(\mathbf{M}, \gamma)$. Let $I' = \mathrm{Cent}_{\mathbf{M}'}(\gamma')^0$. As γ' is $(\mathbf{M}, \mathbf{M}')$-regular, I' is an inner form of I (cf. [K7] 3), so there is a canonical

[5] This definition is consistent with the definition of $\mathfrak{K}(I_0/\mathbb{Q})$ in 1.6: in 1.6, I_0 is the centralizer of a semisimple elliptic element, so $(Z(\widehat{I_0})/Z(\widehat{\mathbf{G}}))^{\mathrm{Gal}(\overline{\mathbb{Q}}/\mathbb{Q})}$ is finite.

isomorphism $Z(\widehat{I}) \simeq Z(\widehat{I'})$. Let $\kappa(\mathbf{M}', s_M, \eta_{M,0}, \gamma')$ be the image of s_M by the morphism $Z(\widehat{\mathbf{M}'}) \subset Z(\widehat{I'}) \simeq Z(\widehat{I})$.

Lemma 2.4.5 *The map* $(\mathbf{M}', s_M, \eta_{M,0}, \gamma') \longmapsto \kappa(\mathbf{M}', s_M, \eta_{M,0}, \gamma')$ *induces a bijection from the set of isomorphism classes of endoscopic* **G***-quadruples for* $(\mathbf{M}, \gamma)$ *to* $\mathfrak{K}_{\mathbf{G}}(I/F)$*. Moreover, the automorphisms of endoscopic* **G***-quadruples for* $(\mathbf{M}, \gamma)$ *are all inner.*

This lemma is lemma 7.1 of [K13]. It is a generalization of lemma 9.7 of [K7] and can be proved in the same way.

Chapter Three

Discrete series

3.1 NOTATION

Let $\mathbf{G}$ be a connected reductive algebraic group over $\mathbb{R}$. In this chapter, we form the L-groups with the Weil group $W_{\mathbb{R}}$. Remember that $W_{\mathbb{R}} = W_{\mathbb{C}} \sqcup W_{\mathbb{C}}\tau$, with $W_{\mathbb{C}} = \mathbb{C}^{\times}$, $\tau^2 = -1 \in \mathbb{C}^{\times}$ and, for every $z \in \mathbb{C}^{\times}$, $\tau z \tau^{-1} = \overline{z}$, and that $W_{\mathbb{R}}$ acts on $\widehat{\mathbf{G}}$ via its quotient $\mathrm{Gal}(\mathbb{C}/\mathbb{R}) \simeq W_{\mathbb{R}}/W_{\mathbb{C}}$. Let $\Pi(\mathbf{G}(\mathbb{R}))$ (resp., $\Pi_{\mathrm{temp}}(\mathbf{G}(\mathbb{R}))$) be the set of equivalence classes of irreductible (resp., irreducible and tempered) admissible representations of $\mathbf{G}(\mathbb{R})$. For every $\pi \in \Pi(\mathbf{G}(\mathbb{R}))$, let Θ_π be the Harish-Chandra character of π (seen as a real analytic function on the set $\mathbf{G}_{\mathrm{reg}}(\mathbb{R})$ of regular elements of $\mathbf{G}(\mathbb{R})$).

Assume that $\mathbf{G}(\mathbb{R})$ has a discrete series. Let $\mathbf{A}_G$ be the maximal ($\mathbb{R}$-)split torus in the center of $\mathbf{G}$ and $\overline{\mathbf{G}}$ be an inner form of $\mathbf{G}$ such that $\overline{\mathbf{G}}/\mathbf{A}_G$ is $\mathbb{R}$-anisotropic.Write $q(G) = \dim(X)/2$, where X is the symmetric space of $\mathbf{G}(\mathbb{R})$. Let $\Pi_{\mathrm{disc}}(\mathbf{G}(\mathbb{R})) \subset \Pi(\mathbf{G}(\mathbb{R}))$ be the set of equivalence classes of representations in the discrete series.

The set $\Pi_{\mathrm{disc}}(\mathbf{G}(\mathbb{R}))$ is the disjoint union of finite subsets called L-packets; L-packets all have the same number of elements and are parametrized by equivalence classes of elliptic Langlands parameters $\varphi : W_{\mathbb{R}} \longrightarrow {}^L\mathbf{G}$, or, equivalently, by isomorphism classes of irreducible representations E of $\overline{\mathbf{G}}(\mathbb{R})$. Let $\Pi(\varphi)$ (resp., $\Pi(E)$) be the L-packet associated to the parameter φ (resp., to the representation E), and let $d(\mathbf{G})$ be the cardinality of a L-packet of $\Pi_{\mathrm{disc}}(\mathbf{G})$.

If $\pi \in \Pi_{\mathrm{disc}}(\mathbf{G}(\mathbb{R}))$, we will write f_π for a pseudo-coefficient of π (cf. [CD]).

For every elliptic Langlands parameter $\varphi : W_{\mathbb{R}} \longrightarrow {}^L\mathbf{G}$, write

$$S\Theta_\varphi = \sum_{\pi \in \Pi(\varphi)} \Theta_\pi .$$

We are going to calculate the integer $d(\mathbf{G})$ for the unitary groups of section 2.1. The following definition will be useful (this notion already appeared in theorem 1.6.6 and in section 2.4).

Definition 3.1.1 Let $\mathbf{G}$ be a connected reductive group over $\mathbb{Q}$. Denote by $\mathbf{A}_G$ the maximal $\mathbb{Q}$-split torus in the center of $\mathbf{G}$. $\mathbf{G}$ is called *cuspidal* if the group $(\mathbf{G}/\mathbf{A}_G)_{\mathbb{R}}$ has a maximal $\mathbb{R}$-torus that is $\mathbb{R}$-anisotropic.

Let $p, q \in \mathbb{N}$ be such that $p \geq q$ and $p + q \geq 1$. Let $\mathbf{G} = \mathbf{GU}(p, q)$, and use the hermitian form $A_{p,q}$ of section 2.2 to define $\mathbf{G}$. Then $\mathbf{A}_G = \mathbb{G}_m I_{p+q}$. Let $\mathbf{T}$ be the diagonal maximal torus of $\mathbf{G}$. Let

$$\mathbf{T}_{\mathrm{ell}} = \left\{ \begin{pmatrix} a_1 & 0 & & & & 0 & b_1 \\ & \ddots & & 0 & & \iddots & \\ 0 & a_q & & & & b_q & 0 \\ & & c_1 & & 0 & & \\ & 0 & & \ddots & & 0 & \\ & & 0 & & c_{p-q} & & \\ 0 & b_q & & & & a_q & 0 \\ & \iddots & & 0 & & \ddots & \\ b_1 & 0 & & & & 0 & a_1 \end{pmatrix} \right\},$$

where $a_i, b_i, c_i \in R_{E/\mathbb{Q}}\mathbb{G}_m$ are such that

$$\begin{cases} a_i\overline{b}_i + \overline{a}_i b_i = 0 \quad \text{for } 1 \leq i \leq q, \\ a_1\overline{a}_1 + b_1\overline{b}_1 = \cdots = a_q\overline{a}_q + b_q\overline{b}_q = c_1\overline{c}_1 = \cdots = c_{p-q}\overline{c}_{p-q}. \end{cases}$$

Then $\mathbf{T}_{\mathrm{ell}}$ is a maximal torus of $\mathbf{G}$, and $\mathbf{T}_{\mathrm{ell}}/\mathbf{A}_G$ is $\mathbb{R}$-anisotropic. So $\mathbf{G}$ is cuspidal. Write

$$u_G = \frac{1}{\sqrt{2}} \begin{pmatrix} I_q & 0 & -(i\otimes i)J_q \\ 0 & (\sqrt{2}\otimes 1)I_{p-q} & 0 \\ (i\otimes i)J_q & 0 & I_q \end{pmatrix} \in \mathbf{SU}(p,q)(\mathbb{C}).$$

Conjugacy by u_G^{-1} is an isomorphism $\alpha : \mathbf{T}_{\mathrm{ell},\mathbb{C}} \xrightarrow{\sim} \mathbf{T}_{\mathbb{C}}$. Use α to identify $\widehat{\mathbf{T}}_{\mathrm{ell}}$ and $\widehat{\mathbf{T}} = \mathbb{C}^\times \times (\mathbb{C}^\times)^{p+q}$. Then the action of $\mathrm{Gal}(\mathbb{C}/\mathbb{R}) = \{1, \tau\}$ on $\widehat{\mathbf{T}}_{\mathrm{ell}}$ is given by

$$\tau((\lambda, (\lambda_1, \dots, \lambda_{p+q}))) = (\lambda\lambda_1 \dots \lambda_{p+q}, (\lambda_1^{-1}, \dots, \lambda_{p+q}^{-1})).$$

Let $\Omega_{\mathbf{G}} = W(\mathbf{T}_{\mathrm{ell}}(\mathbb{C}), \mathbf{G}(\mathbb{C}))$ and $\Omega_{\mathbf{G}(\mathbb{R})} = W(\mathbf{T}_{\mathrm{ell}}(\mathbb{R}), \mathbf{G}(\mathbb{R}))$ be the Weyl groups of $\mathbf{T}_{\mathrm{ell}}$ over $\mathbb{C}$ and $\mathbb{R}$. The group $\Omega_{\mathbf{G}} \simeq W(\mathbf{T}(\mathbb{C}), \mathbf{G}(\mathbb{C})) \simeq \mathfrak{S}_{p+q}$ acts on $\mathbf{T}(\mathbb{C})$ by permuting the diagonal entries. The subgroup $\Omega_{\mathbf{G}(\mathbb{R})}$ of $\Omega_{\mathbf{G}}$ is the group $\mathfrak{S}_p \times \mathfrak{S}_q$ if $p \neq q$, and the union of $\mathfrak{S}_q \times \mathfrak{S}_q$ and of the set of permutations that send $\{1, \dots, q\}$ to $\{q+1, \dots, n\}$ if $p = q$. Hence

$$d(\mathbf{G}) = \begin{cases} \dfrac{(p+q)!}{p!q!} & \text{if } p \neq q, \\ \dfrac{(2q)!}{2(q!)^2} & \text{if } p = q. \end{cases}$$

Remark 3.1.2 The torus $\mathbf{T}_{\mathrm{ell}}$ is isomorphic to $\mathbf{G}(\mathbf{U}(1)^{p+q})$ by the morphism

$$\beta_{\mathbf{G}} : \begin{pmatrix} \mathrm{diag}(a_1, \dots, a_q) & 0 & \mathrm{diag}(b_1, \dots, b_q)J_q \\ 0 & \mathrm{diag}(c_1, \dots, c_{p-q}) & 0 \\ J_q\mathrm{diag}(b_1, \dots, b_q) & 0 & \mathrm{diag}(a_q, \dots, a_1) \end{pmatrix}$$

$$\longmapsto (a_1 - b_1, \dots, a_q - b_q, c_1, \dots, c_{p-q}, a_q + b_q, \dots, a_1 + b_1).$$

These constructions have obvious generalizations to the groups $\mathbf{G}(\mathbf{U}(p_1, q_1) \times \cdots \times \mathbf{U}(p_r, q_r))$. (In particular, these groups are also cuspidal.)

3.2 THE FUNCTIONS $\Phi_M(\gamma, \Theta)$

In this section, we recall a construction of Arthur and Shelstad.

Let $\mathbf{G}$ be a connected reductive group over $\mathbb{R}$. A *virtual character* Θ on $\mathbf{G}(\mathbb{R})$ is a linear combination with coefficients in $\mathbb{Z}$ of functions Θ_π, $\pi \in \Pi(\mathbf{G}(\mathbb{R}))$. The virtual character Θ is called *stable* if $\Theta(\gamma) = \Theta(\gamma')$ for every $\gamma, \gamma' \in \mathbf{G}_{\mathrm{reg}}(\mathbb{R})$ that are stably conjugate.

Let $\mathbf{T}$ be a maximal torus of $\mathbf{G}$. Let $\mathbf{A}$ be the maximal split subtorus of $\mathbf{T}$ and $\mathbf{M} = \mathrm{Cent}_{\mathbf{G}}(\mathbf{A})$ (a Levi subgroup of $\mathbf{G}$). For every $\gamma \in \mathbf{M}(\mathbb{R})$, set

$$D_M^G(\gamma) = \det(1 - \mathrm{Ad}(\gamma), \mathrm{Lie}(\mathbf{G})/\mathrm{Lie}(\mathbf{M})).$$

Lemma 3.2.1 *([A6] 4.1, [GKM] 4.1) Let Θ be a stable virtual character on $\mathbf{G}(\mathbb{R})$. Then the function*

$$\gamma \longmapsto |D_M^G(\gamma)|_{\mathbb{R}}^{1/2}\Theta(\gamma)$$

on $\mathbf{T}_{\mathrm{reg}}(\mathbb{R})$ extends to a continuous function on $\mathbf{T}(\mathbb{R})$, denoted by $\Phi_M(., \Theta)$ or $\Phi_M^G(., \Theta)$.

We will often see $\Phi_M(., \Theta)$ as a function on $\mathbf{M}(\mathbb{R})$, defined as follows : if $\gamma \in \mathbf{M}(\mathbb{R})$ is $\mathbf{M}(\mathbb{R})$-conjugate to a $\gamma' \in \mathbf{T}(\mathbb{R})$, set $\Phi_M(\gamma, \Theta) = \Phi_M(\gamma', \Theta)$; if there is no element of $\mathbf{T}(\mathbb{R})$ conjugate to $\gamma \in \mathbf{M}(\mathbb{R})$, set $\Phi_M(\gamma, \Theta) = 0$.

Remark 3.2.2 The function $\Phi_M(., \Theta)$ on $\mathbf{M}(\mathbb{R})$ is invariant by conjugacy by $\mathrm{Nor}_{\mathbf{G}}(\mathbf{M})(\mathbb{R})$ (because Θ and D_M^G are).

3.3 TRANSFER

We first recall some definitions from [K9] §7.

Let $\mathbf{G}$ be a connected reductive algebraic group over $\mathbb{Q}$. For every maximal torus $\mathbf{T}$ of $\mathbf{G}$, let $\mathcal{B}_G(\mathbf{T})$ be the set of Borel subgroups of $\mathbf{G}_{\mathbb{C}}$ containing $\mathbf{T}$. Assume that $\mathbf{G}$ has a maximal torus $\mathbf{T}_G$ such that $(\mathbf{T}_G/\mathbf{A}_G)_{\mathbb{R}}$ is anisotropic, and let $\overline{\mathbf{G}}$ be an inner form of $\mathbf{G}$ over $\mathbb{R}$ such that $\overline{\mathbf{G}}/\mathbf{A}_{G,\mathbb{R}}$ is anisotropic. Write $\Omega_G = W(\mathbf{T}_G(\mathbb{C}), \mathbf{G}(\mathbb{C}))$. Let $\varphi : W_{\mathbb{R}} \longrightarrow {}^L\mathbf{G}$ be an elliptic Langlands parameter.

Let $(\mathbf{H}, s, \eta_0)$ be an elliptic endoscopic triple for $\mathbf{G}$. Choose an L-morphism $\eta : {}^L\mathbf{H} \longrightarrow {}^L\mathbf{G}$ extending $\eta_0 : \widehat{\mathbf{H}} \longrightarrow \widehat{\mathbf{G}}$ (we assume that such an η exists), and let $\Phi_H(\varphi)$ be the set of equivalence classes of Langlands parameters $\varphi_H : W_{\mathbb{R}} \longrightarrow {}^L\mathbf{H}$ such that $\eta \circ \varphi_H$ and φ are equivalent. Assume that the torus $\mathbf{T}_G$ comes from a maximal torus $\mathbf{T}_H$ of $\mathbf{H}$, and fix an admissible isomorphism $j : \mathbf{T}_H \xrightarrow{\sim} \mathbf{T}_G$. Write $\Omega_H = W(\mathbf{T}_H(\mathbb{C}), \mathbf{H}(\mathbb{C}))$. Then $j_*(\Phi(\mathbf{T}_H, \mathbf{H})) \subset \Phi(\mathbf{T}_G, \mathbf{G})$, so j induces a map $j^* : \mathcal{B}_G(\mathbf{T}_G) \longrightarrow \mathcal{B}_H(\mathbf{T}_H)$ and an injective morphism $\Omega_H \longrightarrow \Omega_G$; we use this morphism to see Ω_H as a subgroup of Ω_G.

Let $\mathbf{B} \in \mathcal{B}_G(\mathbf{T}_G)$, and let $\mathbf{B}_H = j^*(\mathbf{B})$. Set

$$\begin{aligned}\Omega_* &= \{\omega \in \Omega_G | j^*(\omega(\mathbf{B})) = \mathbf{B}_H\} \\ &= \{\omega \in \Omega_G | \omega^{-1}(j_*(\Phi(\mathbf{T}_H, \mathbf{B}_H))) \subset \Phi(\mathbf{T}_G, \mathbf{B})\}.\end{aligned}$$

Then, for every $\omega \in \Omega_G$, there exists a unique pair $(\omega_H, \omega_*) \in \Omega_H \times \Omega_*$ such that $\omega = \omega_H \omega_*$. Moreover, there is a bijection $\Phi_H(\varphi) \xrightarrow{\sim} \Omega_*$ defined as follows: if $\varphi_H \in \Phi_H(\varphi)$, send it to the unique $\omega_*(\varphi_H) \in \Omega_*$ such that $(\omega_*(\varphi_H)^{-1} \circ j, \mathbf{B}, \mathbf{B}_H)$ is aligned with φ_H (in the sense of [K9] p. 184).

The Borel subgroup $\mathbf{B}$ also defines an L-morphism $\eta_B : {}^L\mathbf{T}_G \longrightarrow {}^L\mathbf{G}$, unique up to $\widehat{\mathbf{G}}$-conjugacy (cf. [K9] p. 183).

We will use the normalization of the transfer factors of [K9] §7, which we recall in the next definition.

Definition 3.3.1 For every $\gamma_H \in \mathbf{T}_H(\mathbb{R})$, set (notation as above)

$$\Delta_{j,B}(\gamma_H, \gamma) = (-1)^{q(\mathbf{G})+q(\mathbf{H})} \chi_B(\gamma) \prod_{\alpha \in \Phi(\mathbf{T}_G, \mathbf{B}) - j_*(\Phi(\mathbf{T}_H, \mathbf{B}_H))} (1 - \alpha(\gamma^{-1})),$$

where $\gamma = j(\gamma_H)$ and χ_B is the quasi-character of $\mathbf{T}_G(\mathbb{R})$ associated to the 1-cocyle $a : W_{\mathbb{R}} \longrightarrow \widehat{\mathbf{T}}_G$ such that $\eta \circ \eta_{B_H} \circ \widehat{j}$ and $\eta_B.a$ are conjugate under $\widehat{\mathbf{G}}$.

Remark 3.3.2 (1) Let $\varphi_H \in \Phi_H(\varphi)$ be such that $\omega_*(\varphi_H) = 1$. After replacing φ (resp., φ_H) by a $\widehat{\mathbf{G}}$-conjugate (resp., a $\widehat{\mathbf{H}}$-conjugate), we can write $\varphi = \eta_B \circ \varphi_B$ (resp., $\varphi_H = \eta_{B_H} \circ \varphi_{B_H}$), where φ_B (resp., φ_{B_H}) is a Langlands parameter for $\mathbf{T}_G$ (resp., $\mathbf{T}_H$). Let $\chi_{\varphi,B}$ (resp., χ_{φ_H,B_H}) be the quasi-character of $\mathbf{T}_G(\mathbb{R})$ (resp.,$\mathbf{T}_H(\mathbb{R})$) associated to φ_B (resp., φ_{B_H}). Then $\chi_B = \chi_{\varphi,B}(\chi_{\varphi_H,B_H} \circ j^{-1})^{-1}$.

(2) Let $\omega \in \Omega_G$. Write $\omega = \omega_H \omega_*$, with $\omega_H \in \Omega_H$ and $\omega_* \in \Omega_*$. Then $\Delta_{j,\omega(B)} = \det(\omega_*)\Delta_{j,B}$, where $\det(\omega_*) = \det(\omega_*, X^*(\mathbf{T}_G))$.

Let $p, q \in \mathbb{N}$ be such that $p \geq q$ and that $n := p + q \geq 1$. Fix $n_1, n_2 \in \mathbb{N}^*$ such that n_2 is even and $n_1 + n_2 = n$. Let $\mathbf{G}$ be the group $\mathbf{GU}(p, q)$ and $(\mathbf{H}, s, \eta_0)$ be the elliptic endoscopic triple for $\mathbf{G}$ associated to (n_1, n_2) as in proposition 2.3.1. In section 3.1, we defined elliptic maximal tori $\mathbf{T}_G = \mathbf{T}_{G,\mathrm{ell}}$ and $\mathbf{T}_H = \mathbf{T}_{H,\mathrm{ell}}$ of $\mathbf{G}$ and $\mathbf{H}$, and isomorphisms $\beta_G : \mathbf{T}_G \xrightarrow{\sim} \mathbf{G}(\mathbf{U}(1)^n)$ and $\beta_H : \mathbf{T}_H \xrightarrow{\sim} \mathbf{G}(\mathbf{U}(1)^n)$. Take $j = \beta_G^{-1} \circ \beta_H : \mathbf{T}_H \xrightarrow{\sim} \mathbf{T}_G$. (It is easy to see that this is an admissible isomorphism.) We also defined $u_G \in \mathbf{G}(\mathbb{C})$ such that $\mathrm{Int}(u_G^{-1})$ sends $\mathbf{T}_{G,\mathbb{C}}$ to the diagonal torus $\mathbf{T}$ of $\mathbf{G}_{\mathbb{C}}$. This defines an isomorphism (not compatible with Galois actions in general) $\widehat{\mathbf{T}}_G \simeq \widehat{\mathbf{T}}$. The composition of this isomorphism and of the embedding $\widehat{\mathbf{T}} \subset \widehat{\mathbf{G}}$ defined in section 2.3 gives an embedding $\widehat{\mathbf{T}}_G \subset \widehat{\mathbf{G}}$. Conjugacy by u_G also gives an isomorphism $\Omega_G \simeq \mathfrak{S}_n$. Via this isomorphism, $\Omega_H = \mathfrak{S}_{n_1} \times \mathfrak{S}_{n_2}$, embedded in $\mathfrak{S}_n$ in the obvious way.

Remark 3.3.3 It is easy to give simple descriptions of the subset Ω_* of Ω_G and of the bijection $\Phi_H(\varphi) \xrightarrow{\sim} \Omega_*$ for a particular choice of $\mathbf{B} \in \mathcal{B}_G(\mathbf{T}_G)$. Let

$$\mathbf{B} = \mathrm{Int}(u_G) \begin{pmatrix} * & & * \\ & \ddots & \\ 0 & & * \end{pmatrix}.$$

Then

$$\Omega_* = \{\sigma \in \mathfrak{S}_n | \sigma^{-1}_{|\{1,\dots,n_1\}} \text{ and } \sigma^{-1}_{|\{n_1+1,\dots,n\}} \text{ are nondecreasing}\}.$$

As $W_{\mathbb{C}}$ is commutative, we may assume after replacing η by a $\widehat{\mathbf{G}}$-conjugate that η sends $\{1\} \times W_{\mathbb{C}} \subset {}^L\mathbf{H}$ to $\widehat{\mathbf{T}}_G \times W_{\mathbb{C}} \subset {}^L\mathbf{G}$. As moreover $W_{\mathbb{C}}$ acts trivially on $\widehat{\mathbf{H}}$,

$$\eta((1,z)) = \left(\left(z^a\overline{z}^b, \begin{pmatrix} z^{a_1}\overline{z}^{b_1} I_{n_1} & 0 \\ 0 & z^{a_2}\overline{z}^{\,b_2} I_{n_2} \end{pmatrix}\right), z\right), \quad z \in W_{\mathbb{C}},$$

with $a, b, a_1, a_2, b_1, b_2 \in \mathbb{C}$ such that $a-b, a_1-b_1, a_2-b_2 \in \mathbb{Z}$. Let $\varphi : W_{\mathbb{R}} \longrightarrow {}^L\mathbf{G}$ be an elliptic Langlands parameter. We may assume that $\varphi_{|W_{\mathbb{C}}}$ is

$$z \longmapsto \left(\left(z^{\lambda}\overline{z}^{\mu}, \begin{pmatrix} z^{\lambda_1}\overline{z}^{\mu_1} & & 0 \\ & \ddots & \\ 0 & & z^{\lambda_n}\overline{z}^{\mu_n} \end{pmatrix}\right), z\right),$$

with $\lambda, \mu, \lambda_1, \dots, \lambda_n, \mu_1, \dots, \mu_n \in \mathbb{C}$ such that $\lambda - \mu \in \mathbb{Z}$, $\lambda_i - \mu_i \in \mathbb{Z}$ for every $i \in \{1, \dots, n\}$ and that the λ_i are pairwise distinct. Then there is a commutative diagram

$$\begin{array}{ccc} \Phi_H(\varphi) & \xrightarrow{\quad\sim\quad} & \Omega_* \\ & {}_{\sim}\nwarrow \qquad \swarrow_{\sim} & \\ & \{I \subset \{1, \dots, n\}, |I| = n_1\} & \end{array}$$

where

- the horizontal arrow is $\varphi_H \longmapsto \omega_*(\varphi_H)$;
- the arrow on the right is $\omega_* \longmapsto \omega_*^{-1}(\{1, \dots, n_1\})$;
- if $I \subset \{1, \dots, n\}$ has n_1 elements, write $I = \{i_1, \dots, i_{n_1}\}$ and $\{1, \dots, n\} - I = \{j_1, \dots, j_{n_2}\}$ with $i_1 < \dots < i_{n_1}$ and $j_1 < \dots < j_{n_2}$, and associate to I the unique $\varphi_H \in \Phi_H(\varphi)$ such that, for $z \in W_{\mathbb{C}}$,

$$\varphi_H(z) = \left(\left(z^{\lambda-a}\overline{z}^{\mu-b}, \begin{pmatrix} z^{\lambda_{i_1}-a_1}\overline{z}^{\mu_{i_1}-b_1} & & 0 \\ & \ddots & \\ 0 & & z^{\lambda_{i_{n_1}}-a_1}\overline{z}^{\mu_{i_{n_1}}-b_1} \end{pmatrix},\right.\right.$$

$$\left.\left.\begin{pmatrix} z^{\lambda_{j_1}-a_2}\overline{z}^{\mu_{j_1}-b_2} & & 0 \\ & \ddots & \\ 0 & & z^{\lambda_{j_{n_2}}-a_2}\overline{z}^{\mu_{j_{n_2}}-b_2} \end{pmatrix}\right), z\right).$$

Remember that a Levi subgroup of $\mathbf{G}$ or $\mathbf{H}$ is called *standard* if it is a Levi subgroup of a standard parabolic subgroup and contains the diagonal torus. Let $\mathbf{M}$ be a cuspidal standard Levi subgroup of $\mathbf{G}$, and let $r \in \{1, \dots, q\}$ be such that $\mathbf{M} = \mathbf{M}_{\{1,\dots,r\}} \simeq (R_{E/\mathbb{Q}}\mathbb{G}_m)^r \times \mathbf{GU}(p-r, q-r)$. Let $(\mathbf{M}', s_M, \eta_{M,0})$ be an element of $\mathcal{E}_{\mathbf{G}}(\mathbf{M})$ (cf. 2.4) whose image in $\mathcal{E}(\mathbf{G})$ is $(\mathbf{H}, s, \eta_0)$. There is a conjugacy class of Levi subgroups of $\mathbf{H}$ associated to $(\mathbf{M}', s_M, \eta_{M,0})$; let $\mathbf{M}_H$ be the standard Levi subgroup in this class. If m_1, m_2, r_1, r_2 are defined as in lemma 2.4.3, then

$$\mathbf{M}_H = \mathbf{H} \cap \left(\begin{pmatrix} \mathbf{T}_{r_1} & 0 & 0 \\ 0 & \mathbf{GU}^*(m_1) & 0 \\ 0 & 0 & \mathbf{T}_{r_1} \end{pmatrix}, \begin{pmatrix} \mathbf{T}_{r_2} & 0 & 0 \\ 0 & \mathbf{GU}^*(m_2) & 0 \\ 0 & 0 & \mathbf{T}_{r_2} \end{pmatrix}\right),$$

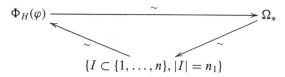

with

$$\mathbf{T}_{r_1} = \begin{pmatrix} * & & 0 \\ & \ddots & \\ 0 & & * \end{pmatrix} \subset R_{E/\mathbb{Q}}\mathbf{GL}_r \quad \text{and} \quad \mathbf{T}_{r_2} = \begin{pmatrix} * & & 0 \\ & \ddots & \\ 0 & & * \end{pmatrix} \subset R_{E/\mathbb{Q}}\mathbf{GL}_{r_2}.$$

Hence $\mathbf{M}_H \simeq \mathbf{G}(\mathbf{U}^*(m_1) \times \mathbf{U}^*(m_2)) \times (R_{E/\mathbb{Q}}\mathbb{G}_m)^{r_1+r_2}$. Set

$$\mathbf{T}_{M_H} = \mathbf{T}_{\mathbf{G}(\mathbf{U}^*(m_1)\times\mathbf{U}^*(m_2)),\mathrm{ell}} \times (R_{E/\mathbb{Q}}\mathbb{G}_m)^{r_1+r_2},$$

$$\mathbf{T}_M = \mathbf{T}_{\mathbf{GU}(p-r,q-r),\mathrm{ell}} \times (R_{E/\mathbb{Q}}\mathbb{G}_m)^r.$$

Then $\mathbf{T}_{M_H}$ (resp., $\mathbf{T}_M$) is an elliptic maximal torus of $\mathbf{M}_H$ (resp., $\mathbf{M}$). We have isomorphisms

$$\beta_{\mathbf{G}(\mathbf{U}^*(m_1)\times\mathbf{U}^*(m_2))} \times \mathrm{id} : \mathbf{T}_{M_H} = \mathbf{T}_{\mathbf{G}(\mathbf{U}^*(m_1)\times\mathbf{U}^*(m_2)),\mathrm{ell}} \times (R_{E/\mathbb{Q}}\mathbb{G}_m)^r \xrightarrow{\sim} \mathbf{G}(\mathbf{U}(1)^{m_1+m_2}) \times (R_{E/\mathbb{Q}}\mathbb{G}_m)^r,$$

$$\beta^{-1}_{\mathbf{GU}(p-r,q-r)} \times \mathrm{id} : \mathbf{G}(\mathbf{U}(1)^{p+q-2r}) \times (R_{E/\mathbb{Q}}\mathbb{G}_m)^r \xrightarrow{\sim} \mathbf{T}_{\mathbf{GU}(p-r,q-r),\mathrm{ell}} \times (R_{E/\mathbb{Q}}\mathbb{G}_m)^r = \mathbf{T}_M.$$

Let

$$j_M : \mathbf{T}_{M_H} \xrightarrow{\sim} \mathbf{T}_M$$

be the composition of these isomorphisms (note that $p + q - 2r = m_1 + m_2$).

As before, the isomorphism j_M is admissible and induces maps $j_{M*} : \Phi(\mathbf{T}_{M_H}, \mathbf{H}) \longrightarrow \Phi(\mathbf{T}_M, \mathbf{G})$ and $j_M^* : \mathcal{B}_G(\mathbf{T}_M) \longrightarrow \mathcal{B}_H(\mathbf{T}_{M_H})$ (and similar maps if we replace $\mathbf{G}$ by $\mathbf{M}$ and $\mathbf{H}$ by $\mathbf{M}_H$). It is easy to see that all the real roots of $\Phi(\mathbf{T}_M, \mathbf{G})$ (resp., $\Phi(\mathbf{T}_M, \mathbf{M})$) are in $j_{M*}(\Phi(\mathbf{T}_{M_H}, \mathbf{H}))$ (resp., $j_{M*}(\Phi(\mathbf{T}_{M_H}, \mathbf{M}_H))$).

Define an element $u \in \mathbf{GU}(p-r, q-r)(\mathbb{C})$ in the same way as the element u_G of 3.1, such that $\mathrm{Int}(u^{-1})$ sends $\mathbf{T}_{\mathbf{GU}(p-r,q-r),\mathrm{ell},\mathbb{C}}$ to the diagonal torus of $\mathbf{GU}(p-r, q-r)_{\mathbb{C}}$. Let $u_M = \mathrm{diag}(I_r, u, I_r)u_G^{-1} \in \mathbf{G}(\mathbb{C})$. Then $\mathrm{Int}(u_M^{-1})$ sends $\mathbf{T}_{M,\mathbb{C}}$ to $\mathbf{T}_{G,\mathbb{C}}$. Similarly, we get $u_{M_H} \in \mathbf{H}(\mathbb{C})$ such that $\mathrm{Int}(u_{M_H}^{-1})$ sends $\mathbf{T}_{M_H}$ to $\mathbf{T}_{H,\mathbb{C}}$. The following diagram is commutative:

$$\begin{array}{ccc} \mathbf{T}_{M,\mathbb{C}} & \xrightarrow{\mathrm{Int}(u_M^{-1})} & \mathbf{T}_{G,\mathbb{C}} \\ {\scriptstyle j_M}\big\downarrow & & \big\downarrow{\scriptstyle j} \\ \mathbf{T}_{M_H,\mathbb{C}} & \xrightarrow[\mathrm{Int}(u_{M_H}^{-1})]{} & \mathbf{T}_{H,\mathbb{C}} \end{array}$$

Use conjugacy by u_M (resp., u_{M_H}) to identify Ω_G (resp., Ω_H) and $W(\mathbf{T}_M(\mathbb{C}), \mathbf{G}(\mathbb{C}))$ (resp., $W(\mathbf{T}_{M_H}(\mathbb{C}), \mathbf{H}(\mathbb{C}))$). If $\mathbf{B} \in \mathcal{B}_G(\mathbf{T}_M)$, we use $\mathrm{Int}(u_M^{-1})(\mathbf{B}) \in \mathcal{B}_G(\mathbf{T}_G)$ to define (as before) a subset Ω_* of Ω_G and a bijection $\Phi_H(\varphi) \xrightarrow{\sim} \Omega_*$.

By [K13] p. 23, the morphism η determines an L-morphism $\eta_M : {}^L\mathbf{M}_H = {}^L\mathbf{M}' \longrightarrow {}^L\mathbf{M}$, unique up to $\widehat{\mathbf{M}}$-conjugacy and extending $\eta_{M,0}$. We use this morphism η_M to define transfer factors $\Delta_{j_M,\mathbf{B}_M}$, for every $\mathbf{B}_M \in \mathcal{B}_M(\mathbf{T}_M)$: if $\gamma_H \in \mathbf{T}_{M_H}(\mathbb{R})$, set

$$\Delta_{j_M,\mathbf{B}_M}(\gamma_H, \gamma) = (-1)^{q(\mathbf{G})+q(\mathbf{H})}\chi_{\mathbf{B}_M}(\gamma) \prod_{\alpha\in\Phi(\mathbf{T}_M,\mathbf{B}_M)-j_{M*}(\Phi(\mathbf{T}_{M_H},\mathbf{B}_{M_H}))} (1 - \alpha(\gamma^{-1}))$$

(note the sign), where $\gamma = j_M(\gamma_H)$, $\mathbf{B}_{M_H} = j_M^*(\mathbf{B}_M)$ and χ_{B_M} is the quasi-character of $\mathbf{T}_M(\mathbb{R})$ associated to the 1-cocyle $a_M : W_{\mathbb{R}} \longrightarrow \widehat{\mathbf{T}}_M$ such that $\eta_M \circ \eta_{B_{M_H}} \circ \widehat{j}_M$ and $\eta_{B_M}.a_M$ are $\widehat{\mathbf{M}}$-conjugate.

The next proposition is a generalization of the calculations of [K9] p. 186.

Proposition 3.3.4 *Fix $\mathbf{B} \in \mathcal{B}_G(\mathbf{T}_M)$ (that determines Ω_* and $\Phi_H(\varphi) \xrightarrow{\sim} \Omega_*$), and let $\mathbf{B}_M = \mathbf{B} \cap \mathbf{M}$. Let $\gamma_H \in \mathbf{T}_{M_H}(\mathbb{R})$ and $\gamma = j_M(\gamma_H)$. Then*

$$\Delta_{j_M,B_M}(\gamma_H,\gamma)\Phi_M(\gamma^{-1}, S\Theta_\varphi) = \sum_{\varphi_H \in \Phi_H(\varphi)} \det(\omega_*(\varphi_H))\Phi_{M_H}(\gamma_H^{-1}, S\Theta_{\varphi_H}).$$

Proof. Both sides of the equality that we want to prove depend on the choice of the L-morphism $\eta : {}^L\mathbf{H} \longrightarrow {}^L\mathbf{G}$ extending $\eta_0 : \widehat{\mathbf{H}} \longrightarrow \widehat{\mathbf{G}}$. Let η' be another such L-morphism. Then the difference between η' and η is given by an element of $\mathrm{H}^1(W_{\mathbb{R}}, Z(\widehat{\mathbf{H}}))$; let χ be the corresponding quasi-character of $\mathbf{H}(\mathbb{R})$. Then, if we replace η by η', the transfer factor $\Delta_{j_M,\mathbf{B}_M}$ is multiplied by χ and the stable characters $S\Theta_{\varphi_H}$, $\varphi_H \in \Phi_H(\varphi)$, are multiplied by χ^{-1}; hence both sides of the equality are multiplied by $\chi(\gamma_H)$. It is therefore enough to prove the proposition for a particular choice of η.

We choose η such that

$$\eta((1,\tau)) = \left(\left(1, \begin{pmatrix} 0 & I_{n_1} \\ (-1)^{n_1} I_{n_2} & 0 \end{pmatrix}\right), \tau\right)$$

and such that, for every $z \in \mathbb{C}^\times = W_{\mathbb{C}}$,

$$\eta((1,z)) = \left(\left(1, \begin{pmatrix} z^{n_2/2}\overline{z}^{-n_2/2} I_{n_1} & 0 \\ 0 & z^{-n_1/2}\overline{z}^{n_1/2} I_{n_2} \end{pmatrix}\right), z\right).$$

We first recall the formulas for $\Phi_M(., S\Theta_\varphi)$ and $\Phi_{M_H}(., S\Theta_{\varphi_H})$. The reference for this is [A6] pp. 272–274.[1]

Let V_φ be the irreducible representation of $\overline{\mathbf{G}}(\mathbb{R})$ corresponding to φ and ξ_φ be the quasi-character by which $\mathbf{A}_G(\mathbb{R})^0$ acts on V_φ. Let $\mathbf{B}_0 = \mathrm{Int}(u_M^{-1})(\mathbf{B})$, Z be the maximal compact subgroup of the center of $\mathbf{G}(\mathbb{R})$ and $\mathfrak{t}_G = \mathrm{Lie}(\mathbf{T}_G)$. Define functions ρ_G and Δ_G on $\mathfrak{t}_G(\mathbb{R})$ by

$$\rho_G = \frac{1}{2} \sum_{\alpha \in \Phi(\mathbf{T}_G,\mathbf{B}_0)} \alpha,$$

$$\Delta_G = \prod_{\alpha \in \Phi(\mathbf{T}_G,\mathbf{B}_0)} (e^{\alpha/2} - e^{-\alpha/2})$$

[1]Note that there is a mistake in this reference. Namely, with the notation used below, the formula of [A6] is correct for elements in the image of the map $Z \times \mathfrak{t}_M(\mathbb{R}) \longrightarrow \mathbf{T}_M(\mathbb{R})$, $(z, X) \longmapsto z\exp(X)$, but it is not true in general, as claimed in [A6], that the stable discrete series characters vanish outside the image of this map. This is not a problem here because the exponential map $\mathfrak{t}_M(\mathbb{R}) \longrightarrow \mathbf{T}_M(\mathbb{R})$ is surjective unless $\mathbf{M}$ is a torus, and, if $\mathbf{M}$ is a torus, elements in $\mathbf{M}(\mathbb{R})$ that are not in the image of the exponential map are also not in $Z(\mathbf{G})(\mathbb{R})\mathbf{G}_{\mathrm{der}}(\mathbb{R})$, so all discrete series characters vanish on these elements. A formula that is correct in the general case can be found in section 4 of [GKM].

(we use the same notation for characters on $\mathbf{T}_G$ and the linear forms on $\mathfrak{t}_G$ that are defined by differentiating these characters). Notice that $\mathbf{T}_G(\mathbb{R}) = Z\exp(\mathfrak{t}_G(\mathbb{R}))$ (this is a general fact; here, $\mathbf{T}_G(\mathbb{R})$ is even equal to $\exp(\mathfrak{t}_G(\mathbb{R}))$). The representation V_φ corresponds to a pair $(\zeta_\varphi, \lambda_\varphi)$, where ζ_φ is a quasi-character of Z and λ_φ is a linear form on $\mathfrak{t}_G(\mathbb{C})$, such that

- λ_φ is regular dominant;
- the morphism $Z\times\mathfrak{t}_G(\mathbb{R}) \longrightarrow \mathbb{C}^\times$, $(z, X) \longmapsto \zeta_\varphi(z)e^{(\lambda_\varphi-\rho_G)(X)}$, factors through the surjective morphism $Z \times \mathfrak{t}_G(\mathbb{R}) \longrightarrow \mathbf{T}_G(\mathbb{R})$, $(z, X) \longmapsto z\exp(X)$, and defines a quasi-character on $\mathbf{T}_G(\mathbb{R})$, whose restriction to $\mathbf{A}_G(\mathbb{R})^0$ is ξ_φ.

Note that the quasi-character $z\exp(X) \longmapsto \zeta_\varphi(z)e^{(\lambda_\varphi-\rho_G)(X)}$ on $\mathbf{T}_G(\mathbb{R})$ is equal to the quasi-character χ_{φ,B_0} defined in remark 3.3.2 (1). Remember the Weyl character formula: if $\gamma \in \mathbf{T}_{G,\mathrm{reg}}(\mathbb{R})$ is such that $\gamma = z\exp(X)$, with $z \in Z$ and $X \in \mathfrak{t}_G(\mathbb{R})$, then

$$\mathrm{Tr}(\gamma, V_\varphi) = (-1)^{q(\mathbf{G})}S\Theta_\varphi(\gamma) = \Delta_G(X)^{-1}\zeta_\varphi(z)\sum_{\omega\in\Omega_G}\det(\omega)e^{(\omega\lambda_\varphi)(X)}.$$

Let R be a root system whose Weyl group $W(R)$ contains -1. Then, to every pair (Q^+, R^+) such that $R^+ \subset R$ and $Q^+ \subset R^\vee$ are positive root systems, we can associate an integer $\overline{c}(Q^+, R^+)$. The definition of $\overline{c}(Q^+, R^+)$ is recalled in [A6] p. 273.

Let R be the set of real roots in $\Phi(\mathbf{T}_M, \mathbf{G})$, $R^+ = R \cap \Phi(\mathbf{T}_M, \mathbf{B})$ and $\mathfrak{t}_M = \mathrm{Lie}(\mathbf{T}_M)$. If X is a regular element of $\mathfrak{t}_M(\mathbb{R})$, let

$$R_X^+ = \{\alpha \in R|\alpha(X) > 0\},$$

$$\varepsilon_R(X) = (-1)^{|R_X^+\cap(-R^+)|}.$$

If ν is a linear form on $\mathfrak{t}_M(\mathbb{C})$ such that $\nu(\alpha^\vee) \neq 0$ for every $\alpha^\vee \in R^\vee$, let

$$Q_\nu^+ = \{\alpha^\vee \in R^\vee|\nu(\alpha^\vee) > 0\}.$$

Define a function Δ_M on $\mathfrak{t}_M(\mathbb{R})$ by

$$\Delta_M = \prod_{\alpha\in\Phi(\mathbf{T}_M,\mathbf{B}_M)}(e^{\alpha/2} - e^{-\alpha/2}).$$

As $\mathrm{Int}(u_M)$ sends $\mathbf{T}_G(\mathbb{C})$ onto $\mathbf{T}_M(\mathbb{C})$, $\mathrm{Ad}(u_M)$ defines an isomorphism $\mathfrak{t}_M(\mathbb{C})^* \xrightarrow{\sim} \mathfrak{t}_G(\mathbb{C})^*$.

Let $\gamma \in \mathbf{T}_{M,\mathrm{reg}}(\mathbb{R})$. If there exist $z \in Z$ and $X \in \mathfrak{t}_M(\mathbb{R})$ such that $\gamma = z\exp(X)$, then (formula (4.8) of [A6])

$$\Phi_M(\gamma, S\Theta_\varphi) = (-1)^{q(\mathbf{G})}\Delta_M(X)^{-1}\varepsilon_R(X)\zeta_\varphi(z)$$
$$\times\sum_{\omega\in\Omega_G}\det(\omega)\overline{c}(Q^+_{\mathrm{Ad}(u_M)\omega\lambda}, R_X^+)e^{(\mathrm{Ad}(u_M)\omega\lambda)(X)}.$$

There are similar objects, defined by replacing $\mathbf{G}$ by $\mathbf{H}$, etc., and similar formulas for the functions $\Phi_{M_H}(., S\Theta_{\varphi_H})$.

Let $\gamma_H \in \mathbf{T}_{M_H}(\mathbb{R})$ and $\gamma = j_M(\gamma_H)$; we want to prove the equality of the proposition. We may assume that γ is regular in $\mathbf{G}$ (because the set of γ_H such that

$j_M(\gamma_H)$ is $\mathbf{G}$-regular is dense in $\mathbf{T}_{M_H}(\mathbb{R})$). Note that, as $\mathbf{T}_M$ and $\mathbf{T}_{M_H}$ are both isomorphic to $R_{E/\mathbb{Q}}\mathbb{G}_m^r \times \mathbf{G}(\mathbf{U}(1)^{m_1+m_2})$, the exponential maps $\mathfrak{t}_M(\mathbb{R}) \longrightarrow \mathbf{T}_M(\mathbb{R})$ and $\mathfrak{t}_{M_H}(\mathbb{R}) \longrightarrow \mathbf{T}_{M_H}(\mathbb{R})$ are surjective unless $\mathbf{M}$ is a torus (i.e., $m_1+m_2=0$). If $\mathbf{M}$ is a torus (so $\mathbf{M}_H$ is also a torus) and γ_H is not in the image of the exponential map, then $c(\gamma_H)=c(\gamma)<0$, so $\gamma \notin Z(\mathbf{G})(\mathbb{R})\mathbf{G}_{\mathrm{der}}(\mathbb{R})$ and $\gamma_H \notin Z(\mathbf{H})(\mathbb{R})\mathbf{H}_{\mathrm{der}}(\mathbb{R})$, and all discrete series characters vanish on γ and γ_H; so the equality of the proposition is obvious.

These remarks show that we may assume that there exists $X_H \in \mathfrak{t}_{M_H}(\mathbb{R})$ such that $\gamma_H^{-1}=\exp(X_H)$. Then $\gamma^{-1}=\exp(X)$, with $X=j_M(X_H)\in\mathfrak{t}_M(\mathbb{R})$. As all the real roots of $\Phi(\mathbf{T}_M,\mathbf{G})$ are in $j_{M*}(\Phi(\mathbf{T}_{M_H},\mathbf{H}))$, R is equal to $j_{M*}(R_H)$. Hence $R_X^+=j_{M*}(R_{H,X_H}^+)$ and $\varepsilon_R(X)=\varepsilon_{R_H}(X_H)$. On the other hand, using the description of the bijection $\Phi_H(\varphi)\xrightarrow{\sim}\Omega_*$ given above and the choice of η, it is easy to see that, for every $\varphi_H\in\Phi_H(\varphi)$, $\zeta_{\varphi_H}=\zeta_\varphi$ and $\lambda_{\varphi_H}=\omega_*(\varphi_H)(\lambda_\varphi)\circ j_M+\rho_H-\rho_G\circ j_M$. As $\rho_H-\rho_G\circ j_M$ is Ω_H-invariant and vanishes on the elements of $R_H^\vee$, this implies that, for every $\varphi_H\in\Phi_H(\varphi)$ and $\omega_H\in\Omega_H$, $Q^+_{\mathrm{Ad}(u_M^{-1})\omega_H\omega_*(\varphi_H)\lambda_\varphi}=j_{M*}(Q^+_{\mathrm{Ad}(u_{M_H}^{-1})\omega_H\lambda_{\varphi_H}})$. So we get

$$(-1)^{q(\mathbf{H})}\sum_{\varphi_H\in\Phi_H(\varphi)}\det(\omega_*(\varphi_H))\Phi_{M_H}(\gamma_H^{-1},S\Theta_{\varphi_H})$$
$$=\Delta_{M_H}(X_H)^{-1}\varepsilon_R(X)\sum_{\omega_*\in\Omega_*}\det(\omega_*)\sum_{\omega_H\in\Omega_H}\det(\omega_H)\overline{c}(Q^+_{\mathrm{Ad}(u_M^{-1})\omega_H\omega_*\lambda_\varphi},R_X^+)$$
$$\times e^{\mathrm{Ad}(u_M^{-1})(\omega_H\omega_*\lambda_\varphi+\rho_H\circ j_M^{-1}-\rho_G)(X)}$$
$$=(-1)^{q(\mathbf{G})}\Delta_{M_H}(X_H)^{-1}\Delta_M(X)e^{\rho_H(X_H)-\rho_G(X)}\Phi_M(\gamma^{-1},S\Theta_\varphi).$$

To finish the proof, it is enough to show that

$$(-1)^{q(\mathbf{G})+q(\mathbf{H})}\Delta_{j_M,\mathbf{B}_M}(\gamma_H,\gamma)=\Delta_{M_H}(X_H)^{-1}\Delta_M(X)e^{\rho_H(X_H)-\rho_G(X)}.$$

Let $\Phi^+=\Phi(\mathbf{T}_M,\mathbf{B})$, $\Phi_H^+=j_{M*}(\Phi(\mathbf{T}_{M_H},j_M^*(\mathbf{B})))$, $\Phi_M^+=\Phi(\mathbf{T}_M,\mathbf{B}_M)$, $\Phi_{M_H}^+=j_{M*}(\Phi(\mathbf{T}_{M_H},j_M^*(\mathbf{B}_M)))$. Then

$$\Delta_{M_H}(X_H)^{-1}\Delta_M(X)e^{\rho_H(X_H)-\rho_G(X)}$$
$$=\prod_{\alpha\in\Phi_M^+-\Phi_{M_H}^+}(e^{\alpha(X)/2}-e^{-\alpha(X)/2})\prod_{\alpha\in\Phi^+-\Phi_H^+}e^{-\alpha(X)/2}$$
$$=\prod_{\alpha\in\Phi_M^+-\Phi_{M_H}^+}(1-\alpha(\gamma^{-1}))\prod_{\alpha\in\Phi^+-(\Phi_H^+\cup\Phi_M^+)}e^{-\alpha(X)/2}.$$

So it is enough to show that

$$\chi_{B_M}(\gamma)=\prod_{\alpha\in\Phi^+-(\Phi_H^+\cup\Phi_M^+)}e^{-\alpha(X)/2}.$$

Remember that χ_{B_M} is the quasi-character of $\mathbf{T}_M(\mathbb{R})$ corresponding to the 1-cocycle $a_M: W_\mathbb{R}\longrightarrow\widehat{\mathbf{T}}_M$ such that $\eta_M\circ\eta_{B_{M_H}}\circ\widehat{j}_M$ and $\eta_{B_M}.a_M$ are $\widehat{\mathbf{M}}$-conjugate, where $\mathbf{B}_{M_H}=j_M^*(\mathbf{B}_M)$. So the equality above is an easy consequence of the definitions of η_{B_M} and $\eta_{B_{M_H}}$ and of the choice of η. □

Remark 3.3.5 As $\Delta_{j_M,B_M}(\gamma_H,\gamma) = 0$ if γ_H is not $(\mathbf{M},\mathbf{M}_H)$-regular, the right-hand side of the equality of the proposition is nonzero only if γ_H is $(\mathbf{M},\mathbf{M}_H)$-regular.

3.4 CALCULATION OF CERTAIN $\Phi_M(\gamma,\Theta)$

As before, let $\mathbf{G}=\mathbf{GU}(p,q)$, with $p,q\in\mathbb{N}$ such that $p\geq q$ and $n:=p+q\geq 1$. Fix $s\in\{1,\dots,q\}$, and set $S=\{1,\dots,s\}$ and $\mathbf{M}=\mathbf{M}_S$ (with notation as in section 2.2). The goal of this section is to calculate $\Phi_M(\gamma,\Theta)$, for Θ the character of a L-packet of the discrete series of $\mathbf{G}(\mathbb{R})$ associated to an algebraic representation of $\mathbf{G}_{\mathbb{C}}$ and certain γ in $\mathbf{M}(\mathbb{R})$.

The linear part of $\mathbf{M}$ (resp., $\mathbf{M}_s$) is $\mathbf{L}_S=(R_{E/\mathbb{Q}}\mathbb{G}_m)^s$ (resp., $\mathbf{L}_s=R_{E/\mathbb{Q}}\mathbf{GL}_s$), and its hermitian part is $\mathbf{G}_s=\mathbf{GU}(p-s,q-s)$. The group $\mathbf{L}_S$ is a minimal Levi subgroup of $\mathbf{L}_s$. The Weyl group $W(\mathbf{L}_S(\mathbb{Q}),\mathbf{L}_s(\mathbb{Q}))$ is obviously isomorphic to $\mathfrak{S}_s$, and we identify these groups; we extend the action of $\mathfrak{S}_s$ on $\mathbf{L}_S$ to an action on $\mathbf{M}=\mathbf{L}_S\times\mathbf{G}_s$, by declaring that the action is trivial on $\mathbf{G}_s$.

For every $r\in\{1,\dots,q\}$, let $t_r=r(r-n)$.

Proposition 3.4.1 *Let E be an irreducible algebraic representation of $\mathbf{G}_{\mathbb{C}}$. Let $m\in\mathbb{Z}$ be such that the central torus $\mathbb{G}_{m,\mathbb{C}}I_n$ of $\mathbf{G}_{\mathbb{C}}$ acts on E by multiplication by the character $z\longmapsto z^m$. For every $r\in\{1,\dots,q\}$, let $t'_r=t_r+m$. Choose an elliptic Langlands parameter $\varphi: W_{\mathbb{R}}\longrightarrow {}^L\mathbf{G}$ corresponding to E (seen as an irreducible representation of $\mathbf{GU}(n)(\mathbb{R})\subset\mathbf{GU}(n)(\mathbb{C})\simeq\mathbf{G}(\mathbb{C})$), and let $\Theta=(-1)^{q(\mathbf{G})}S\Theta_\varphi$. Let $\gamma\in\mathbf{M}(\mathbb{R})$ be semisimple elliptic. Write $\gamma=\gamma_l\gamma_h$, with $\gamma_l=(\lambda_1,\dots,\lambda_s)\in(\mathbb{C}^\times)^s=\mathbf{L}_S(\mathbb{R})$ and $\gamma_h\in\mathbf{G}_s(\mathbb{R})$. Then $c(\gamma)=c(\gamma_h)>0$ unless $\mathbf{M}$ is a torus (i.e., $s=q$). If $c(\gamma)<0$, then $\Phi_M(\gamma,\Theta)=0$. If $c(\gamma)>0$, then*

(i) If $c(\gamma)|\lambda_r|^2\geq 1$ for every $r\in\{1,\dots,s\}$, then $\Phi_M(\gamma,\Theta)$ is equal to

$$2^s\sum_{\substack{S'\subset S\\ S'\ni s}}(-1)^{\dim(A_M/A_{M_{S'}})}|W(\mathbf{L}_S(\mathbb{Q}),\mathbf{L}_{S'}(\mathbb{Q}))|^{-1}$$
$$\times\sum_{\sigma\in\mathfrak{S}_s}|D_{\mathbf{M}}^{\mathbf{M}_{S'}}(\sigma\gamma)|_{\mathbb{R}}^{1/2}\delta_{P_{S'}(\mathbb{R})}^{1/2}(\sigma\gamma)\mathrm{Tr}(\sigma\gamma,R\Gamma(\mathrm{Lie}(\mathbf{N}_{S'}),E)_{<t'_r,r\in S'}).$$

(ii) If $0<c(\gamma)|\lambda_r|^2\leq 1$ for every $r\in\{1,\dots,s\}$, then $\Phi_M(\gamma,\Theta)$ is equal to

$$(-1)^s2^s\sum_{\substack{S'\subset S\\ S'\ni s}}(-1)^{\dim(A_M/A_{M_{S'}})}|W(\mathbf{L}_S(\mathbb{Q}),\mathbf{L}_{S'}(\mathbb{Q}))|^{-1}$$
$$\times\sum_{\sigma\in\mathfrak{S}_s}|D_{\mathbf{M}}^{\mathbf{M}_{S'}}(\sigma\gamma)|_{\mathbb{R}}^{1/2}\delta_{P_{S'}(\mathbb{R})}^{1/2}(\sigma\gamma)\mathrm{Tr}(\sigma\gamma,R\Gamma(\mathrm{Lie}(\mathbf{N}_{S'}),E)_{>t'_r,r\in S'}).$$

The notations $R\Gamma(\mathrm{Lie}(\mathbf{N}_{S'}),E)_{<t'_r,r\in S'}$ and $R\Gamma(\mathrm{Lie}(\mathbf{N}_{S'}),E)_{>t'_r,r\in S'}$ are those of proposition 1.4.5.

Proof. Let $\gamma\in\mathbf{M}(\mathbb{R})$ be semi-simple elliptic. Use the notation of section 3.3, in particular of the proof of proposition 3.3.4. As γ is elliptic in $\mathbf{M}(\mathbb{R})$, we may assume that $\gamma \in \mathbf{T}_M(\mathbb{R})$. The fact that $\Phi_M(\gamma, \Theta) = 0$ if $c(\gamma) < 0$ (and that this can happen only if $\mathbf{M}$ is a torus) has already been noted in the proof of proposition 3.3.4. So we may assume that $c(\gamma) > 0$.

The proofs of (i) and (ii) are similar. Let us prove (ii). Assume that $c(\gamma)|\lambda_r|^2 \leq 1$ for every $r \in S$. As both sides of the equality we want to prove are continuous functions of γ, we may assume that $c(\gamma)|\lambda_r|^2 < 1$ for every $r \in S$ and that γ is regular in $\mathbf{G}$. Let X be an element of $\mathfrak{t}_M(\mathbb{R})$ such that $\gamma = \exp(X)$ (remember that, as the torus $\mathbf{T}_M$ is isomorphic to $(R_{E/\mathbb{Q}}\mathbb{G}_m)^s \times \mathbf{G}(\mathbf{U}(1)^{n-2s})$, such a X exists if and only if $c(\gamma) > 0$). Choose an element $\mathbf{B}$ of $\mathcal{B}_G(\mathbf{T}_M)$ such that $\mathbf{B} \subset \mathbf{P}_S$. There is a pair (ζ, λ_G) associated to E as in section 3.3 (ζ is a quasi-character of Z, and $\lambda_G \in \mathfrak{t}_G(\mathbb{C})^*$). Write $\lambda = \mathrm{Ad}(u_M)(\lambda_G) \in \mathfrak{t}_M(\mathbb{C})^*$ and $\rho_B = \frac{1}{2}\sum_{\alpha \in \Phi(\mathbf{T}_M, \mathbf{B})} \alpha$. Then $\lambda - \rho_B$ is the highest weight of E relative to $(\mathbf{T}_M, \mathbf{B})$.

Let $S' \subset S$ be such that $s \in S'$. We use Kostant's theorem (see, e.g., [GHM] §11) to calculate the trace of γ on $R\Gamma(\mathrm{Lie}(\mathbf{N}_{S'}), E)_{>t'_r, r \in S'}$. Let $\Omega = W(\mathbf{T}_M(\mathbb{C}), \mathbf{G}(\mathbb{C}))$, ℓ be the length function on Ω, $\Omega_{S'} = W(\mathbf{T}_M(\mathbb{C}), \mathbf{M}_{S'}(\mathbb{C}))$ and $\Phi^+ = \Phi(\mathbf{T}_M, \mathbf{B})$. For every $\omega \in \Omega$, let $\Phi^+(\omega) = \{\alpha \in \Phi^+ | \omega^{-1}\alpha \in -\Phi^+\}$. Then $\Omega'_{S'} := \{\omega \in \Omega | \Phi^+(\omega) \subset \Phi(\mathbf{T}_M, \mathbf{N}_{S'})\}$ is a system of representatives of $\Omega_{S'} \setminus \Omega$. Kostant's theorem says that, for every $k \in \mathbb{N}$,

$$\mathrm{H}^k(\mathrm{Lie}(\mathbf{N}_{S'}), E) \simeq \bigoplus_{\omega \in \Omega'_{S'}, \ell(\omega) = k} V_{\omega(\lambda) - \rho_B},$$

where, for every $\omega \in \Omega$, $V_{\omega(\lambda)-\rho_B}$ is the algebraic representation of $\mathbf{M}_{S',\mathbb{C}}$ with highest weight $\omega(\lambda) - \rho_B$ (relative to $(\mathbf{T}_M, \mathbf{B} \cap \mathbf{M}_{S',\mathbb{C}})$).

For every $r \in \{1, \dots, s\}$, let

$$\varpi_r : \mathbb{G}_m \longrightarrow \mathbf{T}_M, \lambda \longrightarrow \begin{pmatrix} \lambda I_r & & 0 \\ & I_{n-2r} & \\ 0 & & \lambda^{-1} I_r \end{pmatrix};$$

we use the same notation for the morphism $\mathrm{Lie}(\mathbb{G}_m) \longrightarrow \mathfrak{t}_M$ obtained by differentiating ϖ_r. Let $k \in \mathbb{N}$. By definition of the truncation,

$$\mathrm{H}^k(\mathrm{Lie}(\mathbf{N}_{S'}), E)_{>t'_r, r \in S'} \simeq \bigoplus_{\omega} V_{\omega(\lambda)-\rho_B},$$

where the sum is taken over the set of $\omega \in \Omega'_{S'}$ of length k and such that, for every $r \in S'$, $\langle \omega(\lambda) - \rho_B, \varpi_r \rangle > t_r$. As $t_r = \langle -\rho_B, \varpi_r \rangle$ for every $r \in \{1, \dots, s\}$, the last condition on ω is equivalent to : $\langle \omega(\lambda), \varpi_r \rangle > 0$, for every $r \in S'$.

On the other hand, by the Weyl character formula, for every $\omega \in \Omega'_{S'}$,

$$\mathrm{Tr}(\gamma, V_{\omega(\lambda)-\rho_B}) = \Delta_{M_{S'}}(X)^{-1} \sum_{\omega_M \in \Omega_{S'}} \det(\omega_M) e^{(\omega_M(\omega(\lambda)-\rho_B+\rho_{S'}))(X)},$$

where $\rho_{S'} = \frac{1}{2}\sum_{\alpha \in \Phi(\mathbf{T}_M, \mathbf{B} \cap \mathbf{M}_{S',\mathbb{C}})} \alpha$. As $\rho_{S'} - \rho_B$ is invariant by $\Omega_{S'}$ and

$$e^{(\rho_{S'} - \rho_B)(X)} = \delta^{-1/2}_{P_{S'}(\mathbb{R})}(\gamma),$$

this formula becomes

$$\mathrm{Tr}(\gamma, V_{\omega(\lambda)-\rho_B}) = \Delta_{M_{S'}}(X)^{-1} \delta^{-1/2}_{P_{S'}(\mathbb{R})}(\gamma) \sum_{\omega_M \in \Omega_{S'}} \det(\omega_M) e^{(\omega_M \omega(\lambda))(X)}.$$

Hence

$$\mathrm{Tr}(\gamma, R\Gamma(\mathrm{Lie}(\mathbf{N}_{S'}), E)_{>t'_r, r\in S'})$$
$$= \Delta_{M_{S'}}(X)^{-1}\delta_{P_{S'}(\mathbb{R})}^{-1/2}(\gamma)\sum_{\omega_M\in\Omega_{S'}}\sum_{\omega}\det(\omega_M\omega)e^{(\omega_M\omega(\lambda))(X)},$$

where the second sum is taken over the set of $\omega\in\Omega'_{S'}$ such that $\langle\omega(\lambda),\varpi_r\rangle>0$ for every $r\in S'$. As the ϖ_r, $r\in S'$, are invariant by $\Omega_{S'}$, for every $\omega_M\in\Omega_{S'}$, $\omega\in\Omega'_{S'}$ and $r\in S'$, $\langle\omega(\lambda),\varpi_r\rangle=\langle\omega_M\omega(\lambda),\varpi_r\rangle$. Hence

$$\mathrm{Tr}(\gamma, R\Gamma(\mathrm{Lie}(\mathbf{N}_{S'}), E)_{>t'_r, r\in S'}) = \Delta_{M_{S'}}(X)^{-1}\delta_{P_{S'}(\mathbb{R})}^{-1/2}(\gamma)\sum_{\omega}\det(\omega)e^{(\omega(\lambda))(X)},$$

where the sum is taken over the set of $\omega\in\Omega$ such that $\langle\omega(\lambda),\varpi_r\rangle>0$ for every $r\in S'$.

Moreover,

$$|D_{\mathbf{M}}^{\mathbf{M}_{S'}}(\gamma)|^{1/2} = |\Delta_{M_{S'}}(X)||\Delta_M(X)|^{-1}.$$

But all the roots of $\mathbf{T}_M$ in $\mathrm{Lie}(\mathbf{M}_{S'})/\mathrm{Lie}(\mathbf{M})$ are complex, so

$$\Delta_{M_{S'}}(X)\Delta_M(X)^{-1} = \prod_{\substack{\alpha\in\Phi(\mathbf{T}_M,\mathrm{Lie}(\mathbf{M}_{S'})/Lie(\mathbf{M}))\\ \alpha>0}}(e^{\alpha(X)/2}-e^{-\alpha(X)/2})\in\mathbb{R}^+,$$

and

$$|D_{\mathbf{M}}^{\mathbf{M}_{S'}}(\gamma)|^{1/2} = \Delta_{M_{S'}}(X)\Delta_M(X)^{-1}.$$

Finally,

$$|D_{\mathbf{M}}^{\mathbf{M}_{S'}}(\gamma)|^{1/2}\delta_{P_{S'}(\mathbb{R})}^{1/2}(\gamma)\mathrm{Tr}(\gamma, R\Gamma(\mathrm{Lie}(\mathbf{N}_{S'}), E)_{>t'_r, r\in S'})$$
$$= \Delta_M(X)^{-1}\sum_{\omega}\det(\omega)e^{(\omega(\lambda))(X)},$$

where the sum is taken as before on the set of $\omega\in\Omega$ such that $\langle\omega(\lambda),\varpi_r\rangle>0$ for every $r\in S'$.

The action of the group $\mathfrak{S}_s$ on $\mathbf{T}_M(\mathbb{C})$ gives an injective morphism $\mathfrak{S}_s\longrightarrow\Omega$. Use this morphism to see $\mathfrak{S}_s$ as a subgroup of Ω. For every $\sigma\in\mathfrak{S}_s$, $\det(\sigma)=1$, and the function Δ_M is invariant by $\mathfrak{S}_s$. Hence

$$\sum_{\sigma\in\mathfrak{S}_s}|D_{\mathbf{M}}^{\mathbf{M}_{S'}}(\sigma\gamma)|^{1/2}\delta_{P_{S'}(\mathbb{R})}(\sigma\gamma)^{1/2}\mathrm{Tr}(\sigma\gamma, R\Gamma(\mathrm{Lie}(\mathbf{N}_{S'}), E)_{>t'_r, r\in S'})$$
$$= \Delta_M(X)^{-1}\sum_{\omega\in\Omega}\det(\omega)e^{(\omega(\lambda))(X)}|\{\sigma\in\mathfrak{S}_s|\langle\sigma\omega(\lambda),\varpi_r\rangle>0\text{ for every }r\in S'\}|.$$

We now use the formula of [A6] pp. 272–274 (recalled in the proof of proposition 3.3.4) to calculate $\Phi_M(\gamma,\Theta)$. Let R be the set of real roots in $\Phi(\mathbf{T}_M,\mathbf{G})$. For every $r\in\{1,\dots,s\}$, let

$$\alpha_r:\mathbf{T}_M\simeq(R_{E/\mathbb{Q}}\mathbb{G}_m)^s\times\mathbf{G}(\mathbf{U}(1)^{n-2s})\longrightarrow\mathbb{G}_m,\ ((\lambda_1,\dots,\lambda_s),g)\longmapsto c(g)\lambda_r\overline{\lambda}_r.$$

Then $R = \{\pm\alpha_1, \dots, \pm\alpha_s\}$, $R^+ := R \cap \Phi^+ = \{\alpha_1, \dots, \alpha_s\}$, and, for every $r \in \{1, \dots, s\}$, the coroot $\alpha_r^\vee$ is the morphism

$$\mathbb{G}_m \longrightarrow \mathbf{T}_M, \lambda \longmapsto ((\underbrace{1, \dots, 1}_{r-1}, \lambda, \underbrace{1, \dots, 1}_{s-r}), 1).$$

Note that $\varpi_r = \alpha_1^\vee + \dots + \alpha_r^\vee$. As $c(\gamma)|\lambda_r|^2 \in]0, 1[$ for every $r \in \{1, \dots, s\}$, $R_X^+ = \{-\alpha_1, \dots, -\alpha_s\} = -R^+$ and $\varepsilon_R(X) = (-1)^s$. Let Q^+ be a positive root system in $R^\vee$. If $Q^+ \neq \{\alpha_1^\vee, \dots, \alpha_s^\vee\}$, then $\overline{c}(Q^+, R_X^+) = 0$ by property (ii) of the function $\overline{c}$ of [A6] p. 273. Suppose that $Q^+ = \{\alpha_1^\vee, \dots, \alpha_s^\vee\}$. Note that R is the product of the root systems $\{\pm\alpha_r\}$, $1 \leq r \leq s$. Hence

$$\overline{c}(Q^+, R_X^+) = \prod_{r=1}^{s} \overline{c}(\{\alpha_r^\vee\}, \{-\alpha_r\}).$$

But it is easy to see that $\overline{c}(\{\alpha_r^\vee\}, \{-\alpha_r\}) = 2$ for every $r \in \{1, \dots, s\}$ (by property (iii) of [A6] p. 273). Hence $\overline{c}(Q^+, R_X^+) = 2^s$. Finally, we find

$$\Phi_M(\gamma, \Theta) = (-1)^s 2^s \Delta_M(X)^{-1} \sum_{\omega} \det(\omega) e^{(\omega(\lambda))(X)},$$

where the sum is taken over the set of $\omega \in \Omega$ such that $\langle \omega(\lambda), \alpha_r^\vee \rangle > 0$ for every $r \in \{1, \dots, s\}$.

To finish the proof, it is enough to show that, if $\omega \in \Omega$ is fixed, then

$$\sum_{\substack{S' \subset S \\ s \in S'}} (-1)^{|S|-|S'|} |W(\mathbf{L}_S(\mathbb{Q}), \mathbf{L}_{S'}(\mathbb{Q}))|^{-1} |\{\sigma \in \mathfrak{S}_s | \langle \sigma\omega(\lambda), \varpi_r \rangle > 0 \text{ for every } r \in S'\}|$$

is equal to 1 if $\langle \omega(\lambda), \alpha_r^\vee \rangle > 0$ for every $r \in \{1, \dots, s\}$ and to 0 otherwise. This is proved in lemma 3.4.2 below. □

Let $n \in \mathbb{N}^*$. Let $S \subset \{1, \dots, n\}$. If $\lambda = (\lambda_1, \dots, \lambda_n) \in \mathbb{R}^n$, we say that $\lambda >_S 0$ if, for every $r \in S$, $\lambda_1 + \dots + \lambda_r > 0$, and we write

$$\mathfrak{S}_S(\lambda) = \{\sigma \in \mathfrak{S}_n | \sigma(\lambda) >_S 0\}.$$

If $S = \{r_1, \dots, r_k\}$ with $r_1 < \dots < r_k$, write

$$w_S = r_1! \prod_{i=1}^{k-1} (r_{i+1} - r_i)!.$$

Lemma 3.4.2 *Let* $\lambda = (\lambda_1, \dots, \lambda_n) \in \mathbb{R}^n$. *Then*

$$\sum_{\substack{S \subset \{1, \dots, n\} \\ S \ni n}} (-1)^{|S|} w_S^{-1} |\mathfrak{S}_S(\lambda)| = \begin{cases} (-1)^n & \text{if } \lambda_r > 0 \text{ for every } r \in \{1, \dots, n\}, \\ 0 & \text{otherwise.} \end{cases}$$

Proof. First we reformulate the problem. Let $\lambda = (\lambda_1, \dots, \lambda_n) \in \mathbb{R}^n$. Write $\lambda > 0$ if $\lambda_1 > 0, \lambda_1 + \lambda_2 > 0, \dots, \lambda_1 + \dots + \lambda_n > 0$. For every $I \subset \{1, \dots, n\}$, let $s_I(\lambda) = \sum_{i \in I} \lambda_i$. Let $\mathcal{P}_{\text{ord}}(n)$ be the set of ordered partitions of $\{1, \dots, n\}$. For every $p = (I_1, \dots, I_k) \in \mathcal{P}_{\text{ord}}(n)$, set $|p| = k$ and $\lambda_p = (s_{I_1}(\lambda), \dots, s_{I_k}(\lambda)) \in \mathbb{R}^k$. Let

$$\mathcal{P}_{\text{ord}}(\lambda) = \{p \in \mathcal{P}_{\text{ord}}(n) | \lambda_p > 0\}.$$

Then it is obvious that :

$$\sum_{\substack{S\subset\{1,\dots,n\}\\ S\ni n}} (-1)^{|S|} w_S^{-1} |\mathfrak{S}_S(\lambda)| = \sum_{p\in\mathcal{P}_{\mathrm{ord}}(\lambda)} (-1)^{|p|}.$$

We show the lemma by induction on the pair $(n, |\mathcal{P}_{\mathrm{ord}}(\lambda)|)$ (we use the lexicographical ordering). If $n = 1$ or if $\mathcal{P}_{\mathrm{ord}}(\lambda) = \varnothing$ (i.e., $\lambda_1 + \cdots + \lambda_n \leq 0$), the result is obvious. Assume that $n \geq 2$, that $\mathcal{P}_{\mathrm{ord}}(\lambda) \neq \varnothing$ and that the result is known for

- all the elements of $\mathbb{R}^m$ if $1 \leq m < n$;
- all the $\lambda' \in \mathbb{R}^n$ such that $|\mathcal{P}_{\mathrm{ord}}(\lambda')| < |\mathcal{P}_{\mathrm{ord}}(\lambda)|$.

Let $\mathcal{E}$ be the set of $I \subset \{1, \dots, n\}$ such that there exists $p = (I_1, \dots, I_k) \in \mathcal{P}_{\mathrm{ord}}(\lambda)$ with $I_1 = I$; $\mathcal{E}$ is nonempty because $\mathcal{P}_{\mathrm{ord}}(\lambda)$ is nonempty. Let ε be the minimum of the $s_I(\lambda)/|I|$, for $I \in \mathcal{E}$. Let $I \in \mathcal{E}$ be an element with minimal cardinality among the elements J of $\mathcal{E}$ such that $s_J(\lambda) = \varepsilon|J|$. Define $\lambda' = (\lambda'_1, \dots, \lambda'_n) \in \mathbb{R}^n$ by

$$\lambda'_i = \begin{cases} \lambda_i & \text{if } i \notin I, \\ \lambda_i - \varepsilon & \text{if } i \in I. \end{cases}$$

Let $\mathcal{P}'$ be the set of $p = (I_1, \dots, I_k) \in \mathcal{P}_{\mathrm{ord}}(\lambda)$ such that there exists $r \in \{1, \dots, k\}$ with $I = I_1 \cup \cdots \cup I_r$. Then $\mathcal{P}' \neq \varnothing$.

It is obvious that $\mathcal{P}_{\mathrm{ord}}(\lambda') \subset \mathcal{P}_{\mathrm{ord}}(\lambda) - \mathcal{P}'$ (because $s_I(\lambda') = 0$). Let us show that $\mathcal{P}_{\mathrm{ord}}(\lambda') = \mathcal{P}_{\mathrm{ord}}(\lambda) - \mathcal{P}'$. Let $p = (I_1, \dots, I_k) \in \mathcal{P}_{\mathrm{ord}}(\lambda) - \mathcal{P}'$, and let us show that $p \in \mathcal{P}_{\mathrm{ord}}(\lambda')$. It is enough to show that $s_{I_1}(\lambda') > 0$, because $(I_1 \cup \cdots \cup I_r, I_{r+1}, \dots, I_k) \in \mathcal{P}_{\mathrm{ord}}(\lambda) - \mathcal{P}'$ for every $r \in \{1, \dots, k\}$. By definition of λ', $s_{I_1}(\lambda') = s_{I_1}(\lambda) - \varepsilon|I \cap I_1|$. If $s_{I_1}(\lambda) > \varepsilon|I_1|$, then $s_{I_1}(\lambda') > \varepsilon(|I_1| - |I \cap I_1|) \geq 0$. If $s_{I_1}(\lambda) = \varepsilon|I_1|$, then $|I_1| \geq |I|$ by definition of I and $I_1 \neq I$ because $p \notin \mathcal{P}'$, so $I_1 \not\subset I$, and $s_{I_1}(\lambda') = \varepsilon(|I_1| - |I \cap I_1|) > 0$.

As $s_I(\lambda') = 0$, there exists $i \in I$ such that $\lambda'_i \leq 0$. By the induction hypothesis, $\sum_{p\in\mathcal{P}_{\mathrm{ord}}(\lambda')} (-1)^{|p|} = 0$. Hence $\sum_{p\in\mathcal{P}_{\mathrm{ord}}(\lambda)} (-1)^{|p|} = \sum_{p\in\mathcal{P}'} (-1)^{|p|}$. As the equality of the lemma does not change if the λ_i are permuted, we may assume that there exists $m \in \{1, \dots, n\}$ such that $I = \{1, \dots, m\}$. Assume first that $m < n$. Let $\mu = (\lambda_1, \dots, \lambda_m)$ and $\nu = (\lambda_{m+1}, \dots, \lambda_n)$. Identify $\{m+1, \dots, n\}$ to $\{1, \dots, n-m\}$ by the map $k \longmapsto k - m$, and define a map $\varphi : \mathcal{P}' \longrightarrow \mathcal{P}_{\mathrm{ord}}(m) \times \mathcal{P}_{\mathrm{ord}}(n-m)$ as follows: if $p = (I_1, \dots, I_k) \in \mathcal{P}'$ and if $r \in \{1, \dots, k\}$ is such that $I_1 \cup \cdots \cup I_r = I$, set $\varphi(p) = ((I_1, \dots, I_r), (I_{r+1}, \dots, I_k))$. The map φ is clearly injective. Let us show that the image of φ is $\mathcal{P}_{\mathrm{ord}}(\mu) \times \mathcal{P}_{\mathrm{ord}}(\nu)$. The inclusion $\mathcal{P}_{\mathrm{ord}}(\mu) \times \mathcal{P}_{\mathrm{ord}}(\nu) \subset \varphi(\mathcal{P}')$ is obvious. Let $p = (I_1, \dots, I_k) \in \mathcal{P}'$, and let $r \in \{1, \dots, k\}$ be such that $I = I_1 \cup \cdots \cup I_r$. We want to show that $\varphi(p) \in \mathcal{P}_{\mathrm{ord}}(\mu) \times \mathcal{P}_{\mathrm{ord}}(\nu)$, i.e., that, for every $s \in \{r+1, \dots, k\}$, $s_{I_{r+1}\cup\dots\cup I_s}(\lambda) > 0$. After replacing p by $(I_1 \cup \cdots \cup I_r, I_{r+1} \cup \cdots \cup I_s, I_{s+1}, \dots, I_k)$, we may assume that $r = 1$ (hence $I = I_1$) and $s = 2$. Then

$$s_{I_2}(\lambda) = s_{I\cup I_2}(\lambda) - s_I(\lambda) = s_{I\cup I_2}(\lambda) - \varepsilon|I| \geq \varepsilon|I \cup I_2| - \varepsilon|I| > 0.$$

Finally :

$$\sum_{p\in\mathcal{P}'} (-1)^{|p|} = \left(\sum_{p\in\mathcal{P}_{\mathrm{ord}}(\mu)} (-1)^{|p|}\right)\left(\sum_{p\in\mathcal{P}_{\mathrm{ord}}(\nu)} (-1)^{|p|}\right).$$

Hence the conclusion of the lemma is a consequence of the induction hypothesis, applied to μ and ν.

We still have to treat the case $I=\{1,\ldots,n\}$. Let us show that there is no partition $\{I_1, I_2\}$ of $\{1,\ldots,n\}$ such that $s_{I_1}(\lambda)>0$ and $s_{I_2}(\lambda)>0$ (in particular, there exists at least one i such that $\lambda_i \leq 0$). If such a partition existed, then, by definition of I, we would have inequalities $s_{I_1}(\lambda)>\varepsilon|I_1|$ and $s_{I_2}(\lambda)>\varepsilon|I_2|$, hence $s_I(\lambda)>\varepsilon|I|$; but that is impossible. Let $\mathcal{P}(n)$ be the set of (unordered) partitions of $\{1,\ldots,n\}$. For every $q=\{I_\alpha, \alpha\in A\}\in\mathcal{P}(n)$, write $|q|=|A|$. Let $q=\{I_\alpha,\alpha\in A\}\in\mathcal{P}(n)$. By lemma 3.4.3, applied to $(s_{I_{\alpha_1}}(\lambda),\ldots,s_{I_{\alpha_k}}(\lambda))$ for a numbering $(\alpha_1,\ldots,\alpha_k)$ of A (the choice of numbering is unimportant), there are exactly $(|q|-1)!$ ways to order q in order to get an element of $\mathcal{P}_{\mathrm{ord}}(\lambda)$. Hence

$$\sum_{p\in\mathcal{P}_{\mathrm{ord}}(\lambda)}(-1)^{|p|}=\sum_{q\in\mathcal{P}(n)}(-1)^{|q|}(|q|-1)!.$$

If q' is a partition of $\{1,\ldots,n-1\}$, we can associate to it a partition q of $\{1,\ldots,n\}$ in one of the following ways:

(i) Adding n to one of the sets of q'. There are $|q'|$ ways of doing this, and we get $|q|=|q'|$.

(ii) Adding to q' the set $\{n\}$. There is only one way of doing this, and we get $|q|=|q'|+1$.

We get every partition of $\{1,\ldots,n\}$ in this way, and we get it only once. Hence (remember that $n\geq 2$),

$$\sum_{q\in\mathcal{P}(n)}(-1)^{|q|}(|q|-1)!=\sum_{q'\in\mathcal{P}(n-1)}(-1)^{|q'|}|q'|(|q'|-1)! + \sum_{q'\in\mathcal{P}(n-1)}(-1)^{|q'|+1}(|q'|)!=0. \qquad \square$$

In the lemma below, $\mathfrak{S}_n$ acts on $\mathbb{R}^n$ in the usual way (permuting the coordinates).

Lemma 3.4.3 *Let $\lambda=(\lambda_1,\ldots,\lambda_n)\in\mathbb{R}^n$. Assume that $\lambda_1+\cdots+\lambda_n>0$ and that there is no partition $\{I_1,I_2\}$ of $\{1,\ldots,n\}$ such that $s_{I_1}(\lambda)>0$ and $s_{I_2}(\lambda)>0$. Then*

$$|\{\sigma\in\mathfrak{S}_n|\sigma(\lambda)>0\}|=(n-1)!.$$

Proof. Let $\mathfrak{S}(\lambda)=\{\sigma\in\mathfrak{S}_n|\sigma(\lambda)>0\}$. Let $\tau\in\mathfrak{S}_n$ be the permutation that sends an element i of $\{1,\ldots,n-1\}$ to $i+1$, and sends n to 1. Let us show that there exists a unique $k\in\{1,\ldots,n\}$ such that $\tau^k\in\mathfrak{S}(\lambda)$. Let $s=\min\{\lambda_1+\ldots\lambda_l, 1\leq l\leq n\}$. Let k be the biggest element of $\{1,\ldots,n\}$ such that $\lambda_1+\cdots+\lambda_k=s$. If $l\in\{k+1,\ldots,n\}$, then $\lambda_1+\cdots+\lambda_l>\lambda_1+\cdots+\lambda_k$, hence $\lambda_{k+1}+\cdots+\lambda_l>0$. If $l\in\{1,\ldots,k\}$, then

$$\begin{aligned}\lambda_{k+1}+\cdots+\lambda_n+\lambda_1+\cdots+\lambda_l &= (\lambda_1+\cdots+\lambda_n)-(\lambda_1+\cdots+\lambda_k)\\ &\quad +(\lambda_1+\cdots+\lambda_l)\\ &> -(\lambda_1+\cdots+\lambda_k)+(\lambda_1+\cdots+\lambda_l)\\ &\geq 0.\end{aligned}$$

This proves that $\tau^k(\lambda) = (\lambda_{k+1}, \ldots, \lambda_n, \lambda_1, \ldots, \lambda_k) > 0$. Suppose that there exists $k, l \in \{1, \ldots, n\}$ such that $k < l$, $\tau^k(\lambda) > 0$ and $\tau^l(\lambda) > 0$. Let $I_1 = \{k+1, \ldots, l\}$ and $I_2 = \{1, \ldots, n\} - I_1$. Then $s_{I_1}(\lambda) = \lambda_{k+1} + \cdots + \lambda_l > 0$ because $\tau^k(\lambda) > 0$, and $s_{I_2}(\lambda) = \lambda_{l+1} + \cdots + \lambda_n + \lambda_1 + \cdots + \lambda_k > 0$ because $\tau^l(\lambda) > 0$. This contradicts the assumption on λ.

Applying the above reasoning to $\sigma(\lambda)$, for $\sigma \in \mathfrak{S}_n$, we see that $\mathfrak{S}_n$ is the disjoint union of the subsets $\tau^k \mathfrak{S}(\lambda)$, $1 \leq k \leq n$. This implies the conclusion of the lemma. □

Chapter Four

Orbital integrals at p

4.1 A SATAKE TRANSFORM CALCULATION (AFTER KOTTWITZ)

Lemma 4.1.1 *(cf. [K3] 2.1.2, [K9] p. 193) Let F be a local or global field and $\mathbf{G}$ be a connected reductive algebraic group over F. For every cocharacter $\mu : \mathbb{G}_{m,F} \longrightarrow \mathbf{G}$, there exists a representation r_μ of ${}^L\mathbf{G}(= \widehat{\mathbf{G}} \rtimes W_F)$, unique up to isomorphism, satisfying the following conditions:*

(a) The restriction of r_μ to $\widehat{\mathbf{G}}$ is irreducible algebraic of highest weight μ.
(b) For every $\mathrm{Gal}(\overline{F}/F)$-fixed splitting of $\widehat{\mathbf{G}}$, the group W_F, embedded in ${}^L\mathbf{G}$ by the section associated to the splitting, acts trivially on the highest weight subspace of r_μ (determined by the same splitting).

Let p be a prime number, $\overline{\mathbb{Q}}_p$ be an algebraic closure of $\mathbb{Q}_p$, $\mathbb{Q}_p^{ur}$ be the maximal unramified extension of $\mathbb{Q}_p$ in $\overline{\mathbb{Q}}_p$, $F \subset \overline{\mathbb{Q}}_p$ be a finite unramified extension of $\mathbb{Q}_p$, W_F be the Weyl group of F, ϖ_F be a uniformizer of F and $n = [F : \mathbb{Q}_p]$. The cardinality of the residual field of F is p^n.

Let $\mathbf{G}$ be a connected reductive algebraic group over F, and assume that $\mathbf{G}$ is unramified. Fix a hyperspecial maximal compact subgroup K of $\mathbf{G}(F)$, and let $\mathcal{H} = \mathcal{H}(\mathbf{G}(F), \mathrm{K}) := C_c^\infty(\mathrm{K} \setminus \mathbf{G}(F)/\mathrm{K})$ be the associated Hecke algebra. For every cocharacter $\mu : \mathbb{G}_{m,F} \longrightarrow \mathbf{G}$ of $\mathbf{G}$, let

$$f_\mu = \frac{\mathbb{1}_{\mathrm{K}\mu(\varpi_F^{-1})\mathrm{K}}}{\mathrm{vol}(\mathrm{K})} \in \mathcal{H}.$$

In this section, the L-group of $\mathbf{G}$ will be ${}^L\mathbf{G} = \widehat{\mathbf{G}} \rtimes W_F$. Let $\varphi \longmapsto \pi_\varphi$ be the bijection between the set of equivalence classes of admissible unramified morphisms $\varphi : W_F \longrightarrow {}^L\mathbf{G}$ and the set of isomorphism classes of spherical representations of $\mathbf{G}(F)$.

Theorem 4.1.2 *([K3] 2.1.3) Let $\mu : \mathbb{G}_{m,F} \longrightarrow \mathbf{G}$ be such that the weights of the representation $\mathrm{Ad} \circ \mu : \mathbb{G}_{m,F} \longrightarrow \mathrm{Lie}(\mathbf{G}_{\overline{\mathbb{Q}}_p})$ are in $\{-1, 0, 1\}$. Fix a maximal torus $\mathbf{T}$ of $\mathbf{G}$ such that μ factors through $\mathbf{T}$, and an ordering on the roots of $\mathbf{T}$ in $\mathbf{G}$ such that μ is dominant. Let ρ be half the sum of the positive roots.*

Then, for every admissible unramified morphism $\varphi : W_F \longrightarrow {}^L\mathbf{G}$,

$$\mathrm{Tr}(\pi_\varphi(f_\mu)) = p^{n\langle\rho,\mu\rangle}\,\mathrm{Tr}(r_{-\mu}(\varphi(\Phi_F))),$$

where $r_{-\mu}$ is the representation defined in lemma 4.1.1 above and $\Phi_F \in W(\mathbb{Q}_p^{ur}/F)$ is the geometric Frobenius.

Remark 4.1.3 In 2.1.3 of [K3], the arithmetic Frobenius is used instead of the geometric Frobenius. The difference comes from the fact that we use here the other normalization of the class field isomorphism (cf. [K9] p. 193).

4.2 EXPLICIT CALCULATIONS FOR UNITARY GROUPS

This section contains explicit descriptions of the Satake isomorphism, the base change map, the transfer map, and the twisted transfer (or unstable base change) map for the spherical Hecke algebras of the unitary groups of section 2.1. These calculations will be useful when proving proposition 4.3.1 and in the applications of chapter 7 and section 8.4.

Let p be a prime number, and let $\overline{\mathbb{Q}}_p$ and $\mathbb{Q}_p^{ur}$ be as in 4.1. Remember that, if $\mathbf{G}$ and $\mathbf{H}$ are unramified groups over $\mathbb{Q}_p$ and if $\eta : {}^L\mathbf{H} := \widehat{\mathbf{H}} \rtimes W_{\mathbb{Q}_p} \longrightarrow {}^L\mathbf{G} := \widehat{\mathbf{G}} \rtimes W_{\mathbb{Q}_p}$ is an unramified L-morphism (i.e., a L-morphism that comes by inflation from a morphism $\widehat{\mathbf{H}} \rtimes W(\mathbb{Q}_p^{ur}/\mathbb{Q}_p) \longrightarrow \widehat{\mathbf{G}} \rtimes W(\mathbb{Q}_p^{ur}/\mathbb{Q}_p)$), then it induces a morphism of algebras $b_\eta : \mathcal{H}_\mathbf{G} \longrightarrow \mathcal{H}_\mathbf{H}$, where $\mathcal{H}_\mathbf{H}$ (resp., $\mathcal{H}_\mathbf{G}$) is the spherical Hecke algebra of $\mathbf{H}$ (resp. $\mathbf{G}$). This construction is recalled in more detail in 9.2 (just before lemma 9.2.5).

Let $E = \mathbb{Q}[\sqrt{-b}]$ be an imaginary quadratic extension where p is unramified, and fix a place $\wp$ of E above p (i.e., an embedding $E \subset \mathbb{Q}_p^{ur}$). Let $n_1, \dots, n_r \in \mathbb{N}^*$ and let $J_1 \in \mathbf{GL}_{n_1}(\mathbb{Z}), \dots, J_r \in \mathbf{GL}_{n_r}(\mathbb{Z})$ be symmetric matrices that are anti-diagonal (i.e., in the subset $\begin{pmatrix} 0 & & * \\ & \cdot^{\cdot^{\cdot}} & \\ * & & 0 \end{pmatrix}$). Let q_i be the floor (integral part) of $n_i/2$, $n = n_1 + \cdots + n_r$, and $\mathbf{G} = \mathbf{G}(\mathbf{U}(J_1) \times \cdots \times \mathbf{U}(J_r))$. Then the group $\mathbf{G}$ is unramified over $\mathbb{Q}_p$. Hence $\mathbf{G}$ extends to a reductive group scheme over $\mathbb{Z}_p$ (i.e., to a group scheme over $\mathbb{Z}_p$ with connected geometric fibers whose special fiber is a reductive group over $\mathbb{F}_p$); we gave an example of such a group scheme in remark 2.1.1. We will still denote this group scheme by $\mathbf{G}$. In this section, the L-group of $\mathbf{G}$ will be the L-group over $\mathbb{Q}_p$, i.e., ${}^L\mathbf{G} = \mathbf{G} \rtimes W_{\mathbb{Q}_p}$.

Satake isomorphism

Let $L \subset \mathbb{Q}_p^{ur}$ be an unramified extension of $\mathbb{Q}_p$. Let $\mathrm{K}_L = \mathbf{G}(\mathcal{O}_L)$; it is a hyperspecial maximal compact subgroup of $\mathbf{G}(L)$. We calculate the Satake isomorphism for $\mathcal{H}(\mathbf{G}(L), \mathrm{K}_L)$.

Suppose first that $\mathbf{G}$ splits over L, that is, $L \supset E$ (if $L = \mathbb{Q}_p$, this means that p splits in E). Then $\mathbf{G}_L \simeq \mathbb{G}_{m,L} \times \mathbf{GL}_{n_1,L} \times \cdots \times \mathbf{GL}_{n_r,L}$. For every $i \in \{1, \dots, r\}$, let $\mathbf{T}_i$ be the diagonal torus of $\mathbf{GU}(J_i)$. Identify $\mathbf{T}_{i,L}$ with the torus $\mathbb{G}_{m,L} \times \mathbb{G}_{m,L}^{n_i}$ by the isomorphism

$$g = \mathrm{diag}(\lambda_1, \dots, \lambda_{n_i}) \longmapsto (c(g), (u(\lambda_1), \dots, u(\lambda_{n_i}))),$$

where u is the morphism $L \otimes_{\mathbb{Q}_p} E \longrightarrow L, x \otimes 1 + y \otimes \sqrt{-b} \longmapsto x + y\sqrt{-b}$. A split maximal torus of $\mathbf{G}_L$ is the diagonal torus $\mathbf{T}_{G,L}$, where

$$\mathbf{T}_G = \{(g_1, \dots, g_r) \in \mathbf{T}_1 \times \cdots \times \mathbf{T}_r | c(g_1) = \cdots = c(g_r)\}.$$

The above isomorphisms give an isomorphism $\mathbf{T}_{G,L} \simeq \mathbb{G}_{m,L} \times \mathbb{G}_{m,L}^n$. Let $\Omega_G(L) = W(\mathbf{T}_G(L), \mathbf{G}(L))$ be the relative Weyl group of $\mathbf{T}_{G,L}$ (as $\mathbf{G}$ splits over L, this group is actually equal to the absolute Weyl group). Then $\Omega_G(L) \simeq \mathfrak{S}_{n_1} \times \cdots \times \mathfrak{S}_{n_r}$. The Satake isomorphism is an isomorphism

$$\mathcal{H}(\mathbf{G}(L), \mathrm{K}_L) \xrightarrow{\sim} \mathbb{C}[X_*(\mathbf{T}_G)]^{\Omega_G(L)}.$$

There is an isomorphism

$$\mathbb{C}[X_*(\mathbf{T}_G)] \simeq \mathbb{C}[X^{\pm 1}, X_{i,j}^{\pm 1}, 1 \leq i \leq r, 1 \leq j \leq n_i]$$

induced by the isomorphism $\mathbf{T}_{G,L} \simeq \mathbb{G}_{m,L} \times \mathbb{G}_{m,L}^n$ defined above. Explicitly

- X corresponds to the cocharacter

$$\lambda \longmapsto \left(\frac{\lambda+1}{2} \otimes 1 + \frac{1-\lambda}{2\sqrt{-b}} \otimes \sqrt{-b}\right)(I_{n_1}, \ldots, I_{n_r}).$$

- Let $i \in \{1, \ldots, r\}$ and $s \in \{1, \ldots, q_i\} \cup \{n_i + 1 - q_i, \ldots, n_i\}$. Then $X_{i,s}$ corresponds to the cocharacter

$$\lambda \longmapsto (I_1, \ldots, I_{i-1}, \mathrm{diag}(a_1(\lambda), \ldots, a_{n_i}(\lambda)), I_{i+1}, \ldots, I_r),$$

  with :

$$a_j(\lambda) = \begin{cases} \dfrac{\lambda+1}{2} \otimes 1 + \dfrac{\lambda-1}{2\sqrt{-b}} \otimes \sqrt{-b} & \text{if} \quad j = s, \\ \dfrac{\lambda^{-1}+1}{2} \otimes 1 + \dfrac{1-\lambda^{-1}}{2\sqrt{-b}} \otimes \sqrt{-b} & \text{if} \quad s = n+1-j, \\ 1 & \text{otherwise.} \end{cases}$$

- If $i \in \{1, \ldots, r\}$ is such that n_i is odd, then $X_{i,(n_i+1)/2}$ corresponds to the cocharacter

$$\lambda \longmapsto \left(I_1, \ldots, I_{i-1}, \begin{pmatrix} I_{q_i} & & 0 \\ & \dfrac{\lambda+\lambda^{-1}}{2} \otimes 1 + \dfrac{\lambda-\lambda^{-1}}{2\sqrt{-b}} \otimes \sqrt{-b} & \\ 0 & & I_{q_i} \end{pmatrix}, I_{i+1}, \ldots, I_r\right).$$

We get an isomorphism

$$\mathbb{C}[X_*(\mathbf{T}_G)]^{\Omega_G(L)} \simeq \mathbb{C}[X^{\pm 1}] \otimes \mathbb{C}[X_{i,j}^{\pm 1}, 1 \leq i \leq r, 1 \leq j \leq n_i]^{\mathfrak{S}_{n_1} \times \cdots \times \mathfrak{S}_{n_r}},$$

where $\mathfrak{S}_{n_i}$ acts by permutations on $X_{i,1}, \ldots, X_{i,n_i}$ and trivially on the $X_{i',j}$ if $i' \neq i$.

Suppose now that $\mathbf{G}$ does not split over L (this implies that p is inert in E). For every $i \in \{1, \ldots, r\}$, a maximal split torus of $\mathbf{GU}(J_i)_L$ is $\mathbf{S}_{i,L}$, where

$$\mathbf{S}_i = \{\mathrm{diag}(\lambda\lambda_1, \ldots, \lambda\lambda_{q_i}, \lambda_{q_i}^{-1}, \ldots, \lambda_1^{-1}), \lambda, \lambda_1, \ldots, \lambda_{q_i} \in \mathbb{G}_{m,\mathbb{Q}_p}\} \simeq \mathbb{G}_{m,\mathbb{Q}_p}^{q_i+1}$$

if n_i is even, and

$$\mathbf{S}_i = \{\mathrm{diag}(\lambda\lambda_1, \ldots, \lambda\lambda_{q_i}, \lambda, \lambda\lambda_{q_i}^{-1}, \ldots, \lambda\lambda_1^{-1}), \lambda, \lambda_1, \ldots, \lambda_{q_i} \in \mathbb{G}_{m,\mathbb{Q}_p}\} \simeq \mathbb{G}_{m,\mathbb{Q}_p}^{q_i+1}$$

if n_i is odd. A maximal split torus of $\mathbf{G}_L$ is $\mathbf{S}_{G,L}$, where

$$\mathbf{S}_G = \{(g_1, \ldots, g_r) \in \mathbf{S}_1 \times \cdots \times \mathbf{S}_r | c(g_1) = \cdots = c(g_r)\}^0 \simeq \mathbb{G}_{m,\mathbb{Q}_p}^{q_1+\cdots+q_r+1}.$$

Let $\Omega_G(L) = W(\mathbf{S}_G(L), \mathbf{G}(L))$ be the relative Weyl group of $\mathbf{S}_G(L)$. Then $\Omega_G(L) \simeq \Omega_1 \times \cdots \times \Omega_r$, where Ω_i is the subgroup of $\mathfrak{S}_{n_i}$ generated by the transposition $(1, n_i)$ and by the image of the morphism

$$\mathfrak{S}_{q_i} \longrightarrow \mathfrak{S}_{n_i},$$
$$\sigma \longmapsto \left(\tau : j \longmapsto \begin{cases} \sigma(j) & \text{if} \quad 1 \leq j \leq q_i, \\ j & \text{if} \quad q_i + 1 \leq j \leq n_i - q_i, \\ n_i + 1 - \sigma(n_i + 1 - j) & \text{if} \quad n_i + 1 - q_i \leq j \leq n_i \end{cases}\right).$$

Hence Ω_i is isomorphic to the semidirect product $\{\pm 1\}^{q_i} \rtimes \mathfrak{S}_{q_i}$, where $\mathfrak{S}_{q_i}$ acts on $\{\pm 1\}^{q_i}$ by $(\sigma, (\varepsilon_1, \ldots, \varepsilon_{q_i})) \longmapsto (\varepsilon_{\sigma^{-1}(1)}, \ldots, \varepsilon_{\sigma^{-1}(q_i)})$. The Satake isomorphism is an isomorphism

$$\mathcal{H}(\mathbf{G}(L), \mathrm{K}_L) \xrightarrow{\sim} \mathbb{C}[X_*(\mathbf{S}_G)]^{\Omega_G(L)}.$$

Assume that n_i is even. Then there is an isomorphism

$$\mathbb{C}[X_*(\mathbf{S}_i)] \simeq \mathbb{C}[X_i'^{\pm 1}, X_{i,1}^{\pm 1}, \ldots, X_{i,q_i}^{\pm 1}]$$

that sends X_i' to the cocharacter

$$\lambda \longmapsto \begin{pmatrix} \lambda I_{q_i} & 0 \\ 0 & I_{q_i} \end{pmatrix}$$

and $X_{i,s}$, $1 \leq s \leq q_i$, to the cocharacter

$$\lambda \longmapsto \mathrm{diag}(a_1(\lambda), \ldots, a_{n_i}(\lambda)),$$

with

$$a_j(\lambda) = \begin{cases} \lambda & \text{if} \quad j = s, \\ \lambda^{-1} & \text{if} \quad j = n_i + 1 - s, \\ 1 & \text{otherwise.} \end{cases}$$

Hence we get an isomorphism

$$\mathbb{C}[X_*(\mathbf{S}_i)]^{\Omega_i} \simeq \mathbb{C}[X_i'^{\pm 1}, X_{i,1}^{\pm 1}, \ldots, X_{i,q_i}^{\pm 1}]^{\{\pm 1\}^{q_i} \rtimes \mathfrak{S}_{q_i}},$$

where $\mathfrak{S}_{q_i}$ acts by permutations on $X_{i,1}, \ldots, X_{i,q_i}$ and trivially on X_i', and $\{\pm 1\}^{q_i}$ acts by $((\varepsilon_1, \ldots, \varepsilon_{q_i}), X_{i,j}) \longmapsto X_{i,j}^{\varepsilon_j}$ and

$$((\varepsilon_1, \ldots, \varepsilon_{q_i}), X_i') \longmapsto X_i' \prod_{j \ st \ \varepsilon_j = -1} X_{i,j}^{-1}.$$

Note that the (Ω_i-invariant) cocharacter $\lambda \longmapsto \lambda I_{n_i}$ corresponds to $X_i := X_i'^2 X_{i,1}^{-1} \ldots X_{i,q_i}^{-1}$.

Assume that n_i is odd. Then there is an isomorphism

$$\mathbb{C}[X_*(\mathbf{S}_i)] \simeq \mathbb{C}[X_i^{\pm 1}, X_{i,1}^{\pm 1}, \ldots, X_{i,q_i}^{\pm 1}]$$

that sends X_i to the cocharacter $\lambda \longmapsto \lambda I_{n_i}$ and $X_{i,s}$, $1 \leq s \leq q_i$, to the cocharacter defined by the same formula as when n_i is even. Hence we get an isomorphism

$$\mathbb{C}[X_*(\mathbf{S}_i)]^{\Omega_i} \simeq \mathbb{C}[X_i^{\pm 1}, X_{i,1}^{\pm 1}, \ldots, X_{i,q_i}^{\pm 1}]^{\{\pm 1\}^{q_i} \rtimes \mathfrak{S}_{q_i}},$$

where $\{\pm 1\}^{q_i} \rtimes \mathfrak{S}_{q_i}$ acts as before on $X_{i,1}, \ldots, X_{i,q_i}$ (and trivially on X_i).

Finally, we get, if all the n_i are even,

$$\mathbb{C}[X_*(\mathbf{S}_G)]^{\Omega_G(L)} \simeq \mathbb{C}[(X'_1 \dots X'_r)^{\pm 1}, X_{i,j}^{\pm 1}, 1 \leq i \leq r, 1 \leq j \leq q_i]^{\Omega_1 \times \dots \times \Omega_r},$$

and, if at least one of the n_i is odd,

$$\mathbb{C}[X_*(\mathbf{S}_G)]^{\Omega_G(L)} \simeq \mathbb{C}[(X_1 \dots X_r)^{\pm 1}, X_{i,j}^{\pm 1}, 1 \leq i \leq r, 1 \leq j \leq q_i]^{\Omega_1 \times \dots \times \Omega_r}.$$

Let $X' = X'_1 \dots X'_r$ and $X = X_1 \dots X_r$.

In order to unify notation later, write, for every $i \in \{1, \dots, r\}$ and $j \in \{n_i + 1 - q_i, \dots, n_i\}$, $X_{i,j} = X_{i,n_i+1-j}^{-1}$ and, for every $i \in \{1, \dots, r\}$ such that n_i is odd, $X_{i,\frac{n_i+1}{2}} = 1$, and $\mathbf{S}_G = \mathbf{T}_G$.

Note that X does not stand for the same cocharacter if $\mathbf{G}$ splits over L or not (neither do the $X_{i,j}$, but this is more obvious). Let $\nu : \mathbb{G}_{m,L} \longrightarrow \mathbf{G}_L$ be the cocharacter corresponding to X. If $\mathbf{G}$ does not split over L, then ν is defined over $\mathbb{Q}$ and $c(\nu(\lambda)) = \lambda^2$ for every $\lambda \in \mathbb{G}_m$. If $\mathbf{G}$ splits over L, then ν is defined over E (and is not defined over $\mathbb{Q}$) and $c(\nu(\lambda)) = \lambda$ for every $\lambda \in \mathbb{G}_{m,E}$.

We end this subsection with an explicit version of the result of 4.1. Assume that $\mathbf{G}$ splits over L (i.e., that L contains $E_\wp$), and fix a uniformizer ϖ_L of L. Set $d = [L : \mathbb{Q}_p]$. Let $s_1, \dots, s_r \in \mathbb{N}$ be such that $s_i \leq n_i$. For every $i \in \{1, \dots, r\}$, there is a cocharacter $\mu_{s_i} : \mathbb{G}_{m,E} \longrightarrow \mathbf{GU}(J_i)_E$, defined in 2.1.2. Let $\mu = (\mu_{s_1}, \dots, \mu_{s_r}) : \mathbb{G}_{m,E} \longrightarrow \mathbf{G}_E$ and

$$\phi = \frac{\mathbb{1}_{\mathrm{K}_L \mu(\varpi_L^{-1}) \mathrm{K}_L}}{vol(\mathrm{K}_L)} \in \mathcal{H}(\mathbf{G}(L), \mathrm{K}_L)$$

(with the notation of section 4.1, $\phi = f_\mu$). Let $r_{-\mu}$ be the representation of ${}^L\mathbf{G}_{E_\wp}$ associated to $-\mu$ as in lemma 4.1.1, and let $\Phi \in W(\mathbb{Q}_p^{ur}/\mathbb{Q}_p)$ be the geometric Frobenius (so Φ^d is a generator of $W(\mathbb{Q}_p^{ur}/L)$).

Proposition 4.2.1 *For every admissible unramified morphism $\varphi : W_L \longrightarrow {}^L\mathbf{G}$,*

$$\mathrm{Tr}(\pi_\varphi(\phi)) = p^{d(s_1(n_1-s_1)/2+\dots+s_r(n_r-s_r)/2)} \, \mathrm{Tr}(r_{-\mu}(\varphi(\Phi^d))).$$

In other words, the Satake transform of ϕ is

$$p^{d(s_1(n_1-s_1)/2+\dots+s_r(n_r-s_r)/2)} X^{-1} \sum_{\substack{I_1 \subset \{1,\dots,n_1\} \\ |I_1|=s_1}} \dots \sum_{\substack{I_r \subset \{1,\dots,n_r\} \\ |I_r|=s_r}} \prod_{i=1}^{r} \prod_{j \in I_i} X_{i,j}^{-1}.$$

Proof. To deduce the first formula from theorem 4.1.2 (i.e., theorem 2.1.3 of [K3]), it is enough to show that

$$\langle \rho, \mu \rangle = s_1(n_1 - s_1)/2 + \dots + s_r(n_r - s_r)/2,$$

where ρ is half the sum of the roots of $\mathbf{T}_G$ in the standard Borel subgroup of $\mathbf{G}$ (i.e., the group of upper triangular matrices). This is an easy consequence of the definition of μ.

In the reformulation below formula (2.3.4) of [K3], theorem 2.1.3 of that article says that the Satake transform of ϕ is

$$p^{d(s_1(n_1-s_1)/2+\dots+s_r(n_r-s_r)/2)} \sum_{\nu \in \Omega_G(L)(-\mu)} \nu.$$

To prove the second formula, it is therefore enough to notice that $-\mu \in X_*(T_G)$ corresponds by the isomorphism $\mathbb{C}[X_*(\mathbf{T}_G)] \simeq \mathbb{C}[X^{\pm 1}] \otimes \mathbb{C}[X_{i,j}^{\pm 1}]$ to $X^{-1}\prod_{i=1}^{r}(X_{i,1} \dots X_{i,s_i})^{-1}$. □

The base change map

In this subsection, L is still an unramified extension of $\mathbb{Q}_p$. Write $\mathrm{K}_0 = \mathbf{G}(\mathbb{Z}_p)$ and $d = [L : \mathbb{Q}_p]$. If L contains $E_\wp$, write $a = [L : E_\wp]$. Let $R = R_{L/\mathbb{Q}_p}\mathbf{G}_L$. Then there is a "diagonal" L-morphism $\eta : {}^L\mathbf{G} \longrightarrow {}^L R$ (cf. example 8.1.1). It induces a morphism $b_\eta : \mathcal{H}(\mathbf{G}(L), \mathrm{K}_L) \longrightarrow \mathcal{H}(\mathbf{G}(\mathbb{Q}_p), \mathrm{K}_0)$, called *base change map* (or *stable base change map*). We want to calculate this morphism. To avoid confusion, when writing the Satake isomorphism for $\mathcal{H}(\mathbf{G}(L), \mathrm{K}_L)$, we will use the letter Z (instead of X) for the indeterminates (and we will still use X when writing the Satake isomorphism for $\mathcal{H}(\mathbf{G}(\mathbb{Q}_p), \mathrm{K}_0)$).

Assume first that $\mathbf{G}$ does not split over L (so that p is inert in E). Then the base change morphism corresponds by the Satake isomorphisms to the morphism induced by

$$\mathbb{C}[Z^{\pm 1}, Z_{i,j}^{\pm 1}, 1 \le i \le r, 1 \le j \le q_i] \longrightarrow \mathbb{C}[X^{\pm 1}, X_{i,j}^{\pm 1}, 1 \le i \le r, 1 \le j \le q_i],$$
$$Z \longmapsto X^d,$$
$$Z_{i,j} \longmapsto X_{i,j}^d.$$

Assume that $\mathbf{G}$ splits over L but not over $\mathbb{Q}_p$. Then $L \supset E_\wp$, p is inert in E and $d = 2a$. The base change morphism corresponds by the Satake isomorphisms to the morphism induced by

$$\mathbb{C}[Z^{\pm 1}, Z_{i,j}^{\pm 1}, 1 \le i \le r, 1 \le j \le n_i] \longrightarrow \mathbb{C}[X^{\pm 1}, X_{i,j}^{\pm 1}, 1 \le i \le r, 1 \le j \le q_i],$$
$$Z \longmapsto X^a,$$
$$Z_{i,j} \longmapsto \begin{cases} X_{i,j}^a & \text{if } 1 \le j \le q_i, \\ 1 & \text{if } q_i + 1 \le j \le n_i - q_i, \\ X_{i,n_i+1-j}^{-a} & \text{if } n_i + 1 - q_i \le j \le n_i. \end{cases}$$

Assume that $\mathbf{G}$ splits over L and $\mathbb{Q}_p$. Then p splits in E and $L \supset E_\wp = \mathbb{Q}_p$, so $d = a$. The base change morphism corresponds by the Satake isomorphisms to the morphism induced by

$$\mathbb{C}[Z^{\pm 1}, Z_{i,j}^{\pm 1}, 1 \le i \le r, 1 \le j \le n_i] \longrightarrow \mathbb{C}[X^{\pm 1}, X_{i,j}^{\pm 1}, 1 \le i \le r, 1 \le j \le n_i],$$
$$Z \longmapsto X^a,$$
$$Z_{i,j} \longmapsto X_{i,j}^a.$$

Notice that, with the conventions of the previous subsection in the case when $\mathbf{G}$ does not split over $\mathbb{Q}_p$, the base change morphism is given by the same formulas in the last two cases (this was the point of the conventions).

Remark 4.2.2 Assume that p is inert in E and that $L = E_\wp = E_p$. Then the image of the base change morphism is $\mathbb{C}[(X_1 \dots X_r)^{\pm 1}, X_{i,j}^{\pm 1}, 1 \leq i \leq r, 1 \leq j \leq q_i]^{\Omega_1 \times \dots \times \Omega_r}$. In particular, the base change morphism is surjective if and only if one of the n_i is odd.

The transfer map

In this subsection and the next, we consider, to simplify notation, the group $\mathbf{G} = \mathbf{GU}(J)$ with $J \in \mathbf{GL}_n(\mathbb{Z})$ symmetric and antidiagonal; all results extend in an obvious way to the groups considered before.

Let $n_1, n_2 \in \mathbb{N}$ be such that n_2 is even and $n = n_1 + n_2$. Let $(\mathbf{H}, s, \eta_0)$ be the elliptic endoscopic triple of $\mathbf{G}$ associated to (n_1, n_2) as in proposition 2.3.1 (note that this endoscopic triple is not always elliptic over $\mathbb{Q}_p$). Let q (resp., q_1, q_2) be the integral part of $n/2$ (resp., $n_1/2, n_2/2$). The group $\mathbf{H}$ is unramified over $\mathbb{Q}_p$. We will write $\mathbf{H}$ for the group scheme over $\mathbb{Z}_p$ extending $\mathbf{H}$ that is defined in remark 2.1.1 and $\mathrm{K}_{H,0}$ for $\mathbf{H}(\mathbb{Z}_p)$.

Any unramified L-morphism $\eta : {}^L\mathbf{H} \longrightarrow {}^L\mathbf{G}$ extending η_0 induces a morphism $b_\eta : \mathcal{H}(\mathbf{G}(\mathbb{Q}_p), \mathrm{K}_0) \longrightarrow \mathcal{H}(\mathbf{H}(\mathbb{Q}_p), \mathrm{K}_{H,0})$, called the *transfer map*. We want to give explicit formulas for this morphism. We will start with a particular case. Let, as before, $\Phi \in W(\mathbb{Q}_p^{ur}/\mathbb{Q}_p)$ be the geometric Frobenius. Let $\eta_{\mathrm{simple}} : {}^L\mathbf{H} \longrightarrow {}^L\mathbf{G}$ be the unramified morphism extending η_0 and such that $\eta(\Phi)$ is equal to $(1, \Phi)$ if p splits in E, and to $((1, A), \Phi)$ if p is inert in E, where A is defined in proposition 2.3.2. Let $b_0 = b_{\eta_{\mathrm{simple}}} : \mathcal{H}(\mathbf{G}(\mathbb{Q}_p), \mathrm{K}_0) \longrightarrow \mathcal{H}(\mathbf{H}(\mathbb{Q}_p), \mathrm{K}_{H,0})$.

Then, if p is inert in E, b_0 corresponds by the Satake isomorphisms to the morphism

$$\mathbb{C}[X_*(\mathbf{S}_G)]^{\Omega_G(\mathbb{Q}_p)} \longrightarrow \mathbb{C}[X_*(\mathbf{S}_H)]^{\Omega_H(\mathbb{Q}_p)}$$

defined by

$$\begin{array}{rcll} X' & \longmapsto & X'_1 X'_2 & \text{if} \quad n \text{ is even}, \\ X & \longmapsto & X_1 X_2 & \text{if} \quad n \text{ is odd}, \\ X_i & \longmapsto & \begin{cases} X_{1,i} & \text{if} \quad 1 \leq i \leq q_1, \\ X_{2,i-q_1} & \text{if} \quad q_1 + 1 \leq i \leq q_2. \end{cases} \end{array}$$

If p splits in E, b_0 corresponds by the Satake isomorphisms to the morphism

$$\mathbb{C}[X_*(\mathbf{T}_G)]^{\Omega_G(\mathbb{Q}_p)} \longrightarrow \mathbb{C}[X_*(\mathbf{T}_H)]^{\Omega_H(\mathbb{Q}_p)}$$

defined by

$$\begin{array}{rcl} X & \longmapsto & X_1 X_2, \\ X_i & \longmapsto & \begin{cases} X_{1,i} & \text{if} \quad 1 \leq i \leq n_1, \\ X_{2,i-n_1} & \text{if} \quad n_1 + 1 \leq i \leq n_2. \end{cases} \end{array}$$

Now let $\eta : {}^L\mathbf{H} \longrightarrow {}^L\mathbf{G}$ be any unramified L-morphism extending η_0. Then $\eta = c\eta_{\mathrm{simple}}$, where $c : W_{\mathbb{Q}_p} \longrightarrow Z(\widehat{\mathbf{H}})$ is a 1-cocycle. Write χ_η for the (unramified) quasi-character of $\mathbf{H}(\mathbb{Q}_p)$ corresponding to the class of c in $\mathrm{H}^1(W_{\mathbb{Q}_p}, Z(\widehat{\mathbf{H}}))$. Then $b_\eta : \mathcal{H}(\mathbf{G}(\mathbb{Q}_p), \mathrm{K}_0) \longrightarrow \mathcal{H}(\mathbf{H}(\mathbb{Q}_p), \mathrm{K}_{H,0})$ is given by the following formula: for every $f \in \mathcal{H}(\mathbf{G}(\mathbb{Q}_p), \mathrm{K}_0)$, $b_\eta(f) = \chi_\eta b_0(f)$.

Following Kottwitz ([K9] p. 181), we use this to define b_η even if η is not unramified. So let $\eta : {}^L\mathbf{H} \longrightarrow {}^L\mathbf{G}$ be a (not necessarily unramified) L-morphism extending η_0. Define a quasi-character χ_η of $\mathbf{H}(\mathbb{Q}_p)$ as before (χ_η can be ramified), and *define* $b_\eta : \mathcal{H}(\mathbf{G}(\mathbb{Q}_p), \mathrm{K}_0) \longrightarrow C_c^\infty(\mathbf{H}(\mathbb{Q}_p))$ by the formula: for every $f \in \mathcal{H}(\mathbf{G}(\mathbb{Q}_p), \mathrm{K}_0)$, $b_\eta(f) = \chi_\eta b_0(f)$.

The twisted transfer map

Keep the notation of the previous subsection. Fix an unramified extension L of $\mathbb{Q}_p$, and write as before $\mathrm{K}_L = \mathbf{G}(\mathcal{O}_L)$, $d = [L : \mathbb{Q}_p]$ and, if $L \supset E_\wp$, $a = [L : E_\wp]$. We will use the same conventions as before when writing the Satake isomorphism for $\mathcal{H}(\mathbf{G}(L), \mathbf{G}(\mathcal{O}_L))$ (i.e., the indeterminates will be Z and the Z_i). Let $\eta : {}^L\mathbf{H} \longrightarrow {}^L\mathbf{G}$ be an unramified L-morphism extending η_0.

Remember the definition of the twisted endoscopic datum associated to $(\mathbf{H}, s, \eta)$ and to the field extension $L/\mathbb{Q}_p$ (cf. [K9] pp. 179–180). Let $\Phi \in W(\mathbb{Q}_p^{ur}/\mathbb{Q}_p)$ be the geometric Frobenius, $R = R_{L/\mathbb{Q}_p}\mathbf{G}_L$, and θ be the automorphism of R corresponding to Φ. Then

$$\widehat{R} = (\widehat{\mathbf{G}})^d,$$

where the ith factor corresponds to the image of Φ^{d-i} in $\mathrm{Gal}(L/\mathbb{Q}_p)$. The group $W_{\mathbb{Q}_p}$ acts on $\widehat{R}$ via its quotient $W(\mathbb{Q}_p^{ur}/\mathbb{Q}_p)$, and Φ acts on $\widehat{R}$ by

$$\Phi(g_1, \dots, g_d) = \widehat{\theta}(\Phi(g_1), \dots, \Phi(g_d)) = (\Phi(g_2), \dots, \Phi(g_d), \Phi(g_1)).$$

In particular, the diagonal embedding $\widehat{\mathbf{G}} \longrightarrow \widehat{R}$ is $W_{\mathbb{Q}_p}$-equivariant, so it extends in an obvious way to an L-morphism ${}^L\mathbf{G} \longrightarrow {}^LR$; let $\eta' : {}^L\mathbf{H} \longrightarrow {}^LR$ denote the composition of $\eta : {}^L\mathbf{H} \longrightarrow {}^L\mathbf{G}$ and of this morphism. Let $t_1, \dots, t_d \in Z(\widehat{\mathbf{H}})^{\mathrm{Gal}(\overline{\mathbb{Q}}_p/\mathbb{Q}_p)}$ be such that $t_1 \dots t_d = s$, and write $t = (t_1, \dots, t_d) \in \widehat{R}$. Define a morphism $\widetilde{\eta} : \widehat{\mathbf{H}} \rtimes W(\mathbb{Q}_p^{\mathrm{ur}}/\mathbb{Q}_p) \longrightarrow \widehat{R} \rtimes W(\mathbb{Q}_p^{\mathrm{ur}}/\mathbb{Q}_p)$ by

- $\widetilde{\eta}_{|\widehat{\mathbf{H}}}$ is the composition of $\eta_0 : \widehat{\mathbf{H}} \longrightarrow \widehat{\mathbf{G}}$ and of the diagonal embedding $\widehat{\mathbf{G}} \longrightarrow \widehat{R} = (\widehat{\mathbf{G}})^d$;
- $\widetilde{\eta}((1, \Phi)) = (t, 1)\eta'(1, \Phi)$.

Then the $\widehat{R}$-conjugacy class of $\widetilde{\eta}$ does not depend on the choice of $t_1, \dots, t_d$, and $(\mathbf{H}, t, \widetilde{\eta})$ is a twisted endoscopic datum for (R, θ). The map

$$b_{\widetilde{\eta}} : \mathcal{H}(\mathbf{G}(L), \mathrm{K}_L) \longrightarrow \mathcal{H}(\mathbf{H}(\mathbb{Q}_p), \mathrm{K}_{H,0})$$

induced by $\widetilde{\eta}$ is called the *twisted transfer map* (or the *unstable base change map*).

Assume first that $\eta = \eta_{\mathrm{simple}}$, and write $\widetilde{b}_0$ for $b_{\widetilde{\eta}}$. If $\mathbf{G}$ does not split over L, then the twisted transfer map $\widetilde{b}_0$ corresponds by the Satake isomorphisms to the morphism induced by

$$\begin{aligned} \mathbb{C}[Z^{\pm 1}] \otimes \mathbb{C}[Z_i^{\pm 1}] &\longrightarrow \mathbb{C}[X^{\pm 1}] \otimes \mathbb{C}[X_{1,1}^{\pm 1}, \dots, X_{1,n_1}^{\pm 1}] \otimes \mathbb{C}[X_{2,1}^{\pm 1}, \dots, X_{2,n_2}^{\pm 1}], \\ Z &\longmapsto X^d, \\ Z_i &\longmapsto \begin{cases} X_{1,i}^d & \text{if} \quad 1 \leq i \leq n_1, \\ -X_{2,i-n_1}^d & \text{if} \quad n_1 + 1 \leq i \leq n. \end{cases} \end{aligned}$$

If $\mathbf{G}$ splits over L (so $L \supset E_\wp$), then the twisted transfer map $\widetilde{b}_0$ corresponds by the Satake isomorphisms to the morphism induced by

$$\begin{aligned}
\mathbb{C}[Z^{\pm 1}] \otimes \mathbb{C}[Z_i^{\pm 1}] &\longrightarrow \mathbb{C}[X^{\pm 1}] \otimes \mathbb{C}[X_{1,1}^{\pm 1}, \dots, X_{1,n_1}^{\pm 1}] \otimes \mathbb{C}[X_{2,1}^{\pm 1}, \dots, X_{2,n_2}^{\pm 1}], \\
Z &\longmapsto X^a, \\
Z_i &\longmapsto \begin{cases} X_{1,i}^a & \text{if} \quad 1 \leq i \leq n_1, \\ -X_{2,i-n_1}^a & \text{if} \quad n_1 + 1 \leq i \leq n. \end{cases}
\end{aligned}$$

Assume now that η is any unramified extension of η_0, and define an unramified quasi-character χ_η of $\mathbf{H}(\mathbb{Q}_p)$ as in the previous subsection. Then, for every $f \in \mathcal{H}(\mathbf{G}(L), \mathrm{K}_L)$, $b_{\widetilde{\eta}}(f) = \chi_\eta \widetilde{b}_0(f)$.

As in the previous subsection, we can use this to define $b_{\widetilde{\eta}}$ for a possibly ramified η (this is just [K9] p. 181). Let $\eta : {}^L\mathbf{H} \longrightarrow {}^L\mathbf{G}$ be any L-morphism extending η_0, and attach to it a (possibly ramified) quasi-character χ_η of $\mathbf{H}(\mathbb{Q}_p)$. Define $b_{\widetilde{\eta}} : \mathcal{H}(\mathbf{G}(L), \mathrm{K}_L) \longrightarrow C_c^\infty(\mathbf{H}(\mathbb{Q}_p))$ by the following formula: for every $f \in \mathcal{H}(\mathbf{G}(L), \mathrm{K}_L)$, $b_{\widetilde{\eta}}(f) = \chi_\eta \widetilde{b}_0(f)$.

4.3 TWISTED TRANSFER MAP AND CONSTANT TERMS

In this section, we consider, to simplify notation, the situation of the last subsection of 4.2, but all results extend in an obvious way to the groups $\mathbf{G}(\mathbf{U}(J_1) \times \cdots \times \mathbf{U}(J_r))$ of the beginning of section 4.2. Assume that $\mathbf{G}$ splits over L (i.e., that L contains $E_\wp$), and fix an L-morphism $\eta : {}^L\mathbf{H} \longrightarrow {}^L\mathbf{G}$ extending η_0; we do not assume that η is unramified.

Let $\mathbf{M}$ be a cuspidal standard Levi subgroup of $\mathbf{G}$, and let $r \in \{1, \dots, q\}$ be such that $\mathbf{M} = \mathbf{M}_{\{1,\dots,r\}} \simeq (R_{E/\mathbb{Q}}\mathbb{G}_m)^r \times \mathbf{GU}^*(m)$, with $m = n - 2r$. Let $(\mathbf{M}', s_M, \eta_{M,0})$ be an element of $\mathcal{E}_\mathbf{G}(\mathbf{M})$ (cf. section 2.4) whose image in $\mathcal{E}(\mathbf{G})$ is $(\mathbf{H}, s, \eta_0)$. Assume that $s_M = s_{A,m_1,m_2}$, with $A \subset \{1, \dots, r\}$ and $m_1 + m_2 = m$, and where notation is as in lemma 2.4.3; define r_1 and r_2 as in this lemma. There is a conjugacy class of Levi subgroups of $\mathbf{H}$ associated to $(\mathbf{M}', s_M, \eta_{M,0})$; let $\mathbf{M}_H$ be the standard Levi subgroup in that class. Then $\mathbf{M}_H = \mathbf{H} \cap (\mathbf{M}_{H,1} \times \mathbf{M}_{H,2})$, with

$$\mathbf{M}_{H,i} = \mathbf{GU}^*(n_i) \cap \begin{pmatrix} R_{E/\mathbb{Q}}\mathbb{G}_m & & 0 & & & 0 \\ & \ddots & & & & \\ 0 & & R_{E/\mathbb{Q}}\mathbb{G}_m & & & \\ & & & \mathbf{GU}^*(m_i) & & \\ & & & & R_{E/\mathbb{Q}}\mathbb{G}_m & 0 \\ & & & & & \ddots \\ 0 & & & & 0 & R_{E/\mathbb{Q}}\mathbb{G}_m \end{pmatrix},$$

where the diagonal blocks are of size r_i, m_i, r_i. On the other hand, $\mathbf{M}_H = \mathbf{M}_{H,l} \times \mathbf{M}_{H,h}$, where $\mathbf{M}_{H,l} = (R_{E/\mathbb{Q}}\mathbb{G}_m)^{r_1} \times (R_{E/\mathbb{Q}}\mathbb{G}_m)^{r_2}$ is the linear part of $\mathbf{M}_H$ and $\mathbf{M}_{H,h} = \mathbf{G}(\mathbf{U}^*(m_1) \times \mathbf{U}^*(m_2))$ is the hermitian part. Similarly, $\mathbf{M} = \mathbf{M}_l \times \mathbf{M}_h$, where $\mathbf{M}_l = \mathbf{M}_{H,l} = (R_{E/\mathbb{Q}}\mathbb{G}_m)^r$ is the linear part of $\mathbf{M}$ and $\mathbf{M}_h = \mathbf{GU}^*(m)$ is the hermitian part. The morphism $\eta : {}^L\mathbf{H} \longrightarrow {}^L\mathbf{G}$ determines an L-morphism

$\eta_M : {}^L\mathbf{M}_H = {}^L\mathbf{M}' \longrightarrow {}^L\mathbf{M}$ extending $\eta_{M,0}$, unique up to $\widehat{\mathbf{M}}$-conjugacy; η_M is unramified if η is unramified.

As in lemma 2.4.3, identify $\widehat{\mathbf{M}}$ with the Levi subgroup

$$\mathbb{C}^\times \times \begin{pmatrix} * & & 0 & & & & \\ & \ddots & & & & 0 & \\ 0 & & * & & & & \\ & & & \mathbf{GL}_m(\mathbb{C}) & & & \\ & & & & * & & 0 \\ & 0 & & & & \ddots & \\ & & & & 0 & & * \end{pmatrix}$$

(blocks of size r, m, r) of $\widehat{\mathbf{G}}$. Identify $\widehat{\mathbf{M}}_H$ to a Levi subgroup of $\widehat{\mathbf{H}}$ in a similar way. Let s'_M be the element of $Z(\widehat{\mathbf{M}}_H) \subset \widehat{\mathbf{M}} \simeq \mathbb{C}^\times \times (\mathbb{C}^\times)^r \times \mathbf{GL}_m(\mathbb{C}) \times (\mathbb{C}^\times)^r$ equal to $(1, (1, \dots, 1), s_{M_h}, (1, \dots, 1))$, where s_{M_h} is the image of s_M by the projection $\widehat{\mathbf{M}} \longrightarrow \mathbf{GL}_m(\mathbb{C})$ (in the notation of lemma 2.4.3, $s'_M = s_{\varnothing, m_1, m_2}$). Then $(\mathbf{M}_H, s'_M, \eta_{M,0})$ is an elliptic endoscopic triple for $\mathbf{M}$, isomorphic to $(\mathbf{M}', s_M, \eta_{M,0})$ as an endoscopic $\mathbf{M}$-triple (but not as an endoscopic $\mathbf{G}$-triple).

Write $b_{s_M}, b_{s'_M} : \mathcal{H}(\mathbf{M}(L), \mathbf{M}(\mathcal{O}_L)) \longrightarrow C_c^\infty(\mathbf{M}_H(\mathbb{Q}_p))$ for the twisted transfer maps associated to $(\mathbf{M}', s_M, \eta_M)$ and $(\mathbf{M}_H, s'_M, \eta_M)$, and $f \longmapsto f_{\mathbf{M}}$ (resp., $f \longmapsto f_{\mathbf{M}_H}$) for the constant term map $\mathcal{H}(\mathbf{G}(L), \mathrm{K}_L) \longrightarrow \mathcal{H}(\mathbf{M}(L), \mathbf{M}(\mathcal{O}_L))$ or $\mathcal{H}(\mathbf{G}(\mathbb{Q}_p), \mathrm{K}_0) \longrightarrow \mathcal{H}(\mathbf{M}(\mathbb{Q}_p), \mathbf{M}(\mathbb{Z}_p))$ (resp., $\mathcal{H}(\mathbf{H}(\mathbb{Q}_p), \mathrm{K}_{H,0}) \longrightarrow \mathcal{H}(\mathbf{M}_H(\mathbb{Q}_p), \mathbf{M}_H(\mathbb{Z}_p))$). Then it is easy to see from the definitions that, if η is unramified, then, for every $f \in \mathcal{H}(\mathbf{G}(L), \mathrm{K}_L)$, $b_{s_M}(f_{\mathbf{M}}) = (b_{\widetilde{\eta}}(f))_{\mathbf{M}_H}$. There is a similar formula for a general η. Let χ_η be the quasi-character of $\mathbf{H}(\mathbb{Q}_p)$ associated to η as in the last two subsections of 4.2, and write $b_{s_M,0}$ for the twisted transfer map defined by s_M in the case $\eta = \eta_{\mathrm{simple}}$ (and $\widetilde{b}_0 = b_{\widetilde{\eta}_{\mathrm{simple}}}$, as before). Then, for every $f \in \mathcal{H}(\mathbf{G}(L), \mathrm{K}_L)$, $b_{s_M}(f_{\mathbf{M}}) = \chi_{\eta|\mathbf{M}_H(\mathbb{Q}_p)} b_{s_M,0}(f_{\mathbf{M}}) = \chi_{\eta|\mathbf{M}_H(\mathbb{Q}_p)}(\widetilde{b}_0(f))_{\mathbf{M}_H}$.

Later, we will use the twisted transfer map $b_{s'_M}$ and not b_{s_M}, so we need to compare it to b_{s_M} (or to $b_{\widetilde{\eta}}$), at least on certain elements of $\mathcal{H}(\mathbf{M}(L), \mathbf{M}(\mathcal{O}_L))$. First we give explicit formulas for it in the case $\eta = \eta_{\mathrm{simple}}$. Write $\Omega_M(L) = W(\mathbf{T}_G(L), \mathbf{M}(L))$ and $\Omega_{M_H}(\mathbb{Q}_p) = W(\mathbf{T}_H(\mathbb{Q}_p), \mathbf{M}_H(\mathbb{Q}_p))$. Then we get Satake isomorphisms $\mathcal{H}(\mathbf{M}(L), \mathbf{M}(\mathcal{O}_L)) \simeq \mathbb{C}[X_*(\mathbf{T}_G)]^{\Omega_M(L)}$ and $\mathcal{H}(\mathbf{M}_H(\mathbb{Q}_p), \mathbf{M}_H(\mathbb{Z}_p)) \simeq \mathbb{C}[X_*(\mathbf{T}_H)]^{\Omega_{M_H}(\mathbb{Q}_p)}$, and, if $\eta = \eta_{\mathrm{simple}}$, then the twisted transfer map $b_{s'_M}$ is induced by the morphism:

$$\begin{aligned}
\mathbb{C}[Z^{\pm 1}] \otimes \mathbb{C}[Z_1^{\pm}, \dots, Z_n^{\pm 1}] &\longrightarrow \mathbb{C}[X^{\pm 1}] \otimes \mathbb{C}[X_{1,1}^{\pm 1}, \dots, X_{1,n_1}^{\pm 1}] \otimes \mathbb{C}[X_{2,1}^{\pm 1}, \dots, X_{2,n_2}^{\pm 1}], \\
Z &\longmapsto X^a, \\
Z_{i_k} &\longmapsto X_{1,k}^a, \\
Z_{n+1-i_k} &\longmapsto X_{1,n_1+1-k}^a, \\
Z_{j_l} &\longmapsto X_{2,l}^a, \\
Z_{n+1-j_l} &\longmapsto X_{2,n_2+1-l}^a, \\
Z_i &\longmapsto \begin{cases} X_{1,i-r_2}^a & \text{if} \quad r+1 \leq i \leq r+m_1, \\ -X_{2,i-(r_1+m_1)}^a & \text{if} \quad r+m_1+1 \leq i \leq r+m. \end{cases}
\end{aligned}$$

where we write $\{1, \dots, r\} - A = \{i_1, \dots, i_{r_1}\}$ and $A = \{j_1, \dots, j_{r_2}\}$ with $i_1 < \cdots < i_{r_1}$ and $j_1 < \cdots < j_{r_2}$.

Let $\alpha \in \mathbb{N}$ such that $n - q \leq \alpha \leq n$. Write $\mu = \mu_\alpha$, where $\mu_\alpha : \mathbb{G}_{m,L} \longmapsto \mathbf{GU}^*(n_i)_L$ is the cocharacter defined in section 2.1.2. This cocharacter factors through $\mathbf{M}_L$, and we denote by μ_M the cocharacter of $\mathbf{M}_L$ that it induces. Set

$$\phi = \frac{\mathbb{1}_{\mathrm{K}_L \mu(\varpi_L^{-1})\mathrm{K}_L}}{vol(\mathrm{K}_L)} \in \mathcal{H}(\mathbf{G}(L), \mathrm{K}_L)$$

and

$$\phi^{\mathbf{M}} = \frac{\mathbb{1}_{\mathbf{M}(\mathcal{O}_L)\mu_M(\varpi_L^{-1})\mathbf{M}(\mathcal{O}_L)}}{vol(\mathbf{M}(\mathcal{O}_L))} \in \mathcal{H}(\mathbf{M}(L), \mathbf{M}(\mathcal{O}_L)).$$

Note that, if $\alpha \leq n - r$, then $\phi^{\mathbf{M}}$ is the product of a function in $\mathcal{H}(\mathbf{M}_h(L), \mathbf{M}_h(\mathcal{O}_L))$ and the unit element of $\mathcal{H}(\mathbf{M}_l(L), \mathbf{M}_l(\mathcal{O}_L))$ (because the image of μ_M is included in $\mathbf{M}_{h,L}$ in that case).

The Satake transform of ϕ has been calculated in proposition 4.2.1; it is equal to

$$p^{d\alpha(n-\alpha)/2} Z^{-1} \sum_{\substack{I \subset \{1,\dots,n\} \\ |I| = \alpha}} \prod_{i \in I} Z_i^{-1}.$$

Identify $\mathcal{H}(\mathbf{G}(L), \mathrm{K}_L)$ with a subalgebra of $\mathcal{H}(\mathbf{M}(L), \mathbf{M}(\mathcal{O}_L))$ using the constant term morphism (via the Satake isomorphisms, this corresponds to the obvious inclusion $\mathbb{C}[X_*(\mathbf{T}_G)]^{\Omega_G(L)} \subset \mathbb{C}[X_*(\mathbf{T}_G)]^{\Omega_M(L)}$). If $\alpha \geq n - r + 1$, then the Satake transform of $\phi^{\mathbf{M}}$ is simply $(Z Z_1 \dots Z_\alpha)^{-1}$. If $\alpha \leq n - r$, then, by proposition 4.2.1 (applied to $\mathbf{M}_h$), the Satake transform of $\phi^{\mathbf{M}}$ is

$$p^{d(\alpha-r)(n-\alpha-r)/2} (Z Z_1 \dots Z_r)^{-1} \sum_{\substack{I \subset \{r+1,\dots,n-r\} \\ |I| = \alpha - r}} \prod_{i \in I} Z_i^{-1}.$$

Let $f^{\mathbf{H}} = b_{\widetilde{\eta}}(\phi) \in C_c^\infty(\mathbf{H}(\mathbb{Q}_p))$, $f^{\mathbf{M}_H} = b_{s'_M}(\phi^{\mathbf{M}}) \in C_c^\infty(\mathbf{M}_H(\mathbb{Q}_p))$ and $\psi^{\mathbf{M}_H} = b_{s'_M}(\phi_{\mathbf{M}}) \in C_c^\infty(\mathbf{M}_H(\mathbb{Q}_p))$. By the definition of $b_{\widetilde{\eta}}$, there exists a quasi-character χ_η of $\mathbf{H}(\mathbb{Q}_p)$ such that $\chi_\eta^{-1} f^{\mathbf{H}} \in \mathcal{H}(\mathbf{H}(\mathbb{Q}_p), \mathrm{K}_{H,0})$; write $f^{\mathbf{H}}_{\mathbf{M}_H} \in C_c^\infty(\mathbf{H}(\mathbb{Q}_p))$ for $\chi_{\eta|\mathbf{M}_H(\mathbb{Q}_p)} (\chi_\eta^{-1} f^{\mathbf{H}})_{\mathbf{M}_H}$. (Of course, because we used $b_{s'_M}$ and not b_{s_M} to define $\psi^{\mathbf{M}_H}$, the functions $\psi^{\mathbf{M}_H}$ and $f^{\mathbf{H}}_{\mathbf{M}_H}$ are different in general.)

Let Ω^{M_H} be the subgroup of $\Omega_H(\mathbb{Q}_p) \subset \mathfrak{S}_{n_1} \times \mathfrak{S}_{n_2}$ generated by the transpositions $((j, n_1 + 1 - j), 1)$, $1 \leq j \leq r_1$, and $(1, (j, n_2 + 1 - j))$, $1 \leq j \leq r_2$. Then $\Omega^{M_H} \simeq \{\pm 1\}^{r_1} \times \{\pm 1\}^{r_2}$ (actually, Ω^{M_H} is even a subgroup of the relative Weyl group over $\mathbb{Q}$, $W(\mathbf{T}_H(\mathbb{Q}), \mathbf{H}(\mathbb{Q}))$).

Proposition 4.3.1 *Let $\gamma_H \in \mathbf{M}_H(\mathbb{Q})$ be such that $O_{\gamma_H}(f^{\mathbf{H}}_{\mathbf{M}_H}) \neq 0$. Write $\gamma_H = \gamma_l \gamma_h$, with $\gamma_h \in \mathbf{M}_{H,h}(\mathbb{Q})$ and $\gamma_l = ((\lambda_{1,1}, \dots, \lambda_{1,r_1}), (\lambda_{2,1}, \dots, \lambda_{2,r_2})) \in \mathbf{M}_{H,l}(\mathbb{Q}) = (E^\times)^{r_1} \times (E^\times)^{r_2}$, and let*

$$N_{s_M}(\gamma_H) = \frac{1}{2a} \sum_{i=1}^{r_2} \mathrm{val}_p(|\lambda_{2,i} \overline{\lambda}_{2,i}|_p)$$

(where val_p is the p-adic valuation). Then $|c(\gamma_H)|_p = p^d$, $\frac{1}{2a}\mathrm{val}_p(|\lambda_{i,j}\overline{\lambda}_{i,j}|_p)$ is an integer for every i, j (in particular, $N_{s_M}(\gamma_H) \in \mathbb{Z}$), and one and only one of the

following two assertions is true:

(A) *There exists* $\omega \in \Omega^{M_H}$, *uniquely determined by* γ_H, *such that* $\omega(\gamma_H) \in \mathbf{M}_{H,l}(\mathbb{Z}_p)\mathbf{M}_{H,h}(\mathbb{Q}_p)$.

(B) *There exist* $i \in \{1,2\}$ *and* $j \in \{1,\dots,r_i\}$ *such that* $\frac{1}{2a}\mathrm{val}_p(|\lambda_{i,j}\overline{\lambda}_{i,j}|_p)$ *is odd.*

Besides, case (A) can occur only if $\alpha \leq n-r$.

Choose an element $\gamma \in \mathbf{M}(\mathbb{Q}_p)$ *coming from* γ_H *(such a* γ *always exists because* $\mathbf{M}$ *is quasi-split over* $\mathbb{Q}_p$, *cf. [K1]). Then*

$$O_{\gamma_H}(f^{\mathbf{H}}_{\mathbf{M}_H}) = \langle \mu, s\rangle\,\langle \mu, s'_M\rangle \varepsilon_{s_M}(\gamma) O_{\gamma_H}(\psi^{\mathbf{M}_H}),$$

where $\varepsilon_{s_M}(\gamma) = (-1)^{N_{s_M}(\gamma)}$. *If moreover* $\gamma_l \in \mathbf{M}_{H,l}(\mathbb{Z}_p)$, *(this can happen only if* $\alpha \leq n-r$*), then*

$$O_{\gamma_H}(f^{\mathbf{H}}_{\mathbf{M}_H}) = \langle \mu, s\rangle\langle \mu, s'_M\rangle \delta^{1/2}_{P(\mathbb{Q}_p)}(\gamma) O_{\gamma_H}(f^{\mathbf{M}_H}),$$

where $\mathbf{P}$ *is the standard parabolic subgroup of* $\mathbf{G}$ *with Levi subgroup* $\mathbf{M}$.

Let γ_0 be the component of γ in $\mathbf{M}_h(\mathbb{Q}_p)$. For every $\delta \in \mathbf{M}(L)$ σ-semisimple, let δ_h be the component of δ in $\mathbf{M}_h(L)$ and define $\alpha_p(\gamma,\delta)$ and $\alpha_p(\gamma_0,\delta_h)$ as in as [K9] p. 180 (cf. also subsections A.2.3 and A.3.5 of the appendix).

From the definition of s'_M, it is clear that

$$\langle \alpha_p(\gamma,\delta), s'_M\rangle = \langle \alpha_p(\gamma_0,\delta_h), s'_M\rangle = \langle \alpha_p(\gamma_0,\delta_h), s_M\rangle.$$

After applying corollary 9.5.3, we get:[1]

Corollary 4.3.2 *There is an equality*

$$SO_{\gamma_H}(f^{\mathbf{H}}_{\mathbf{M}_H}) = \langle \mu, s\rangle\langle \mu, s'_M\rangle \varepsilon_{s_M}(\gamma) \times \sum_{\delta} \langle \alpha_p(\gamma_0,\delta_h), s_M\rangle \Delta^{\mathbf{M}}_{\mathbf{M}_H,s_M,p}(\gamma_H,\gamma) e(\delta) TO_\delta(\phi_{\mathbf{M}}).$$

If moreover $\gamma \in \mathbf{M}_l(\mathbb{Z}_p)\mathbf{M}_h(\mathbb{Q}_p)$, *then*

$$SO_{\gamma_H}(f^{\mathbf{H}}_{\mathbf{M}_H}) = \langle \mu, s\rangle\langle \mu, s'_M\rangle \delta^{1/2}_{\mathbf{P}(\mathbb{Q}_p)}(\gamma) \times \sum_{\delta} \langle \alpha_p(\gamma_0,\delta_h), s_M\rangle \Delta^{\mathbf{M}}_{\mathbf{M}_H,s_M,p}(\gamma_H,\gamma) e(\delta) TO_\delta(\phi^{\mathbf{M}}).$$

The two sums above are taken over the set of σ*-conjugacy classes* δ *in* $\mathbf{M}(L)$ *such that* γ *and* $N\delta$ *(defined in section 1.6, after theorem 1.6.1) are* $\mathbf{M}(\overline{\mathbb{Q}}_p)$*-conjugate; for every such* δ, *we write* $e(\delta) = e(R_{\delta\theta})$, *where* $R_{\delta\theta}$ *is the* θ*-centralizer of* δ *in* R *(denoted by* $I(p)$ *in 1.6) and* e *is the sign of [K2].*

Proof of the proposition. It is obvious from the definition of the twisted transfer maps that it suffices to prove the proposition for $\eta = \eta_{\mathrm{simple}}$. So we assume this throughout the proof.

[1] Corollary 9.5.3 applies only to a particular choice of η, i.e., to $\eta = \eta_{\mathrm{simple}}$. But it is explained in [K9] (after formula (7.3)) why this is enough to prove the next corollary for any choice of η.

It is easy to see that the element $\omega \in \Omega^{M_H}$ in (A) is necessarily unique (if we already know that $|c(\gamma_H)|_p = p^d$). This comes from the fact that, for every $\gamma_H \in \mathbf{M}_{H,l}(\mathbb{Z}_p)\mathbf{M}_{H,h}(\mathbb{Q}_p)$ such that $|c(\gamma_H)|_p \neq 1$ and for every $\omega \in \Omega^{M_H}$, $\omega(\gamma_H) \neq \gamma_H$.

We know that the Satake transform of ϕ is

$$p^{d\alpha(n-\alpha)/2} Z^{-1} \sum_{\substack{I\subset\{1,\dots,n\}\\|I|=\alpha}} \prod_{i\in I} Z_i^{-1}.$$

For every $I \subset \{1,\dots,n\}$, write

$$n(I) = |I \cap \{n_1+1,\dots,n\}|,$$

and

$$a_I = X^{-a} \prod_{i\in I\cap\{1,\dots,n_1\}} X_{1,i}^{-a} \prod_{i\in I\cap\{n_1+1,\dots,n\}} X_{2,i-n_1}^{-a}.$$

Then the Satake transform of $f^{\mathbf{H}}_{M_H}$ (that is equal to the Satake transform of $f^{\mathbf{H}}$) is

$$S := p^{d\alpha(n-\alpha)/2} \sum_{\substack{I\subset\{1,\dots,n\}\\|I|=\alpha}} (-1)^{n(I)} a_I.$$

As S is the product of X^{-a} and a polynomial in the $X_{i,j}^{-1}$, if $O_{\gamma_H}(f^{\mathbf{H}}_{\mathbf{M}_H}) \neq 0$, then $|c(\gamma_H)|_p$ must be equal to p^d.

Let $A_{l,1} = \{1,\dots,r_1\}\cup\{n_1+1-r_1,\dots,n_1\}$, $A_{l,2} = \{n_1+1,\dots,n_1+r_2\}\cup\{n+1-r_2,\dots,n\}$, $A_l = A_{l,1}\cup A_{l,2}$, $A_{h,1} = \{r_1+1,\dots,n_1-r_1\}$, $A_{h,2} = \{n_1+r_2+1,\dots,n-r_2\}$, and $A_h = A_{h,1}\cup A_{h,2}$. For every $I_l \subset A_l$, write $n_l(I_l) = |I_l \cap A_{l,2}|$ and

$$b_{I_l} = \prod_{j\in I_l\cap A_{l,1}} X_{1,j}^{-a} \prod_{j\in I_l\cap A_{l,2}} X_{2,j-n_1}^{-a}.$$

For every $I_h \subset A_h$, write $n_h(I_h) = |I_h \cap A_{h,2}|$, and

$$c_{I_h} = X^{-a} \prod_{j\in I_h\cap A_{h,1}} X_{1,j}^{-a} \prod_{j\in I_h\cap A_{h,2}} X_{2,j-n_1}^{-a}.$$

As $a_I = b_{I\cap A_l} c_{I\cap A_h}$ and $n(I) = n_l(I\cap A_l) + n_h(I\cap A_h)$ for every $I \subset \{1,\dots,n\}$,

$$S = p^{d\alpha(n-\alpha)/2} \sum_{k=0}^{\alpha} \left(\sum_{I_l\subset A_l, |I_l|=k} (-1)^{n_l(I_l)} b_{I_l} \right) \left(\sum_{I_h\subset A_h, |I_h|=\alpha-k} (-1)^{n_h(I_h)} c_{I_h} \right).$$

For every $k \in \{0,\dots,\alpha\}$, the polynomial $\sum_{I_h\subset A_h,|I_h|=\alpha-k} (-1)^{n_h(I_h)} c_{I_h}$ is the Satake transform of a function in $\mathcal{H}(\mathbf{M}_{H,h}(\mathbb{Q}_p), \mathbf{M}_{H,h}(\mathbb{Z}_p))$, that will be denoted by $\psi_{h,k}$. For every $I_l \subset A_l$, the monomial $(-1)^{n_l(I_l)} b_{I_l}$ is the Satake transform of a function in $\mathcal{H}(\mathbf{M}_{H,l}(\mathbb{Q}_p), \mathbf{M}_{H,l}(\mathbb{Z}_p))$, that will be denoted by ψ_{I_l}. Then

$$f^{\mathbf{H}}_{\mathbf{M}_H} = p^{d\alpha(n-\alpha)/2} \sum_{k=0}^{\alpha} \psi_{h,k} \left(\sum_{I_l\subset A_l, |I_l|=k} \psi_{I_l} \right).$$

As $O_{\gamma_H}(f^{\mathbf{H}}_{\mathbf{M}_H}) \neq 0$, there exist $k \in \{0,\dots,\alpha\}$ and $I_l \subset A_l$ such that $|I_l| = k$ and $O_{\gamma_H}(\psi_{I_l}\psi_{h,k}) \neq 0$. Write $\gamma_H = \gamma_l\gamma_h$, with $\gamma_h \in \mathbf{M}_{H,h}(\mathbb{Q}_p)$ and $\gamma_l = ((\lambda_{1,1},\dots,\lambda_{1,r_1}),(\lambda_{2,1},\dots,\lambda_{2,r_2})) \in \mathbf{M}_{H,l}(\mathbb{Q}_p)$. Then $O_{\gamma_H}(\psi_{I_l}\psi_{h,k}) = O_{\gamma_l}(\psi_{I_l}) O_{\gamma_h}(\psi_{h,k})$. We have $O_{\gamma_l}(\psi_{I_l}) \neq 0$ if and only if γ_l is in the product of

$\mathbf{M}_{H,l}(\mathbb{Z}_p)$ and the image of p by the cocharacter corresponding to the monomial $(X_{1,1}\dots X_{1,r_1}X_{2,1}\dots X_{2,r_2})^a b_{I_l}$, and this implies that

- for every $j\in\{1,\dots,r_1\}$,

$$|\lambda_{1,j}\overline{\lambda}_{1,j}|_p=\begin{cases}1 & \text{if} \quad j\in I_l \text{ and } n_1+1-j\notin I_l,\\ p^{-2a} & \text{if} \quad j,n_1+1-j\in I_l \text{ or } j,n_1+1-j\notin I_l,\\ p^{-4a} & \text{if} \quad j\notin I_l \text{ and } n_1+1-j\in I_l;\end{cases}$$

- for every $j\in\{1,\dots,r_2\}$,

$$|\lambda_{2,j}\overline{\lambda}_{2,j}|_p=\begin{cases}1 & \text{if} \quad n_1+j\in I_l \text{ and } n+1-j\notin I_l,\\ p^{-2a} & \text{if} \quad n_1+j,n+1-j\in I_l \text{ or } n_1+j,n+1-j\notin I_l,.\\ p^{-4a} & \text{if} \quad n_1+j\notin I_l \text{ and } n+1-j\in I_l.\end{cases}$$

This implies in particular that $\frac{1}{2a}\mathrm{val}_p(|\lambda_{i,j}\overline{\lambda}_{i,j}|_p)\in\mathbb{Z}$ for every i,j.

On the other hand, $c(\gamma_H)=c(\gamma_h)$, so $|c(\gamma_h)|_p=p^d$.

There are three cases to consider:

(1) Assume that, for every $j\in\{1,\dots,r_1\}$, either $j\in I_l$ or $n_1+1-j\in I_l$, and that, for every $j\in\{1,\dots,r_2\}$, either $n_1+j\in I_l$ or $n+1-j\in I_l$. Let $\omega=((\omega_{1,1},\dots,\omega_{1,r_1}),(\omega_{2,1},\dots,\omega_{2,r_2}))\in\Omega^{M_H}$ be such that, for every $j\in\{1,\dots,r_1\}$, $\omega_{1,j}=1$ if $j\notin I_l$ and $\omega_{1,j}=-1$ if $j\in I_l$ and, for every $j\in\{1,\dots,r_2\}$, $\omega_{2,j}=1$ if $n_1+j\notin I_l$ and $\omega_{2,j}=-1$ if $n_1+j\in I_l$. It is easy to see that $\omega(\gamma)\in\mathbf{M}_{H,l}(\mathbb{Z}_p)\mathbf{M}_{H,h}(\mathbb{Q}_p)$, and it is clear that $\frac{1}{2a}\mathrm{val}_p(|\lambda_{i,j}\overline{\lambda}_{i,j}|_p)$ is even for every i,j.
On the other hand, $k=|I_l|=r_1+r_2=r$, so, for $\psi_{h,k}=\psi_{h,r}$ to be non-zero, we must have $\alpha-r\le n-2r$, i.e., $\alpha\le n-r$.

(2) Assume that there exists $j\in\{1,\dots,r_1\}$ such that $j,n_1+1-j\in I_l$ or $j,n_1+1-j\notin I_l$. Then $|\lambda_{1,j}\overline{\lambda}_{1,j}|_p=p^{-2a}$, hence $\frac{1}{2a}\mathrm{val}_p(|\lambda_{1,j}\overline{\lambda}_{1,j}|_p)=-1$.

(3) Assume that there exists $j\in\{1,\dots,r_2\}$ such that $n_1+j,n+1-j_2\in I_l$ or $n_1+j,n+1-j\notin I_l$. As in case (2), we see that $\frac{1}{2a}\mathrm{val}_p(|\lambda_{2,j}\overline{\lambda}_{2,j}|_p)=-1$.

We now show the last two statements of the proposition. First note that $\langle\mu,s\rangle$ $\langle\mu,s_M\rangle=(-1)^{r_2}$. For every $I\subset\{1,\dots,n\}$, write

$$m(I)=|I\cap\{n_1+r_2+1,\dots,n_1+r_2+m_2\}|.$$

The Satake transform of $\phi_{\mathbf{M}}$ is equal to the Satake transform of ϕ, so the Satake transform of $\psi^{\mathbf{M}_H}$ is

$$p^{d\alpha(n-\alpha)/2}\sum_{\substack{I\subset\{1,\dots,n\}\\|I|=\alpha}}(-1)^{m(I)}a_I.$$

Hence

$$\psi^{\mathbf{M}_H}=p^{d\alpha(n-\alpha)/2}\sum_{k=0}^{\alpha}\psi_{h,k}\left(\sum_{I_l\subset A_l,|I_l|=k}(-1)^{n_l(I_l)}\psi_{I_l}\right).$$

To show the first equality of the proposition, it is therefore enough to see that, for every $I_l\subset A_l$ such that $O_{\gamma_l}(\psi_{I_l})\neq 0$, $n_l(I_l)=r_2+\varepsilon_{s_M}(\gamma)$ modulo 2; but this is an easy consequence of the nonvanishing condition for $O_{\gamma_l}(\psi_{I_l})$ that we wrote above.

Assume that $\gamma_l \in \mathbf{M}_{H,l}(\mathbb{Z}_p)$. Then the only $I_l \subset A_l$ such that $O_{\gamma_l}(\psi_{I_l}) \neq 0$ is $I_l = \{1, \dots, r_1\} \cup \{n_1 + 1, \dots, n_1 + r_2\}$, and $|I_l| = r$ and $n_l(I_l) = r_2$. We have already seen that $\psi_{h,r} = 0$ unless $\alpha \leq n - r$, so we may assume this. Then $O_{\gamma_H}(f^{\mathbf{H}}_{\mathbf{M}_H}) = O_{\gamma_H}(\psi')$, where $\psi' \in \mathcal{H}(\mathbf{M}_H(\mathbb{Q}_p), \mathbf{M}_H(\mathbb{Z}_p))$ is the function with Satake transform

$$p^{d\alpha(n-\alpha)/2}(-1)^{r_2}\psi_{h,r}\prod_{j=1}^{r_1} X_{1,j}^{-a}\prod_{j=1}^{r_2} X_{2,j}^{-a}.$$

Applying the calculation of the twisted transfer morphism to $\mathbf{M}_h$ and $\mathbf{M}_{H,h}$ instead of $\mathbf{G}$ and $\mathbf{H}$, we find that the Satake transform of $f^{\mathbf{M}_H}$ is

$$p^{d(\alpha-r)(n-\alpha-r)/2}\psi_{h,r}\prod_{j=1}^{r_1} X_{1,j}^{-a}\prod_{j=1}^{r_2} X_{2,j}^{-a}.$$

So, to finish the proof of the proposition, it is enough to show that

$$\delta_{P(\mathbb{Q}_p)}^{-1/2}(\gamma) = p^{d(\alpha-r)(n-\alpha-r)/2-d\alpha(n-\alpha)/2}.$$

As γ comes from γ_H, $c(\gamma) = c(\gamma_H)$. On the other hand, $\gamma_l \in \mathbf{M}_l(\mathbb{Z}_p)\mathbf{M}_h(\mathbb{Q}_p)$, so

$$\delta_{P(\mathbb{Q}_p)}(\gamma_l) = 1.$$

As the image of $\gamma_h \in \mathbf{GU}^*(m)(\mathbb{Q}_p)$ in $\mathbf{G}(\mathbb{Q}_p)$ is

$$\gamma_h = \begin{pmatrix} c(\gamma_h)I_r & 0 & 0 \\ 0 & \gamma_h & 0 \\ 0 & 0 & I_r \end{pmatrix},$$

we get

$$\delta_{P(\mathbb{Q}_p)}(\gamma) = \delta_{P(\mathbb{Q}_p)}(\gamma_h) = |c(\gamma_h)|_p^{r(r+m)} = |c(\gamma)|_p^{r(r+m)} = p^{dr(n-r)}.$$

To conclude, it suffices to notice that

$$\alpha(n-\alpha) - (\alpha - r)(n - \alpha - r) = r(n-r). \qquad \square$$

Remark 4.3.3 From the proof above, it is easy to see that the set of (i, j) such that $\frac{1}{2a}\mathrm{val}_p(|\lambda_{i,j}\overline{\lambda}_{i,j}|_p)$ is odd has an even number of elements. In particular, the sign $\varepsilon_{s_M}(\gamma)$ does not change if $N_{s_M}(\gamma_H)$ is replaced by $\frac{1}{2a}\sum_{i=1}^{r_1}\mathrm{val}_p(|\lambda_{1,i}\overline{\lambda}_{1,i}|_p)$.

Chapter Five

The geometric side of the stable trace formula

5.1 NORMALIZATION OF THE HAAR MEASURES

We use the following rules to normalize the Haar measures.

(1) In the situation of theorem 1.6.1, use the normalizations of this theorem.

(2) Let $\mathbf{G}$ be a connected reductive group over $\mathbb{Q}$. We always take Haar measures on $\mathbf{G}(\mathbb{A}_f)$ such that the volumes of open compact subgroups are rational numbers. Let p be a prime number such that $\mathbf{G}$ is unramified over $\mathbb{Q}_p$, and let L be a finite unramified extension of $\mathbb{Q}_p$; then we use the Haar measure on $\mathbf{G}(L)$ such that the volume of hyperspecial maximal compact subgroups is 1. If a Haar measure dg_f on $\mathbf{G}(\mathbb{A}_f)$ is fixed, then we use the Haar measure dg_∞ on $\mathbf{G}(\mathbb{R})$ such that $dg_f dg_\infty$ is the Tamagawa measure on $\mathbf{G}(\mathbb{A})$ (cf. [O]).

(3) (cf. [K7] 5.2) Let F be a local field of characteristic 0, $\mathbf{G}$ be a connected reductive group on F and $\gamma \in \mathbf{G}(F)$ be semi-simple. Write $I = \mathbf{G}_\gamma$ $(= \mathrm{Cent}_{\mathbf{G}}(\gamma)^0)$, and choose Haar measures on $\mathbf{G}(F)$ and $I(F)$. If $\gamma' \in \mathbf{G}(F)$ is stably conjugate to γ, then $I' := \mathbf{G}_{\gamma'}$ is an inner form of I, so the measure on $I(F)$ gives a measure on $I'(F)$. When we take the stable orbital integral at γ of a function in $C_c^\infty(\mathbf{G}(F))$, we use these measures on the centralizers of elements in the stable conjugacy class of γ.

(4) Let F be a local field of characteristic 0, $\mathbf{G}$ be a connected reductive group over F and $(\mathbf{H}, s, \eta_0)$ be an endoscopic triple for $\mathbf{G}$. Let $\gamma_H \in \mathbf{H}(F)$ be semi-simple and $(\mathbf{G}, \mathbf{H})$-regular. Assume that there exists an image $\gamma \in \mathbf{G}(F)$ of γ_H. Then $I := \mathbf{G}_\gamma$ is an inner form of $I_H := \mathbf{H}_{\gamma_H}$ ([K7] 3.1). We always choose Haar measures on $I(F)$ and $I_H(F)$ that correspond to each other.

(5) Let $\mathbf{G}$ be a connected reductive group over $\mathbb{Q}$ as in section 1.6, and let $(\gamma_0; \gamma, \delta)$ be a triple satisfying conditions (C) of 1.6 and such that the invariant $\alpha(\gamma_0; \gamma, \delta)$ is trivial. We associate to $(\gamma_0; \gamma, \delta)$ a group I (connected and reductive over $\mathbb{Q}$) as in 1.6. In particular, $I_{\mathbb{R}}$ is an inner form of $I(\infty) := \mathbf{G}_{\mathbb{R},\gamma_0}$. If we already chose a Haar measure on $I(\mathbb{R})$ (for example, using rule (2), if we have a Haar measure on $I(\mathbb{A}_f)$), then we take the corresponding Haar measure on $I(\infty)(\mathbb{R})$.

5.2 NORMALIZATION OF THE TRANSFER FACTORS

The properties of transfer factors that we will use here are stated in [K7]. Note that transfer factors have been defined in all generality (for ordinary endoscopy) by Langlands and Shelstad, cf. [LS1] and [LS2]. The formula of [K7] 5.6 is proved in [LS1] 4.2, and conjecture 5.3 of [K7] is proved in proposition 1 (section 3) of [K8].

Let $\mathbf{G}$ be one of the unitary groups of section 2.1, and let $(\mathbf{H}, s, \eta_0)$ be an elliptic endoscopic triple for $\mathbf{G}$. Choose an L-morphism $\eta : {}^L\mathbf{H} \longrightarrow {}^L\mathbf{G}$ extending η_0. The local transfer factors associated to η are defined only up to a scalar.

At the infinite place, normalize the transfer factor as in [K9] pp. 184–185 (this is recalled in section 3.3), using the morphism j of 3.3 and the Borel subgroup of 3.3.3.

Let p be a prime number unramified in E (so $\mathbf{G}$ and $\mathbf{H}$ are unramified over $\mathbb{Q}_p$). Normalize the transfer factor at p as in [K9] pp. 180–181. If η is unramified at p, then this normalization is the one given by the $\mathbb{Z}_p$-structures on $\mathbf{G}$ and $\mathbf{H}$ (it has been defined by Hales in [H1] II 7, cf. also [Wa3] 4.6).

Choose the transfer factors at other places such that condition 6.10 (b) of [K7] is satisfied. We write $\Delta_{\mathbf{H},v}^{\mathbf{G}}$ for the transfer factors normalized in this way.

Let $\mathbf{M}$ be a cuspidal standard Levi subgroup of $\mathbf{G}$, let $(\mathbf{M}', s_M, \eta_{M,0}) \in \mathcal{E}_{\mathbf{G}}(\mathbf{M})$, and let $(\mathbf{H}, s, \eta_0)$ be its image in $\mathcal{E}(\mathbf{G})$. As in 3.3 and 4.3, associate to $(\mathbf{M}', s_M, \eta_{M,0})$ a cuspidal standard Levi subgroup $\mathbf{M}_H \simeq \mathbf{M}'$ of $\mathbf{H}$ and a L-morphism $\eta_M : {}^L\mathbf{M}_H = {}^L\mathbf{M}' \longrightarrow {}^L\mathbf{M}$ extending $\eta_{M,0}$ (in 3.3 and 4.3, we took $\mathbf{G} = \mathbf{GU}(p,q)$, but the general case is similar). We want to define a normalization of the transfer factors for η_M associated to this data.

At the infinite place, normalize the transfer factor as in 3.3, for the Borel subgroup of $\mathbf{M}$ related to the Borel subgroup of $\mathbf{G}$ fixed above as in 3.3.

If v is a finite place of $\mathbb{Q}$, choose the transfer factor at v that satisfies the condition

$$\Delta_v(\gamma_H, \gamma)_{\mathbf{M}_H}^{\mathbf{M}} = |D_{\mathbf{M}_H}^{\mathbf{H}}(\gamma_H)|_v^{1/2} |D_{\mathbf{M}}^{\mathbf{G}}(\gamma)|_v^{-1/2} \Delta_v(\gamma_H, \gamma)_{\mathbf{H}}^{\mathbf{G}},$$

for every $\gamma_H \in \mathbf{M}_H(\mathbb{Q}_v)$ semi-simple $\mathbf{G}$-regular and every image $\gamma \in \mathbf{M}(\mathbb{Q}_v)$ of γ_H (cf. [K13] lemma 7.5).

We write $\Delta_{\mathbf{M}_H,s_M,v}^{\mathbf{M}}$ for the transfer factors normalized in this way. Note that, if p is unramified in E, then $\Delta_{\mathbf{M}_H,s_M,p}^{\mathbf{M}}$ depends only on the image of $(\mathbf{M}', s_M, \eta_{M,0})$ in $\mathcal{E}(\mathbf{M})$ (because it is simply the transfer factor with the normalization of [K9] pp. 180–181, i.e., if η_M is unramified at p, with the normalization given by the $\mathbb{Z}_p$-structures on $\mathbf{M}$ and $\mathbf{M}_H$). However, the transfer factors $\Delta_{\mathbf{M}_H,s_M,v}^{\mathbf{M}}$ at other places may depend on $(\mathbf{M}', s_M, \eta_{M,0}) \in \mathcal{E}_{\mathbf{G}}(\mathbf{M})$, and not only on its image in $\mathcal{E}(\mathbf{M})$.

5.3 FUNDAMENTAL LEMMA AND TRANSFER CONJECTURE

We state here the forms of the fundamental lemma and of the transfer conjecture that we will need in chapter 6. The local and adelic stable orbital integrals are defined in [K7] 5.2 and 9.2.

Let $\mathbf{G}$ be one of the groups of 2.1, $(\mathbf{H}, s, \eta_0)$ be an elliptic endoscopic triple for $\mathbf{G}$ and $\eta : {}^L\mathbf{H} \longrightarrow {}^L\mathbf{G}$ be an L-morphism extending η_0. The transfer conjecture is stated in [K7] 5.4 and 5.5. It says that, for every place v of $\mathbb{Q}$, for every function $f \in C_c^\infty(\mathbf{G}(\mathbb{Q}_v))$, there exists a function $f^{\mathbf{H}} \in C_c^\infty(\mathbf{H}(\mathbb{Q}_v))$ such that, if $\gamma_H \in \mathbf{H}(\mathbb{Q}_v)$ is semisimple $(\mathbf{G}, \mathbf{H})$-regular, then

$$SO_{\gamma_H}(f^{\mathbf{H}}) = \sum_{\gamma} \Delta_v(\gamma_H, \gamma) e(\mathbf{G}_\gamma) O_\gamma(f),$$

where the sum is taken over the set of conjugacy classes γ in $\mathbf{G}(\mathbb{Q}_v)$ that are images of γ_H (so, if γ_H has no image in $\mathbf{G}(\mathbb{Q}_v)$, we want $SO_{\gamma_H}(f^{\mathbf{H}}) = 0$), $\mathbf{G}_\gamma = \mathrm{Cent}_{\mathbf{G}}(\gamma)^0$ and e is the sign of [K2]. We say that the function $f^{\mathbf{H}}$ is a transfer of f.

The fundamental lemma says that, if v is a finite place where $\mathbf{G}$ and $\mathbf{H}$ are unramified, if η is unramified at v and if $b : \mathcal{H}(\mathbf{G}(\mathbb{Q}_v), \mathbf{G}(\mathbb{Z}_v)) \longrightarrow \mathcal{H}(\mathbf{H}(\mathbb{Q}_v), \mathbf{H}(\mathbb{Z}_v))$ is the morphism induced by η (this morphism is made explicit in section 4.2), then, for every $f \in \mathcal{H}(\mathbf{G}(\mathbb{Q}_v), \mathbf{G}(\mathbb{Z}_v))$, $b(f)$ is a transfer of f.

If $v = \infty$, the transfer conjecture was proved by Shelstad (cf. [Sh1]).

For unitary groups, the fundamental lemma and the transfer conjecture were proved by Laumon-Ngo, Waldspurger and Hales (cf. [LN], [Wa1], [Wa2], and [H2]).

We will also need another fundamental lemma. Let p be a finite place where $\mathbf{G}$ and $\mathbf{H}$ are unramified, $\wp$ be the place of E above p determined by the fixed embedding $E \longrightarrow \overline{\mathbb{Q}}_p$, $j \in \mathbb{N}^*$, and L be the unramified extension of degree j of $E_\wp$ in $\overline{\mathbb{Q}}_p$. Assume that η is unramified at p; then it defines a morphism $b : \mathcal{H}(\mathbf{G}(L), \mathbf{G}(\mathcal{O}_L)) \longrightarrow \mathcal{H}(\mathbf{H}(\mathbb{Q}_p), \mathbf{H}(\mathbb{Z}_p))$ (cf. section 4.2 for the definition of b and its description). The fundamental lemma corresponding to this situation says that, for every $\phi \in \mathcal{H}(\mathbf{G}(L), \mathbf{G}(\mathcal{O}_L))$ and every $\gamma_H \in \mathbf{H}(\mathbb{Q}_p)$ semisimple and $(\mathbf{G}, \mathbf{H})$-regular,

$$SO_{\gamma_H}(b(\phi)) = \sum_{\delta} \langle \alpha_p(\gamma_0; \delta), s \rangle \Delta_p(\gamma_H, \gamma_0) e(\mathbf{G}_{\delta\sigma}) TO_\delta(\phi), \qquad (*)$$

where the sum is taken over the set of σ-conjugacy classes δ in $\mathbf{G}(L)$ such that $N\delta$ is $\mathbf{G}(\overline{\mathbb{Q}}_p)$-conjugate to an image $\gamma_0 \in \mathbf{G}(\mathbb{Q}_p)$ of γ_H, $\mathbf{G}_{\delta\sigma}$ is the σ-centralizer of δ in $R_{L/\mathbb{Q}_p}\mathbf{G}_L$ and $\alpha_p(\gamma_0; \delta)$ is defined in [K9] §7 p. 180 (see [K9] §7 pp. 180–181 for more details). This conjecture (modulo a calculation of transfer factors) is proved in [Wa3] when ϕ is the unit element of the Hecke algebra $\mathcal{H}(\mathbf{G}(L), \mathbf{G}(\mathcal{O}_L))$. The reduction of the general case to this case is done in chapter 9, and the necessary transfer factor calculation is done in the appendix by Kottwitz.

5.4 A RESULT OF KOTTWITZ

We recall here a theorem of Kottwitz about the geometric side of the stable trace formula for a function that is stable cuspidal at infinity. The reference for this result is [K13].

Let $\mathbf{G}$ be a connected reductive algebraic group over $\mathbb{Q}$. Assume that $\mathbf{G}$ is cuspidal (cf. definition 3.1.1) and that the derived group of $\mathbf{G}$ is simply connected. Let K_∞ be a maximal compact subgroup of $\mathbf{G}(\mathbb{R})$. Let $\mathbf{G}^*$ be a quasi-split inner form of $\mathbf{G}$ over $\mathbb{Q}$, $\overline{\mathbf{G}}$ be an inner form of $\mathbf{G}$ over $\mathbb{R}$ such that $\overline{\mathbf{G}}/\mathbf{A}_{G,\mathbb{R}}$ is $\mathbb{R}$-anisotropic, and $\mathbf{T}_e$ be a maximal elliptic torus of $\mathbf{G}_{\mathbb{R}}$. Write

$$\overline{v}(\mathbf{G}) = e(\overline{\mathbf{G}}) \operatorname{vol}(\overline{\mathbf{G}}(\mathbb{R})/\mathbf{A}_G(\mathbb{R})^0)$$

($e(\overline{\mathbf{G}})$ is the sign associated to $\overline{\mathbf{G}}$ in [K2]), and

$$k(\mathbf{G}) = |\mathrm{Im}(\mathbf{H}^1(\mathbb{R}, \mathbf{T}_e \cap \mathbf{G}^{der}) \longrightarrow \mathbf{H}^1(\mathbb{R}, \mathbf{T}_e))|.$$

For every Levi subgroup $\mathbf{M}$ of $\mathbf{G}$, set

$$n_M^G = |(\mathrm{Nor}_{\mathbf{G}}(\mathbf{M})/\mathbf{M})(\mathbb{Q})|.$$

Let ν be a quasi-character of $\mathbf{A}_G(\mathbb{R})^0$. Let $\Pi_{\mathrm{temp}}(\mathbf{G}(\mathbb{R}), \nu)$ (resp., $\Pi_{\mathrm{disc}}(\mathbf{G}(\mathbb{R}), \nu)$) be the subset of π in $\Pi_{\mathrm{temp}}(\mathbf{G}(\mathbb{R}))$ (resp., $\Pi_{\mathrm{disc}}(\mathbf{G}(\mathbb{R}))$) such that the restriction to $\mathbf{A}_G(\mathbb{R})^0$ of the central character of π is equal to ν. Let $C_c^\infty(\mathbf{G}(\mathbb{R}), \nu^{-1})$ be the set of functions $f_\infty : \mathbf{G}(\mathbb{R}) \longrightarrow \mathbb{C}$ smooth, with compact support modulo $\mathbf{A}_G(\mathbb{R})^0$ and such that, for every $(z, g) \in \mathbf{A}_G(\mathbb{R})^0 \times \mathbf{G}(\mathbb{R})$, $f_\infty(zg) = \nu^{-1}(z) f_\infty(g)$.

We say that $f_\infty \in C_c^\infty(\mathbf{G}(\mathbb{R}), \nu^{-1})$ is *stable cuspidal* if f_∞ is left and right K_∞-finite and if the function

$$\Pi_{\mathrm{temp}}(\mathbf{G}(\mathbb{R}), \nu) \longrightarrow \mathbb{C}, \pi \longmapsto \mathrm{Tr}(\pi(f_\infty))$$

vanishes outside $\Pi_{\mathrm{disc}}(\mathbf{G}(\mathbb{R}))$ and is constant on the L-packets of $\Pi_{\mathrm{disc}}(\mathbf{G}(\mathbb{R}), \nu)$.

Let $f_\infty \in C_c^\infty(\mathbf{G}(\mathbb{R}), \nu^{-1})$. For every L-packet Π of $\Pi_{\mathrm{disc}}(\mathbf{G}(\mathbb{R}), \nu)$, write $\mathrm{Tr}(\Pi(f_\infty)) = \sum_{\pi\in\Pi} \mathrm{Tr}(\pi(f_\infty))$ and $\Theta_\Pi = \sum_{\pi\in\Pi} \Theta_\pi$. For every cuspidal Levi subgroup $\mathbf{M}$ of $\mathbf{G}$, define a function $S\Phi_M(., f_\infty) = S\Phi_M^G(., f_\infty)$ on $\mathbf{M}(\mathbb{R})$ by the formula :

$$\begin{aligned} S\Phi_M(\gamma, f_\infty) = {} & (-1)^{\dim(\mathbf{A}_M/\mathbf{A}_G)} k(\mathbf{M}) k(\mathbf{G})^{-1} \overline{v}(\mathbf{M}_\gamma)^{-1} \\ & \times \sum_{\Pi} \Phi_M(\gamma^{-1}, \Theta_\Pi) \mathrm{Tr}(\Pi(f_\infty)), \end{aligned}$$

where the sum is taken over the set of L-packets Π in $\Pi_{\mathrm{disc}}(\mathbf{G}(\mathbb{R}), \nu)$ and $\mathbf{M}_\gamma = \mathrm{Cent}_{\mathbf{M}}(\gamma)$. Of course, $S\Phi_M(\gamma, f_\infty) = 0$ unless γ is semisimple and elliptic in $\mathbf{M}(\mathbb{R})$. If $\mathbf{M}$ is a Levi subgroup of $\mathbf{G}$ that is not cuspidal, set $S\Phi_M^G = 0$.

Let $f : \mathbf{G}(\mathbb{A}) \longrightarrow \mathbb{C}$. Assume that $f = f^\infty f_\infty$, with $f^\infty \in C_c^\infty(\mathbf{G}(\mathbb{A}_f))$ and $f_\infty \in C_c^\infty(\mathbf{G}(\mathbb{R}), \nu^{-1})$. For every Levi subgroup $\mathbf{M}$ of $\mathbf{G}$, set

$$ST_M^G(f) = \tau(\mathbf{M}) \sum_\gamma SO_\gamma(f_{\mathbf{M}}^\infty) S\Phi_M(\gamma, f_\infty),$$

where the sum is taken over the set of stable conjugacy classes γ in $\mathbf{M}(\mathbb{Q})$ that are semisimple and elliptic in $\mathbf{M}(\mathbb{R})$, and $f_{\mathbf{M}}^\infty$ is the constant term of f^∞ at $\mathbf{M}$ (the constant term depends on the choice of a parabolic subgroup of $\mathbf{G}$ with Levi subgroup $\mathbf{M}$, but its integral orbitals do not). Set

$$ST^G(f) = \sum_{\mathbf{M}} (n_M^G)^{-1} ST_M^G(f),$$

where the sum is taken over the set of $\mathbf{G}(\mathbb{Q})$-conjugacy classes $\mathbf{M}$ of Levi subgroups of $\mathbf{G}$.

Let T^G be the distribution of Arthur's invariant trace formula [A3]. For every $(\mathbf{H}, s, \eta_0) \in \mathcal{E}(\mathbf{G})$ (cf. 2.4), fix an L-morphism $\eta : {}^L\mathbf{H} \longrightarrow {}^L\mathbf{G}$ extending η_0, and let

$$\iota(\mathbf{G}, \mathbf{H}) = \tau(\mathbf{G})\tau(\mathbf{H})^{-1}|\Lambda(\mathbf{H}, s, \eta_0)|^{-1}.$$

Kottwitz proved the following theorem in [K13] (theorem 5.1).

Theorem 5.4.1 *Let $f = f^\infty f_\infty$ be as above. Assume that f_∞ is stable cuspidal and that, for every $(\mathbf{H}, s, \eta_0) \in \mathcal{E}(\mathbf{G})$, there exists a transfer $f^{\mathbf{H}}$ of f. Then*

$$T^G(f) = \sum_{(\mathbf{H},s,\eta_0)\in\mathcal{E}(\mathbf{G})} \iota(\mathbf{G},\mathbf{H}) ST^H(f^{\mathbf{H}}).$$

We calculate $k(\mathbf{G})$ for $\mathbf{G}$ a unitary group.

Lemma 5.4.2 *Let $p_1, \dots, p_r, q_1, \dots, q_r \in \mathbb{N}$ such that $p_i + q_i \geq 1$ for $1 \leq i \leq r$; write $n_i = p_i + q_i$, $n = n_1 + \dots + n_r$ and $\mathbf{G} = \mathbf{G}(\mathbf{U}(p_1, q_1) \times \dots \times \mathbf{U}(p_r, q_r))$. Then*

$$k(\mathbf{G}) = 2^{n-r-1}$$

if all the n_i are even, and

$$k(\mathbf{G}) = 2^{n-r}$$

otherwise.

In particular, $k(R_{E/\mathbb{Q}}\mathbb{G}_m) = k(\mathbf{GU}(1)) = 1$.

Proof. Write $\Gamma(\infty) = \mathrm{Gal}(\mathbb{C}/\mathbb{R})$. In 3.1, we defined an elliptic maximal torus $\mathbf{T}_e$ of $\mathbf{G}$ and an isomorphism $\mathbf{T}_e \xrightarrow{\sim} \mathbf{G}(\mathbf{U}(1)^n)$. Tate-Nakayama duality induces an isomorphism between the dual of $\mathrm{Im}(\mathrm{H}^1(\mathbb{R}, \mathbf{T}_e \cap \mathbf{G}^{der}) \longrightarrow \mathrm{H}^1(\mathbb{R}, \mathbf{T}_e))$ and $\mathrm{Im}(\pi_0(\widehat{\mathbf{T}}_e^{\Gamma(\infty)}) \longrightarrow \pi_0((\widehat{\mathbf{T}}_e/Z(\widehat{\mathbf{G}})^{\Gamma(\infty)})))$ (cf. [K4] 7.9). Moreover, there is an exact sequence

$$(X_*(\widehat{\mathbf{T}}_e/Z(\widehat{\mathbf{G}})))^{\Gamma(\infty)} \longrightarrow \pi_0(Z(\widehat{\mathbf{G}})^{\Gamma(\infty)}) \longrightarrow \pi_0(\widehat{\mathbf{T}}_e^{\Gamma(\infty)}) \longrightarrow \pi_0((\widehat{\mathbf{T}}_e/Z(\widehat{\mathbf{G}}))^{\Gamma(\infty)})$$

(cf. [K4] 2.3), and $(X_*(\widehat{\mathbf{T}}_e/Z(\widehat{\mathbf{G}})))^{\Gamma(\infty)} = 0$ because $\mathbf{T}_e$ is elliptic; hence

$$k(\mathbf{G}) = |\pi_0(\widehat{\mathbf{T}}_e^{\Gamma(\infty)})||\pi_0(Z(\widehat{\mathbf{G}})^{\Gamma(\infty)})|^{-1}.$$

Of course, $\widehat{\mathbf{T}}_e^{\Gamma(\infty)} = \widehat{\mathbf{T}}_e^{\mathrm{Gal}(E/\mathbb{Q})}$ et $Z(\widehat{\mathbf{G}})^{\Gamma(\infty)} = Z(\widehat{\mathbf{G}})^{\mathrm{Gal}(E/\mathbb{Q})}$. We already calculated these groups in (i) of lemma 2.3.3. This implies the result. □

Remark 5.4.3 Note that $k(\mathbf{G})\tau(\mathbf{G}) = 2^{n-1}$.

Chapter Six

Stabilization of the fixed point formula

To simplify the notation, we suppose in this chapter that the group $\mathbf{G}$ is $\mathbf{GU}(p, q)$, but all the results generalize in an obvious way to the groups $\mathbf{G}(\mathbf{U}(p_1, q_1) \times \cdots \times \mathbf{U}(p_r, q_r))$.

6.1 PRELIMINARY SIMPLIFICATIONS

We first rewrite the fixed point formula using proposition 3.4.1.

The notation is as in chapters 1 (especially section 1.7), 2 and 3. Fix $p, q \in \mathbb{N}$ such that $n := p + q \geq 1$, and let $\mathbf{G} = \mathbf{GU}(p, q)$. We may and will assume that $p \geq q$. Let V be an irreducible algebraic representation of $\mathbf{G}_{\mathbb{C}}$ and $\varphi : W_{\mathbb{R}} \longrightarrow {}^L\mathbf{G}$ be an elliptic Langlands parameter corresponding to V^* as in proposition 3.4.1. Let $K \subset \mathbb{C}$ be a number field such that V is defined over K. Set

$$\Theta = (-1)^{q(\mathbf{G})} S\Theta_{\varphi}.$$

Fix $g \in \mathbf{G}(\mathbb{A}_f)$, $\mathrm{K}, \mathrm{K}' \subset \mathbf{G}(\mathbb{A}_f)$, $j \in \mathbb{N}^*$, prime numbers p and ℓ, and a place λ of K above ℓ as in sections 1.4 and 1.5. We get a cohomological correspondence

$$\overline{u}_j : (\Phi^j \overline{T}_g)^* IC^{\mathrm{K}} V \longrightarrow \overline{T}_1^! IC^{\mathrm{K}} V.$$

Proposition 6.1.1 *Write $\mathbf{G}_0 = \mathbf{M}_{\varnothing} = \mathbf{P}_{\varnothing} = \mathbf{G}$ and $\mathbf{L}_{\varnothing} = \{1\}$. For every $s \in \{0, \ldots, q\}$, set*

$$\mathrm{Tr}_s = (-1)^s m_s (n_{M_S}^G)^{-1} \chi(\mathbf{L}_S) \sum_{\gamma_L \in \mathbf{L}_S(\mathbb{Q})} \sum_{(\gamma_0; \gamma, \delta) \in C'_{\mathbf{G}_S, j}} c(\gamma_0; \gamma, \delta) O_{\gamma_L}(\mathbb{1}_{\mathbf{L}_S(\mathbb{Z}_p)})$$
$$\times O_{\gamma_L \gamma}\left(f_{\mathbf{M}_S}^{\infty, p}\right) \delta_{P_S(\mathbb{Q}_p)}^{1/2}(\gamma_0) T O_{\delta}(\phi_j^{\mathbf{G}_s}) \Phi_{M_S}^G((\gamma_L \gamma_0)^{-1}, \Theta),$$

where $S = \{1, 2, \ldots, s\}$, $m_s = 1$ if $s < n/2$, and $m_{n/2} = |\mathcal{X}_{n/2}| = 2$ if n is even. Then, if j is big enough :

$$\mathrm{Tr}(\overline{u}_j, R\Gamma(M^{\mathrm{K}}(\mathbf{G}, \mathcal{X})^*_{\mathbb{F}}, (IC^{\mathrm{K}} V)_{\mathbb{F}})) = \sum_{s=0}^{q} \mathrm{Tr}_s .$$

If $g = 1$ and $\mathrm{K} = \mathrm{K}'$, then the above formula is true for every $j \in \mathbb{N}^*$.

Proof. Let $m \in \mathbb{Z}$ be the weight of V as a representation of $\mathbf{G}$ (cf. section 1.3). For every $r \in \{1, \dots, q\}$, set $t_r = r(r-n)$. By proposition 1.4.3, there is a canonical isomorphism

$$IC^{\mathrm{K}}V \simeq W^{\geq t_1+1, \dots, \geq t_q+1}V.$$

Write

$$\mathrm{Tr} = \mathrm{Tr}(\overline{u}_j, R\Gamma(M^{\mathrm{K}}(\mathbf{G}, \mathcal{X})^*_{\mathbb{F}}, (IC^{\mathrm{K}}V)_{\mathbb{F}})).$$

If j is big enough, then, by theorem 1.7.1,

$$\mathrm{Tr} = \mathrm{Tr}_G + \sum_{\mathbf{P}} \mathrm{Tr}_P,$$

where the sum is taken over the set of standard parabolic subgroups of $\mathbf{G}$. Set $\mathrm{Tr}'_0 = \mathrm{Tr}_G$ and, for every $s \in \{1, \dots, q\}$,

$$\mathrm{Tr}'_s = \sum_{\substack{S' \subset \{1, \dots, s\} \\ S' \ni s}} \mathrm{Tr}_{P_{S'}}.$$

We want to show that $\mathrm{Tr}'_s = \mathrm{Tr}_s$. For $s = 0$, this comes from the fact that, for every semisimple $\gamma_0 \in \mathbf{G}(\mathbb{Q})$ that is elliptic in $\mathbf{G}(\mathbb{R})$,

$$\mathrm{Tr}(\gamma_0, V) = \Theta(\gamma_0^{-1}) = \Phi_G^G(\gamma_0^{-1}, \Theta).$$

Let $s \in \{1, \dots, q\}$; write $S = \{1, \dots, s\}$. Let $S' \subset S$ be such that $s \in S'$. Then, up to $\mathbf{L}_{S'}(\mathbb{Q})$-conjugacy, the only cuspidal Levi subgroup of $\mathbf{L}_{S'}$ is $\mathbf{L}_S = (R_{E/\mathbb{Q}}\mathbb{G}_m)^s$. Hence

$$\begin{aligned}\mathrm{Tr}_{P_{S'}} = &(-1)^{\dim(\mathbf{A}_{L_S}/\mathbf{A}_{L_{S'}})} m_{P_{S'}} (n_{L_S}^{L_{S'}})^{-1} \chi(\mathbf{L}_S) \\ &\times \sum_{\gamma_L \in \mathbf{L}_S(\mathbb{Q})} |D_{L_S}^{L_{S'}}(\gamma_L)|^{1/2} \sum_{(\gamma_0;\gamma,\delta) \in C'_{\mathbf{G}_s,j}} c(\gamma_0;\gamma,\delta) O_{\gamma_L\gamma}((f^{\infty,p})_{\mathbf{M}_S}) \\ &\times O_{\gamma_L}(\mathbb{1}_{\mathbf{L}_S(\mathbb{Z}_p)}) \delta^{1/2}_{P_{S'}(\mathbb{Q}_p)}(\gamma_0) T O_\delta(\phi_j^{\mathbf{G}_s}) \delta^{1/2}_{P_{S'}(\mathbb{R})}(\gamma_L\gamma_0) L_{S'}(\gamma_L\gamma_0),\end{aligned}$$

where

$$L_{S'}(\gamma_L\gamma_0) = \mathrm{Tr}(\gamma_L\gamma_0, R\Gamma(\mathrm{Lie}(\mathbf{N}_{S'}), V)_{>t_r+m, r \in S'}).$$

As γ_0 is in $\mathbf{G}_s(\mathbb{Q})$,

$$\delta_{P_{S'}(\mathbb{Q}_p)}(\gamma_0) = \delta_{P_S(\mathbb{Q}_p)}(\gamma_0).$$

Hence

$$\begin{aligned}\mathrm{Tr}'_s = &m_S \chi(\mathbf{L}_S) \sum_{\gamma_L \in \mathbf{L}_S(\mathbb{Q})} \sum_{(\gamma_0;\gamma,\delta) \in C'_{\mathbf{G}_s,j}} c(\gamma_0;\gamma,\delta) O_{\gamma_L\gamma}((f^{\infty,p})_{\mathbf{M}_S}) O_{\gamma_L}(\mathbb{1}_{\mathbf{L}_S(\mathbb{Z}_p)}) \\ &\times \delta^{1/2}_{P_S(\mathbb{Q}_p)}(\gamma_0) T O_\delta(\phi_j^{\mathbf{G}_s}) \sum_{\substack{S' \subset S \\ S' \ni s}} (-1)^{\dim(\mathbf{A}_{L_S}/\mathbf{A}_{L_{S'}})} (n_{L_S}^{L_{S'}})^{-1} \\ &\times |D_{L_S}^{L_{S'}}(\gamma_L)|^{1/2} \delta^{1/2}_{P_{S'}(\mathbb{R})}(\gamma_L\gamma_0) L_{S'}(\gamma_L\gamma_0).\end{aligned}$$

Consider the action of the group $\mathfrak{S}_s$ on $\mathbf{M}_S$ defined in section 3.4 (so $\mathfrak{S}_s$ acts on $\mathbf{L}_S = (R_{E/\mathbb{Q}}\mathbb{G}_m)^s$ by permuting the factors, and acts trivially on $\mathbf{G}_s$). Then

$$c(\gamma_0;\gamma,\delta)O_{\gamma_L\gamma}((f^{\infty,p})_{\mathbf{M}_S})O_{\gamma_L}(\mathbb{1}_{\mathbf{L}_S(\mathbb{Z}_p)})\delta^{1/2}_{P_S(\mathbb{Q}_p)}(\gamma_0)TO_\delta(\phi_j^{\mathbf{G}_s})$$

is invariant by the action of $\mathfrak{S}_s$. Let $\gamma_M \in \mathbf{M}_S(\mathbb{Q})$ be semisimple and elliptic in $\mathbf{M}_S(\mathbb{R})$. Write $\gamma_M = \gamma_L\gamma_0$, with $\gamma_L \in \mathbf{L}_S(\mathbb{Q})$ and $\gamma_0 \in \mathbf{G}_s(\mathbb{Q})$. Note that, for every $S' \subset S$ such that $s \in S'$, $\dim(\mathbf{A}_{L_{S'}}/\mathbf{A}_{L_S}) = \dim(\mathbf{A}_{M_{S'}}/\mathbf{A}_{M_S})$ and $D^{M_{S'}}_{M_S}(\gamma_M) = D^{L_{S'}}_{L_S}(\gamma_L)$. If γ_M satisfies the condition of part (ii) of proposition 3.4.1, then, by this proposition,

$$\sum_{\gamma'\in\mathfrak{S}_s.\gamma_M}\sum_{\substack{S'\subset S\\ S'\ni s}}(-1)^{\dim(\mathbf{A}_{L_S}/\mathbf{A}_{L_{S'}})}(n^{L_{S'}}_{L_S})^{-1}|D^{M_{S'}}_{M_S}(\gamma')|^{1/2}\delta^{1/2}_{P_{S'}(\mathbb{R})}(\gamma')L_{S'}(\gamma')$$

$$= (-1)^s(n^G_{M_S})^{-1}\sum_{\gamma'\in\mathfrak{S}_s.\gamma_M}\Phi^G_{M_S}(\gamma'^{-1},\Theta),$$

because the function $\Phi^G_{M_S}(.,\Theta)$ is invariant by the action of $\mathfrak{S}_s$, and $n^G_{M_S} = 2^s s! = 2^s|\mathfrak{S}_s|$. Moreover, also by proposition 3.4.1, for every $\gamma_L \in \mathbf{L}_S(\mathbb{R})$ and $(\gamma_0;\gamma,\delta) \in C_{\mathbf{G}_s,j} - C'_{\mathbf{G}_s,j}$, $\Phi^G_{M_S}((\gamma_L\gamma_0)^{-1},\Theta) = 0$ (because $c(\gamma_L\gamma_0) = c(\gamma_0) < 0$).

To finish the proof of the proposition, it is enough to show that, if j is big enough, then, for every $\gamma_L \in \mathbf{L}_S(\mathbb{Q})$ and $(\gamma_0;\gamma,\delta) \in C'_{\mathbf{G}_s,j}$ such that $O_{\gamma_L}(\mathbb{1}_{\mathbf{L}_S(\mathbb{Z}_p)})TO_\delta(\phi_j^{\mathbf{G}_s})$ $O_{\gamma_L\gamma}((f^{\infty,p})_{\mathbf{M}_S}) \neq 0$, the element $\gamma_L\gamma_0$ of $\mathbf{M}_S(\mathbb{Q})$ satisfies the condition of part (ii) of proposition 3.4.1.

Let Σ be the set of $(\gamma_L,\gamma_0) \in \mathbf{M}_S(\mathbb{Q}) = (E^\times)^s \times \mathbf{G}_s(\mathbb{Q})$ such that there exists $(\gamma,\delta) \in \mathbf{G}_s(\mathbb{A}_f^p) \times \mathbf{G}_s(L)$ with $(\gamma_0;\gamma,\delta) \in C'_{\mathbf{G}_s,j}$ and $O_{\gamma_L}(\mathbb{1}_{\mathbf{L}_S(\mathbb{Z}_p)})TO_\delta(\phi_j^{\mathbf{G}_s}) \times O_{\gamma_L\gamma}((f^{\infty,p})_{\mathbf{M}_S}) \neq 0$. By remark 1.7.5, the function $\gamma_M \longmapsto O_{\gamma_M}((f^{\infty,p})_{\mathbf{M}_S})$ on $\mathbf{M}_S(\mathbb{A}_f^p)$ has compact support modulo conjugacy. So there exist $C_1, C_2 \in \mathbb{R}^{+*}$ such that, for every $(\gamma_L = (\lambda_1,\dots,\lambda_s),\gamma_0) \in \Sigma$,

$$|c(\gamma_0)|_{\mathbb{A}_f^p} \geq C_1,$$

$$\sup_{1\leq r\leq s}|\lambda_r|^2_{\mathbb{A}_f^p} \geq C_2.$$

On the other hand, if $(\gamma_L = (\lambda_1,\dots,\lambda_s),\gamma_0) \in \Sigma$, then $|\lambda_r|_{\mathbb{Q}_p} = 1$ for $1 \leq r \leq s$ (because $\gamma_L \in \mathbf{L}_S(\mathbb{Z}_p)$), and there exists $\delta \in \mathbf{G}_s(L)$ such that $TO_\delta(\phi_j^{\mathbf{G}_s}) \neq 0$ and $N\delta$ is $\mathbf{G}_s(\overline{\mathbb{Q}}_p)$-conjugate to γ_0; this implies that

$$|c(\gamma_0)|_{\mathbb{Q}_p} = |c(\delta)|_L = p^d \geq p^j,$$

because $d = j$ or $2j$. (If $TO_\delta(\phi_j^{\mathbf{G}_s}) \neq 0$, then δ is σ-conjugate to an element δ' of $\mathbf{G}(\mathcal{O}_L)\mu_{\mathbf{G}_s}(\varpi_L^{-1})\mathbf{G}(\mathcal{O}_L)$, so $|c(\delta)|_L = |c(\delta')|_L = p^d$.) Finally, if $(\gamma_L = (\lambda_1, \dots, \lambda_s), \gamma_0) \in \Sigma$, then

$$|c(\gamma_0)|_\infty \sup_{1 \leq r \leq s} |\lambda_r|_\infty^2 = |c(\gamma_0)|_{\mathbb{Q}_p}^{-1} |c(\gamma_0)|_{\mathbb{A}_f^p}^{-1} \sup_{1 \leq r \leq s} |\lambda_r|_{\mathbb{A}_f^p}^{-2} \leq p^{-j} C_1^{-1} C_2^{-1}.$$

Moreover, if $(\gamma_0; \gamma, \delta) \in C'_{\mathbf{G}_s, j}$, then $c(\gamma_0) > 0$. Hence, if j is such that $p^j C_1 C_2 \geq 1$, then all the elements of Σ satisfy the condition of part (ii) of proposition 3.4.1.

Assume that $g = 1$ and $\mathrm{K} = \mathrm{K}'$. Then theorem 1.7.1 is true for every $j \in \mathbb{N}^*$. Moreover, by remark 1.7.5, the support of the function $\gamma_M \longmapsto O_{\gamma_M}((f^{\infty,p})_{\mathbf{M}_S})$ is contained in the union of the conjugates of a finite union of open compact subgroups of $\mathbf{M}_S(\mathbb{A}_f^p)$, so we may take $C_1 = C_2 = 1$, and every $j \in \mathbb{N}^*$ satisfies $p^j C_1 C_2 \geq 1$. □

6.2 STABILIZATION OF THE ELLIPTIC PART, AFTER KOTTWITZ

In [K9], Kottwitz stabilized the elliptic part of the fixed point formula (i.e., with the notation used here, the term Tr_0). We will recall his result in this section, and, in the next section, apply his method to the terms Tr_s, $s \in \{1, \dots, q\}$.

For every $(\mathbf{H}, s, \eta_0) \in \mathcal{E}(\mathbf{G})$, fix an L-morphism $\eta : {}^L\mathbf{H} \longrightarrow {}^L\mathbf{G}$ extending η_0 (in this section, we make the L-groups with $W_\mathbb{Q}$).

Theorem 6.2.1 *([K9] 7.2) There is an equality*

$$\mathrm{Tr}_0 = \sum_{(\mathbf{H},s,\eta_0)\in\mathcal{E}(\mathbf{G})} \iota(\mathbf{G}, \mathbf{H})\tau(\mathbf{H}) \sum_{\gamma_H} SO_{\gamma_H}(f_{\mathbf{H}}^{(j)}),$$

where the second sum is taken over the set of semisimple stable conjugacy classes in $\mathbf{H}(\mathbb{Q})$ that are elliptic in $\mathbf{H}(\mathbb{R})$.

We have to explain the notation. Let $(\mathbf{H}, s, \eta_0) \in \mathcal{E}(\mathbf{G})$. As in section 5.4, write

$$\iota(\mathbf{G}, \mathbf{H}) = \tau(\mathbf{G})\tau(\mathbf{H})^{-1} |\Lambda(\mathbf{H}, s, \eta_0)|^{-1}.$$

The function $f_{\mathbf{H}}^{(j)}$ is a function in $C^\infty(\mathbf{H}(\mathbb{A}))$ such that $f_{\mathbf{H}}^{(j)} = f_{\mathbf{H}}^{\infty,p} f_{\mathbf{H},p}^{(j)} f_{\mathbf{H},\infty}$, with $f_{\mathbf{H}}^{\infty,p} \in C_c^\infty(\mathbf{H}(\mathbb{A}_f^p))$, $f_{\mathbf{H},p}^{(j)} \in C_c^\infty(\mathbf{H}(\mathbb{Q}_p))$ and $f_{\mathbf{H},\infty} \in C^\infty(\mathbf{H}(\mathbb{R}))$.

The first function $f_{\mathbf{H}}^{\infty,p}$ is simply a transfer of $f^{\infty,p} \in C_c^\infty(\mathbf{G}(\mathbb{A}_f^p))$.

Use the notation of the last subsection of section 4.2, and set $\eta_p = \eta_{|\widehat{\mathbf{H}} \rtimes W_{\mathbb{Q}_p}}$. Then $f_{\mathbf{H},p}^{(j)} \in C_c^\infty(\mathbf{H}(\mathbb{Q}_p))$ is equal to the twisted transfer $b_{\widetilde{\eta}_p}(\phi_j^{\mathbf{G}})$.

For every elliptic Langlands parameter $\varphi_H : W_\mathbb{R} \longrightarrow \widehat{\mathbf{H}} \rtimes W_\mathbb{R}$, set

$$f_{\varphi_H} = d(\mathbf{H})^{-1} \sum_{\pi \in \Pi(\varphi_H)} f_\pi$$

(with the notation of section 3.1).

Let $\mathbf{B}$ be the standard Borel subgroup of $\mathbf{G}_{\mathbb{C}}$ (i.e., the subgroup of upper triangular matrices). It determines as in section 3.3 a subset $\Omega_* \subset \Omega_G$ and a bijection $\Phi_H(\varphi) \xrightarrow{\sim} \Omega_*$, $\varphi_H \longmapsto \omega_*(\varphi_H)$. Take

$$f_{\mathbf{H},\infty} = \langle \mu_{\mathbf{G}}, s\rangle(-1)^{q(\mathbf{G})} \sum_{\varphi_H \in \Phi_H(\varphi)} \det(\omega_*(\varphi_H)) f_{\varphi_H},$$

where $\mu_{\mathbf{G}}$ is the cocharacter of $\mathbf{G}_{\mathbb{C}}$ associated to the Shimura datum as in 2.1. Note that, as suggested by the notation, $f_{\mathbf{H},p}^{(j)}$ is the only part of $f_{\mathbf{H}}^{(j)}$ that depends on j.

Remark 6.2.2 In theorem 7.2 of [K9], the second sum is taken over the set of semisimple elliptic stable conjugacy classes that are $(\mathbf{G}, \mathbf{H})$-regular. But proposition 3.3.4 implies that $SO_{\gamma_H}(f_{\mathbf{H},\infty}) = 0$ if $\gamma_H \in \mathbf{H}(\mathbb{R})$ is not $(\mathbf{G}, \mathbf{H})$-regular (cf. remark 3.3.5).

6.3 STABILIZATION OF THE OTHER TERMS

The stabilization process that we follow here is mainly due to Kottwitz, and is explained (in a more general situation) in [K13]. The differences between the stabilization of the trace formula (in [K13]) and the stabilization of the fixed point formula considered here are concentrated at the places p and ∞. In particular, the vanishing of part of the contribution of the linear part of Levi subgroups is particular to the stabilization of the fixed point formula.

We will use freely the definitions and notation of section 2.4.

Theorem 6.3.1 (i) *Let $r \in \{1, \dots, q\}$. Write $\mathbf{M} = \mathbf{M}_{\{1,\dots,r\}}$. Then*

$$\mathrm{Tr}_r = (n_M^G)^{-1} \sum_{(\mathbf{M}', s_M, \eta_{M,0}) \in \mathcal{E}_{\mathbf{G}}(\mathbf{M})} \tau(\mathbf{G})\tau(\mathbf{H})^{-1} |\Lambda_{\mathbf{G}}(\mathbf{M}', s_M, \eta_{M,0})|^{-1} ST_{M'}^H(f_{\mathbf{H}}^{(j)}),$$

where, for every $(\mathbf{M}', s_M, \eta_{M,0}) \in \mathcal{E}_{\mathbf{G}}(\mathbf{M})$, $(\mathbf{H}, s, \eta_0)$ is the corresponding element of $\mathcal{E}(\mathbf{G})$.

(ii) *Write $\mathbf{G}^* = \mathbf{GU}^*(n)$. Let $r \in \mathbb{N}$ such that $r \leq n/2$. Denote by $\mathbf{M}^*$ the standard Levi subgroup of $\mathbf{G}^*$ that corresponds to $\{1, \dots, r\}$ (as in section 2.2). If $r \geq q + 1$, then*

$$\sum_{(\mathbf{M}', s_M, \eta_{M,0}) \in \mathcal{E}_{\mathbf{G}^*}(\mathbf{M}^*)} \tau(\mathbf{H})^{-1} |\Lambda_{\mathbf{G}^*}(\mathbf{M}', s_M, \eta_{M,0})|^{-1} ST_{M'}^H(f_{\mathbf{H}}^{(j)}) = 0.$$

Corollary 6.3.2 *If j is big enough, then*

$$\mathrm{Tr}(\overline{u}_j, R\Gamma(M^{\mathrm{K}}(\mathbf{G}, \mathcal{X})^*_{\mathbb{F}}, IC^{\mathrm{K}} V_{\mathbb{F}})) = \sum_{(\mathbf{H}, s, \eta_0) \in \mathcal{E}(\mathbf{G})} \iota(\mathbf{G}, \mathbf{H}) ST^H(f_{\mathbf{H}}^{(j)}).$$

If $g = 1$ and $\mathrm{K} = \mathrm{K}'$, then this formula is true for every $j \in \mathbb{N}^*$.

Remark 6.3.3 Let $\mathcal{H}_K = \mathcal{H}(\mathbf{G}(\mathbb{A}_f), \mathrm{K})$. Define an object W_λ of the Grothendieck group of representations of $\mathcal{H}_K \times \mathrm{Gal}(\overline{\mathbb{Q}}/F)$ in a finite-dimensional K_λ-vector space by

$$W_\lambda = \sum_{i\geq 0}(-1)^i[\mathrm{H}^i(M^{\mathrm{K}}(\mathbf{G},\mathcal{X})^*_{\overline{\mathbb{Q}}}, IC^{\mathrm{K}}V_{\overline{\mathbb{Q}}})].$$

Then, for every $j \in \mathbb{N}^*$,

$$\mathrm{Tr}(\overline{u}_j, R\Gamma(M^{\mathrm{K}}(\mathbf{G},\mathcal{X})^*_{\mathbb{F}}, IC^{\mathrm{K}}V_{\mathbb{F}})) = \mathrm{Tr}(\Phi^j_\wp h, W_\lambda),$$

where $\Phi_\wp \in \mathrm{Gal}(\overline{\mathbb{Q}}/F)$ is a lifting of the geometric Frobenius at $\wp$ (the fixed place of F above p) and $h = \mathrm{vol}(\mathrm{K})^{-1}\mathbb{1}_{\mathrm{K}g\mathrm{K}}$.

So the corollary implies that, for every $f^\infty \in \mathcal{H}_{\mathrm{K}}$ such that $f^\infty = f^{\infty,p}\mathbb{1}_{\mathbf{G}(\mathbb{Z}_p)}$ and for every j big enough (in a way that may depend on f^∞),

$$\mathrm{Tr}(\Phi^j_\wp f^\infty, W_\lambda) = \sum_{(\mathbf{H},s,\eta_0)\in\mathcal{E}(\mathbf{G})} \iota(\mathbf{G},\mathbf{H})ST^H(f^{(j)}_{\mathbf{H}}),$$

where, for every $(\mathbf{H}, s, \eta_0) \in \mathcal{E}(\mathbf{G})$, $f^{(j)}_{\mathbf{H},p}$ and $f_{\mathbf{H},\infty}$ are defined as before, and $f^{\infty,p}_{\mathbf{H}}$ is a transfer of $f^{\infty,p}$.

Proof. The corollary is an immediate consequence of theorem 6.2.1, the above theorem, proposition 6.1.1 and lemma 2.4.2. □

Proof of the theorem. As j is fixed, we omit the superscripts (j) in this proof.

We prove (i). As $\mathbf{M}$ is a proper Levi subgroup of $\mathbf{G}$, $|\Lambda_{\mathbf{G}}(\mathbf{M}', s_M, \eta_{M,0})| = 1$ for every $(\mathbf{M}', s_M, \eta_{M,0}) \in \mathcal{E}_{\mathbf{G}}(\mathbf{M})$ (lemma 2.4.3). Write, with the notation of the theorem,

$$\mathrm{Tr}'_M = \sum_{(\mathbf{M}',s_M,\eta_{M,0})\in\mathcal{E}_{\mathbf{G}}(\mathbf{M})} \tau(\mathbf{H})^{-1}ST^H_{M'}(f_{\mathbf{H}}).$$

By the definition of $ST^H_{M'}$ in section 5.4,

$$\mathrm{Tr}'_M = \sum_{(\mathbf{M}',s_M,\eta_{M,0})\in\mathcal{E}_{\mathbf{G}}(\mathbf{M})} \tau(\mathbf{H})^{-1}\tau(\mathbf{M}')\sum_{\gamma'} SO_{\gamma'}((f^\infty_{\mathbf{H}})_{\mathbf{M}'})S\Phi^H_{M'}(\gamma', f_{\mathbf{H},\infty}),$$

where the second sum is taken over the set of semisimple stable conjugacy classes of $\mathbf{M}'(\mathbb{Q})$ that are elliptic in $\mathbf{M}'(\mathbb{R})$. By proposition 3.3.4 (and remark 3.3.5), the terms indexed by a stable conjugacy class γ' that is not $(\mathbf{M},\mathbf{M}')$-regular all vanish.

By lemma 2.4.5,

$$\mathrm{Tr}'_M = \sum_{\gamma_M} \sum_{\kappa \in \mathfrak{K}_{\mathbf{G}}(I/\mathbb{Q})_e} \tau(\mathbf{M}')\tau(\mathbf{H})^{-1}\psi(\gamma_M, \kappa),$$

where

- The first sum is taken over the set of semisimple stable conjugacy classes γ_M in $\mathbf{M}(\mathbb{Q})$ that are elliptic in $\mathbf{M}(\mathbb{R})$.
- $I = \mathbf{M}_{\gamma_M}$, and $\mathfrak{K}_{\mathbf{G}}(I/\mathbb{Q})$ is defined above lemma 2.4.5.
- Let γ_M be as above and $\kappa \in \mathfrak{K}_{\mathbf{G}}(I/\mathbb{Q})$. Let $(\mathbf{M}', s_M, \eta_{M,0}, \gamma')$ be an endoscopic $\mathbf{G}$-quadruple associated to κ by lemma 2.4.5. The subset $\mathfrak{K}_{\mathbf{G}}(I/\mathbb{Q})_e$ of $\mathfrak{K}_{\mathbf{G}}(I/\mathbb{Q})$ is the set of κ such that $(\mathbf{M}', s_M, \eta_{M,0}) \in \mathcal{E}_{\mathbf{G}}(\mathbf{M})$. If $\kappa \in \mathfrak{K}_{\mathbf{G}}(I/\mathbb{Q})_e$, set

$$\psi(\gamma_M, \kappa) = SO_{\gamma'}((f_{\mathbf{H}}^{\infty})_{\mathbf{M}'}) S\Phi^H_{M'}(\gamma', f_{\mathbf{H},\infty}),$$

where $(\mathbf{H}, s, \eta_0)$ is as before the image of $(\mathbf{M}', s_M, \eta_{M,0})$ in $\mathcal{E}(\mathbf{G})$.

Fix γ_M, $\kappa \in \mathfrak{K}_{\mathbf{G}}(I/\mathbb{Q})_e$, and $(\mathbf{M}', s_M, \eta_{M,0}, \gamma')$ as above. Let $(\mathbf{H}, s, \eta_0)$ be the element of $\mathcal{E}(\mathbf{G})$ associated to $(\mathbf{M}', s_M, \eta_{M,0})$, and define s'_M as in section 4.3. Let γ_0 be the component of γ_M in the hermitian part $\mathbf{G}_r(\mathbb{Q})$ of $\mathbf{M}(\mathbb{Q})$. We want to calculate $\psi(\gamma_M, \kappa)$. This number is the product of three terms:

(a) Outside p and ∞: by (ii) of lemma 6.3.4,

$$SO_{\gamma'}((f_{\mathbf{H}}^{p,\infty})_{\mathbf{M}'}) = \sum_{\gamma} \Delta^{\mathbf{M},\infty,p}_{\mathbf{M}',s_M}(\gamma', \gamma) e(\gamma) O_\gamma(f_{\mathbf{M}}^{\infty,p}),$$

where the sum is taken over the set of semisimple conjugacy classes $\gamma = (\gamma_v)_{v \neq p,\infty}$ in $\mathbf{M}(\mathbb{A}_f^p)$ such that γ_v is stably conjugate to γ_M for every v, and $e(\gamma) = \prod_{v\neq p,\infty} e(\mathbf{M}_{\mathbb{Q}_v,\gamma_v})$. By [K7] 5.6, this sum is equal to

$$\Delta^{\mathbf{M},\infty,p}_{\mathbf{M}',s_M}(\gamma', \gamma_M) \sum_{\gamma} \langle \alpha(\gamma_M, \gamma), \kappa\rangle e(\gamma) O_\gamma(f_{\mathbf{M}}^{\infty,p}),$$

where the sum is over the same set as before and $\alpha(\gamma_M, \gamma)$ is the invariant denoted by $\mathrm{inv}(\gamma_M, \gamma)$ in [K7] (the article [K7] is only stating a conjecture, but this conjecture has been proved since; see section 5.3 for explanations). Moreover, as the linear part of $\mathbf{M}$ is isomorphic to $(R_{E/\mathbb{Q}}\mathbb{G}_m)^r$, we may replace $\alpha(\gamma_M, \gamma)$ with $\alpha(\gamma_0, \gamma_h)$, where γ_h is the component of γ in $\mathbf{G}_r(\mathbb{A}_f^p)$.

(b) At p: by corollary 4.3.2 (and with the notation of section 4.3), $SO_{\gamma'}((f_{\mathbf{H},p})_{\mathbf{M}'})$ is equal to

$$\langle \mu, s\rangle \langle \mu, s'_M\rangle \varepsilon_{s_M}(\gamma_M) \sum_{\delta} \langle \alpha_p(\gamma_0, \delta_h), s_M\rangle$$
$$\times \Delta^{\mathbf{M}}_{\mathbf{M}',s_M,p}(\gamma', \gamma_M) e(\delta) TO_\delta((\phi_j^{\mathbf{G}})_{\mathbf{M}}),$$

and, if $\gamma_M \in \mathbf{L}_r(\mathbb{Z}_p)\mathbf{G}_r(\mathbb{Q}_p)$, to

$$\langle\mu, s\rangle\langle\mu, s'_M\rangle\delta^{1/2}_{\mathbf{P}_r(\mathbb{Q}_p)}(\gamma_M)\sum_{\delta}\langle\alpha_p(\gamma_0, \delta_h), s_M\rangle$$
$$\times\Delta^{\mathbf{M}}_{\mathbf{M}', s_M, p}(\gamma', \gamma_M)e(\delta)TO_\delta(\mathbb{1}_{\mathbf{L}_r(\mathcal{O}_L)} \times \phi_j^{\mathbf{G}_r}).$$

Both sums are taken over the set of σ-conjugacy classes δ in $\mathbf{M}(L)$ such that $\gamma_M \in \mathcal{N}\delta$ (this notation is explained after example 8.1.1), and δ_h is the component of δ in $\mathbf{G}_r(L)$.

(c) At ∞ : by the definitions and proposition 3.3.4, $S\Phi^H_{M'}(\gamma', f_{\mathbf{H},\infty})$ is equal to

$$\langle\mu, s\rangle(-1)^{\dim(\mathbf{A}_M/\mathbf{A}_G)}k(\mathbf{M}')k(\mathbf{H})^{-1}\overline{v}(I)^{-1}\Delta^{\mathbf{M}}_{\mathbf{M}', s_M, \infty}(\gamma', \gamma_M)\Phi^G_M(\gamma_0^{-1}, \Theta).$$

(Note that $\mathbf{A}_{M'} \simeq \mathbf{A}_M$ and $\mathbf{A}_H \simeq \mathbf{A}_G$ because the endoscopic data $(\mathbf{H}, s, \eta_0)$ and $(\mathbf{M}', s_M, \eta_{M,0})$ are elliptic, and that $\overline{v}(I) = \overline{v}(\mathbf{M}'_{\gamma'})$ because I is an inner form of $\mathbf{M}'_{\gamma'}$ by [K7] 3.1.)

Finally, we find that $\psi(\gamma_M, \kappa)$ is equal to

$$(-1)^{\dim(\mathbf{A}_M/\mathbf{A}_G)}\varepsilon_{s_M}(\gamma_M)k(\mathbf{M}')k(\mathbf{H})^{-1}$$

$$\sum_{(\gamma,\delta)}\langle\alpha(\gamma_0; \gamma_h, \delta_h), \kappa\rangle e(\gamma)e(\delta)\overline{v}(I)^{-1}O_\gamma(f^{\infty,p}_{\mathbf{M}})TO_\delta((\phi_j^{\mathbf{G}})_{\mathbf{M}})\Phi^G_M(\gamma_M^{-1}, \Theta),$$

where the sum is taken over the set of equivalence classes of $(\gamma, \delta) \in \mathbf{M}(\mathbb{A}^p_f) \times \mathbf{M}(L)$ such that $(\gamma_M; \gamma, \delta)$ satisfies conditions (C) of section 1.6 and, if (γ_h, δ_h) is the component of (γ, δ) in $\mathbf{G}_r(\mathbb{A}^p_f) \times \mathbf{G}_r(L)$, then $\alpha(\gamma_0; \gamma_h, \delta_h) \in \mathfrak{K}_{\mathbf{M}}(I/\mathbb{Q})^D$ is the invariant associated to $(\gamma_0; \gamma_h, \delta_h)$ by Kottwitz in [K9] §2 (it is easy to check that $(\gamma_0; \gamma_h, \delta_h)$ also satisfies conditions (C) of section 1.6). We say that (γ_1, δ_1) and (γ_2, δ_2) are equivalent if γ_1 and γ_2 are $\mathbf{M}(\mathbb{A}^p_f)$-conjugate and δ_1 and δ_2 are σ-conjugate in $\mathbf{M}(L)$. In particular, $\psi(\gamma_M, \kappa)$ is the product of a term depending only on the image of κ in $\mathfrak{K}_{\mathbf{M}}(I/\mathbb{Q})$ and of $k(\mathbf{M}')k(\mathbf{H})^{-1}\varepsilon_{s_M}(\gamma_M)$.

Moreover, if $\gamma_M \in \mathbf{L}_r(\mathbb{Z}_p)\mathbf{G}_r(\mathbb{Q}_p)$, then $\psi(\gamma_M, \kappa)$ is equal to

$$(-1)^{\dim(\mathbf{A}_M/\mathbf{A}_G)}k(\mathbf{M}')k(\mathbf{H})^{-1}\delta^{1/2}_{\mathbf{P}_r(\mathbb{Q}_p)}(\gamma_M)$$

$$\sum_{(\gamma,\delta)}\langle\alpha(\gamma_0; \gamma_h, \delta_h), \kappa\rangle e(\gamma)e(\delta)\overline{v}(I)^{-1}O_\gamma(f^{\infty,p}_{\mathbf{M}})TO_\delta(\mathbb{1}_{\mathbf{L}_r(\mathcal{O}_L)} \times \phi_j^{\mathbf{G}_r})\Phi^G_M(\gamma_M^{-1}, \Theta),$$

where the sum is taken over the same set. Note that, as $\alpha(\gamma_0; \gamma_h, \delta_h) \in \mathfrak{K}_{\mathbf{M}}(I/\mathbb{Q})^D$, the number $\langle\alpha(\gamma_0; \gamma_h, \delta_h), \kappa\rangle$ depends only on the image of κ in $\mathfrak{K}_{\mathbf{M}}(I/\mathbb{Q})$.

Let Σ_L be the set of $(\lambda_1, \dots, \lambda_r) \in \mathbf{L}_r(\mathbb{Q}) = (E^\times)^r$ such that

- for every $i \in \{1, \dots, r\}$, $|\lambda_i\overline{\lambda}_i|_p \in p^{2a\mathbb{Z}}$;
- there exists $i \in \{1, \dots, r\}$ such that $\frac{1}{2a}\mathrm{val}_p(|\lambda_i\overline{\lambda}_i|_p)$ is odd.

Remember that we defined in section 4.3 a subgroup Ω^M of the group of automorphisms of $\mathbf{M}$ (it was called Ω^{M_H} in section 4.3); if we write $\mathbf{M} = (R_{E/\mathbb{Q}}\mathbb{G}_m)^r \times \mathbf{G}_r$, then Ω^M is the group generated by the involutions

$$((\lambda_1, \dots, \lambda_r), g) \longmapsto ((\lambda_1, \dots, \lambda_{i-1}, \lambda_i^{-1}c(g)^{-1}, \lambda_{i+1}, \dots, \lambda_r), g),$$

with $1 \leq i \leq r$. The order of the group Ω^M is 2^r. On the other hand, there is an action of $\mathfrak{S}_r$ on $\mathbf{M}$, given by the formula

$$(\sigma, ((\lambda_1, \dots, \lambda_r), g)) \longmapsto ((\lambda_{\sigma^{-1}(1)}, \dots, \lambda_{\sigma^{-1}(r)}), g).$$

The subset $\Sigma_L \mathbf{G}_r(\mathbb{Q})$ of $\mathbf{M}_r(\mathbb{Q})$ is of course stable by $\mathfrak{S}_r$. By proposition 4.3.1, if $\gamma_M \in \mathbf{M}(\mathbb{Q})$ is such that $\psi(\gamma_M, \kappa) \neq 0$, then either there exists $\omega \in \Omega^M$ (uniquely determined) such that $\omega(\gamma_M) \in \mathbf{L}_r(\mathbb{Z}_p)\mathbf{G}_r(\mathbb{Q})$, or $\gamma_M \in \Sigma_L \mathbf{G}_r(\mathbb{Q})$. By (i) of lemma 6.3.6, if $\gamma_M \in \Sigma_L \mathbf{G}_r(\mathbb{Q})$, then, for every $\kappa_M \in \mathfrak{K}_{\mathbf{M}}(I/\mathbb{Q})$,

$$\sum_{\kappa \mapsto \kappa_M} \sum_{\gamma'_M \in \mathfrak{S}_r \gamma_M} \varepsilon_{s_M}(\gamma'_M)\tau(\mathbf{M}')k(\mathbf{M}')\tau(\mathbf{H})^{-1}k(\mathbf{H})^{-1} = 0.$$

Hence

$$\sum_{\gamma_M} \sum_{\kappa} \tau(\mathbf{M}')\tau(\mathbf{H})^{-1}\psi(\gamma_M, \kappa) = 0,$$

where the first sum is taken over the set of semisimple stable conjugacy classes γ_M in $\mathbf{M}(\mathbb{Q})$ that are elliptic in $\mathbf{M}(\mathbb{R})$ and such that there is no $\omega \in \Omega^M$ such that $\omega(\gamma_M) \in \mathbf{L}_r(\mathbb{Z}_p)\mathbf{G}_r(\mathbb{Q})$. As Ω^M is a subgroup of $\mathrm{Nor}_{\mathbf{G}}(\mathbf{M})(\mathbb{Q})$, the first expression above for $\psi(\gamma_M, \kappa)$ shows that $\psi(\omega(\gamma_M), \kappa) = \psi(\gamma_M, \kappa)$ for every $\omega \in \Omega^M$. So we get

$$\mathrm{Tr}'_M = (-1)^{\dim(\mathbf{A}_M/\mathbf{A}_G)} 2^r \sum_{\gamma_M} \delta^{1/2}_{\mathbf{P}_r(\mathbb{Q}_p)}(\gamma_M)$$

$$\times \sum_{(\gamma,\delta)} e(\gamma)e(\delta)\overline{v}(I)^{-1} O_\gamma(f_{\mathbf{M}}^{\infty,p}) T O_\delta(\mathbb{1}_{\mathbf{L}_r(\mathcal{O}_L)} \times \phi_j^{\mathbf{G}_r})$$

$$\Phi^G_M(\gamma_M^{-1}, \Theta) \sum_{\kappa_M \in \mathfrak{K}_{\mathbf{M}}(I/\mathbb{Q})} \langle \alpha(\gamma_0; \gamma_h, \delta_h), \kappa_M \rangle \sum_{\substack{\kappa \in \mathfrak{K}_{\mathbf{G}}(I/\mathbb{Q})_e \\ \kappa \mapsto \kappa_M}} \tau(\mathbf{M}')k(\mathbf{M}')\tau(\mathbf{H})^{-1}k(\mathbf{H})^{-1},$$

where the first sum is taken over the set of semisimple stable conjugacy classes γ_M in $\mathbf{M}(\mathbb{Q})$ that are elliptic in $\mathbf{M}(\mathbb{R})$ and the second sum is taken over the set of equivalence classes (γ, δ) in $\mathbf{M}(\mathbb{A}_f^p) \times \mathbf{M}(L)$ such that $(\gamma_M; \gamma, \delta)$ satisfies conditions (C) of section 1.6. By (ii) of lemma 6.3.6, for every $\kappa_M \in \mathfrak{K}_{\mathbf{M}}(I/\mathbb{Q})$,

$$\tau(\mathbf{G})\tau(\mathbf{M})^{-1} \sum_{\kappa \mapsto \kappa_M} \tau(\mathbf{M}')k(\mathbf{M}')\tau(\mathbf{H})^{-1}k(\mathbf{H})^{-1} = \begin{cases} 2^{-r} & \text{if } r < n/2 \\ 2^{-r+1} & \text{if } r = n/2, \end{cases}$$

In particular, this sum is independant of κ_M. So, by the reasoning of [K9] §4 and by the definition of $\overline{v}(I)$ in section 5.4 and lemma 6.3.7 (and the fact that $\mathbf{L}_r$ is a torus), we get

$$\tau(\mathbf{G})\,\mathrm{Tr}'_M = (-1)^r m_r \chi(\mathbf{L}_r) \sum_{\gamma_l \in \mathbf{L}_r(\mathbb{Q})} \sum_{(\gamma_0;\gamma_h,\delta_h)\in C'_{\mathbf{G}_r,j}} c(\gamma_0;\gamma_h,\delta_h)\delta^{1/2}_{\mathbf{P}_r(\mathbb{Q}_p)}(\gamma_0)$$

$$\times O_{\gamma_l\gamma_h}(f^{\infty,p}_{\mathbf{M}})O_{\gamma_l}(\mathbb{1}_{\mathbf{L}_r(\mathbb{Z}_p)})TO_{\delta_h}(\phi_j^{\mathbf{G}_r})\Phi^G_M((\gamma_l\gamma_0)^{-1},\Theta),$$

where the integer m_r is as in proposition 6.1.1 (equal to 2 if $r = n/2$ and to 1 if $r < n/2$). This (combined with proposition 6.1.1) finishes the proof of the equality of (i).

We now prove (ii). The proof is very similar to that of (i). Assume that $r \geq q+1$, and write

$$\mathrm{Tr}'_r = \sum_{(\mathbf{M}',s_M,\eta_{M,0})\in\mathcal{E}_{\mathbf{G}^*}(\mathbf{M}^*)} \tau(\mathbf{H})^{-1}|\Lambda_{\mathbf{G}^*}(\mathbf{M}',s_M,\eta_{M,0})|^{-1}ST^H_{M'}(f^{(j)}_{\mathbf{H}}).$$

Then, as in the prooof of (i), we see that

$$\mathrm{Tr}'_r = \sum_{\gamma_M} \sum_{\kappa\in\mathfrak{K}_{\mathbf{G}^*}(I/\mathbb{Q})_e} \tau(\mathbf{M}')\tau(\mathbf{H})^{-1}\psi(\gamma_M,\kappa),$$

where

- The first sum is taken over the set of semisimple stable conjugacy classes γ_M in $\mathbf{M}^*(\mathbb{Q})$ that are elliptic in $\mathbf{M}^*(\mathbb{R})$.
- $I = \mathbf{M}^*_{\gamma_M}$, and $\mathfrak{K}_{\mathbf{G}^*}(I/\mathbb{Q})$ is defined above lemma 2.4.5.
- The subset $\mathfrak{K}_{\mathbf{G}^*}(I/\mathbb{Q})_e$ of $\mathfrak{K}_{\mathbf{G}^*}(I/\mathbb{Q})$ is defined as above. If $\kappa \in \mathfrak{K}_{\mathbf{G}^*}(I/\mathbb{Q})_e$ and $(\mathbf{M}', s_M, \eta_{M,0}, \gamma')$ is an endoscopic $\mathbf{G}^*$-quadruple associated to κ by lemma 2.4.5, then

$$\psi(\gamma_M,\kappa) = SO_{\gamma'}((f^{\infty}_{\mathbf{H}})_{\mathbf{M}'})S\Phi^H_{M'}(\gamma', f_{\mathbf{H},\infty}),$$

  where $(\mathbf{H}, s, \eta_0)$ is the image of $(\mathbf{M}', s_M, \eta_{M,0})$ in $\mathcal{E}(\mathbf{G}^*) = \mathcal{E}(\mathbf{G})$.

If there exists a place $v \neq p, \infty$ of $\mathbb{Q}$ such that $\mathbf{M}^*_{\mathbb{Q}_v}$ does not transfer to $\mathbf{G}_{\mathbb{Q}_v}$ (see lemma 6.3.5), then $\mathrm{Tr}'_r = 0$ by (ii) of lemma 6.3.5. So we may assume that $\mathbf{M}^*_{\mathbb{Q}_v}$ transfers to $\mathbf{G}_{\mathbb{Q}_v}$ for every $v \neq p, \infty$. Then, reasoning exactly as in the proof of (i) (and using (i) of lemma 6.3.5), we see that, for every γ_M and every κ as above, $\psi(\gamma_M,\kappa)$ is the product of a term depending only on the image of κ in $\mathfrak{K}_{\mathbf{M}^*}(I/\mathbb{Q})$ and of $k(\mathbf{M}')k(\mathbf{H})^{-1}\varepsilon_{s_M}(\gamma_M)$. Now, to show that $\mathrm{Tr}'_r = 0$, we can use (i) of lemma 6.3.6 (and proposition 4.3.1) as in the proof of (i). □

The next two lemmas are results of [K13].

Lemma 6.3.4 *(cf. [K13] 7.10 and lemma 7.6) Fix a place v of $\mathbb{Q}$. Let $\mathbf{M}$ be a Levi subgroup of $\mathbf{G}_{\mathbb{Q}_v}$, $(\mathbf{M}', s_M, \eta_{M,0}) \in \mathcal{E}_{\mathbf{G}_{\mathbb{Q}_v}}(\mathbf{M})$ and $(\mathbf{H}, s, \eta_0)$ be the image of $(\mathbf{M}', s_M, \eta_{M,0})$ in $\mathcal{E}(\mathbf{G}_{\mathbb{Q}_v})$. As in lemma 2.4.2, we identify $\mathbf{M}'$ with a Levi subgroup of $\mathbf{H}$. Choose compatible extensions $\eta : {}^L\mathbf{H} \longrightarrow {}^L\mathbf{G}_{\mathbb{Q}_v}$ and $\eta_M : {}^L\mathbf{M}' \longrightarrow {}^L\mathbf{M}$ of η_0 and $\eta_{M,0}$, and normalize transfer factors as in section 5.2.*

(i) *Let* $f \in C_c^\infty(\mathbf{G}(\mathbb{Q}_v))$. *Then, for every* $\gamma \in \mathbf{M}(\mathbb{Q}_v)$ *semisimple and* $\mathbf{G}$*-regular,*

$$SO_\gamma(f_{\mathbf{M}}) = |D_M^G(\gamma)|_v^{1/2} SO_\gamma(f).$$

(Remember that $D_M^G(\gamma) = \det(1 - \mathrm{Ad}(\gamma), \mathrm{Lie}(\mathbf{G})/\mathrm{Lie}(\mathbf{M})$.*)*

(ii) *Let* $f \in C_c^\infty(\mathbf{G}(\mathbb{Q}_v))$, *and let* $f^{\mathbf{H}} \in C_c^\infty(\mathbf{H}(\mathbb{Q}_v))$ *be a transfer of* f. *Then, for every* $\gamma_H \in \mathbf{M}'(\mathbb{Q}_v)$ *semisimple and* $(\mathbf{M}, \mathbf{M}')$*-regular,*

$$SO_{\gamma_H}((f^{\mathbf{H}})_{\mathbf{M}'}) = \sum_\gamma \Delta_{\mathbf{M}',s_M}^{\mathbf{M}}(\gamma_H, \gamma) e(\mathbf{M}_\gamma) O_\gamma(f_{\mathbf{M}}),$$

where the sum is taken over the set of semisimple conjugacy classes γ *in* $\mathbf{M}(\mathbb{Q}_v)$ *that are images of* γ_H.

Proof. (cf. [K13]) We show (i). Let $\gamma \in \mathbf{M}(\mathbb{Q}_v)$ be semisimple and $\mathbf{G}$-regular. By the descent formula ([A2] corollary 8.3),

$$O_\gamma(f_{\mathbf{M}}) = |D_M^G(\gamma)|_v^{1/2} O_\gamma(f).$$

On the other hand, as $\mathbf{M}_\gamma = \mathbf{G}_\gamma$ and as the morphism $\mathbf{H}^1(\mathbb{Q}_v, \mathbf{M}) \longrightarrow \mathbf{H}^1(\mathbb{Q}_v, \mathbf{G})$ is injective,[1] the obvious map $\mathrm{Ker}(\mathbf{H}^1(\mathbb{Q}_v, \mathbf{M}_\gamma) \longrightarrow \mathbf{H}^1(\mathbb{Q}_v, \mathbf{M})) \longrightarrow \mathrm{Ker}(\mathbf{H}^1(\mathbb{Q}_v, \mathbf{G}_\gamma) \longrightarrow \mathbf{H}^1(\mathbb{Q}_v, \mathbf{G}))$ is a bijection. In other words, there is a natural bijection from the set of conjugacy classes in the stable conjugacy class of γ in $\mathbf{M}(\mathbb{Q}_v)$ to the set of conjugacy classes in the stable conjugacy class of γ in $\mathbf{G}(\mathbb{Q}_v)$. This proves the equality of (i).

We show (ii). By lemma 2.4.A of [LS2], it is enough to show the equality for γ_H regular in $\mathbf{H}$. We may even assume that all the images of γ_H in $\mathbf{M}(\mathbb{Q}_v)$ are regular in $\mathbf{G}$; in that case, all the signs $e(\mathbf{M}_\gamma)$ in the equality that we are trying to prove are trivial. Applying (i) to $f^{\mathbf{H}}$, we get

$$SO_{\gamma_H}((f^{\mathbf{H}})_{\mathbf{M}'}) = |D_{M'}^H(\gamma_H)|_v^{1/2} SO_{\gamma_H}(f^{\mathbf{H}}).$$

By definition of the transfer, this implies that

$$SO_{\gamma_H}((f^{\mathbf{H}})_{\mathbf{M}'}) = |D_{M'}^H(\gamma_H)|_v^{1/2} \sum_\gamma \Delta_{\mathbf{H}}^{\mathbf{G}}(\gamma_H, \gamma) O_\gamma(f),$$

where the sum is taken over the set of conjugacy classes γ in $\mathbf{G}(\mathbb{Q}_v)$ that are images of γ_H. Such a conjugacy class has a nonempty intersection with $\mathbf{M}(\mathbb{Q}_v)$, so the equality of (ii) is a consequence of the descent formula and of the normalization of the transfer factors. □

[1]This is explained in [K13] A.1, and is true for any reductive group over a field of characteristic 0. Choose a parabolic subgroup $\mathbf{P}$ of $\mathbf{G}$ with Levi subgroup $\mathbf{M}$. Then the map $\mathbf{H}^1(\mathbb{Q}_v, \mathbf{M}) \longrightarrow \mathbf{H}^1(\mathbb{Q}_v, \mathbf{P})$ is bijective, and the map $\mathbf{H}^1(\mathbb{Q}_v, \mathbf{P}) \longrightarrow \mathbf{H}^1(\mathbb{Q}_v, \mathbf{G})$ is injective (the second map has a trivial kernel by theorem 4.13(c) of [BT], and it is not hard to deduce from this that it is injective).

Lemma 6.3.5 *(cf. [K13] lemma 7.4 and A.2) Write as before* $\mathbf{G}^* = \mathbf{GU}^*(n)$. *Fix a place* v *of* $\mathbb{Q}$. *Let* $\mathbf{M}^*$ *be a Levi subgroup of* $\mathbf{G}^*_{\mathbb{Q}_v}$. *As in [K13] A.2, we say that* $\mathbf{M}^*$ transfers to $\mathbf{G}_{\mathbb{Q}_v}$ *if there exists an inner twisting* $\psi : \mathbf{G}^* \longrightarrow \mathbf{G}$ *such that the restriction of* ψ *to* $\mathbf{A}_{M^*}$ *is defined over* $\mathbb{Q}_v$.

(i) Assume that $\mathbf{M}^*$ *transfers to* $\mathbf{G}_{\mathbb{Q}_v}$, *and let* $\psi : \mathbf{G}^* \longrightarrow \mathbf{G}$ *be an inner twisting such that* $\psi_{|\mathbf{A}_{M^*}}$ *is defined over* $\mathbb{Q}_v$. *Then* $\mathbf{M} := \psi(\mathbf{M}^*)$ *is a Levi subgroup of* $\mathbf{G}_{\mathbb{Q}_v}$, *and* $\psi_{|\mathbf{M}^*} : \mathbf{M}^* \longrightarrow \mathbf{M}$ *is an inner twisting.*

(ii) Assume that $\mathbf{M}^*$ *does not transfer to* $\mathbf{G}_{\mathbb{Q}_v}$. *Let* $(\mathbf{M}', s_M, \eta_{M,0}) \in \mathcal{E}_{\mathbf{G}^*_{\mathbb{Q}_v}}(\mathbf{M}^*)$, *and let* $(\mathbf{H}, s, \eta_0)$ *be the image of* $(\mathbf{M}', s_M, \eta_{M,0})$ *in* $\mathcal{E}(\mathbf{G}^*_{\mathbb{Q}_v}) = \mathcal{E}(\mathbf{G}_{\mathbb{Q}_v})$. *As in lemma 2.4.2, we identify* $\mathbf{M}'$ *with a Levi subgroup of* $\mathbf{H}$. *Let* $f \in C_c^\infty(\mathbf{G}(\mathbb{Q}_v))$, *and let* $f^{\mathbf{H}} \in C_c^\infty(\mathbf{H}(\mathbb{Q}_v))$ *be a transfer of* f. *Then, for every* $\gamma_H \in \mathbf{M}'(\mathbb{Q}_v)$ *semisimple and* $(\mathbf{M}, \mathbf{M}')$*-regular,*

$$SO_{\gamma_H}((f^{\mathbf{H}})_{\mathbf{M}'}) = 0.$$

Proof. Point (i) follows from the fact that $\mathbf{M} = \mathrm{Cent}_{\mathbf{G}_{\mathbb{Q}_v}}(\psi(\mathbf{A}_{M^*}))$. We prove (ii). By lemma 2.4.A of [LS2], we may assume that γ_H is regular in $\mathbf{H}$; by continuity, we may even assume that γ_H is $\mathbf{G}^*$-regular. Let $\mathbf{T}_H$ be the centralizer of γ_H in $\mathbf{H}$. It is a maximal torus of $\mathbf{M}_H$ and $\mathbf{H}$, and it transfers to $\mathbf{M}^*$ and $\mathbf{G}^*$ because γ_H is $\mathbf{G}^*$-regular. By (i) of lemma 6.3.4 and the definition of the transfer, to show that $SO_{\gamma_H}((f^{\mathbf{H}})_{\mathbf{M}'}) = 0$, it is enough to show that $\mathbf{T}_H$, seen as a maximal torus in $\mathbf{G}^*$, does not transfer to $\mathbf{G}$. Assume that $\mathbf{T}_H$ transfers to $\mathbf{G}$. Then there exists an inner twisting $\psi : \mathbf{G}^* \longrightarrow \mathbf{G}$ such that $\psi_{|\mathbf{T}_H}$ is defined over $\mathbb{Q}_v$; but $\mathbf{A}_{M^*} \subset \mathbf{T}_H$, so $\psi_{|\mathbf{A}_{M^*}}$ is defined over $\mathbb{Q}_v$, and this contradicts the fact that $\mathbf{M}^*$ does not transfer to $\mathbf{G}_{\mathbb{Q}_v}$. □

Lemma 6.3.6 *We use the notation of the proof of theorem 6.3.1. Let* $\kappa_M \in \mathfrak{K}_M(I/\mathbb{Q})$.

(i) If $\gamma_M \in \Sigma_L\mathbf{G}_r(\mathbb{Q})$, *then*

$$\sum_{\substack{\kappa\in\mathfrak{K}_{\mathbf{G}}(I/\mathbb{Q})_e \\ \kappa\longmapsto\kappa_M}} \sum_{\gamma'_M\in\mathfrak{S}_r\gamma_M} \varepsilon_{s_M}(\gamma'_M)\tau(\mathbf{M}')k(\mathbf{M}')\tau(\mathbf{H})^{-1}k(\mathbf{H})^{-1} = 0.$$

(ii)

$$\tau(\mathbf{G})\tau(\mathbf{M})^{-1} \sum_{\substack{\kappa\in\mathfrak{K}_{\mathbf{G}}(I/\mathbb{Q})_e \\ \kappa\longmapsto\kappa_M}} \tau(\mathbf{M}')k(\mathbf{M}')\tau(\mathbf{H})^{-1}k(\mathbf{H})^{-1}$$

$$= \begin{cases} 2^{-r} & \text{if } r < n/2 \\ 2^{-r+1} & \text{if } r = n/2. \end{cases}$$

Proof. Remember that $\mathbf{M} = \mathbf{M}_{\{1,\dots,r\}} \simeq (R_{E/\mathbb{Q}}\mathbb{G}_m)^r \times \mathbf{GU}(p-r, q-r)$. By remark 5.4.3, $\tau(\mathbf{H})k(\mathbf{H}) = 2^{n-1}$ for every $(\mathbf{H}, s, \eta_0) \in \mathcal{E}(\mathbf{G})$ and, for every $(\mathbf{M}', s_M, \eta_{M,0}) \in \mathcal{E}_{\mathbf{G}}(\mathbf{M})$,

$$k(\mathbf{M}')\tau(\mathbf{M}') = k(\mathbf{M})\tau(\mathbf{M}) = \begin{cases} 2^{n-2r-1} & \text{if } r < n/2 \\ 1 & \text{if } r = n/2. \end{cases}$$

In particular, the term $\tau(\mathbf{M}')k(\mathbf{M}')\tau(\mathbf{H})^{-1}k(\mathbf{H})^{-1}$ in the two sums of the lemma does not depend on κ; it is equal to 2^{-2r} if $r < n/2$, and to $2^{1-n} = 2^{-2r+1}$ if $r = n/2$. Besides, by lemma 2.3.3, $\tau(\mathbf{G})\tau(\mathbf{M})^{-1}$ is equal to 1 if $r < n/2$, and to 2 if $r = n/2$.

We calculate $\mathfrak{K}_{\kappa_M} := \{\kappa \in \mathfrak{K}_{\mathbf{G}}(I/\mathbb{Q})_e | \kappa \longmapsto \kappa_M\}$. Write $\Gamma = \mathrm{Gal}(\overline{\mathbb{Q}}/\mathbb{Q})$, and choose an embedding $\widehat{\mathbf{M}} \subset \widehat{\mathbf{G}}$ as in lemma 2.4.3. Then we get isomorphisms $Z(\widehat{\mathbf{G}}) \simeq \mathbb{C}^\times \times \mathbb{C}^\times$ and $Z(\widehat{\mathbf{M}}) \simeq \mathbb{C}^\times \times (\mathbb{C}^\times)^r \times \mathbb{C}^\times \times (\mathbb{C}^\times)^r$ such that the embedding $Z(\widehat{\mathbf{G}}) \subset Z(\widehat{\mathbf{M}})$ is $(\lambda, \mu) \longmapsto (\lambda, (\mu, \dots, \mu), \mu, (\mu, \dots, \mu))$ and that the action of $\mathrm{Gal}(E/\mathbb{Q})$ on $Z(\widehat{\mathbf{G}})$ and $Z(\widehat{\mathbf{M}})$ is given by the following formulas

$$\tau(\lambda, \mu) = (\lambda\mu^n, \mu^{-1})$$

$$\tau(\lambda, (\lambda_1, \dots, \lambda_r), \mu, (\lambda'_r, \dots, \lambda'_1)) = (\lambda\mu^{n-2r}\lambda_1 \dots \lambda_r\lambda'_1 \dots \lambda'_r, \\ (\lambda_1'^{-1}, \dots, \lambda_r'^{-1}), \mu^{-1}, (\lambda_1^{-1}, \dots, \lambda_r^{-1}))$$

(remember that τ is the non-trivial element of $\mathrm{Gal}(E/\mathbb{Q})$). This implies that $(Z(\widehat{\mathbf{M}})/Z(\widehat{\mathbf{G}}))^\Gamma \simeq (\mathbb{C}^\times)^r$ is connected (this is a general fact, cf. [K13] A.5). By the exact sequence of [K4] 2.3, the morphism $\mathbf{H}^1(\mathbb{Q}, Z(\widehat{\mathbf{G}})) \longrightarrow \mathbf{H}^1(\mathbb{Q}, Z(\widehat{\mathbf{M}}))$ is injective. By lemma 2.3.3 and [K4] (4.2.2), $\mathrm{Ker}^1(\mathbb{Q}, Z(\widehat{\mathbf{G}})) = \mathrm{Ker}^1(\mathbb{Q}, Z(\widehat{\mathbf{M}})) = 1$; so the following commutative diagram has exact rows:

$$\begin{array}{ccccccc} 1 & \longrightarrow & \mathfrak{K}_{\mathbf{G}}(I/\mathbb{Q}) & \longrightarrow & (Z(\widehat{I})/Z(\widehat{\mathbf{G}}))^\Gamma & \longrightarrow & \mathbf{H}^1(\mathbb{Q}, Z(\widehat{\mathbf{G}})) \\ & & \downarrow & & \downarrow & & \downarrow \\ 1 & \longrightarrow & \mathfrak{K}_{\mathbf{M}}(I/\mathbb{Q}) & \longrightarrow & (Z(\widehat{I})/Z(\widehat{\mathbf{M}}))^\Gamma & \longrightarrow & \mathbf{H}^1(\mathbb{Q}, Z(\widehat{\mathbf{M}})) \end{array}$$

Let $x \in \mathfrak{K}_{\mathbf{M}}(I/\mathbb{Q})$. Then x, seen as an element of $(Z(\widehat{I})/Z(\widehat{\mathbf{M}}))^\Gamma$, has a trivial image in $\mathbf{H}^1(\mathbb{Q}, Z(\widehat{\mathbf{M}}))$, so it is in the image of $Z(\widehat{I})^\Gamma \longrightarrow (Z(\widehat{I})/Z(\widehat{\mathbf{M}}))^\Gamma$. In particular, there exists $y \in (Z(\widehat{I})/Z(\widehat{\mathbf{G}}))^\Gamma$ that is sent to x. As the map $\mathbf{H}^1(\mathbb{Q}, Z(\widehat{\mathbf{G}})) \longrightarrow \mathbf{H}^1(\mathbb{Q}, Z(\widehat{\mathbf{M}}))$ is injective, y has a trivial image in $\mathbf{H}^1(\mathbb{Q}, Z(\widehat{\mathbf{G}}))$, so y is in $\mathfrak{K}_{\mathbf{G}}(I/\mathbb{Q})$. This proves that the map $\alpha : \mathfrak{K}_{\mathbf{G}}(I/\mathbb{Q}) \longrightarrow \mathfrak{K}_{\mathbf{M}}(I/\mathbb{Q})$ in the diagram above is surjective. We want to determine its kernel. There is an obvious injection $\mathrm{Ker}(\alpha) \longrightarrow (Z(\widehat{\mathbf{M}})/Z(\widehat{\mathbf{G}}))^\Gamma$. By the injectivity of $\mathbf{H}^1(\mathbb{Q}, Z(\widehat{\mathbf{G}})) \longrightarrow \mathbf{H}^1(\mathbb{Q}, Z(\widehat{\mathbf{M}}))$, the image of $(Z(\widehat{\mathbf{M}})/Z(\widehat{\mathbf{G}}))^\Gamma$ in $(Z(\widehat{I})/Z(\widehat{\mathbf{G}}))^\Gamma$ is included in $\mathfrak{K}_{\mathbf{G}}(I/\mathbb{Q})$; this implies that $\mathrm{Ker}(\alpha) = (Z(\widehat{\mathbf{M}})/Z(\widehat{\mathbf{G}}))^\Gamma$. Finally, we get an exact sequence

$$1 \longrightarrow (Z(\widehat{\mathbf{M}})/Z(\widehat{\mathbf{G}}))^\Gamma \longrightarrow \mathfrak{K}_{\mathbf{G}}(I/\mathbb{Q}) \longrightarrow \mathfrak{K}_{\mathbf{M}}(I/\mathbb{Q}) \longrightarrow 1$$

(it is the exact sequence of [K13] (7.2.1)).

If $\kappa \in \mathfrak{K}(\kappa_M)$ and $\gamma'_M \in \mathfrak{S}_r\gamma_M$, write $\varepsilon_\kappa(\gamma'_M)$ instead of $\varepsilon_{s_M}(\gamma'_M)$ (this sign depends only on κ, cf. remark 4.3.3). As I is the centralizer of an elliptic element of $\mathbf{M}(\mathbb{Q})$, it is easy to see from lemma 2.4.3 that $\mathfrak{K}(\kappa_M)$ is non-empty and that we can choose $\kappa_0 \in \mathfrak{K}(\kappa_M)$ such that $\varepsilon_{\kappa_0}(\gamma'_M) = 1$ for every $\gamma'_M \in \mathbf{M}(\mathbb{Q})$. Fix such a κ_0. For every $A \subset \{1, \dots, r\}$, let s_A be the image in $Z(\widehat{\mathbf{M}})/Z(\widehat{\mathbf{G}})$ of the element $(1, (s_1, \dots, s_r), 1, (s_r, \dots, s_1))$ of $Z(\widehat{\mathbf{M}})$, where $s_i = 1$ if $i \notin A$ and $s_i = -1$ if $i \in A$. Lemma 2.4.3 implies that $\mathfrak{K}(\kappa_M) = \{\kappa_0 + s_A, A \subset \{1, \dots, r\}\}$. If $r < n/2$, then the s_A are pairwise distinct, so $|\mathfrak{K}(\kappa_M)| = 2^r$. If $r = n/2$, then $s_A = s_{A'}$ if and only if $\{1, \ldots, r\} = A \sqcup A'$, so $|\mathfrak{K}(\kappa_M)| = 2^{r-1}$. This finishes the proof of (ii).

We now prove (i). Let $\gamma_M \in \Sigma_L \mathbf{G}_r(\mathbb{Q})$. We want to show that

$$\sum_{\kappa \in \mathfrak{K}(\kappa_M)} \sum_{\gamma'_M \in \mathfrak{S}_r \gamma_M} \varepsilon_\kappa(\gamma'_M) = 0.$$

Write $\gamma_M = ((\lambda_1, \ldots, \lambda_r), g) \in (E^\times)^r \times \mathbf{G}_r(\mathbb{Q})$, and let B the set of $i \in \{1, \ldots, r\}$ such that $\frac{1}{2a}\mathrm{val}_p(|\lambda_i \overline{\lambda}_i|_p)$ is odd.

It is easy to see from the definition of ε_{s_M} that, for every $\sigma \in \mathfrak{S}_r$ and $A \subset \{1, \ldots, r\}$, $\varepsilon_{\kappa_0 + s_A}(\sigma(\gamma_M)) = (-1)^{|A \cap \sigma(B)|}$. (If $r = n/2$, this sign is the same for A and $\{1, \ldots, r\} - A$ because $|B|$ is even by remark 4.3.3.) So it is enough to show that, for every $\sigma \in \mathfrak{S}_r$, $\sum\limits_{A \subset \{1,\ldots,r\}} (-1)^{|A \cap \sigma(B)|} = 0$. But

$$\sum_{A \subset \{1,\ldots,r\}} (-1)^{|A \cap \sigma(B)|} = 2^{n-|B|} \sum_{A \subset \sigma(B)} (-1)^{|A|},$$

and this is equal to 0 because B is nonempty by the hypothesis on γ_M. □

Lemma 6.3.7 *Let $s \in \{1, \ldots, q\}$. Write $S = \{1, \ldots, s\}$. Then*

$$\chi(\mathbf{L}_S) = \mathrm{vol}(\mathbf{A}_{L_S}(\mathbb{R})^0 \setminus \mathbf{L}_S(\mathbb{R}))^{-1}$$

(where $\mathbf{A}_{L_S}$ is the maximal split subtorus of $\mathbf{L}_S$, i.e., $\mathbb{G}_m^s$).

Proof. By [GKM] 7.10 and the fact that $\mathbf{L}_S/\mathbf{A}_{L_S}$ is $\mathbb{R}$-anisotropic, we get

$$\chi(\mathbf{L}_S) = (-1)^{q(\mathbf{L}_S)} \tau(\mathbf{L}_S) \,\mathrm{vol}(\mathbf{A}_{L_S}(\mathbb{R})^0 \setminus \mathbf{L}_S(\mathbb{R}))^{-1} d(\mathbf{L}_S).$$

As $\mathbf{L}_S$ is a torus, $q(\mathbf{L}_S) = 0$ and $d(\mathbf{L}_S) = 1$. Moreover, by lemma 2.3.3, $\tau(\mathbf{L}_S) = 1$. □

Chapter Seven

Applications

This chapter contains a few applications of corollary 6.3.2. First we show how corollary 6.3.2 implies a variant of theorem 5.4.1 for the unitary groups of section 2.1. The only reason we do this is to make the other applications in this chapter logically independent of the unpublished [K13] (this independence is of course only formal, as the whole stabilization of the fixed point formula in this book was inspired by [K13]). Then we give applications to the calculation of the (Hecke) isotypical components of the intersection cohomology and to the Ramanujan-Petersson conjecture.

7.1 STABLE TRACE FORMULA

The simplest way to apply corollary 6.3.2 is to use theorem 5.4.1 (i.e., the main result of [K13]) to calculate the terms $ST^H(f_{\mathbf{H}}^{(i)})$. In this section, we show how to avoid this reference to the unpublished [K13].

Notation is as in chapter 6, but with $\mathbf{G}$ any of the unitary groups defined in 2.1. As in 6.2, fix, for every $(\mathbf{H}, s, \eta_0) \in \mathcal{E}(\mathbf{G})$, an L-morphism $\eta : {}^L\mathbf{H} \longrightarrow {}^L\mathbf{G}$ extending η_0.

Definition 7.1.1 Let $f_\infty \in C^\infty(\mathbf{G}(\mathbb{R}))$. Suppose that $f_\infty = \sum_\varphi c_\varphi f_\varphi$, where the sum is taken over the set of equivalence classes of elliptic Langlands parameters $\varphi : W_\mathbb{R} \longrightarrow \widehat{\mathbf{G}} \rtimes W_\mathbb{R}$ and the c_φ are complex numbers that are almost all 0 (f_φ is defined at the beginning of section 6.2). Then, for every $(\mathbf{H}, s, \eta_0) \in \mathcal{E}(\mathbf{G})$, set

$$(f_\infty)_\mathbf{H} = \langle \mu_\mathbf{G}, s\rangle \sum_\varphi c_\varphi \sum_{\varphi_H \in \Phi_H(\varphi)} \det(\omega_*(\varphi_H)) f_{\varphi_H},$$

where the bijection $\Phi_H(\varphi) \xrightarrow{\sim} \Omega_*$, $\varphi_H \longmapsto \omega_*(\varphi_H)$ is as in section 6.2 determined by the standard Borel subgroup of $\mathbf{G}_\mathbb{C}$.

Remark 7.1.2 By the trace Paley-Wiener theorem of Clozel and Delorme ([CD], cf. the beginning of section 3 of [A6]), if $f_\infty \in C^\infty(\mathbf{G}(\mathbb{R}))$ is stable cuspidal, then f_∞ satisfies the condition of definition 7.1.1, so $(f_\infty)_\mathbf{H}$ is defined.

Definition 7.1.3 Let $a_1, b_1, \dots, a_r, b_r \in \mathbb{N}$ be such that $a_i \geq b_i$ for all i and $\mathbf{G} = \mathbf{G}(\mathbf{U}(a_1, b_1) \times \cdots \times \mathbf{U}(a_r, b_r))$. Write $n_i = a_i + b_i$; then the quasi-split inner form of $\mathbf{G}$ is $\mathbf{G}(\mathbf{U}^*(n_1) \times \cdots \times \mathbf{U}^*(n_r))$. Fix $n_1^+, n_1^-, \dots, n_r^+, n_r^- \in \mathbb{N}$ such that $n_i = n_i^+ + n_i^-$

for every $i \in \{1, \dots, r\}$ and that $n_1^- + \dots + n_r^-$ is even. Let $(\mathbf{H}, s, \eta_0)$ be the elliptic endoscopic triple for $\mathbf{G}$ associated to these integers as in proposition 2.3.1. For every $i \in \{1, \dots, r\}$, if $I_i \subset \{1, \dots, n_i\}$, set $n_i(I_i) = |I \cap \{n_i^+ + 1, \dots, n_i\}|$. We define a rational number $\iota_{\mathbf{G},\mathbf{H}}$ by

$$\iota_{\mathbf{G},\mathbf{H}} = \iota(\mathbf{G}, \mathbf{H}) |\pi_0(\mathcal{X})|^{-1} \sum_{\substack{I_1 \subset \{1,\dots,n_1\} \\ |I_1| = a_1}} \cdots \sum_{\substack{I_r \subset \{1,\dots,n_r\} \\ |I_r| = a_r}} (-1)^{n_1(I_1) + \dots + n_r(I_r)},$$

where $\mathcal{X}$ is the symmetric space appearing in the Shimura data of section 2.1 for $\mathbf{G}$.

Proposition 7.1.4 *Let $f = f^\infty f_\infty$ be as in theorem 5.4.1. Assume that f_∞ is stable cuspidal and that, for every $(\mathbf{H}, s, \eta_0) \in \mathcal{E}(\mathbf{G})$, there exists a transfer $(f^\infty)^{\mathbf{H}}$ of f^∞. Then*

$$T^G(f) = \sum_{(\mathbf{H}, s, \eta_0) \in \mathcal{E}} \iota_{\mathbf{G},\mathbf{H}} ST^H((f^\infty)^{\mathbf{H}} (f_\infty)_{\mathbf{H}}).$$

Remark 7.1.5 It is not very hard to see that proposition 7.1.4 is a consequence of theorem 5.4.1. The goal here is to prove it directly.

To prove this proposition, we first need an extension of corollary 6.3.2 (proposition 7.1.7 below).

Fix a prime number p where $\mathbf{G}$ is unramified. Remember that we defined, for every $m \in \mathbb{N}^*$, a function $\phi_m^{\mathbf{G}}$ on $\mathbf{G}(L)$, where L is an unramified extension of $\mathbb{Q}_p$; if $(\mathbf{H}, s, \eta_0) \in \mathcal{E}(\mathbf{G})$, write $f_{\mathbf{H},p}^{(m)}$ for the function in $C_c^\infty(\mathbf{H}(\mathbb{Q}_p))$ obtained by twisted transfer from $\phi_m^{\mathbf{G}}$ as in section 4.3. In the proof of proposition 4.3.1, we calculated the Satake transform S of $f_{\mathbf{H},p}^{(m)}$, or more precisely of $\chi_{\eta_p}^{-1} f_{\mathbf{H},p}^{(m)}$, where χ_{η_p} is the quasi-character of $\mathbf{H}(\mathbb{Q}_p)$ associated to $\eta_{|\widehat{\mathbf{H}} \rtimes W_{\mathbb{Q}_p}} : \widehat{\mathbf{H}} \rtimes W_{\mathbb{Q}_p} \longrightarrow \widehat{\mathbf{G}} \rtimes W_{\mathbb{Q}_p}$ as in the last two subsections of 4.2. Note that the expression for the Satake transform S makes sense for any $m \in \mathbb{Z}$.

Definition 7.1.6 If $m \in \mathbb{Z}$, we define $f_{\mathbf{H},p}^{(m)} \in C_c^\infty(\mathbf{H}(\mathbb{Q}_p))$ in the following way: $\chi_{\eta_p}^{-1} f_{\mathbf{H},p}^{(m)} \in \mathcal{H}(\mathbf{H}(\mathbb{Q}_p), \mathbf{H}(\mathbb{Z}_p))$, and its Satake transform is given by the polynomial S in the proof of proposition 4.3.1 (where of course the integer a in the definition of S is replaced by m).

Fix $f^{\infty,p} \in C_c^\infty(\mathbf{G}(\mathbb{A}_f^p))$ and an irreducible algebraic representation V of $\mathbf{G}_{\mathbb{C}}$. For every $(\mathbf{H}, s, \eta_0) \in \mathcal{E}(\mathbf{G})$ and $m \in \mathbb{Z}$, let $f_{\mathbf{H}}^{(m)} = f_{\mathbf{H}}^{p,\infty} f_{\mathbf{H},p}^{(m)} f_{\mathbf{H},\infty} \in C^\infty(\mathbf{H}(\mathbb{A}))$, where $f_{\mathbf{H}}^{p,\infty}$ and $f_{\mathbf{H},\infty}$ are as in 6.2.

Proposition 7.1.7 *Assume that p is inert in E.*[1] *Then, with the notation of section 6.3, for every $m \in \mathbb{Z}$,*

$$\mathrm{Tr}(\Phi_\wp^m f^\infty, W_\lambda) = \sum_{(\mathbf{H}, s, \eta_0) \in \mathcal{E}(\mathbf{G})} \iota(\mathbf{G}, \mathbf{H}) ST^H(f_{\mathbf{H}}^{(m)}),$$

where $f^\infty = f^{\infty,p} \mathbb{1}_{\mathbf{G}(\mathbb{Z}_p)}$.

---

[1]This is not really necessary, but it makes the proof slightly simpler.

Notice that, for $m >> 0$, this is simply corollary 6.3.2 (cf. the remark following this corollary).

Proof. Fix an (arbitrary) embedding $\iota : K_\lambda \subset \mathbb{C}$ and write $W = \iota_*(W_\lambda)$. Then W is a virtual complex representation of $\mathcal{H}_K \times \mathrm{Gal}(\overline{\mathbb{Q}}/F)$. As the actions of $\Phi_\wp$ and f^∞ on W commute, there exist a finite set I_0 and families of complex numbers $(c_i)_{i\in I_0}$ and $(\alpha_i)_{i\in I_0}$ such that, for every $m \in \mathbb{Z}$,

$$\mathrm{Tr}(\Phi_\wp^m f^\infty, W) = \sum_{i\in I_0} c_i \alpha_i^m .$$

We now want to find a similar expression for the right-hand side of the equality of the proposition. Remember from the definitions in section 5.4 that

$$ST^H(f_{\mathbf{H}}^{(m)}) = \sum_{\mathbf{M}_H} (n_{M_H}^H)^{-1} \tau(\mathbf{M}_H) \sum_{\gamma_H} SO_{\gamma_H}((f_{\mathbf{H}}^{\infty,p})_{\mathbf{M}_H}) \times SO_{\gamma_H}((f_{\mathbf{H},p}^{(m)})_{\mathbf{M}_H}) S\Phi_{M_H}^H(\gamma_H, f_{H,\infty}),$$

where the first sum is taken over the set of conjugacy classes of cuspidal Levi subgroups $\mathbf{M}_H$ of $\mathbf{H}$ and the second sum over the set of semisimple stable conjugacy classes $\gamma_H \in \mathbf{M}_H(\mathbb{Q})$ that are elliptic in $\mathbf{M}_H(\mathbb{R})$. Note that the first sum is finite. In the second sum, all but finitely many terms are zero, but the set of γ_H such that the term associated to γ_H is nonzero may depend on m.

Fix $(\mathbf{H}, s, \eta_0) \in \mathcal{E}(\mathbf{G})$ and a cuspidal Levi subgroup $\mathbf{M}_H$ of $\mathbf{H}$. By the Howe conjecture, proved by Clozel in [Cl1], the space of linear forms $\mathcal{H}(\mathbf{M}_H(\mathbb{Q}_p), \mathbf{M}_H(\mathbb{Z}_p)) \longrightarrow \mathbb{C}$ generated by the elements $h \longmapsto SO_{\gamma_H}(h)$, for $\gamma_H \in \mathbf{M}_H(\mathbb{Q}_p)$ semisimple elliptic, is finite-dimensional. As p is inert in E, any semisimple $\gamma_H \in \mathbf{M}_H(\mathbb{Q})$ that is elliptic in $\mathbf{M}_H(\mathbb{R})$ is also elliptic in $\mathbf{M}_H(\mathbb{Q}_p)$.[2] So we find that the space D of linear forms on $\mathcal{H}(\mathbf{M}_H(\mathbb{Q}_p), \mathbf{M}_H(\mathbb{Z}_p))$ generated by the $h \longmapsto SO_{\gamma_H}(h)$, for $\gamma_H \in \mathbf{M}_H(\mathbb{Q})$ semisimple and elliptic in $\mathbf{M}_H(\mathbb{R})$, is finite-dimensional.

On the other hand, by Kazhdan's density theorem ([Ka] theorem 0), every distribution $h \longmapsto SO_{\gamma_H}(h)$ on $\mathcal{H}(\mathbf{M}_H(\mathbb{Q}_p), \mathbf{M}_H(\mathbb{Z}_p))$ is a finite linear combination of distributions of the type $h \longmapsto \mathrm{Tr}(\pi(h))$, for π a smooth irreducible representation of $\mathbf{M}_H(\mathbb{Z}_p)$ (which we may assume to be unramified). So the space D is generated by a finite number of distributions of that type.

Using the form of the Satake transform of $\chi_{\eta_p}^{-1} f_{\mathbf{H},p}^{(m)}$, it is easy to see that this implies that there exist a finite set $I_{M_H,H}$ and a family of complex numbers $(\beta_{M_H,H,i})_{i\in I_{M_H,H}}$ such that, for every $\gamma_H \in \mathbf{M}_H(\mathbb{Q})$ semisimple and elliptic in $\mathbf{M}_H(\mathbb{R})$, there exists a family of complex numbers $(d_{M_H,H,i}(\gamma_H))_{i\in I_{M_H,H}}$ with

$$SO_{\gamma_H}((f_{\mathbf{H}}^{\infty,p})_{\mathbf{M}_H}) SO_{\gamma_H}((f_{\mathbf{H},p}^{(m)})_{\mathbf{M}_H}) S\Phi_{M_H}^H(\gamma_H, f_{H,\infty}) = \sum_{i\in I_{M_H,H}} d_{M_H,H,i}(\gamma_H)\beta_{M_H,H,i}^m$$

for every $m \in \mathbb{Z}$.

[2]Such a γ_H is elliptic in $\mathbf{M}_H(\mathbb{R})$ (resp. $\mathbf{M}_H(\mathbb{Q}_p)$) if and only if its centralizer is contained in no Levi subgroup of $\mathbf{M}_H(\mathbb{R})$ (resp. $\mathbf{M}_H(\mathbb{Q}_p)$). But the Levi subgroups of $\mathbf{M}_H(\mathbb{R})$ and $\mathbf{M}_H(\mathbb{Q}_p)$ are all defined over $\mathbb{Q}$.

Let $m_0 \in \mathbb{Z}$. We want to prove the equality of the proposition for $m = m_0$. Let $N \in \mathbb{N}$ such that the equality of the proposition is true for $m \geq N$ (such an N exists by corollary 6.3.2). We may assume that $m_0 \leq N$. Let

$$M = |I_0| + \sum_{(\mathbf{H},s,\eta_0)\in\mathcal{E}(\mathbf{G})} \sum_{\mathbf{M}_H} |I_{M_H,H}|,$$

where the second sum is taken over the set of conjugacy classes of cuspidal Levi subgroups of $\mathbf{H}$. For every $\mathbf{H}$ and $\mathbf{M}_H$ as before, let $\Gamma_{M_H,H}$ be the set of semisimple stable conjugacy classes $\gamma_H \in \mathbf{M}_H(\mathbb{Q})$ that are elliptic in $\mathbf{M}_H(\mathbb{R})$ and such that there exists $m \in \mathbb{Z}$ with $m_0 \leq m \leq N + M - 1$ and

$$SO_{\gamma_H}((f_{\mathbf{H}}^{\infty,p})_{\mathbf{M}_H}) SO_{\gamma_H}((f_{\mathbf{H},p}^{(m)})_{\mathbf{M}_H}) S\Phi_{\mathbf{M}_H}^{\mathbf{H}}(\gamma_H, f_{H,\infty}) \neq 0.$$

This set is finite. So, by the above calculations, there exist families of complex numbers $(d_{M_H,H,i})_{i\in I_{M_H,H}}$, for all $\mathbf{M}_H$ and $\mathbf{H}$ as before, such that, for every $m \in \mathbb{Z}$ with $m_0 \leq m \leq N + M - 1$,

$$\sum_{(\mathbf{H},s,\eta_0)\in\mathcal{E}(\mathbf{G})} \iota(\mathbf{G},\mathbf{H}) ST^H(f_{\mathbf{H}}^{(m)}) = \sum_{(\mathbf{H},s,\eta_0)\in\mathcal{E}(\mathbf{G})} \sum_{\mathbf{M}_H} \sum_{i\in I_{M_H,H}} d_{M_H,H,i} \beta_{M_H,H,i}^m.$$

All the sums above are finite. So we can reformulate this as: there exist a finite set J (with $|J| = \sum_H \sum_{M_H} |I_{M_H,H}|$) and families of complex numbers $(d_j)_{j\in J}$ and $(\beta_j)_{j\in J}$ such that, if $m_0 \leq m \leq N + M - 1$, then

$$\sum_{(\mathbf{H},s,\eta_0)\in\mathcal{E}(\mathbf{G})} \iota(\mathbf{G},\mathbf{H}) ST^H(f_{\mathbf{H}}^{(m)}) = \sum_{j\in J} d_j \beta_j^m.$$

So the result that we want to prove is that the equality

$$\sum_{i\in I_0} c_i \alpha_i^m = \sum_{j\in J} d_j \beta_j^m$$

holds for $m = m_0$. But we know that this equality holds if $N \leq m \leq N + M - 1$, and $M = |I_0| + |J|$, so this equality holds for all $m \in \mathbb{Z}$. □

Proof of proposition 7.1.4. We may assume that f^∞ is a product $\bigotimes_p f_p$, with $f_p = \mathbb{1}_{\mathbf{G}(\mathbb{Z}_p)}$ for almost all p. Let $\mathrm{K} = \prod_p \mathrm{K}_p$ be a neat open compact subgroup of $\mathbf{G}(\mathbb{A}_f)$ such that $f^\infty \in \mathcal{H}(\mathbf{G}(\mathbb{A}_f), \mathrm{K})$. Fix a prime number p that is inert in E and such that $\mathbf{G}$ is unramified at p, $f_p = \mathbb{1}_{\mathbf{G}(\mathbb{Z}_p)}$ and $\mathrm{K}_p = \mathbf{G}(\mathbb{Z}_p)$. Define a virtual representation W of $\mathcal{H}(\mathbf{G}(\mathbb{A}_f), \mathrm{K}) \times \mathrm{Gal}(\overline{\mathbb{Q}}/F)$ as in the proof of proposition 7.1.7. Then, by formula (3.5) and theorem 6.1 of [A6], and by theorem 7.14.B and paragraph (7.19) of [GKM][3]:

$$\mathrm{Tr}(f^\infty, W) = |\pi_0(\mathcal{X})| T^G(f).$$

On the other hand, using proposition 7.1.7 at the place p and for $m = 0$, we find

$$\mathrm{Tr}(f^\infty, W) = \sum_{(\mathbf{H},s,\eta_0)\in\mathcal{E}(\mathbf{G})} \iota(\mathbf{G},\mathbf{H}) ST^H(f_{\mathbf{H}}^{(0)}).$$

[3] In the articles [A6] and [GKM], the authors consider only connected symmetric spaces, i.e., they use $\pi_0(\mathbf{G}(\mathbb{R})) \setminus \mathcal{X}$ instead of $\mathcal{X}$ (in the cases considered here, $\mathbf{G}(\mathbb{R})$ acts transitively on $\mathcal{X}$, so $\pi_0(\mathbf{G}(\mathbb{R})) \setminus \mathcal{X}$ is connected). When we pass from $\pi_0(\mathbf{G}(\mathbb{R})) \setminus \mathcal{X}$ to $\mathcal{X}$, the trace of Hecke operators is multiplied by $|\pi_0(\mathcal{X})|$.

But it is obvious from the definitions of $f^{(m)}_{\mathbf{H},p}$ and $\iota_{\mathbf{G},\mathbf{H}}$ that

$$f^{(0)}_{\mathbf{H},p} = \frac{\iota_{\mathbf{G},\mathbf{H}}}{\iota(\mathbf{G},\mathbf{H})}|\pi_0(\mathcal{X})|\chi_{\eta_p}\mathbb{1}_{\mathbf{H}(\mathbb{Z}_p)},$$

and we know that $\chi_{\eta_p}\mathbb{1}_{\mathbf{H}(\mathbb{Z}_p)}$ is a transfer of $f_p = \mathbb{1}_{\mathbf{G}(\mathbb{Z}_p)}$ by the fundamental lemma (cf. section 5.3). This finishes the proof. □

7.2 ISOTYPICAL COMPONENTS OF THE INTERSECTION COHOMOLOGY

The notation is still as in section 7.1, and we assume that $\mathbf{G} = \mathbf{GU}(p,q)$, with $n = p + q$ (for the other unitary groups of section 2.1, everything would work the same way, but with more complicated notation). In particular, V is an irreducible algebraic representation of $\mathbf{G}$ defined over a number field K, λ is a place of K over ℓ, and $\varphi : W_{\mathbb{R}} \longrightarrow \widehat{\mathbf{G}} \rtimes W_{\mathbb{R}}$ is an elliptic Langlands parameter corresponding to the contragredient V^* of V (as in proposition 3.4.1).

Let $\mathcal{H}_K = \mathcal{H}(\mathbf{G}(\mathbb{A}_f), \mathrm{K})$. Define, as in section 6.3, an object W_λ of the Grothendieck group of representations of $\mathcal{H}_K \times \mathrm{Gal}(\overline{\mathbb{Q}}/F)$ in a finite dimensional K_λ-vector space by

$$W_\lambda = \sum_{i\geq 0}(-1)^i[\mathrm{H}^i(M^{\mathrm{K}}(\mathbf{G},\mathcal{X})^*_{\overline{\mathbb{Q}}}, IC^{\mathrm{K}}V_{\overline{\mathbb{Q}}})].$$

Let $\iota : K_\lambda \longrightarrow \mathbb{C}$ be an embedding. Then there is an isotypical decomposition of $\iota_*(W_\lambda)$ as a $\mathcal{H}_{\mathrm{K}}$-module:

$$\iota_*(W_\lambda) = \sum_{\pi_f}\iota_*(W_\lambda)(\pi_f)\otimes\pi_f^{\mathrm{K}},$$

where the sum is taken over the set of isomorphism classes of irreducible admissible representations π_f of $\mathbf{G}(\mathbb{A}_f)$ such that $\pi_f^{\mathrm{K}} \neq 0$, and where the $\iota_*(W_\lambda)(\pi_f)$ are virtual representations of $\mathrm{Gal}(\overline{\mathbb{Q}}/F)$ in finite dimensional $\mathbb{C}$-vector spaces. As there are only a finite number of π_f such that $\iota_*(W_\lambda)(\pi_f) \neq 0$, we may assume, after replacing K_λ by a finite extension, that there exist virtual representations $W_\lambda(\pi_f)$ of $\mathrm{Gal}(\overline{\mathbb{Q}}/F)$ in finite-dimensional K_λ-vector spaces such that $\iota_*(W_\lambda(\pi_f)) = \iota_*(W_\lambda)(\pi_f)$. So we get

$$W_\lambda = \sum_{\pi_f} W_\lambda(\pi_f)\otimes\pi_f^{\mathrm{K}}.$$

Notation 7.2.1 Let $\mathbf{H}$ be a connected reductive group over $\mathbb{Q}$ and ξ be a quasi-character of $\mathbf{A}_H(\mathbb{R})^0$. We write $\Pi(\mathbf{H}(\mathbb{A}),\xi)$ for the set of isomorphism classes of irreducible admissible representations of $\mathbf{H}(\mathbb{A})$ on which $\mathbf{A}_H(\mathbb{R})^0$ acts by ξ. For every $\pi \in \Pi(\mathbf{H}(\mathbb{A}),\xi)$, let $m_{\mathrm{disc}}(\pi)$ be the multiplicity of π in the discrete part of $L^2(\mathbf{H}(\mathbb{Q})\setminus\mathbf{H}(\mathbb{A}),\xi)$ (cf. [A6], §2).

Let ξ_G be the quasi-character by which the group $\mathbf{A}_G(\mathbb{R})^0$ acts on the contragredient of V.

For every $(\mathbf{H}, s, \eta_0) \in \mathcal{E}(\mathbf{G})$, fix an L-morphism $\eta : {}^L\mathbf{H} \longrightarrow {}^L\mathbf{G}$ extending η_0 as in proposition 2.3.2. Let $\mathcal{E}^0(\mathbf{G})$ be the set of $(\mathbf{H}, s, \eta_0) \in \mathcal{E}(\mathbf{G})$ such that $\mathbf{H}$ is not an inner form of $\mathbf{G}$. If $n_1, \dots, n_r \in \mathbb{N}^*$ and $\mathbf{H} = \mathbf{G}(\mathbf{U}^*(n_1) \times \dots \times \mathbf{U}^*(n_r))$, we define in the same way a subset $\mathcal{E}^0(\mathbf{H})$ of $\mathcal{E}(\mathbf{H})$ and fix, for every $(\mathbf{H}', s, \eta_0) \in \mathcal{E}(\mathbf{H})$, an L-morphism $\eta : {}^L\mathbf{H}' \longrightarrow {}^L\mathbf{H}$ extending η_0 as in proposition 2.3.2.

Let $\mathcal{F}_\mathbf{G}$ be the set of sequences $(e_1, \dots, e_r)$ of variable length $r \in \mathbb{N}^*$, with $e_1 = (\mathbf{H}_1, s_1, \eta_1) \in \mathcal{E}^0_\mathbf{G}$ and, for every $i \in \{2, \dots, r\}$, $e_i = (\mathbf{H}_i, s_i, \eta_i) \in \mathcal{E}^0_{\mathbf{H}_{i-1}}$.

Let $\underline{e} = (e_1, \dots, e_r) \in \mathcal{F}_\mathbf{G}$. Write $e_i = (\mathbf{H}_i, s_i, \eta_i)$ and $\mathbf{H}_0 = \mathbf{G}$. Set $\ell(\underline{e}) = r$, $\mathbf{H}_{\underline{e}} = \mathbf{H}_r$, $\eta_{\underline{e}} = \eta_1 \circ \dots \circ \eta_r : {}^L\mathbf{H}_{\underline{e}} \longrightarrow {}^L\mathbf{G}$,

$$\iota(\underline{e}) = \iota_{\mathbf{G},\mathbf{H}_1} \iota_{\mathbf{H}_1,\mathbf{H}_2} \dots \iota_{\mathbf{H}_{r-1},\mathbf{H}_r}$$

and

$$\iota'(\underline{e}) = \iota(\mathbf{G}, \mathbf{H}_1) \iota_{\mathbf{H}_1,\mathbf{H}_2} \dots \iota_{\mathbf{H}_{r-1},\mathbf{H}_r}.$$

For every finite set S of prime numbers, write $\mathbb{A}_S = \prod_{p \in S} \mathbb{Q}_p$ and $\mathbb{A}_f^S = \prod'_{p \notin S} \mathbb{Q}_p$; if $\pi_f = \bigotimes' \pi_p$ is an irreducible admissible representation of $\mathbf{G}(\mathbb{A}_f)$, write $\pi_S = \bigotimes_{p \in S} \pi_p$ and $\pi^S = \bigotimes'_{p \notin S} \pi_p$; if $\mathbf{G}$ is unramified at every $p \notin S$, write $\mathrm{K}^S = \prod_{p \notin S} \mathbf{G}(\mathbb{Z}_p)$. If $f^S \in C_c^\infty(\mathbf{G}(\mathbb{A}_f^S))$ and $f_S \in C_c^\infty(\mathbf{G}(\mathbb{A}_S))$, define functions $(f^S)^{\underline{e}} \in C_c^\infty(\mathbf{H}_{\underline{e}}(\mathbb{A}_f^S))$ and $(f_S)^{\underline{e}} \in C_c^\infty(\mathbf{H}_{\underline{e}}(\mathbb{A}_S))$ by

$$(f^S)^{\underline{e}} = (\dots((f^S)^{\mathbf{H}_1})^{\mathbf{H}_2} \dots)^{\mathbf{H}_r},$$

$$(f_S)^{\underline{e}} = (\dots((f_S)^{\mathbf{H}_1})^{\mathbf{H}_2} \dots)^{\mathbf{H}_r}.$$

Define a function $f_\infty^{\underline{e}}$ on $\mathbf{H}_{\underline{e}}(\mathbb{R})$ by

$$f_\infty^{\underline{e}} = (\dots((f_\infty)_{\mathbf{H}_1})_{\mathbf{H}_2} \dots)_{\mathbf{H}_r},$$

where $f_\infty = (-1)^{q(\mathbf{G})} f_\varphi$ (this function is defined in section 6.2). The function $f_\infty^{\underline{e}}$ is stable cuspidal by definition.

Let $k \in \{1, \dots, r\}$. Consider the morphism

$$\varphi_k : W_\mathbb{R} \xrightarrow{j} {}^L\mathbf{H}_{k,\mathbb{R}} \xrightarrow{\eta_\infty} {}^L\mathbf{H}_{k-1,\mathbb{R}} \xrightarrow{p} {}^L(\mathbf{A}_{H_{k-1}})_\mathbb{R},$$

where j is the obvious inclusion, η_∞ is induced by η_k and p is the dual of the inclusion $\mathbf{A}_{H_{k-1}} \longrightarrow \mathbf{H}_{k-1}$. The morphism φ_k is the Langlands parameter of a quasi-character on $\mathbf{A}_{H_{k-1}}(\mathbb{R})$, and we write χ_k for the restriction of this quasi-character to $\mathbf{A}_{H_{k-1}}(\mathbb{R})^0$. As $\mathbf{A}_{H_r} = \dots = \mathbf{A}_{H_1} = \mathbf{A}_G$ (because $(\mathbf{H}_k, s_k, \eta_{k,0})$ is an elliptic endoscopic datum for $\mathbf{H}_{k-1}$ for every $k \in \{1, \dots, r\}$), we may define a quasi-character $\xi_{\underline{e}}$ on $\mathbf{A}_{H_{\underline{e}}}(\mathbb{R})^0$ by the formula

$$\xi_{\underline{e}} = \xi_\mathbf{G} \chi_1^{-1} \dots \chi_r^{-1}.$$

This quasi-character satisfies the following property: if $\varphi_{H_{\underline{e}}} : W_\mathbb{R} \longrightarrow {}^L\mathbf{H}_{\underline{e},\mathbb{R}}$ is a Langlands parameter corresponding to an L-packet of representations of $\mathbf{H}_{\underline{e}}(\mathbb{R})$ with central character $\xi_{\underline{e}}$ on $\mathbf{A}_{H_{\underline{e}}}(\mathbb{R})^0$, then $\eta_{\underline{e}} \circ \varphi_{H_{\underline{e}}} : W_\mathbb{R} \longrightarrow {}^L\mathbf{G}_\mathbb{R}$ corresponds to an L-packet of representations of $\mathbf{G}(\mathbb{R})$ with central character $\xi_\mathbf{G}$ on $\mathbf{A}_G(\mathbb{R})^0$. (This is the construction of [K13] 5.5.) Write $\Pi_{\underline{e}} = \Pi(\mathbf{H}_{\underline{e}}(\mathbb{A}), \xi_{\underline{e}})$. Let $R_{\underline{e}}(V)$ be the set of $\pi_\infty \in \Pi(\mathbf{H}_{\underline{e}}(\mathbb{R}))$ such that there exists an elliptic Langlands parameter

$\varphi_{\underline{e}}\colon W_{\mathbb{R}} \longrightarrow {}^L\mathbf{H}_{\underline{e},\mathbb{R}}$ satisfying the following properties: $\eta_{\underline{e}} \circ \varphi_{\underline{e}}$ is $\widehat{\mathbf{G}}$-conjugate to φ, and $\mathrm{Tr}(\pi_\infty(f_{\varphi_{\underline{e}}})) \neq 0$ (remember that $f_{\varphi_{\underline{e}}}$ is defined in 6.2). Then $R_{\underline{e}}(V)$ is finite.

If p is a prime number unramified in E, let $\eta_{\underline{e},p} = \eta_{\underline{e}|\widehat{\mathbf{H}}_{\underline{e}} \rtimes W_{\mathbb{Q}_p}}$ and write $\eta_{\underline{e},p,\mathrm{simple}} : {}^L\mathbf{H}_{\underline{e}} \longrightarrow {}^L\mathbf{G}$ for the L-morphism extending $\eta_{1,0} \circ \cdots \circ \eta_{r,0}$ and equal to the composition of the analogs of the morphism η_{simple} of the last two subsections of 4.2. Write $\eta_{\underline{e},p} = c\eta_{\underline{e},p,\mathrm{simple}}$, where $c : W_{\mathbb{Q}_p} \longrightarrow Z(\widehat{\mathbf{H}}_{\underline{e}})$ is a 1-cocycle. Let $\chi_{\underline{e},p} = \chi_{\eta_{\underline{e},p}}$ be the quasi-character of $\mathbf{H}_{\underline{e}}(\mathbb{Q}_p)$ corresponding to the class of c in $\mathrm{H}^1(W_{\mathbb{Q}_p}, Z(\widehat{\mathbf{H}}_{\underline{e}}))$.

Suppose that $(\mathbf{H}_1, s_1, \eta_{1,0})$ is the elliptic endoscopic triple for $\mathbf{G}$ defined by a pair $(n^+, n^-) \in \mathbb{N}^2$ as in proposition 2.3.1 (so $n = n^+ + n^-$ and n^- is even). Write

$$\mathbf{H}_{\underline{e}} = \mathbf{G}(\mathbf{U}^*(n_1^+) \times \cdots \times \mathbf{U}^*(n_r^+) \times \mathbf{U}^*(n_1^-) \times \cdots \times \mathbf{U}^*(n_s^-)),$$

where the identification is chosen such that $\eta_2 \circ \cdots \circ \eta_r$ sends $\widehat{\mathbf{U}^*(n_1^+)} \times \cdots \times \widehat{\mathbf{U}^*(n_r^+)}$ (resp., $\widehat{\mathbf{U}^*(n_1^-)} \times \cdots \times \widehat{\mathbf{U}^*(n_s^-)}$) in $\widehat{\mathbf{U}^*(n^+)}$ (resp., $\widehat{\mathbf{U}^*(n^-)}$).

If $p_1^+, \dots, p_r^+, p_1^-, \dots, p_s^- \in \mathbb{N}$ are such that $1 \leq p_i^+ \leq n_i^+$ and $1 \leq p_j^- \leq n_j^-$ for every $i \in \{1, \dots, r\}$ and $j \in \{1, \dots, s\}$, write

$$\mu = \mu_{p_1^+,\dots,p_r^+,p_1^-,\dots,p_s^-} = (\mu_{p_1^+}, \dots, \mu_{p_r^+}, \mu_{p_1^-}, \dots, \mu_{p_s^-}) : \mathbb{G}_{m,E} \longrightarrow \mathbf{H}_{\underline{e},E}$$

(cf. 2.1.2 for the definition of μ_p), and

$$s(\mu) = (-1)^{p_1^- + \cdots + p_s^-}.$$

Let $M_{\underline{e}}$ be the set of cocharacters $\mu_{p_1^+,\dots,p_r^+,p_1^-,\dots,p_s^-}$ with $p = p_1^+ + \cdots + p_r^+ + p_1^- + \cdots + p_s^-$. For every $\mu \in M_{\underline{e}}$ and every finite place $\wp$ of F where $\mathbf{H}_{\underline{e}}$ is unramified, we get a representation $r_{-\mu}$ of ${}^L\mathbf{H}_{\underline{e},F_\wp}$, defined in 4.1.1.

For every irreducible admissible representation $\pi_{\underline{e},f}$ of $\mathbf{H}_{\underline{e}}(\mathbb{A}_f)$, let

$$c_{\underline{e}}(\pi_{\underline{e}}) = \sum_{\substack{\pi_{\underline{e},\infty} \in \Pi(\mathbf{H}_{\underline{e}}(\mathbb{R})),\\ \pi_{\underline{e},f} \otimes \pi_{\underline{e},\infty} \in \Pi_{\underline{e}}}} m_{\mathrm{disc}}(\pi_{\underline{e},f} \otimes \pi_{\underline{e},\infty}) \, \mathrm{Tr}(\pi_{\underline{e},\infty}(f_\infty^{\underline{e}}))$$

(as $\mathrm{Tr}(\pi_{\underline{e},\infty}(f_\infty^{\underline{e}})) = 0$ unless $\pi_{\underline{e},\infty} \in R_{\underline{e}}(V)$, this sum has only a finite number of nonzero terms).

Write $\Pi_{\mathbf{G}} = \Pi(\mathbf{G}(\mathbb{A}), \xi_G)$. For every irreducible admissible representation π_f of $\mathbf{G}(\mathbb{A}_f)$, let

$$c_{\mathbf{G}}(\pi_f) = \sum_{\substack{\pi_\infty \in \Pi(\mathbf{G}(\mathbb{R})),\\ \pi_\infty \otimes \pi_f \in \Pi_{\mathbf{G}}}} m_{\mathrm{disc}}(\pi_f \otimes \pi_\infty) \, \mathrm{Tr}(\pi_\infty(f_\infty))$$

(this sum has only a finite number of non-zero terms because there are only finitely many π_∞ in $\Pi(\mathbf{G}(\mathbb{R}))$ such that $\mathrm{Tr}(\pi_\infty(f_\infty)) \neq 0$). Remember that there is a cocharacter $\mu_G : \mathbb{G}_{m,E} \longrightarrow \mathbf{G}_E$ associated to the Shimura datum (cf. 2.1); this cocharacter gives a representation $r_{-\mu_G}$ of ${}^L\mathbf{G}_{F_\wp}$, for every finite place $\wp$ of F where $\mathbf{G}$ is unramified.

Let $\pi_f = \bigotimes'_p \pi_p$ be an irreducible admissible representation of $\mathbf{G}(\mathbb{A}_f)$ such that $\pi_f^{\mathrm{K}} \neq 0$, and let $\underline{e} \in \mathcal{F}_{\mathbf{G}}$. Write $R_{\underline{e}}(\pi_f)$ for the set of equivalence classes of irreducible admissible representations $\pi_{\underline{e},f} = \bigotimes'_p \pi_{\underline{e},p}$ of $\mathbf{H}_{\underline{e}}(\mathbb{A}_f)$ such that, for almost every prime number p where π_f and $\pi_{\underline{e},f}$ are unramified, the morphism

$\eta_{\underline{e}} : {}^L\mathbf{H}_{\underline{e}} \longrightarrow {}^L\mathbf{G}$ sends a Langlands parameter of $\pi_{\underline{e},p}$ to a Langlands parameter of π_p.

Let p be a prime number. Remember that we fixed embeddings $F \subset \overline{\mathbb{Q}} \subset \overline{\mathbb{Q}}_p$, that determine a place $\wp$ of F above p and a morphism $\mathrm{Gal}(\overline{\mathbb{Q}}_p/F_\wp) \longrightarrow \mathrm{Gal}(\overline{\mathbb{Q}}/F)$. Let $\Phi_\wp \in \mathrm{Gal}(\overline{\mathbb{Q}}_p/F_\wp)$ be a lift of the geometric Frobenius, and use the same notation for its image in $\mathrm{Gal}(\overline{\mathbb{Q}}/F)$. If $\mathbf{H}$ is a reductive unramified group over $\mathbb{Q}_p$ and π_p is an unramified representation of $\mathbf{H}(\mathbb{Q}_p)$, denote by $\varphi_{\pi_p} : W_{\mathbb{Q}_p} \longrightarrow {}^L\mathbf{H}_{\mathbb{Q}_p}$ a Langlands parameter of π_p.

Theorem 7.2.2 *Let π_f be an irreducible admissible representation of $\mathbf{G}(\mathbb{A}_f)$ such that $\pi_f^{\mathrm{K}} \neq 0$. Then there exists a function $f^\infty \in C_c^\infty(\mathbf{G}(\mathbb{A}_f))$ such that, for almost every prime number p and for every $m \in \mathbb{Z}$,*

$$\begin{aligned}\mathrm{Tr}(\Phi_\wp^m, W_\lambda(\pi_f)) &= (N\wp)^{md/2} c_{\mathbf{G}}(\pi_f)\dim(\pi_f^{\mathrm{K}})\,\mathrm{Tr}(r_{-\mu_G}\circ\varphi_{\pi_p}(\Phi_\wp^m))\\ &+(N\wp)^{md/2}\sum_{\underline{e}\in\mathcal{F}_{\mathbf{G}}}(-1)^{\ell(\underline{e})}\iota(\underline{e})\sum_{\pi_{\underline{e},f}\in R_{\underline{e}}(\pi_f)} c_{\underline{e}}(\pi_{\underline{e},f})\,\mathrm{Tr}(\pi_{\underline{e}}((f^\infty)^{\underline{e}}))\\ &\times\sum_{\mu\in M_{\underline{e}}}\Big(1-(-1)^{s(\mu)}\frac{\iota'(\underline{e})}{\iota(\underline{e})}\Big)\,\mathrm{Tr}(r_{-\mu}\circ\varphi_{\pi_{\underline{e},p}\otimes\chi_{\underline{e},p}}(\Phi_\wp^m)),\end{aligned}$$

where the second sum in the right-hand side is taken only over those $\pi_{\underline{e},f}$ such that $\pi_{\underline{e},p}\otimes\chi_{\underline{e},p}$ is unramified, $d = \dim(M^{\mathrm{K}}(\mathbf{G},\mathcal{X}))$ and $N\wp = \#(\mathcal{O}_{F_\wp}/\wp)$.

Remark 7.2.3 The lack of control over the set of "good" prime numbers in the theorem above comes from the fact that we do not have a strong multiplicity one theorem for $\mathbf{G}$ (and not from a lack of information about the integral models of Shimura varieties). If π_f extends to an automorphic representation of $\mathbf{G}(\mathbb{A})$ whose base change to $\mathbf{G}(\mathbb{A}_E)$ is cuspidal (cf. section 8.5), then it is possible to do better by using corollary 8.5.3.

Proof. It is enough to prove the equality of the theorem for m big enough (where the meaning of "big enough" can depend on p).

Let R' be the set of isomorphism classes of irreducible admissible representations π'_f of $\mathbf{G}(\mathbb{A}_f)$ satisfying the following properties:

- $\pi'_f \not\simeq \pi_f$;
- $(\pi'_f)^{\mathrm{K}} \neq 0$;
- $W_\lambda(\pi'_f) \neq 0$ or $c_{\mathbf{G}}(\pi'_f) \neq 0$.

Then R' is finite, so there exists $h \in \mathcal{H}_{\mathrm{K}} = \mathcal{H}(\mathbf{G}(\mathbb{A}_f),\mathrm{K})$ such that $\mathrm{Tr}(\pi_f(h)) = \mathrm{Tr}(\pi_f(\mathbb{1}_{\mathrm{K}}))$ and $\mathrm{Tr}(\pi'_f(h)) = 0$ for every $\pi'_f \in R'$.

Let T be a finite set of prime numbers such that all the representations in R' are unramified outside T, $\mathbf{G}$ is unramified at every $p \notin T$, $\mathrm{K} = \mathrm{K}_T\mathrm{K}^T$ with $\mathrm{K}_T \subset \mathbf{G}(\mathbb{A}_T)$, and $h = h_T\mathbb{1}_{\mathrm{K}^T}$ with $h_T \in \mathcal{H}(\mathbf{G}(\mathbb{A}_T),\mathrm{K}_T)$. Then, for every function g^T in $\mathcal{H}(\mathbf{G}(\mathbb{A}_f^T),\mathrm{K}^T)$, $\mathrm{Tr}(\pi_f(h_Tg^T)) = \mathrm{Tr}(\pi_T(\mathbb{1}_{\mathrm{K}_T}))\,\mathrm{Tr}(\pi^T(g^T))$ and $\mathrm{Tr}(\pi'_f(h_Tg^T)) = 0$ if $\pi'_f \in R'$.

For every $\underline{e} \in \mathcal{F}_{\mathbf{G}}$, let $R'_{\underline{e}}$ be the set of isomorphism classes of irreducible admissible representations ρ_f of $\mathbf{H}_{\underline{e}}(\mathbb{A}_f)$ such that $\rho_f \notin R_{\underline{e}}(\pi_f)$, $\mathrm{Tr}(\rho_f(h^{\underline{e}})) \neq 0$

and $c_{\underline{e}}(\rho_f) \neq 0$. As $\mathcal{F}_{\mathbf{G}}$ is finite and $R'_{\underline{e}}$ is finite for every $\underline{e} \in \mathcal{F}_{\mathbf{G}}$, there exists $g^T \in \mathcal{H}(\mathbf{G}(\mathbb{A}_f^T), \mathrm{K}^T)$ such that

- $\mathrm{Tr}(\pi^T(g^T)) = 1$;
- for every $\underline{e} \in \mathcal{F}_{\mathbf{G}}$ and $\rho_f \in R'_{\underline{e}}$, if k^T is the function on $\mathbf{H}_{\underline{e}}(\mathbb{A}_f^T)$ obtained from g^T by the base change morphism associated to $\eta_{\underline{e}}$, then $\mathrm{Tr}(\rho^T(k^T)) = 0$.

Let $S \supset T$ be a finite set of prime numbers such that $g^T = g_{S-T}\mathbb{1}_{\mathrm{K}^S}$, with g_{S-T} a function on $\mathbf{G}(\mathbb{A}_{S-T})$. Set

$$f^\infty = h_T g^T.$$

Let $p \notin S$ be a prime number big enough for corollary 6.3.2 to be true. Then $f^\infty = f^{\infty,p}\mathbb{1}_{\mathbf{G}(\mathbb{Z}_p)}$, and there are functions $(f^{\infty,p})^{\underline{e}}$ and $(f^\infty)^{\underline{e}}$, defined as above, for every $\underline{e} \in \mathcal{F}_{\mathbf{G}}$.

Let $m \in \mathbb{Z}$. Consider the functions

$$f^{(m)} = f^{\infty,p} f_p^{(m)} f_\infty \in C_c^\infty(\mathbf{G}(\mathbb{A}_f^p)) C_c^\infty(\mathbf{G}(\mathbb{Q}_p)) C^\infty(\mathbf{G}(\mathbb{R}))$$

and

$$f_{\mathbf{H}}^{(m)} = (f^{\infty,p})^{\mathbf{H}} f_{\mathbf{H},p}^{(m)} f_{\mathbf{H},\infty} \in C_c^\infty(\mathbf{H}(\mathbb{A}_f^p)) C_c^\infty(\mathbf{H}(\mathbb{Q}_p)) C^\infty(\mathbf{H}(\mathbb{R}))$$

for every $(\mathbf{H}, s, \eta_0) \in \mathcal{E}(\mathbf{G})$, where

- $f_p^{(m)} \in \mathcal{H}(\mathbf{G}(\mathbb{Q}_p), \mathbf{G}(\mathbb{Z}_p))$ is the function obtained by base change from the function $\phi_m^{\mathbf{G}}$ of theorem 1.6.1;
- $f_{\mathbf{H},p}^{(m)} \in C_c^\infty(\mathbf{H}(\mathbb{Q}_p))$ is the function obtained by twisted transfer from $\phi_m^{\mathbf{G}}$.

Then, by corollary 6.3.2 and the choice of f^∞, for m big enough,

$$\mathrm{Tr}(\Phi_\wp^m f^\infty, W_\lambda(\pi_f)) = \mathrm{Tr}(\Phi_\wp^m f^\infty, W_\lambda) = \sum_{(\mathbf{H},s,\eta_0)\in\mathcal{E}(\mathbf{G})} \iota(\mathbf{G},\mathbf{H}) ST^H(f_{\mathbf{H}}^{(m)}).$$

By proposition 7.1.4 and the fact that $f_{\mathbf{H}}^{(m)}$ is simply a transfer of $f^{(m)}$ if $\mathbf{H} = \mathbf{G}^*$ (the quasi-split inner form of $\mathbf{G}$), we get

$$\mathrm{Tr}(\Phi_\wp^m f^\infty, W_\lambda(\pi_f)) = T^G(f^{(m)}) + \sum_{\underline{e}\in\mathcal{F}_{\mathbf{G}}} (-1)^{\ell(\underline{e})} \iota(\underline{e}) T^{H_{\underline{e}}}(f^{(m),\underline{e}})$$
$$+ \sum_{\underline{e}\in\mathcal{F}_{\mathbf{G}}} (-1)^{\ell(\underline{e})-1} \iota'(\underline{e}) T^{H_{\underline{e}}}(f_{\underline{e}}^{(m)}),$$

where, for every $\underline{e} = ((\mathbf{H}_1, s_1, \eta_1), \dots, (\mathbf{H}_r, s_r, \eta_r)) \in \mathcal{F}_{\mathbf{G}}$, we write $\mathbf{H}_{\underline{e}} = \mathbf{H}_r$,

$$f^{(m),\underline{e}} = (f^{\infty,p})^{\underline{e}} (f_p^{(m)})^{\underline{e}} f_\infty^{\underline{e}}$$

and

$$f_{\underline{e}}^{(m)} = (f^{\infty,p})^{\underline{e}} f_{\underline{e},p}^{(m)} f_\infty^{\underline{e}},$$

with

$$f_{\underline{e},p}^{(m)} = (\dots (f_{\mathbf{H}_1,p}^{(m)})^{\mathbf{H}_2} \dots)^{\mathbf{H}_r}.$$

By the calculation of [A6] pp. 267–268:

$$T^G(f^{(m)}) = \sum_{\rho\in\Pi_{\mathbf{G}}} m_{\mathrm{disc}}(\rho)\,\mathrm{Tr}(\rho(f^{(m)}))$$

$$T^{H_{\underline{e}}}(f^{(m),\underline{e}}) = \sum_{\rho\in\Pi_{\underline{e}}} m_{\mathrm{disc}}(\rho)\,\mathrm{Tr}(\rho(f^{(m),\underline{e}}))$$

$$T^{H_{\underline{e}}}(f_{\underline{e}}^{(m)}) = \sum_{\rho\in\Pi_{\underline{e}}} m_{\mathrm{disc}}(\rho)\,\mathrm{Tr}(\rho(f_{\underline{e}}^{(m)})).$$

Let $\underline{e} = ((\mathbf{H}_1, s_1, \eta_1), \dots, (\mathbf{H}_r, s_r, \eta_r)) \in \mathcal{F}_{\mathbf{G}}$ and $\rho = \rho^{\infty,p} \otimes \rho_p \otimes \rho_\infty = \rho_f \otimes \rho_\infty \in \Pi_{\underline{e}}$. Then

$$\mathrm{Tr}(\rho(f^{(m),\underline{e}})) = \mathrm{Tr}(\rho^{\infty,p}((f^{\infty,p})^{\underline{e}}))\,\mathrm{Tr}(\rho_p(f_p^{(m)})^{\underline{e}}))\,\mathrm{Tr}(\rho_\infty(f_\infty^{\underline{e}})).$$

As $\chi_{\underline{e},p}^{-1}(f_p^{(m)})^{\underline{e}} \in \mathcal{H}(\mathbf{H}_r(\mathbb{Q}_p), \mathbf{H}_r(\mathbb{Z}_p))$, the trace above is 0 unless $\rho_p \otimes \chi_{\underline{e},p}$ is unramified. So

$$\mathrm{Tr}(\rho(f^{(m),\underline{e}})) = \mathrm{Tr}(\rho_f((f^\infty)^{\underline{e}}))\,\mathrm{Tr}(\rho_p((f_p^{(m)})^{\underline{e}}))\,\mathrm{Tr}(\rho_\infty(f_\infty^{\underline{e}})),$$

because both sides are zero if unless $\rho_p \otimes \chi_{\underline{e},p}$ is unramified and, if $\rho_p \otimes \chi_{\underline{e},p}$ is unramified, then $\mathrm{Tr}(\rho_p(\chi_{\underline{e},p}\mathbb{1}_{\mathbf{H}_r(\mathbb{Z}_p)})) = \dim((\rho_p \otimes \chi_{\underline{e},p})^{\mathbf{H}_r(\mathbb{Z}_p)}) = 1$ (and, by the fundamental lemma, we may assume that $(f_p^{(m)})^{\underline{e}}$ is equal to $\chi_{\underline{e},p}\mathbb{1}_{\mathbf{H}_{\underline{e}}(\mathbb{Z}_p)}$). Assume that $\rho_p \otimes \chi_{\underline{e},p}$ is unramified, and let $\varphi_{\rho_p\otimes\chi_{\underline{e},p}} : W_{\mathbb{Q}_p} \longrightarrow {}^L\mathbf{H}_{r,\mathbb{Q}_p}$ be a Langlands parameter of $\rho_p\otimes\chi_{\underline{e},p}$. Then, by proposition 4.2.1 and the calculation of the transfer of a function in the spherical Hecke algebra in section 4.2, we get

$$\mathrm{Tr}(\rho_p((f_p^{(m)})^{\underline{e}})) = (N\wp)^{md/2} \sum_{\mu\in M_{\underline{e}}} \mathrm{Tr}(r_{-\mu} \circ \varphi_{\rho_p\otimes\chi_{\underline{e},p}}(\Phi_\wp^m)).$$

Similarly, using the calculation of the twisted transfer in section 4.2, we see that $\mathrm{Tr}(\rho(f_{\underline{e}}^{(m)}))$ is equal to 0 if $\rho_p \otimes \chi_{\underline{e},p}$ is ramified, and to

$$\mathrm{Tr}(\rho_f((f^\infty)^{\underline{e}}))\,\mathrm{Tr}(\rho_\infty(f_\infty^{\underline{e}}))(N\wp)^{md/2} \sum_{\mu\in M_{\underline{e}}} (-1)^{s(\mu)}\,\mathrm{Tr}(r_{-\mu} \circ \varphi_{\rho_p\otimes\chi_{\underline{e},p}}(\Phi_\wp^m))$$

if $\rho_p \otimes \chi_{\underline{e},p}$ is unramified.

Moreover, by the choice of f^∞, if $\rho_f \notin R_{\underline{e}}(\pi_f)$, then

$$c_{\underline{e}}(\rho_f)\,\mathrm{Tr}(\rho_f((f^\infty)^{\underline{e}})) = 0.$$

A similar (but simpler) calculation shows that, for every $\rho = \rho_f \otimes \rho_\infty \in \Pi_{\mathbf{G}}$: $c_{\mathbf{G}}(\rho_f)\,\mathrm{Tr}(\rho_f(f^{\infty,p} f_p^{(m)})) = 0$ if ρ is ramified at p or if $\rho_f \not\simeq \pi_f$, and, if $\rho_f \simeq \pi_f$ (so ρ is unramified at p), then

$$\mathrm{Tr}(\rho(f^{(m)})) = \dim(\pi_f^{\mathrm{K}})\,\mathrm{Tr}(\rho_\infty(f_\infty))(N\wp)^{md/2}\,\mathrm{Tr}(r_{-\mu_{\mathbf{G}}} \circ \varphi_{\pi_p}(\Phi_\wp^m)).$$

These calculations imply the equality of the theorem. □

Remark 7.2.4 Take any f^∞ in $C_c^\infty(\mathbf{G}(\mathbb{A}_f))$. Then the calculations in the proof of the theorem show that, for every prime number p unramified in E and such that $f^\infty = f^{\infty,p}\mathbb{1}_{\mathbf{G}(\mathbb{Z}_p)}$, and for every $m \in \mathbb{Z}$,

$$\sum_{(\mathbf{H},s,\eta_0)\in\mathcal{E}(\mathbf{G})} \iota(\mathbf{G},\mathbf{H})ST^H(f_{\mathbf{H}}^{(m)})$$
$$= (N\wp)^{md/2}\sum_{\pi_f} c_{\mathbf{G}}(\pi_f)\,\mathrm{Tr}(\pi_f(f^\infty))\,\mathrm{Tr}(r_{-\mu_G}\circ\varphi_{\pi_p}(\Phi_\wp^m))$$
$$+(N\wp)^{md/2}\sum_{\underline{e}\in\mathcal{F}_{\mathbf{G}}}(-1)^{\ell(\underline{e})}\iota(\underline{e})\sum_{\pi_{\underline{e},f}} c_{\underline{e}}(\pi_{\underline{e},f})\,\mathrm{Tr}(\pi_{\underline{e},f}((f^\infty)^{\underline{e}}))$$
$$\times\sum_{\mu\in M_{\underline{e}}}\left(1-(-1)^{s(\mu)}\frac{\iota'(\underline{e})}{\iota(\underline{e})}\right)\mathrm{Tr}(r_{-\mu}\circ\varphi_{\pi_{\underline{e},p}\otimes\chi_{\underline{e},p}}(\Phi_\wp^m)),$$

where the first (resp., third) sum on the right-hand side is taken over the set of isomorphism classes of irreducible admissible respresentations π_f (resp., $\pi_{\underline{e},f}$) of $\mathbf{G}(\mathbb{A}_f)$ (resp., $\mathbf{H}_{\underline{e}}(\mathbb{A}_f)$) such that π_p is unramified (resp., $\pi_{\underline{e},p}\otimes\chi_{\underline{e},p}$ is unramified), and the function $f_{\mathbf{H},p}^{(m)}$ for $m \leq 0$ is defined in definition 7.1.6.

This implies that corollary 6.3.2 is true for every $j \in \mathbb{Z}$, and not just for j big enough (because that corollary can be rewritten as an equality $\sum_{i\in I} c_i\alpha_i^j = \sum_{k\in K} d_k\beta_k^j$, where $(c_i)_{i\in I}$, $(\alpha_i)_{i\in I}$, $(d_k)_{k\in K}$ and $(\beta_k)_{k\in K}$ are finite families of complex numbers). This is the statement of proposition 7.1.7 (if p is inert), but note that proposition 7.1.7 was used in the proof of this remark.

For every $i \in \mathbb{Z}$, consider the representation $\mathrm{H}^i(M^{\mathrm{K}}(\mathbf{G},\mathcal{X})^*_{\overline{\mathbb{Q}}}, IC^{\mathrm{K}}V_{\overline{\mathbb{Q}}})$ of $\mathcal{H}_{\mathrm{K}} \times \mathrm{Gal}(\overline{\mathbb{Q}}/F)$. After making K_λ bigger, we may assume that all the $\mathcal{H}_{\mathrm{K}}$-isotypical components of this representation are defined over K_λ. Write W_λ^i for the semisimplification of this representation, and let

$$W_\lambda^i = \bigoplus_{\pi_f} W_\lambda^i(\pi_f)\otimes\pi_f^{\mathrm{K}}$$

be its isotypical decomposition as a $\mathcal{H}_{\mathrm{K}}$-module (so, as before, the sum is taken over the set of isomorphism classes of irreducible admissible representations π_f of $\mathbf{G}(\mathbb{A}_f)$ such that $\pi_f^{\mathrm{K}} \neq 0$). Of course, $W_\lambda = \sum_{i\in\mathbb{Z}}(-1)^i[W_\lambda^i]$ and $W_\lambda(\pi_f) = \sum_{i\in\mathbb{Z}}(-1)^i[W_\lambda^i(\pi_f)]$ for every π_f.

Then, just as in Kottwitz's article [K10] (see also section 5.2 of Clozel's article [Cl5]), we get the following characterization of the representations π_f that appear in W_λ.

Remark 7.2.5 Let π_f be an irreducible admissible representation of $\mathbf{G}(\mathbb{A}_f)$ such that $\pi_f^{\mathrm{K}} \neq 0$. Then the following conditions are equivalent.

(1) $W_\lambda(\pi_f) \neq 0$.
(2) There exists $i \in \mathbb{Z}$ such that $W_\lambda^i(\pi_f) \neq 0$.
(3) There exist $\pi_\infty \in \Pi(\mathbf{G}(\mathbb{R}))$ and $i \in \mathbb{Z}$ such that $m_{\mathrm{disc}}(\pi_f\otimes\pi_\infty) \neq 0$ and $\mathrm{H}^i(\mathfrak{g}, \mathrm{K}'_\infty; \pi_\infty\otimes V) \neq 0$.

The notation used in condition (3) is that of the proof of lemma 7.3.5. Moreover, all these conditions are implied by

(4) $c_{\mathbf{G}}(\pi_f) \neq 0$.

Assume that, for every $\underline{e} \in \mathcal{F}_{\mathbf{G}}$ and every $\pi_{e,f} \in R_{\underline{e}}(\pi_f)$, $c_{\underline{e}}(\pi_{\underline{e},f}) = 0$. Then (1) implies (4).

Proof. It is obvious that (1) implies (2).

By lemma 3.2 of [K10], there exists a positive integer N such that, for every $\pi_\infty \in \Pi(\mathbf{G}(\mathbb{R}))$,

$$\mathrm{Tr}(\pi_\infty(f_\infty)) = N^{-1} \sum_{i \in \mathbb{Z}} (-1)^i \dim(\mathrm{H}^i(\mathfrak{g}, \mathrm{K}'_\infty; \pi_\infty \otimes V)).$$

This shows in particular that (4) implies (3).

By Matsushima's formula (generalized by Borel and Casselman) and Zucker's conjecture (proved by Looijenga, Saper-Stern, Looijenga-Rapoport), for every $i \in \mathbb{Z}$, there is an isomorphism of $\mathbb{C}$-vector spaces:

$$\iota(W^i_\lambda(\pi_f)) = \bigoplus_{\pi_\infty \in \Pi(\mathbf{G}(\mathbb{R}))} m_{\mathrm{disc}}(\pi_f \otimes \pi_\infty)\, \mathrm{H}^i(\mathfrak{g}, \mathrm{K}'_\infty; \pi_\infty \otimes V)$$

(remember that $\iota : K_\lambda \longrightarrow \mathbb{C}$ is an embedding that was fixed at the begining of this section). This is explained in the proof of lemma 7.3.5. The equivalence of (2) and (3) follows from this formula.

We show that (2) implies (1). This is done just as in section 6 of [K10]. Let m be the weight of V in the sense of section 1.3. Then the local system $\mathcal{F}^{\mathrm{K}}V$ defined by V is pure of weight $-m$ (cf. 1.3), so the intersection complex $IC^{\mathrm{K}}V$ is also pure of weight $-m$. Hence, for every $i \in \mathbb{Z}$, $W^i_\lambda(\pi_f)$ is pure of weight $-m+i$ as a representation of $\mathrm{Gal}(\overline{\mathbb{Q}}/F)$ (i.e., it is unramified and pure of weight $-m+i$ at almost all places of F). In particular, $W^i_\lambda(\pi_f)$ and $W^j_\lambda(\pi_f)$ cannot have isomorphic irreducible subquotients if $i \neq j$, so there are no cancellations in the sum $W_\lambda(\pi_f) = \sum_{i \in \mathbb{Z}} (-1)^i [W^i_\lambda(\pi_f)]$. This shows that (2) implies (1).

We now prove the last statement. By the assumption on π_f and theorem 7.2.2, for almost every prime number p and every $m \in \mathbb{Z}$,

$$\mathrm{Tr}(\Phi^m_\wp, W_\lambda(\pi_f)) = (N\wp)^{md/2} c_{\mathbf{G}}(\pi_f) \dim(\pi_f^{\mathrm{K}}) \,\mathrm{Tr}(r_{-\mu_G} \circ \varphi_{\pi_p}(\Phi^m_\wp)).$$

Fix p big enough for this equality to be true. If $W_\lambda(\pi_f) \neq 0$, then there exists $m \in \mathbb{Z}$ such that $\mathrm{Tr}(\Phi^m_\wp, W_\lambda(\pi_f)) \neq 0$, so $c_{\mathbf{G}}(\pi_f) \neq 0$. □

7.3 APPLICATION TO THE RAMANUJAN-PETERSSON CONJECTURE

We keep the notation of 7.2, but we take here $\mathbf{G} = \mathbf{G}(\mathbf{U}(p_1, q_1) \times \cdots \times \mathbf{U}(p_r, q_r))$, with $p_1, q_1, \ldots, p_r, q_r \in \mathbb{N}$ such that, for every $i \in \{1, \ldots, r\}$, $n_i := p_i + q_i \geq 1$. Assume that, for every $i \in \{1, \ldots, r\}$, if $n_i \geq 2$, then $q_i \geq 1$. Write $n = n_1 + \cdots + n_r$ and $d = \dim M^{\mathrm{K}}(\mathbf{G}, \mathcal{X})$ (so $d = p_1 q_1 + \cdots + p_r q_r$). Let $\mathbf{T}$ be the diagonal torus of $\mathbf{G}$.

Theorem 7.3.1 *Let π_f be an irreducible admissible representation of $\mathbf{G}(\mathbb{A}_f)$ such that there exists an irreducible representation π_∞ of $\mathbf{G}(\mathbb{R})$ with $\mathrm{Tr}(\pi_\infty(f_\infty)) \neq 0$ and $m_{\mathrm{disc}}(\pi_f \otimes \pi_\infty) \neq 0$. For every prime number p where π_f is unramified, let*

$$(z^{(p)}, ((z^{(p)}_{1,1}, \dots, z^{(p)}_{1,n_1}), \dots, (z^{(p)}_{r,1}, \dots, z^{(p)}_{r,n_r}))) \in \widehat{\mathbf{T}}^{\mathrm{Gal}(\overline{\mathbb{Q}}_p/\mathbb{Q}_p)}$$

be the Langlands parameter of π_p. Assume that V is pure of weight 0 in the sense of section 1.3 (i.e., $\mathbb{G}_m$, seen as a subgroup of the center of $\mathbf{G}$, acts trivially on V). Then, for every p where π_f is unramified,

$$|z^{(p)}| = |z^{(p)}_{1,1} \dots z^{(p)}_{1,n_1}| = \dots = |z^{(p)}_{r,1} \dots z^{(p)}_{r,n_r}| = 1.$$

Moreover:

(i) Assume that the highest weight of V is regular. Then, for p big enough, for every $i \in \{1, \dots, r\}$ and $j \in \{1, \dots, n_i\}$, $|z^{(p)}_{i,j}| = 1$.

(ii) Assume that $r = 1$, that $W_\lambda(\pi_f) \neq 0$ and that, for every $\underline{e} \in \mathcal{F}_{\mathbf{G}}$ and every $\pi_{e,f} \in R_{\underline{e}}(\pi_f)$, $c_{\underline{e}}(\pi_{\underline{e},f}) = 0$. Then, for p big enough,

- *if p splits in E, then, for every $j \in \{1, \dots, n_1\}$, $\log_p |z^{(p)}_{1,j}| \in \frac{1}{\gcd(p_1,q_1)}\mathbb{Z}$;*
- *if p is inert in E, then, for every $j \in \{1, \dots, n_1\}$, $\log_p |z^{(p)}_{1,j}| \in \frac{1}{\gcd(2,p_1,q_1)}\mathbb{Z}$.*

Proof. Let K be a neat open compact subgroup of $\mathbf{G}(\mathbb{A}_f)$ such that $\pi_f^{\mathrm{K}} \neq \{0\}$.

The center Z of $\mathbf{G}$ is isomorphic in an obvious way to $\mathbf{G}(\mathbf{U}(1)^r)$. As π_f is irreducible, $Z(\mathbb{A}_f)$ acts on the space of π_f by a character $\chi : Z(\mathbb{A}_f) \longrightarrow \mathbb{C}^\times$, that is unramified wherever π_f is and trivial on $\mathrm{K} \cap Z(\mathbb{A}_f)$. The character χ is also trivial on $Z(\mathbb{Q})$, because there exists a representation π_∞ of $\mathbf{G}(\mathbb{R})$ such that $\pi_\infty \otimes \pi_f$ is a direct factor of $L^2(\mathbf{G}(\mathbb{Q}) \backslash \mathbf{G}(\mathbb{A}), 1)$ (where 1 is the trivial character of $\mathbf{A}_G(\mathbb{R})^0$, i.e., the character by which $\mathbf{A}_G(\mathbb{R})^0$ acts on V). Hence χ is trivial on $Z(\mathbb{Q})(\mathrm{K} \cap Z(\mathbb{A}_f))$; as $Z(\mathbb{Q})(\mathrm{K} \cap Z(\mathbb{A}_f))$ is a subgroup of finite index of $Z(\mathbb{A}_f)$, χ is of finite order. (As $Z(\mathbb{R})/\mathbf{A}_G(\mathbb{R})^0$ is compact, this implies in particular that the central character of $\pi_\infty \otimes \pi_f$ is unitary.)

Use section 2.3 to identify $\widehat{Z}$ and $\mathbb{C}^\times \times (\mathbb{C}^\times)^r$. For every p where π_f is unramified, let $(y^{(p)}, (y_1^{(p)}, \dots, y_r^{(p)})) \in \widehat{Z}^{\mathrm{Gal}(\overline{\mathbb{Q}}_p/\mathbb{Q}_p)}$ be the Langlands parameter of χ_p. As χ is of finite order, $|y^{(p)}| = |y_1^{(p)}| = \dots = |y_r^{(p)}| = 1$.

The morphism $\widehat{\mathbf{G}} = \mathbb{C}^\times \times \mathbf{GL}_{n_1}(\mathbb{C}) \times \dots \times \mathbf{GL}_{n_r}(\mathbb{C}) \longrightarrow \widehat{Z} = \mathbb{C}^\times \times (\mathbb{C}^\times)^r$, $(z, (g_1, \dots, g_r)) \longmapsto (z, \det(g_1), \dots, \det(g_r))$ is dual to the inclusion $Z \subset \mathbf{G}$. So, for every p where π_f is unramified, $z^{(p)} = y^{(p)}$ and $z^{(p)}_{i,1} \dots z^{(p)}_{i,n_i} = y_i^{(p)}$ for every $i \in \{1, \dots, r\}$. This proves the first statement of the theorem.

We show (i). Assume that the highest weight of V is regular. Let R_∞ be the set of $\pi_\infty \in \Pi(\mathbf{G}(\mathbb{R}))$ such that $\pi_\infty \otimes \pi_f \in \Pi_G$ and $\mathrm{Tr}(\pi_\infty(f_\infty)) \neq 0$. By the proof of lemma 6.2 of [A6], all the representations $\pi_\infty \in \Pi(\mathbf{G}(\mathbb{R}))$ such that $\mathrm{Tr}(\pi_\infty(f_\infty)) \neq 0$ are in the discrete series. So R_∞ is contained in the discrete series L-packet associated to the contragredient of V. In particular, the function $\pi_\infty \longmapsto \mathrm{Tr}(\pi_\infty(f_\infty))$ is constant on R_∞, so

$$c_{\mathbf{G}}(\pi_f) = \sum_{\pi_\infty \in R_\infty} m_{\mathrm{disc}}(\pi_\infty \otimes \pi_f)\,\mathrm{Tr}(\pi_\infty(f_\infty)) \neq 0.$$

We prove the result by induction on the set of $(n_1, \dots, n_r) \in (\mathbb{N}^*)^r$ such that $n_1 + \cdots + n_r = n$, with the ordering: $(n'_1, \dots, n'_{r'}) < (n_1, \dots, n_r)$ if and only if $r' > r$.

Assume first that, for every $\underline{e} \in \mathcal{F}_{\mathbf{G}}$ and every $\pi_{\underline{e},f} \in \Pi_{\underline{e}}(\pi_f)$, $c_{\underline{e}}(\pi_{\underline{e},f}) = 0$. Let p be a prime number big enough for theorem 7.2.2 to be true. Then, for every $m \in \mathbb{Z}$,

$$\mathrm{Tr}(\Phi_{\wp}^m, W_\lambda(\pi_f)) = (N\wp)^{md/2} c_{\mathbf{G}}(\pi_f) \dim(\pi_f^{\mathrm{K}}) \,\mathrm{Tr}(r_{-\mu_G} \circ \varphi_{\pi_p}(\Phi_{\wp}^m)).$$

Let $x_1, \dots, x_s \in \mathbb{C}$ be the eigenvalues of $\Phi_\wp$ acting on $W_\lambda(\pi_f)$, and $a_1, \dots, a_s \in \mathbb{Z}$ be their multiplicities. For every $m \in \mathbb{Z}$,

$$\mathrm{Tr}(\Phi_{\wp}^m, W_\lambda(\pi_f)) = \sum_{i=1}^{s} a_i x_i^m.$$

By lemma 7.3.5, the cohomology of $IC^{\mathrm{K}}V$ is concentrated in degree d. As $IC^{\mathrm{K}}V$ is pure of weight 0 (because V is pure of weight 0), $\log_p |x_1| = \dots \log_p |x_r| = n(\wp)d/2$, where $n(\wp) = \log_p(N\wp)$.

On the other hand, by lemma 7.3.2, for every $m \in \mathbb{Z}$:

$$\begin{aligned}&\mathrm{Tr}(r_{-\mu_G} \otimes \varphi_{\pi_p}(\Phi_{\wp}^{2m}))\\ &\quad= (z^{(p)})^{-2m} \sum_{\substack{J_1 \subset \{1,\dots,n_1\}\\ |J_1| = p_1}} \cdots \sum_{\substack{J_r \subset \{1,\dots,n_r\}\\ |J_r| = p_r}} \prod_{i=1}^{r} \prod_{j \in J_i} (z_{i,j}^{(p)})^{-2m[F_\wp:\mathbb{Q}_p]}.\end{aligned}$$

As $|z^{(p)}| = 1$, this implies that, for all $J_1 \subset \{1, \dots, n_1\}, \dots, J_r \subset \{1, \dots, n_r\}$ such that $|J_1| = p_1, \dots, |J_r| = p_r$,

$$\sum_{i=1}^{r} \sum_{j \in J_i} \log_p |z_{i,j}^{(p)}| = 0.$$

By the first statement of the theorem and lemma 7.3.3, we get $\log_p |z_{i,j}^{(p)}| = 0$, i.e., $|z_{i,j}^{(p)}| = 1$, for every $i \in \{1, \dots, r\}$ and $j \in \{1, \dots, n_i\}$.

Assume now that there exist $\underline{e} \in \mathcal{F}_{\mathbf{G}}$ and $\pi_{\underline{e},f} \in \Pi_{\underline{e}}(\pi_f)$ such that $c_{\underline{e}}(\pi_{\underline{e},f}) \neq 0$. Write (with the notation of section 7.2)

$$\mathbf{H}_{\underline{e}} = \mathbf{G}(\mathbf{U}^*(n'_1) \times \cdots \times \mathbf{U}^*(n'_{r'})).$$

Of course, $(n'_1, \dots, n'_{r'}) < (n_1, \dots, n_r)$.

Let $\mathbf{T}_H$ be the diagonal torus of $\mathbf{H}_{\underline{e}}$. If p is a prime number where $\pi_{\underline{e},f}$ is unramified, let

$$(t^{(p)}, ((t_{1,1}^{(p)}, \dots, t_{1,n'_1}^{(p)}), \dots, (t_{r',1}^{(p)}, \dots, t_{r',n'_{r'}}^{(p)}))) \in \widehat{\mathbf{T}}_H^{\mathrm{Gal}(\overline{\mathbb{Q}}_p/\mathbb{Q}_p)}$$

be the Langlands parameter of $\pi_{\underline{e},p}$. By the definition of $R_{\underline{e}}(\pi_f)$ (and the fact that, in proposition 2.3.2, we chose a unitary character μ), up to a permutation of the $t_{i,j}^{(p)}$, there is an equality

$$\begin{aligned}&(z^{(p)}, u_{1,1}^{(p)} z_{1,1}^{(p)}, \dots, u_{1,n_1}^{(p)} z_{1,n_1}^{(p)}, \dots, u_{r,1}^{(p)} z_{r,1}^{(p)}, \dots, u_{r,n_r}^{(p)} z_{r,n_r}^{(p)})\\ &\quad= (t^{(p)}, t_{1,1}^{(p)}, \dots, t_{1,n'_1}^{(p)}, \dots, t_{r',1}^{(p)}, \dots, t_{r',n'_{r'}}^{(p)})\end{aligned}$$

for almost every p, where the $u_{i,j}^{(p)}$ are complex numbers with absolute value 1. So it is enough to show that $|t_{i,j}^{(p)}| = 1$ for all i, j, if p is big enough.

As $c_{\underline{e}}(\pi_{\underline{e},f}) \neq 0$, there exist $\pi_{\underline{e},\infty} \in \Pi(\mathbf{H}_{\underline{e}}(\mathbb{R}))$ and an elliptic Langlands parameter $\varphi_H : W_{\mathbb{R}} \longrightarrow {}^L\mathbf{H}_{\underline{e},\mathbb{R}}$ such that $m_{\mathrm{disc}}(\pi_{\underline{e},\infty} \otimes \pi_{\underline{e},f}) \neq 0$, $\eta_{\underline{e}} \circ \varphi_H$ is $\widehat{\mathbf{G}}$-conjugate to φ and

$$\mathrm{Tr}(\pi_{\underline{e},\infty}(f_{\varphi_H})) \neq 0,$$

where f_{φ_H} is the stable cuspidal function associated to φ_H defined at the end of 6.2. By lemma 7.3.4, φ_H is the Langlands parameter of an L-packet of the discrete series of $\mathbf{H}_{\underline{e}}(\mathbb{R})$ associated to an irreducible algebraic representation of $\mathbf{H}_{\underline{e},\mathbb{C}}$ with regular highest weight and pure of weight 0. So the representation $\pi_{\underline{e},f}$ of $\mathbf{H}_{\underline{e}}(\mathbb{A}_f)$ satisfies all the conditions of point (i) of the theorem, and we can apply the induction hypothesis to finish the proof.

We show (ii). Without the assumption on the highest weight of V, the complex $IC^{\mathrm{K}}V$ is still pure of weight 0, but its cohomology is not necessarily concentrated in degree d. By the hypothesis on π_f, for p big enough and for every $m \in \mathbb{Z}$, there is an equality

$$(N\wp)^{md/2} c_{\mathbf{G}}(\pi_f) \dim(\pi_f^{\mathrm{K}}) \,\mathrm{Tr}(r_{-\mu_G} \circ \varphi_{\pi_p}(\Phi_{\wp}^m)) = \mathrm{Tr}(\Phi_{\wp}^m, W_\lambda(\pi_f)) = \sum_{i=1}^{s} a_i x_i^m,$$

where, as in (i), $x_1, \dots, x_s \in \mathbb{C}$ are the eigenvalues of $\Phi_\wp$ acting on $W_\lambda(\pi_f)$ and $a_1, \dots, a_s \in \mathbb{Z}$ are their multiplicities. In particular, all the a_i have the same sign (the sign of $c_{\mathbf{G}}(\pi_f)$), so $W_\lambda(\pi_f)$ is concentrated either in odd degree or in even degree, and the weights of $W_\lambda(\pi_f)$ are either all even or all odd. By applying the same reasoning as above, we find, for p big enough, a linear system

$$\sum_{j \in J} \log_p |z_{1,j}^{(p)}| = \frac{1}{2} w_J, \quad J \subset \{1, \dots, n_1\}, \quad |J| = p_1,$$

where the w_J are in $\mathbb{Z}$ and all have the same parity. As $p_1 < n_1$ if $n_1 \geq 2$, this implies that $\log_p |z_{1,j}^{(p)}| - \log_p |z_{1,j'}^{(p)}| \in \mathbb{Z}$ for every $j, j' \in \{1, \dots, n_1\}$. On the other hand, we know that $\sum_{j=1}^{n_1} \log_p |z_{1,j}^{(p)}| = 0$. So, for every $J \subset \{1, \dots, n_1\}$ such that $|J| = q_1$, $\sum_{j \in J} \log_p |z_{1,j}^{(p)}| \in \mathbb{Z}$.

Let $\alpha \in \mathbb{R}$ be such that $\log_p |z_{1,1}^{(p)}| - \alpha \in \mathbb{Z}$. Then $\log_p |z_{1,j}^{(p)}| - \alpha \in \mathbb{Z}$ for every $j \in \{1, \dots, n_1\}$, so $p_1\alpha, q_1\alpha \in \mathbb{Z}$, and $\gcd(p_1, q_1)\alpha \in \mathbb{Z}$. Assume that p is inert in E. Then the fact that $(z^{(p)}, (z_{1,1}^{(p)}, \dots, z_{1,n_1}^{(p)}))$ is $\mathrm{Gal}(\overline{\mathbb{Q}}_p/\mathbb{Q}_p)$-invariant implies that, for every $j \in \{1, \dots, n_1\}$, $\log_p |z_{1,j}^{(p)}| + \log_p |z_{1,n_1+1-j}^{(p)}| = 0$. So $2\alpha \in \mathbb{Z}$. □

Lemma 7.3.2 *Use the notation of theorem 7.3.1 above. Fix a prime number p where π_f is unramified and $m \in \mathbb{Z}$. Then*

$$\mathrm{Tr}(r_{-\mu_G} \circ \varphi_{\pi_p}(\Phi_{\wp}^m)) = (z^{(p)})^{-m} \sum_{J_1} \cdots \sum_{J_r} \prod_{i=1}^{r} \prod_{j \in J_i} (\pm z_{i,j}^{(p)})^{-m[F_\wp:\mathbb{Q}_p]},$$

where

(i) *if* $F = \mathbb{Q}$, *p is inert in E and m is odd, then, for every* $i \in \{1, \dots, r\}$, *the ith sum is taken over the set of subsets* J_i *of* $\{1, \dots, n_i\}$ *such that*

$$\{1, \dots, n_i\} - J_i = \{n_i + 1 - j, j \in J_i\};$$

(ii) *in all other cases, the ith sum is taken over the set of subsets* J_i *of* $\{1, \dots, n_i\}$ *such that* $|J_i| = p_i$, *and all the signs are equal to 1.*

Proof. To make notation simpler, we assume that $r = 1$. (The proof is exactly the same in the general case.) We first determine the representation $r_{-\mu_G}$ of ${}^L\mathbf{G}_F$. As $r_{-\mu_G}$ is the contragredient of r_{μ_G}, it is enough to calculate r_{μ_G}. Remember that $\mathbf{T}$ is the diagonal torus of $\mathbf{G}$, and that $\widehat{\mathbf{T}} = \mathbb{C}^\times \times (\mathbb{C}^\times)^{n_1} \subset \widehat{\mathbf{G}} = \mathbb{C}^\times \times \mathbf{GL}_{n_1}(\mathbb{C})$. The cocharacter μ_G of $\mathbf{T}$ corresponds to the following character of $\widehat{\mathbf{T}}$:

$$(\lambda, (\lambda_i)_{1 \le i \le n_1}) \longmapsto \lambda \prod_{i=1}^{p_1} \lambda_i.$$

So the space of r_{μ_G} is $V_\mu = \bigwedge^{p_1} \mathbb{C}^{n_1}$, where $\mathbf{GL}_{n_1}(\mathbb{C})$ acts by $\bigwedge^{p_1}$ of the standard representation, and $\mathbb{C}^\times$ acts by the character $z \longmapsto z$. Let $(e_1, \dots, e_{n_1})$ be the canonical basis of $\mathbb{C}^{n_1}$. Then the family $(e_{i_1} \wedge \dots \wedge e_{i_{p_1}})_{1 \le i_1 < \dots < i_{p_1} \le n_1}$ is a basis of V_μ. From the definition of r_{μ_G} (cf. lemma 4.1.1), it is easy to see that W_E acts trivially on V_μ and that, if $F = \mathbb{Q}$ (so n_1 is even and $p_1 = n_1/2$), then an element of $W_\mathbb{Q} - W_E$ sends $e_{i_1} \wedge \dots \wedge e_{i_{p_1}}$, where $1 \le i_1 < \dots < i_{p_1} \le n_1$, to $\pm e_{j_1} \wedge \dots \wedge e_{j_{p_1}}$, with $1 \le j_1 < \dots < j_{p_1} \le n_1$ and $\{n_1 + 1 - j_1, \dots, n_1 + 1 - j_{p_1}\} = \{1, \dots, n_1\} - \{i_1, \dots, i_{p_1}\}$. By definition of the Langlands parameter, we may assume that $\varphi_{\pi_p}(\Phi_\wp) = (((z^{(p)})^{[F_\wp:\mathbb{Q}_p]}, ((z_1^{(p)})^{[F_\wp:\mathbb{Q}_p]}, \dots, (z_{n_1}^{(p)})^{[F_\wp:\mathbb{Q}_p]})), \Phi_\wp)$ (remember that $\Phi_\wp$ is a lift in $\mathrm{Gal}(\overline{\mathbb{Q}}_p/F_\wp)$ of the geometric Frobenius). If $F = E$, p is split in E, or m is even, then the image of $\Phi_\wp^m$ in $W_\mathbb{Q}$ is an element of W_E, so $\Phi_\wp^m$ acts trivially on V_μ (if p is split in E, this comes from the fact that the image of $W_{\mathbb{Q}_p}$ in $W_\mathbb{Q}$ is included in W_E). If $F = \mathbb{Q}$ and p is inert in E, then $\mathrm{Gal}(E_p/\mathbb{Q}_p) \xrightarrow{\sim} \mathrm{Gal}(E/\mathbb{Q})$, and the image of $\Phi_\wp$ in $\mathrm{Gal}(E_p/\mathbb{Q}_p)$ generates $\mathrm{Gal}(E_p/\mathbb{Q}_p)$, so $\Phi_\wp^m \notin W_E$ for m odd. The formula of the lemma is a consequence of these remarks and of the explicit description of r_{μ_G}. □

Lemma 7.3.3

(i) *Let* $n, p \in \mathbb{N}$ *be such that* $1 \le p \le \max(1, n-1)$. *Then there exist* $J_1, \dots, J_n \subset \{1, \dots, n\}$ *such that* $|J_1| = \dots = |J_n| = p$ *and that the only solution of the system of linear equations*

$$\sum_{j \in J_i} X_j = 0, \quad 1 \le i \le n,$$

is the zero solution.

(ii) *Let* $r \in \mathbb{N}^*$, $n_1, \dots, n_r \ge 2$ *and* $p_1, \dots, p_r \in \mathbb{N}$ *be such that* $1 \le p_i \le n_i - 1$ *for* $1 \le i \le r$. *For every* $i \in \{1, \dots, r\}$, *choose subsets* $J_{i,1}, \dots, J_{i,n_i}$ *of*

$\{1, \dots, n_i\}$ *of cardinality* p_i *and satisfying the property of (i). Then the only solution of the system of linear equations*

$$\begin{cases} \sum_{j=1}^{n_i} X_{i,j} = 0, & 1 \le i \le r, \\ \sum_{i=1}^{r} \sum_{j \in J_{i,k_i}} X_{i,j} = 0, & (k_1, \dots, k_r) \in \{1, \dots, n_1\} \times \cdots \times \{1, \dots, n_r\}, \end{cases}$$

is the zero solution.

Proof. We show (i) by induction on n. If $n = 1$, the result is obvious. Suppose that $n \ge 2$, and let $p \in \{1, \dots, n-1\}$. Assume first that $p \le n - 2$. Then, by the induction hypothesis, there exist $J_2, \dots, J_n \subset \{2, \dots, n\}$ of cardinality p such that the only solution of the system of linear equations (with unknowns $X_2, \dots, X_n$)

$$\sum_{j \in J_i} X_j = 0, \qquad 2 \le i \le n,$$

is the zero solution. Take $J_1 = \{1, 2, \dots, p\}$. It is clear that $J_1, \dots, J_n$ satisfy the condition of (i). Assume now that $p = n - 1$. For every $i \in \{1, \dots, n\}$, let $J_i = \{1, \dots, n\} - \{i\}$. To show that $J_1, \dots, J_n$ satisfy the condition of (i), it is enough to show that $\det(A - I_n) \ne 0$, where $A \in M_n(\mathbb{Z})$ is the matrix all of whose entries are equal to 1. But it is clear that the kernel of A is of dimension $n - 1$ and that n is an eigenvalue of A, so A has no eigenvalue $\lambda \notin \{0, n\}$. In particular, $\det(A - I_n) \ne 0$.

We show (ii) by induction on r. The case $r = 1$ is obvious, so we assume that $r \ge 2$. Let (S) be the system of linear equations of (ii). For $2 \le i \le r$, fix $k_i \in \{1, \dots, n_i\}$. Then, by the case $r = 1$, the system (S′):

$$\sum_{i=1}^{r} \sum_{j \in J_{i,k_i}} X_{i,j} = 0, \qquad k_1 \in \{1, \dots, n_1\},$$

has a unique solution in $(X_{1,1}, \dots, X_{1,n_1})$, that is equal to the obvious solution

$$X_{1,1} = \cdots = X_{1,n_1} = -\frac{1}{p_1} \sum_{i=2}^{r} \sum_{j \in J_{i,k_i}} X_{i,j}.$$

So the system (S′) and the equation $\sum_{j=1}^{n_1} X_{1,j} = 0$ imply

$$X_{1,1} = \cdots = X_{1,n_1} = \sum_{i=2}^{r} \sum_{j \in J_{i,k_i}} X_{i,j} = 0.$$

To finish the proof, apply the induction hypothesis to the system analogous to (S) but with $2 \le i \le r$. □

We now take $\mathbf{G} = \mathbf{G}(\mathbf{U}(p_1, q_1) \times \cdots \times \mathbf{U}(p_r, q_r))$. Then $\widehat{\mathbf{G}} = \mathbb{C}^\times \times \mathbf{GL}_{n_1}(\mathbb{C}) \times \cdots \times \mathbf{GL}_{n_r}(\mathbb{C})$, with the action of $W_\mathbb{Q}$ described in section 2.3 ($n_i = p_i + q_i$).

Let $\mathbf{T}$ be the elliptic maximal torus of $\mathbf{G}$ defined in section 3.1, and $u_G \in \mathbf{G}(\mathbb{C})$ be the element defined in 3.1, so that $u_G^{-1}\mathbf{T}u_G$ is the diagonal torus of $\mathbf{G}$. Let $\mathbf{B} \supset \mathbf{T}$ be the Borel subgroup of $\mathbf{G}_\mathbb{C}$ image by $\mathrm{Int}(u_G)$ of the group of upper triangular matrices (we identify $\mathbf{G}_\mathbb{C}$ to $\mathbb{G}_{m,\mathbb{C}} \times \mathbf{GL}_{n_1,\mathbb{C}} \times \cdots \times \mathbf{GL}_{n_r}(\mathbb{C})$ as in 2.3). Identify $\mathbf{T}_\mathbb{C}$ to $\mathbb{G}_{m,\mathbb{C}} \times \mathbb{G}_{m,\mathbb{C}}^{n_1} \times \cdots \times \mathbb{G}_{m,\mathbb{C}}^{n_r}$ and $\widehat{\mathbf{T}}$ to $\mathbb{C}^\times \times (\mathbb{C}^\times)^{n_1} \times \cdots \times (\mathbb{C}^\times)^{n_r}$ as in 3.1.

Let V be an irreducible algebraic representation of $\mathbf{G}_\mathbb{C}$.

Let $\underline{a} = (a, (a_{i,j})_{1\leq i\leq r, 1\leq j\leq n_i}) \in X^*(\mathbf{T})$ be the highest weight of V relative to $(\mathbf{T}, \mathbf{B})$; the notation means that $\underline{a}$ is the character

$$(z, (z_{i,j})_{1\leq i\leq r, 1\leq j\leq n_i}) \longmapsto z^a \prod_{i=1}^{r}\prod_{j=1}^{n_i} z_{i,j}^{a_{i,j}}.$$

By definition of the highest weight, $a, a_{i,j} \in \mathbb{Z}$ and $a_{i,1} \geq a_{i,2} \geq \cdots \geq a_{i,n_i}$ for every $i \in \{1, \ldots, r\}$. Note also that the weight of V, in the sense of section 1.3, is $2a + \sum_{i=1}^{r}\sum_{j=1}^{n_i} a_{i,j}$.

Let $(\mathbf{H}, s, \eta_0)$ be the elliptic endoscopic triple for $\mathbf{G}$ associated to $((n_1^+, n_1^-), \ldots, (n_r^+, n_r^-))$ as in proposition 2.3.1. Then $\mathbf{H} = \mathbf{G}(\mathbf{U}^*(n_1^+) \times \mathbf{U}^*(n_1^-) \times \cdots \times \mathbf{U}^*(n_r^+) \times \mathbf{U}^*(n_r^-))$, and we define an elliptic maximal torus $\mathbf{T}_H$ of $\mathbf{H}$ and a Borel subgroup $\mathbf{B}_H \supset \mathbf{T}_H$ of $\mathbf{H}_\mathbb{C}$ in the same way as $\mathbf{T}$ and $\mathbf{B}$. Let

$$\begin{aligned}\Omega_* = \{\omega = (\omega_1, \ldots, \omega_r) \in \mathfrak{S}_{n_1} \times \cdots \times \mathfrak{S}_{n_r} \mid \\ \forall i, \omega^{-1}_{i|\{1,\ldots,n_i^+\}} \text{ and } \omega^{-1}_{i|\{n_i^+ +1,\ldots,n_i\}} \text{ are increasing}\}.\end{aligned}$$

Ω_* is the set of representatives of $\Omega(\mathbf{T}_H(\mathbb{C}), \mathbf{H}(\mathbb{C})) \setminus \Omega(\mathbf{T}(\mathbb{C}), \mathbf{G}(\mathbb{C}))$ determined by $\mathbf{B}$ and $\mathbf{B}_H$ as in 3.3.

Lemma 7.3.4 *Let $\varphi : W_\mathbb{R} \longrightarrow {}^L\mathbf{G}_\mathbb{R}$ be a Langlands parameter of the L-packet of the discrete series of $\mathbf{G}(\mathbb{R})$ associated to V and $\eta : {}^L\mathbf{H}_\mathbb{R} \longrightarrow {}^L\mathbf{G}_\mathbb{R}$ be an L-morphism extending η_0 as in proposition 2.3.2. Remember that we wrote $\Phi_H(\varphi)$ for the set of equivalence classes of Langlands parameters $\varphi_H : W_\mathbb{R} \longrightarrow {}^L\mathbf{H}_\mathbb{R}$ such that $\eta \circ \varphi_H$ and φ are equivalent.*

Then every $\varphi_H \in \Phi_H(\varphi)$ is the parameter of an L-packet of the discrete series of $\mathbf{H}(\mathbb{R})$ corresponding to an algebraic representation of $\mathbf{H}_\mathbb{C}$; this algebraic representation has a regular highest weight if $\underline{a}$ is regular, and its weight in the sense of 1.3 is equal to the weight of V.

Proof. We may assume that

$$\varphi(\tau) = ((1, (\Phi_{n_1}^{-1}, \ldots, \Phi_{n_r}^{-1})), \tau)$$

and that, for every $z \in \mathbb{C}^\times$,

$$\varphi(z) = ((z^a \overline{z}^{a+S}, (B_1(z), \ldots, B_r(z))), z),$$

where

$$S = \sum_{i=1}^{r}\sum_{j=1}^{n_i} a_{i,j}$$

and

$$B_i(z) = \mathrm{diag}\left(z^{\frac{n_i-1}{2}+a_{i,1}}\overline{z}^{\frac{1-n_i}{2}-a_{i,1}}, z^{\frac{n_i-3}{2}+a_{i,2}}\overline{z}^{\frac{3-n_i}{2}-a_{i,2}}, \ldots, z^{\frac{1-n_i}{2}+a_{i,n_i}}\overline{z}^{\frac{n_i-1}{2}-a_{i,n_i}}\right).$$

(Remember that $W_{\mathbb{R}} = W_{\mathbb{C}} \sqcup W_{\mathbb{C}}\tau$, with $W_{\mathbb{C}} = \mathbb{C}^\times$, $\tau^2 = -1$ and, for every $z \in \mathbb{C}^\times$, $\tau z \tau^{-1} = \overline{z}$.)

Let C be the odd integer associated to η as below proposition 2.3.2. Let $\omega = (\omega_1, \dots, \omega_r) \in \Omega_*$, and let φ_H be the element of $\Phi_H(\varphi)$ associated to ω as in section 3.3. Write, for every $i \in \{1, \dots, r\}$, $j_{i,s} = \omega_i^{-1}(s)$ if $1 \leq s \leq n_i^+$ and $k_{i,t} = \omega_i^{-1}(t + n_i^+)$ if $1 \leq t \leq n_i^-$. Then we may assume that

$$\varphi_H(\tau) = ((1, (\Phi_{n_1^+}^{-1}, \Phi_{n_1^-}^{-1}, \dots, \Phi_{n_r^+}^{-1}, \Phi_{n_r^-}^{-1})), \tau)$$

and that, for every $z \in \mathbb{C}^\times$,

$$\varphi_H(z) = ((z^a \overline{z}^{a+S}, (B_1^+(z), B_1^-(z), \dots, B_r^+(z), B_r^-(z))), z),$$

with

$$B_i^+(z) = \operatorname{diag}\left(z^{\frac{n_i^+-1}{2}+a_{i,1}^+} \overline{z}^{\frac{1-n_i^+}{2}-a_{i,1}^+}, z^{\frac{n_i^+-3}{2}+a_{i,2}^+} \overline{z}^{\frac{3-n_i^+}{2}-a_{i,2}^+}, \dots, z^{\frac{1-n_i^+}{2}+a_{i,n_i^+}^+} \overline{z}^{\frac{n_i^+-1}{2}-a_{i,n_i^+}^+} \right),$$

where $a_{i,s}^+ = a_{i,j_{i,s}} + s - j_{i,s} + n_i^-(1 - C)/2 \in \mathbb{Z}$, and with

$$B_i^-(z) = \operatorname{diag}\left(z^{\frac{n_i^--1}{2}+a_{i,1}^-} \overline{z}^{\frac{1-n_i^-}{2}-a_{i,1}^-}, z^{\frac{n_i^--3}{2}+a_{i,2}^-} \overline{z}^{\frac{3-n_i^-}{2}-a_{i,2}^-}, \dots, z^{\frac{1-n_i^-}{2}+a_{i,n_i^-}^-} \overline{z}^{\frac{n_i^--1}{2}-a_{i,n_i^-}^-} \right),$$

where $a_{i,t}^- = a_{i,k_{i,t}} + t - k_{i,t} + n_i^+(1 + C)/2 \in \mathbb{Z}$. Let $i \in \{1, \dots, r\}$. For all $s \in \{1, \dots, n_i^+ - 1\}$ and $t \in \{1, \dots, n_1^- - 1\}$,

$$a_{i,s}^+ - a_{i,s+1}^+ = (a_{i,j_{i,s}} - a_{i,j_{i,s+1}}) + (j_{i,s+1} - j_{i,s} - 1),$$

$$a_{i,t}^- - a_{i,t+1}^- = (a_{i,k_{i,t}} - a_{i,k_{i,t+1}}) + (k_{i,t+1} - k_{i,t} - 1),$$

so $a_{i,s}^+ \geq a_{i,s+1}^+$ and $a_{i,t}^- \geq a_{i,t+1}^-$, and the inequalities are strict if $\underline{a}$ is regular. Note also that

$$S_H := \sum_{i=1}^{r} \sum_{s=1}^{n_i^+} a_{i,s}^+ + \sum_{i=1}^{r} \sum_{t=1}^{n_i^-} a_{i,t}^- = S.$$

So φ_H is the parameter of the discrete series of $\mathbf{H}(\mathbb{R})$ associated to the algebraic representation of $\mathbf{H}_{\mathbb{C}}$ of highest weight $(a, ((a_{i,s}^+)_{1 \leq s \leq n_i^+}, (a_{i,t}^-)_{1 \leq t \leq n_i^-})_{1 \leq i \leq r})$. This representation has a regular highest weight if $\underline{a}$ is regular by the above calculations, and its weight in the sense of 1.3 is the same as the weight of V because $2a + S_H = 2a + S$. □

We use again the notations of the beginning of this section.

Lemma 7.3.5 *If the highest weight of V is regular, then, for every neat open compact subgroup K of $\mathbf{G}(\mathbb{A}_f)$, the cohomology of the complex $IC^{\mathrm{K}}V$ is concentrated in degree d.*

Proof. By Zucker's conjecture (proved by Looijenga [Lo], Looijenga-Rapoport [LoR], and Saper-Stern [SS]), the intersection cohomology of $M^{\mathrm{K}}(\mathbf{G}, \mathcal{X})^*(\mathbb{C})$ with coefficients in $\mathcal{F}^{\mathrm{K}}V$ is isomorphic to the L^2-cohomology of $M^{\mathrm{K}}(\mathbf{G}, \mathcal{X})(\mathbb{C})$ with coefficients in $\mathcal{F}^{\mathrm{K}}V$. By a result of Borel and Casselman (theorem 4.5 of [BC]), the $\mathbf{H}^q$ of this L^2-cohomology is isomorphic (as a representation of $C_c^\infty(\mathrm{K}\setminus\mathbf{G}(\mathbb{A}_f)/\mathrm{K})$) to

$$\bigoplus_{\pi} m_{\mathrm{disc}}(\pi)(\mathbf{H}^q(\mathfrak{g},\mathrm{K}'_\infty;\pi_\infty\otimes V)\otimes\pi_f^{\mathrm{K}}),$$

where the sum is taken over the set of isomorphism classes of irreducible admissible representations of $\mathbf{G}(\mathbb{A})$, $\mathfrak{g}=\mathrm{Lie}(\mathbf{G}(\mathbb{R}))\otimes\mathbb{C}$ and $\mathrm{K}'_\infty=\mathrm{K}_\infty\mathbf{A}_G(\mathbb{R})^0$, with K_∞ a maximal compact subgroup of $\mathbf{G}(\mathbb{R})$. By the proof of lemma 6.2 of [A6], if π_∞ is an irreducible admissible representation of $\mathbf{G}(\mathbb{R})$ such that $\mathbf{H}^*(\mathfrak{g},\mathrm{K}'_\infty;\pi_\infty\otimes V)\neq 0$, then π_∞ is in the discrete series of $\mathbf{G}(\mathbb{R})$ (this is the only part where we use the fact that the highest weight of V is regular). By theorem II.5.3 of [BW], if π_∞ is in the discrete series of $\mathbf{G}(\mathbb{R})$, then $\mathbf{H}^q(\mathfrak{g},\mathrm{K}'_\infty;\pi_\infty\otimes V)=0$ for every $q\neq d$. □

Chapter Eight

The twisted trace formula

8.1 NONCONNECTED GROUPS

We first recall some definitions from section 1 of [A4].

Let $\widetilde{\mathbf{G}}$ be a reductive group (not necessarily connnected) over a field K. Fix a connected component $\mathbf{G}$ of $\widetilde{\mathbf{G}}$, and assume that $\mathbf{G}$ generates $\widetilde{\mathbf{G}}$ and that $\mathbf{G}(K) \neq \varnothing$. Let $\mathbf{G}^0$ be the connected component of $\widetilde{\mathbf{G}}$ that contains 1.

Consider the polynomial

$$\det((t+1) - \mathrm{Ad}(g), \mathrm{Lie}(\mathbf{G}^0)) = \sum_{k \geq 0} D_k(g) t^k$$

on $\mathbf{G}(K)$. The smallest integer k for which D_k does not vanish identically is called the *rank* of $\mathbf{G}$; we will denote by r. An element g of $\mathbf{G}(K)$ is called *regular* if $D_r(g) \neq 0$.

A *parabolic subgroup* of $\widetilde{\mathbf{G}}$ is the normalizer in $\widetilde{\mathbf{G}}$ of a parabolic subgroup of $\mathbf{G}^0$. A *parabolic subset* of $\mathbf{G}$ is a nonempty subset of $\mathbf{G}$ that is equal to the intersection of $\mathbf{G}$ with a parabolic subgroup of $\widetilde{\mathbf{G}}$. If $\mathbf{P}$ is a parabolic subset of $\mathbf{G}$, write $\widetilde{\mathbf{P}}$ for the subgroup of $\widetilde{\mathbf{G}}$ generated by $\mathbf{P}$ and $\mathbf{P}^0$ for the intersection $\widetilde{\mathbf{P}} \cap \mathbf{G}^0$ (then $\widetilde{\mathbf{P}} = \mathrm{Nor}_{\widetilde{\mathbf{G}}}(\mathbf{P}^0)$ and $\mathbf{P} = \widetilde{\mathbf{P}} \cap \mathbf{G}$).

Let $\mathbf{P}$ be a parabolic subset of $\mathbf{G}$. The *unipotent radical* $\mathbf{N}_P$ of $\mathbf{P}$ is by definition the unipotent radical of $\mathbf{P}^0$. A *Levi component* $\mathbf{M}$ of $\mathbf{P}$ is a subset of $\mathbf{P}$ that is equal to $\widetilde{\mathbf{M}} \cap \mathbf{G}$, where $\widetilde{\mathbf{M}}$ is the normalizer in $\widetilde{\mathbf{G}}$ of a Levi component $\mathbf{M}^0$ of $\mathbf{P}^0$. If $\mathbf{M}$ is a Levi component of $\mathbf{P}$, then $\mathbf{P} = \mathbf{M}\mathbf{N}_P$.

A *Levi subset* of $\mathbf{G}$ is a Levi component of a parabolic subset of $\mathbf{G}$. Let $\mathbf{M}$ be a Levi subset of $\mathbf{G}$. Let $\widetilde{\mathbf{M}}$ be the subgroup of $\widetilde{\mathbf{G}}$ generated by $\mathbf{M}$, $\mathbf{M}^0 = \mathbf{G}^0 \cap \widetilde{\mathbf{M}}$, $\mathbf{A}_M$ be the maximal split subtorus of the centralizer of $\mathbf{M}$ in $\mathbf{M}^0$ (so $\mathbf{A}_M \subset \mathbf{A}_{M^0}$), $X^*(\mathbf{M})$ be the group of characters of $\widetilde{\mathbf{M}}$ that are defined over K, $\mathfrak{a}_M = \mathrm{Hom}(X^*(\mathbf{M}), \mathbb{R})$ and

$$n_M^G = |\,\mathrm{Nor}_{\mathbf{G}^0(\mathbb{Q})}(\mathbf{M})/\mathbf{M}^0(\mathbb{Q})\,|.$$

Fix a minimal parabolic subgroup $\mathbf{P}_0$ of $\mathbf{G}^0$ and a Levi subgroup $\mathbf{M}_0$ of $\mathbf{P}_0$. Write $\mathbf{A}_0 = \mathbf{A}_{M_0}$ and $\mathfrak{a}_0 = \mathfrak{a}_{\mathbf{M}_0}$. If $\mathbf{P}$ is a parabolic subset of $\mathbf{G}$ such that $\mathbf{P}^0 \supset \mathbf{P}_0$, then $\mathbf{P}$ has a unique Levi component $\mathbf{M}$ such that $\mathbf{M}^0 \supset \mathbf{M}_0$; write $\mathbf{M}_P = \mathbf{M}$. Let $\Phi(\mathbf{A}_{M_P}, \mathbf{P})$ be the set of roots of $\mathbf{A}_{M_P}$ in $\mathrm{Lie}(\mathbf{N}_P)$.

Let W_0^G be the set of linear automorphisms of $\mathfrak{a}_0$ induced by elements of $\mathbf{G}(K)$ that normalize $\mathbf{A}_0$, and $W_0 = W_0^{G^0}$. The group W_0 acts on W_0^G on the left and on the right, and both these actions are simply transitive.

Here, we will be interested in the case where $\widetilde{\mathbf{G}} = \mathbf{G}^0 \rtimes \langle\theta\rangle$ and $\mathbf{G} = \mathbf{G}^0 \rtimes \theta$, where $\mathbf{G}^0$ is a connected reductive group over K and θ is an automorphism of finite order of $\mathbf{G}^0$.

In this situation, we say that an element $g \in \mathbf{G}^0(K)$ is *θ-semisimple* (resp., *θ-regular*, resp., *strongly θ-regular*) if $g\theta \in \mathbf{G}(K)$ is semisimple (resp., regular, resp., strongly regular) in $\widetilde{\mathbf{G}}$ (an element of γ of $\widetilde{\mathbf{G}}(K)$ is called strongly regular if its centralizer is a torus). Let $\mathbf{G}^0_{\theta-\mathrm{reg}}$ be the open subset of θ-regular elements in $\mathbf{G}^0$. We say that $g_1, g_2 \in \mathbf{G}^0(K)$ are *θ-conjugate* if $g_1\theta, g_2\theta \in \mathbf{G}(K)$ are conjugate under $\widetilde{\mathbf{G}}(K)$. If $g \in \mathbf{G}^0(K)$, let $\mathrm{Cent}_{\mathbf{G}^0}(g\theta)$ be the centralizer of $g\theta \in \mathbf{G}(K)$ in $\mathbf{G}^0$; we call this group the *θ-centralizer* of g. Write $\mathbf{G}^0_{g\theta}$ for the connected component of 1 in $\mathrm{Cent}_{\mathbf{G}^0}(g\theta)$. Finally, we say that an element $g \in \mathbf{G}^0(K)$ is *θ-elliptic* if $\mathbf{A}_{G^0_{g\theta}} = \mathbf{A}_G$.

Assume that there exist a θ-stable minimal parabolic subgroup $\mathbf{P}_0$ of $\mathbf{G}^0$ and a θ-stable Levi subgroup $\mathbf{M}_0$ of $\mathbf{P}_0$. We say that a parabolic subset $\mathbf{P}$ of $\mathbf{G}$ is *standard* if $\widetilde{\mathbf{P}} \supset \mathbf{P}_0 \rtimes \langle\theta\rangle$, and that a Levi subset $\mathbf{M}$ of $\mathbf{G}$ is *standard* if there exists a standard parabolic subset $\mathbf{P}$ such that $\mathbf{M} = \mathbf{M}_P$ (so $\widetilde{\mathbf{M}} \supset \mathbf{M}_0 \rtimes \langle\theta\rangle$). Then the map $\mathbf{P} \longmapsto \mathbf{P}^0$ is a bijection from the set of standard parabolic subsets of $\mathbf{G}$ onto the set of θ-stable standard parabolic subgroups of $\mathbf{G}^0$. If $\mathbf{P}$ is a standard parabolic subset of $\mathbf{G}$, then $\widetilde{\mathbf{P}} = \mathbf{P}^0 \rtimes \langle\theta\rangle$, $\mathbf{P} = \mathbf{P}^0\theta$, $\widetilde{\mathbf{M}}_P = \mathbf{M}^0_P \rtimes \langle\theta\rangle$ and $\mathbf{M}_P = \mathbf{M}^0_P\theta$. It is easy to see that the centralizer of $\mathbf{M}_P$ in $\mathbf{M}^0_P$ is $Z(\mathbf{M}^0_P)^\theta$; so $\mathbf{A}_{M_P}$ is the maximal split subtorus of $Z(\mathbf{M}^0_P)^\theta$.

Example 8.1.1 Let $\mathbf{H}$ be a connected reductive quasi-split group over K and E/K be a cyclic extension. Let $\mathbf{G}^0 = R_{E/K}\mathbf{H}_E$, θ be the isomorphism of $\mathbf{G}^0$ induced by a fixed generator of $\mathrm{Gal}(E/K)$, $\widetilde{\mathbf{G}} = \mathbf{G}^0 \rtimes \langle\theta\rangle$ and $\mathbf{G} = \mathbf{G}^0 \rtimes \theta$. Fix a Borel subgroup $\mathbf{B}_H$ of $\mathbf{H}$ and a Levi subrgoup $\mathbf{T}_H$ of $\mathbf{B}_H$. Then $\mathbf{B}^0 := R_{E/K}\mathbf{B}_{H,E}$ is a θ-stable Borel subgroup of $\mathbf{G}^0$, and $\mathbf{T}^0 := R_{E/K}\mathbf{T}_{H,E}$ is a θ-stable maximal torus of $\mathbf{G}^0$. The standard parabolic subsets of $\mathbf{G}$ are in bijection with the θ-stable standard parabolic subgroups of $\mathbf{G}^0$, that is, with the standard parabolic subgroups of $\mathbf{H}$. If $\mathbf{P}$ corresponds to $\mathbf{P}_H$, then $\mathbf{P}_H = \mathbf{P}^0 \cap \mathbf{H}$, $\mathbf{P}^0 = R_{E/K}\mathbf{P}_{H,K}$ and $\mathbf{A}_{M_P} = \mathbf{A}_{M_{P_H}}$.

Assume that K is local or global. Then we can associate to $\mathbf{H}$ an endoscopic datum $(\mathbf{H}^*, \mathcal{H}, s, \xi)$ for $(\mathbf{G}^0, \theta, 1)$ in the sense of [KS] 2.1. If $r = [E : K]$, then $\widehat{\mathbf{G}}^0 \simeq \widehat{\mathbf{H}}^r$, with $\hat{\theta}(x_1, \dots, x_r) = (x_2, \dots, x_r, x_1)$. The diagonal embedding $\widehat{\mathbf{H}} \longrightarrow \widehat{\mathbf{G}}^0$ is W_K-equivariant, hence extends in an obvious way to an L-morphism $\xi : \mathcal{H} := {}^L\mathbf{H} \longrightarrow {}^L\mathbf{G}^0$. Finally, take $s = 1$.

Assume that we are in the situation of the example above. In [La3] 2.4, Labesse defines the norm $\mathcal{N}\gamma$ of a θ-semisimple element γ of $\mathbf{G}^0(K)$ (and shows that it exists); $\mathcal{N}\gamma$ is a stable conjugacy class in $\mathbf{H}(K)$ that depends only on the stable θ-conjugacy class of γ, and every element of $\mathcal{N}\gamma$ is stably conjugate to $N\gamma := \gamma\theta(\gamma)\dots\theta^{[E:K]-1}(\gamma) \in \mathbf{G}^0(K) = \mathbf{H}(E)$.

If $\mathbf{M}$ is a Levi subset of $\mathbf{G}$, write, for every θ-semi-simple $\gamma \in \mathbf{M}^0(K)$,

$$D^G_M(\gamma) = \det(1 - \mathrm{Ad}(\gamma) \circ \theta, \mathrm{Lie}(\mathbf{G}^0)/\mathrm{Lie}(\mathbf{M}^0)).$$

If $\mathbf{M}$ is a standard Levi subset of $\mathbf{G}$ (or, more generally, any Levi subset of $\mathbf{G}$ such that $\theta \in \mathbf{M}$), set $\mathbf{M}_H = (\mathbf{M}^0)^\theta = \mathbf{M}^0 \cap \mathbf{H}$; then $\mathbf{M}_H$ is a Levi subgroup of $\mathbf{H}$.

Lemma 8.1.2 *Let* $\mathbf{M}$ *be a standard Levi subset of* $\mathbf{G}$. *Then, for every* θ*-semisimple* $\gamma \in \mathbf{M}^0(K)$;

$$D_M^G(\gamma) = D_{M_H}^H(\mathcal{N}\gamma).$$

Assume now that $K = \mathbb{R}$ and $E = \mathbb{C}$. We will recall results of Clozel, Delorme and Labesse about θ-stable tempered representations of $\mathbf{G}^0(\mathbb{R})$.

Remember that an admissible representation π of $\mathbf{G}^0(\mathbb{R})$ is called θ-stable if $\pi \simeq \pi \circ \theta$. In that case, there exists an intertwining operator $A_\pi : \pi \xrightarrow{\sim} \pi \circ \theta$. We say that A_π is *normalized* if $A_\pi^2 = 1$. The data of a normalized intertwining operator on π are equivalent to those of a representation of $\widetilde{\mathbf{G}}(\mathbb{R})$ extending π. If π is irreducible and θ-stable, then, by Schur's theorem, it always has a normalized intertwining operator.

For ξ a quasi-character of $\mathbf{A}_G(\mathbb{R})^0$, let $C_c^\infty(\mathbf{G}^0(\mathbb{R}), \xi)$ be the set of functions $f \in C^\infty(\mathbf{G}^0(\mathbb{R}))$ that have compact support modulo $\mathbf{A}_G(\mathbb{R})^0$ and such that $f(zg) = \xi(z) f(g)$ for every $(z, g) \in \mathbf{A}_G(\mathbb{R})^0 \times \mathbf{G}^0(\mathbb{R})$.

The following theorem is due to Clozel (cf. [Cl2] 4.1, 5.12, 8.4).

Theorem 8.1.3 *Let* π *be an irreducible admissible* θ*-stable representation of* $\mathbf{G}^0(\mathbb{R})$ *and* A_π *be a normalized intertwining operator on* π. *Let* ξ *be the quasi-character through which* $\mathbf{A}_G(\mathbb{R})^0$ *acts on the space of* π. *Then the map*

$$C_c^\infty(\mathbf{G}^0(\mathbb{R}), \xi^{-1}) \longrightarrow \mathbb{C}, \quad f \longmapsto \mathrm{Tr}(\pi(f) A_\pi)$$

extends to a distribution on $\mathbf{G}^0(\mathbb{R})$ *that is invariant under* θ*-conjugacy; this distribution is tempered if* π *is tempered. Call this distribution the* twisted character *of* π *and denote it by* Θ_π.

Let $\varphi : W_\mathbb{R} \longrightarrow {}^L\mathbf{H}$ *be a tempered Langlands parameter; it defines an* L*-packet* Π_H *of tempered representations of* $\mathbf{H}(\mathbb{R})$. *Write* $\Theta_{\Pi_H} = \sum_{\pi_H \in \Pi_H} \Theta_{\pi_H}$, *where, for every* $\pi_H \in \Pi_H$, Θ_{π_H} *is the character of* π_H. *Then the representation* π *of* $\mathbf{G}^0(\mathbb{R}) = \mathbf{H}(\mathbb{C})$ *associated to* $\varphi_{|W_\mathbb{C}}$ *is tempered and* θ*-stable, and, if* A_π *is a normalized intertwining operator on* π, *there exists* $\varepsilon \in \{\pm 1\}$ *such that, for every* θ*-regular* $g \in \mathbf{G}^0(\mathbb{R})$,

$$\Theta_\pi(g) = \varepsilon \Theta_{\Pi_H}(\mathcal{N}g).$$

In particular, Θ_π is invariant under stable θ-conjugacy.

Remark 8.1.4 Let π be an irreducible tempered θ-stable representation of $\mathbf{G}^0(\mathbb{R})$. If the infinitesimal character of π is equal to that of a finite-dimensional θ-stable representation of $\mathbf{G}^0(\mathbb{R})$, then there exists a tempered Langlands parameter $\varphi : W_\mathbb{R} \longrightarrow {}^L\mathbf{H}$ such that π is associated to the parameter $\varphi_{|W_\mathbb{C}}$ (cf. [J] (5.16)).

Assume from now on that $\mathbf{H}(\mathbb{R})$ has a discrete series. Let K'_∞ be the set of fixed points of a Cartan involution of $\mathbf{G}^0(\mathbb{R})$ that commutes with θ. Write $\mathfrak{g} = \mathrm{Lie}(\mathbf{G})(\mathbb{C})$. For every admissible θ-stable representation ρ of $\mathbf{G}^0(\mathbb{R})$, let

$$\mathrm{ep}(\theta, \rho) := \sum_{i \geq 0} (-1)^i \, \mathrm{Tr}(\theta, \mathbf{H}^i(\mathfrak{g}, \mathrm{K}'_\infty; \rho))$$

be the twisted Euler-Poincaré characteristic of ρ. It depends on the choice of a normalized intertwining operator on ρ. An admissible representation of $\mathbf{G}^0(\mathbb{R})$ is called *θ-discrete* if it is irreducible tempered θ-stable and is not a subquotient of a representation induced from an admissible θ-stable representation of a proper θ-stable Levi subgroup (cf. [AC] I.2.3).

The following theorem is due to Labesse (cf. [La2] proposition 12).

Theorem 8.1.5 *Let π be a θ-discrete representation of $\mathbf{G}^0(\mathbb{R})$, and let ξ be the quasi-character through which $\mathbf{A}_G(\mathbb{R})^0$ acts on the space of π. Assume that π is associated to a Langlands parameter $\varphi_\pi : W_\mathbb{C} \longrightarrow {}^L\mathbf{H}$ satisfying $\varphi_\pi = \varphi_{|W_\mathbb{C}}$, where $\varphi : W_\mathbb{R} \longrightarrow {}^L\mathbf{H}$ is a Langlands parameter of the L-packet of the discrete series of $\mathbf{H}(\mathbb{R})$ associated to the contragredient of an irreducible algebraic representation V of $\mathbf{H}$. As in section 3 of [Cl2], we associate to V a θ-stable algebraic representation W of $\mathbf{G}^0$ and a normalized intertwining operator A_W on W.*

Then there exists a function $\phi \in C_c^\infty(\mathbf{G}^0(\mathbb{R}), \xi^{-1})$, K'_∞-finite on the right and on the left modulo $\mathbf{A}_G(\mathbb{R})^0$, such that, for every admissible θ-stable representation ρ of $\mathbf{G}^0(\mathbb{R})$ that is of finite length and such that $\rho_{|\mathbf{A}_G(\mathbb{R})^0}$ is a sum of copies of ξ and every normalized intertwining operator A_ρ on ρ,

$$\mathrm{Tr}(\rho(\phi)A_\rho) = \mathrm{ep}(\theta, \rho \otimes W).$$

Such a function ϕ is called *twisted pseudo-coefficient* of π. This name is justified by the next remark.

Remark 8.1.6 Let π and ϕ be as in the above theorem, and let A_π be a normalized intertwining operator on π. By the proof of proposition 3.6 of [Cl5] and by theorem 2 (p. 217) of [De],

$$\mathrm{Tr}(\pi(\phi)A_\pi) \neq 0$$

and, for every irreducible θ-stable tempered representation ρ of $\mathbf{G}^0(\mathbb{R})$ and every normalized intertwining operator A_ρ on ρ,

$$\mathrm{Tr}(\rho(\phi)A_\rho) = 0$$

if $\rho \not\simeq \pi$. In particular, the function ϕ is cuspidal (the definition of "cuspidal" is recalled, for example, at the beginning of section 7 of [A3]).

Definition 8.1.7 Let $\mathbf{T}_e$ be a torus of $\mathbf{G}^0_\mathbb{R}$ such that $\mathbf{T}_e(\mathbb{R})$ is a maximal torus of K'_∞. Set

$$d(\mathbf{G}) = |\,\mathrm{Ker}(\mathbf{H}^1(\mathbb{R}, \mathbf{T}_e) \longrightarrow \mathbf{H}^1(\mathbb{R}, \mathbf{G}^0))|,$$

$$k(\mathbf{G}) = |\mathrm{Im}(\mathbf{H}^1(\mathbb{R}, (\mathbf{T}_e)^{\mathrm{sc}}) \longrightarrow \mathbf{H}^1(\mathbb{R}, \mathbf{T}_e))|,$$

with $(\mathbf{T}_e)^{\mathrm{sc}}$ the inverse image of $\mathbf{T}_e$ by the morphism $(\mathbf{G}^0)^{\mathrm{sc}} \longrightarrow \mathbf{G}^0$ (where $(\mathbf{G}^0)^{\mathrm{sc}} \longrightarrow (\mathbf{G}^0)^{\mathrm{der}}$ is the simply connected covering of $(\mathbf{G}^0)^{\mathrm{der}}$).[1]

[1]Cf. [CL] A.1 for the definition of $d(\mathbf{G})$; Clozel and Labesse use a maximal $\mathbb{R}$-anisotropic torus instead of a maximal $\mathbb{R}$-elliptic torus, but this does not give the correct result if $\mathbf{A}_G \neq \{1\}$ (cf. [K13] 1.1 for the case of connected groups).

Remark 8.1.8 As $\mathbf{G}^0$ comes from a complex group by restriction of scalars, $\mathbf{H}^1(\mathbb{R}, \mathbf{G}^0) = \{1\}$, so $d(\mathbf{G}) = |\mathbf{H}^1(\mathbb{R}, \mathbf{T}_e)|$. For example, if $\mathbf{H} = \mathbf{G}(\mathbf{U}^*(n_1) \times \cdots \times \mathbf{U}^*(n_r))$ (cf. section 2.1 for the definition of this group) with $n := n_1 + \cdots + n_r \geq 1$, then $\mathbf{T}_e = \mathbf{G}(\mathbf{U}(1)^n)$, so $d(\mathbf{G}) = 2^{n-1}$.

On the other hand, if the derived group of $\mathbf{G}^0$ is simply connected, then $k(\mathbf{G}) = |\mathrm{Im}(\mathbf{H}^1(\mathbb{R}, \mathbf{T}_e \cap (\mathbf{G}^0)^{\mathrm{der}}) \longrightarrow \mathbf{H}^1(\mathbb{R}, \mathbf{T}_e))|$.

Remember that a θ-semisimple element g of $\mathbf{G}^0(\mathbb{R})$ is called θ-elliptic if $\mathbf{A}_{G^0_{g\theta}} = \mathbf{A}_G (= \mathbf{A}_H)$.

Lemma 8.1.9 *Let $g \in \mathbf{G}^0(\mathbb{R})$ be θ-semisimple. Then g is θ-elliptic if and only if $\mathcal{N} g$ is elliptic. Moreover, if g is not θ-elliptic, then there exists a proper Levi subset $\mathbf{M}$ of $\mathbf{G}$ such that $g\theta \in \mathbf{M}(\mathbb{R})$ and $\mathbf{G}^0_{g\theta} \subset \mathbf{M}^0$.*

Proof. Let $g \in \mathbf{G}^0(\mathbb{R})$ be θ-semisimple. As $h := g\theta(g)$ is $\mathbf{G}^0(\mathbb{R})$-conjugate to an element of $\mathbf{H}(\mathbb{R})$ (cf. [Cl2] p. 55), we may assume, after replacing g by a θ-conjugate, that $h \in \mathbf{H}(\mathbb{R})$. Let $\mathbf{L} = \mathbf{G}^0_h$. Then $\mathbf{L}$ is stable by the morphism (of algebraic groups over $\mathbb{R}$) $\theta' : x \longmapsto \theta(g)\theta(x)\theta(g)^{-1}$, and it is easy to see that $\theta'_{|\mathbf{L}}$ is an involution and that $\mathbf{G}^0_{g\theta} = \mathbf{L}^{\theta'}$ and $\mathbf{H}_h = \mathbf{L}^{\theta}$. This implies that $Z(\mathbf{H}_h) = Z(\mathbf{L})^{\theta}$ and that $Z(\mathbf{G}^0_{g\theta}) = Z(\mathbf{L})^{\theta'}$. But $\theta(g) \in \mathbf{L}(\mathbb{R})$ (we assumed that $g\theta(g) \in \mathbf{H}(\mathbb{R})$, so g and $\theta(g)$ commute), so $\theta_{|Z(\mathbf{L})} = \theta'_{|Z(\mathbf{L})}$. Hence $\mathbf{A}_{G^0_{g\theta}} = \mathbf{A}_{H_h}$, and this proves that g is θ-elliptic if and only if h is elliptic.

Suppose that g is not θ-elliptic. Then h is not elliptic, so there exists a proper Levi subgroup $\mathbf{M}_H$ of $\mathbf{H}$ such that $\mathbf{H}_h \subset \mathbf{M}_H$. Let $\mathbf{M}^0 = R_{\mathbb{C}/\mathbb{R}} \mathbf{M}_{H,\mathbb{C}}$; it is a θ-stable proper Levi subgroup of $\mathbf{G}^0$. Let $\mathbf{M}$ be the Levi subset of $\mathbf{G}$ associated to $\mathbf{M}^0$. As $\mathbf{G}^0_h \subset \mathbf{M}^0$, $\mathbf{G}^0_{g\theta} \subset \mathbf{M}^0$; moreover, $g \in \mathbf{G}^0_h(\mathbb{R})$, so $g\theta \in \mathbf{M}(\mathbb{R})$. □

Note that, by the above proof, for every θ-semisimple $g \in \mathbf{G}^0(\mathbb{R})$, if $h \in \mathcal{N}(g)$, then $\mathbf{G}^0_{g\theta}$ is an inner form of $\mathbf{H}_h$. Later, when we calculate orbital integrals at $g\theta$ and h, we always choose Haar measures on $\mathbf{G}^0_{g\theta}$ and $\mathbf{H}_h$ that correspond to each other.

Finally, we calculate the twisted orbital integrals of some of the twisted pseudo-coefficients defined above at θ-semisimple elements. To avoid technical complications, assume that $\mathbf{G}^{\mathrm{der}}$ is simply connected.

If $\phi \in C_c^\infty(\mathbf{G}^0(\mathbb{R}), \xi^{-1})$ (ξ is a quasi-character on $\mathbf{A}_G(\mathbb{R})^0$) and $g \in \mathbf{G}^0(\mathbb{R})$, the twisted orbital integral of ϕ at g (also called orbital integral of ϕ at $g\theta$) is by definition

$$O_{g\theta}(\phi) = \int_{\mathbf{G}^0_{g\theta}(\mathbb{R}) \backslash \mathbf{G}^0(\mathbb{R})} \phi(x^{-1} g\theta(x)) dx$$

(of course, it depends of the choice of Haar measures on $\mathbf{G}^0(\mathbb{R})$ and $\mathbf{G}^0_{g\theta}(\mathbb{R})$).

Let $\mathbf{G}'$ be a reductive connected algebraic group over $\mathbb{R}$. If $\mathbf{G}'$ has an inner form $\overline{\mathbf{G}}'$ that is anisotropic modulo its center, set

$$v(\mathbf{G}') = (-1)^{q(\mathbf{G}')} \operatorname{vol}(\overline{\mathbf{G}}'(\mathbb{R}) / \mathbf{A}_{G_{\mathbb{R}}}(\mathbb{R})^0) d(\mathbf{G}')^{-1},$$

where $d(\mathbf{G}')$ is defined in section 3.1.

Lemma 8.1.10 *Let V be an irreducible algebraic representation of $\mathbf{H}$, $\varphi : W_{\mathbb{R}} \longrightarrow {}^L\mathbf{H}$ be a Langlands parameter of the L-packet of the discrete series of $\mathbf{H}(\mathbb{R})$ associated to V^*, π_{V^*} be the representation of $\mathbf{H}(\mathbb{C}) = \mathbf{G}^0(\mathbb{R})$ corresponding to $\varphi_{|W_{\mathbb{C}}}$ (so π_{V^*} is θ-discrete) and ϕ_{V^*} be a twisted pseudo-coefficient of π_{V^*}. Let $g \in \mathbf{G}^0(\mathbb{R})$ be θ-semisimple. Then*

$$O_{g\theta}(\phi_{V^*}) = v(\mathbf{G}^0_{g\theta})^{-1}\Theta_{\pi^\vee_{V^*}}(g)$$

if g is θ-elliptic, and

$$O_{g\theta}(\phi_{V^*}) = 0$$

if g is not θ-elliptic.

The proof of this lemma is inspired by the proof of theorem 2.12 of [CCl].

Proof. To simplify the notation, we will write $\pi = \pi_{V^*}$ and $\phi = \phi_{V^*}$.

If V is the trivial representation, then the twisted orbital integrals of ϕ are calculated in theorem A.1.1 of [CL]; write $\phi_0 = \phi$. (Note that Clozel and Labesse choose the Haar measure on $\mathbf{G}^0_{g\theta}(\mathbb{R})$ for which $\mathrm{vol}(\overline{\mathbf{G}}'(\mathbb{R})/\mathbf{A}_{G'}(\mathbb{R})^0) = 1$, where $\overline{\mathbf{G}}'$ is an inner form of $\mathbf{G}^0_{g\theta}$ that is anisotropic modulo its center.)

Assume that V is any irreducible algebraic representation of $\mathbf{H}$. Let W be the θ-stable algebraic representation of $\mathbf{G}^0$ associated to V as in theorem 8.1.5, with the normalized intertwining operator A_W fixed in that theorem. Write $\phi' = \Theta_W\phi_0$. As Θ_W is invariant by θ-conjugacy, proposition 3.4 of [Cl2] implies that, for every θ-semisimple $g \in \mathbf{G}^0(\mathbb{R})$, $O_{g\theta}(\phi') = O_{g\theta}(\phi_0)\Theta_{\pi^\vee}(g)$. So it is enough to show that ϕ and ϕ' have the same orbital integrals. By theorem 1 of [KRo], in order to prove this, it suffices to show that, for every θ-stable tempered representation ρ of $\mathbf{G}^0(\mathbb{R})$ and every normalized intertwining operator A_ρ on ρ,

$$\mathrm{Tr}(\rho(\phi)A_\rho) = \mathrm{Tr}(\rho(\phi')A_\rho).$$

Fix such a representation ρ. Then it is easy to see that

$$\mathrm{Tr}(\rho(\phi')A_\rho) = \mathrm{Tr}((\rho \otimes W)(\phi_0)(A_\rho \otimes A_W)).$$

Hence

$$\mathrm{Tr}(\rho(\phi')A_\rho) = \mathrm{ep}(\theta, \rho \otimes W) = \mathrm{Tr}(\rho(\phi)A_\rho).$$

□

Corollary 8.1.11 *Use the notation of lemma 8.1.10 above.*

(i) *The function ϕ_{V^*} is stabilizing ("stabilisante") in the sense of [La3] 3.8.2.*
(ii) *Let*

$$f_{V^*} = \frac{1}{|\Pi(\varphi)|}\sum_{\pi_H \in \Pi(\varphi)} f_{\pi_H},$$

where $\Pi(\varphi)$ is the discrete series L-packet of $\mathbf{H}(\mathbb{R})$ associated to $\varphi : W_{\mathbb{R}} \longrightarrow {}^L\mathbf{H}$ and, for every representation π_H in the discrete series of $\mathbf{H}(\mathbb{R})$, f_{π_H} is a pseudo-coefficient of π_H. Then the functions ϕ_{V^*} and $d(\mathbf{G})f_{V^*}$ are associated in the sense of [La1] 3.2.

Proof. The result follows from lemma 8.1.10 and the proof of theorem A.1.1 and lemma A.1.2 of [CL]. □

8.2 THE INVARIANT TRACE FORMULA

Note first that, thanks to the work of Delorme-Mezo ([DeM]) and Kottwitz-Rogawski ([KRo]), Arthur's invariant trace formula (see, e.g., [A3]) is now available for nonconnected groups as well as for connected groups.

In [A6], Arthur gave a simple form of the invariant trace formula (on a connected group) for a function that is stable cuspidal at infinity (this notion is defined at the beginning of section 4 of [A6] and recalled in section 5.4). The goal of this section is to give a similar formula for a (very) particular class of nonconnected groups.

Let $\mathbf{H}$ be a connected reductive quasi-split group over $\mathbb{Q}$; fix a Borel subgroup of $\mathbf{H}$ and a Levi subgroup of this Borel subgroup. Fix an imaginary quadratic extension E of $\mathbb{Q}$, and take $\mathbf{G}^0 = R_{E/\mathbb{Q}}\mathbf{H}_E$. Assume that the derived group of $\mathbf{H}$ is simply connected and that $\mathbf{H}$ is cuspidal (in the sense of definition 3.1.1). A Levi subset $\mathbf{M}$ of $\mathbf{G}$ is called cuspidal if it is conjugate to a standard Levi subset $\mathbf{M}'$ such that $\mathbf{M}'_H$ is cuspidal.

We first define the analogs of the functions $\Phi^G_M(.,\Theta)$ of section 3.2.

Lemma 8.2.1 *Let π be a θ-stable irreducible tempered representation of $\mathbf{G}^0(\mathbb{R})$. Fix a normalized intertwining operator A_π on π. Assume that there exists a Langlands parameter $\varphi : W_{\mathbb{R}} \longrightarrow {}^L\mathbf{H}$ such that π is associated to $\varphi_{|W_{\mathbb{C}}}$. Let $\mathbf{M}$ be a standard cuspidal Levi subset of $\mathbf{G}$, and let $\mathbf{T}_H$ be a maximal torus of $\mathbf{M}_{H,\mathbb{R}}$ that is anisotropic modulo $\mathbf{A}_{M_H}$. Write D for the set of $\gamma \in \mathbf{G}^0(\mathbb{R})$ such that $\gamma\theta(\gamma) \in \mathbf{T}_H(\mathbb{R})$. Then the function*

$$D \cap \mathbf{G}^0(\mathbb{R})_{\theta\text{-reg}} \longrightarrow \mathbb{C}, \quad \gamma \longmapsto |D^G_M(\gamma)|^{1/2}\Theta_\pi(\gamma)$$

extends to a continuous function $D \longrightarrow \mathbb{C}$, that will be denoted by $\Phi^G_M(.,\Theta_\pi)$.

Extend $\Phi^G_M(.,\Theta_\pi)$ to a function on $\mathbf{M}^0(\mathbb{R})$ in the following way: if $\gamma \in \mathbf{M}^0(\mathbb{R})$ is θ-elliptic (in $\mathbf{M}^0(\mathbb{R})$), then it is θ-conjugate to $\gamma' \in D$, and we set $\Phi^G_M(\gamma,\Theta_\pi) = \Phi^G_M(\gamma',\Theta_\pi)$; otherwise, we set $\Phi^G_M(\gamma,\Theta_\pi) = 0$. The function $\Phi^G_M(.,\Theta_\pi)$ is clearly invariant by stable θ-conjugacy. As every Levi subset of $\mathbf{G}$ is $\mathbf{G}^0(\mathbb{R})$-conjugate to a standard Levi subset, we can define in the same way a function $\Phi^G_M(.,\Theta_\pi)$ for any cuspidal Levi subset $\mathbf{M}$.

The lemma above follows from the similar lemma for connected groups (lemma 3.2.1, due to Arthur and Shelstad), from theorem 8.1.3 (due to Clozel) and from lemma 8.1.2.

Let $\Pi_{\theta\text{-disc}}(\mathbf{G}^0(\mathbb{R}))$ be the set of isomorphism classes of θ-discrete representations of $\mathbf{G}^0(\mathbb{R})$. For every $\pi \in \Pi_{\theta\text{-disc}}(\mathbf{G}^0(\mathbb{R}))$, fix a normalizing operator A_π on π.

Definition 8.2.2 Let $\mathbf{M}$ be a cuspidal Levi subset of $\mathbf{G}$. Let ξ be a θ-stable quasi-character of $\mathbf{A}_{G^0}(\mathbb{R})^0$. For every function $\phi \in C^\infty_c(\mathbf{G}^0(\mathbb{R}),\xi^{-1})$ that is left and right K'_∞-finite modulo $\mathbf{A}_{G^0}(\mathbb{R})^0$ and every $\gamma \in \mathbf{M}^0(\mathbb{R})$, write

$$\Phi^G_M(\gamma,\phi) = (-1)^{\dim(\mathbf{A}_M/\mathbf{A}_G)} v(\mathbf{M}^0_{\gamma\theta})^{-1} \sum_{\pi \in \Pi_{\theta\text{-disc}}(\mathbf{G}^0(\mathbb{R}))} \Phi^G_M(\gamma,\Theta_{\pi^\vee})\,\mathrm{Tr}(\pi(\phi)A_\pi),$$

and

$$S\Phi_M^G(\gamma,\phi)=(-1)^{\dim(\mathbf{A}_M/\mathbf{A}_G)}k(\mathbf{M})k(\mathbf{G})^{-1}\overline{v}(\mathbf{M}_{\gamma\theta}^0)^{-1}$$
$$\sum_{\pi\in\Pi_{\theta\text{-disc}}(\mathbf{G}^0(\mathbb{R}))}\Phi_M^G(\gamma,\Theta_{\pi^\vee})\operatorname{Tr}(\pi(\phi)A_\pi).$$

The notation k and $\overline{v}$ is defined in section 5.4, and the notation v is that of section 8.1.

Let $\mathbf{M}_0$ be the minimal θ-stable Levi subgroup of $\mathbf{G}^0$ corresponding to the fixed minimal Levi subgroup of $\mathbf{H}$ ($\mathbf{M}_0$ is a torus because $\mathbf{H}$ is quasi-split) and ξ be a θ-stable quasi-character of $\mathbf{A}_{G^0}(\mathbb{R})^0$. Define an action of the group $\widetilde{\mathbf{G}}(\mathbb{A})$ on $L^2(\mathbf{G}^0(\mathbb{Q})\backslash\mathbf{G}^0(\mathbb{A}),\xi)$ in the following way: the subgroup $\mathbf{G}^0(\mathbb{A})$ acts in the usual way, and θ acts by $\phi\longmapsto\phi\circ\theta$. For every irreducible θ-stable representation π of $\mathbf{G}^0(\mathbb{A})$ such that $m_{disc}(\pi)\neq 0$ (i.e., such that π is a direct factor of $L^2(\mathbf{G}^0(\mathbb{Q})\backslash\mathbf{G}^0(\mathbb{A}),\xi)$, seen as a representation of $\mathbf{G}^0(\mathbb{A})$), fix a normalized intertwining operator A_π on π. If π and π' are such that $\pi_\infty\simeq\pi'_\infty$, choose intertwining operators that are compatible at infinity. For the θ-stable irreducible admissible representations of $\mathbf{G}^0(\mathbb{R})$ that do not appear in this way, use any normalized intertwining operator. If π is as above, let $\widetilde{\pi}^+$ (resp., $\widetilde{\pi}^-$) be the representation of $\widetilde{\mathbf{G}}(\mathbb{A})$ defined by π and A_π (resp., $-A_\pi$), and let $m_{disc}^+(\pi)$ (resp., $m_{disc}^-(\pi)$) be the multiplicity of $\widetilde{\pi}^+$ (resp., $\widetilde{\pi}^-$) in $L^2(\mathbf{G}^0(\mathbb{Q})\backslash\mathbf{G}^0(\mathbb{A}),\xi)$.

We write $C_c^\infty(\mathbf{G}^0(\mathbb{A}),\xi^{-1})$ for the vector space of functions $\phi:\mathbf{G}^0(\mathbb{A})\longrightarrow\mathbb{C}$ that are finite linear combinations of functions of the form $\phi^\infty\otimes\phi_\infty$, with $\phi^\infty\in C_c^\infty(\mathbf{G}^0(\mathbb{A}_f))$ and $\phi_\infty\in C_c^\infty(\mathbf{G}^0(\mathbb{R}),\xi^{-1})$.

Let $\mathbf{M}$ be a Levi subset of $\mathbf{G}$. If $\mathbf{M}$ is cuspidal, then, for every function $\phi=\phi^\infty\otimes\phi_\infty\in C^\infty(\mathbf{G}(\mathbb{A})),\xi^{-1})$, write

$$T_{M,\mathrm{geom}}^G(\phi)=\sum_\gamma \operatorname{vol}(\mathbf{M}_{\gamma\theta}^0(\mathbb{Q})\mathbf{A}_M(\mathbb{R})^0\backslash\mathbf{M}_{\gamma\theta}^0(\mathbb{A}))O_{\gamma\theta}(\phi_M^\infty)\Phi_M^G(\gamma,\phi_\infty),$$

where the sum is taken over the set of θ-conjugacy classes of θ-semisimple elements of $\mathbf{M}^0(\mathbb{Q})$ and ϕ_M^∞ is the constant term of ϕ^∞ at $\mathbf{M}$ (defined in exactly the same way as in the case of connected groups).

If $\mathbf{M}$ is not cuspidal, set $T_{M,\mathrm{geom}}^G=0$.

For every $t\geq 0$, define $\Pi_{\mathrm{disc}}(\mathbf{G},t)$ and the function $a_{\mathrm{disc}}=a_{\mathrm{disc}}^G:\Pi_{\mathrm{disc}}(\mathbf{G},t)\longrightarrow\mathbb{C}$ as in section 4 of [A3] (p. 515–517).

Let T^G be the distribution of the θ-twisted invariant trace formula on $\mathbf{G}^0(\mathbb{A})$. The following proposition is the analog of theorem 6.1 of [A6] (and of the formula below (3.5) of that article).

Proposition 8.2.3[2] *Let $\phi=\phi^\infty\phi_\infty\in C_c^\infty(\mathbf{G}^0(\mathbb{A}),\xi^{-1})$. Assume that there exists an irreducible algebraic representation V of $\mathbf{H}$ such that, if $\varphi:W_\mathbb{R}\longrightarrow{}^L\mathbf{H}$ is the Langlands parameter of the discrete series L-packet Π_V of $\mathbf{H}(\mathbb{R})$ associated to V and π_∞ is the θ-discrete representation of $\mathbf{G}^0(\mathbb{R})$ with Langlands parameter $\varphi_{|W_\mathbb{C}}$,*

[2] I thank Sug Woo Shin for pointing out that I had forgotten terms on the spectral side of this proposition.

then ϕ_∞ is a twisted pseudo-coefficient of π_∞. Then

$$T^G(\phi) = \sum_{\mathbf{M}} (n_M^G)^{-1} T^G_{M,\mathrm{geom}}(\phi) = \sum_{t\geq 0} \sum_{\pi\in\Pi_{\mathrm{disc}}(\mathbf{G},t)} a_{\mathrm{disc}}(\pi)\,\mathrm{Tr}(\pi(\phi)A_\pi),$$

where the first sum is taken over the set of $\mathbf{G}^0(\mathbb{Q})$-conjugacy classes of Levi subsets $\mathbf{M}$ of $\mathbf{G}$.

Remarks 8.2.4

(1) If π is a cuspidal θ-stable representation of $\mathbf{G}^0(\mathbb{A})$, then $a_{\mathrm{disc}}(\pi) = m^+_{\mathrm{disc}}(\pi) + m^-_{\mathrm{disc}}(\pi)$. (This is an easy consequence of the definition of a_{disc}, cf. [A3] (4.3) and (4.4).)

(2) The spectral side of the formula of [A6] (formula above (3.5)) is simpler, because only the discrete automorphic representations of $\mathbf{G}^0(\mathbb{A})$ can contribute. However, in the twisted case, it is not possible to eliminate the contributions from the discrete spectrum of proper Levi subsets (because there might be representations of $\mathbf{M}(\mathbb{A})$ that are fixed by a regular element of W_0^G and still have a regular archimedean infinitesimal character).

(3) In theorem 3.3 of [A3], the sum is taken over all Levi subsets $\mathbf{M}$ of $\mathbf{G}$ such that $\mathbf{M}^0$ contains $\mathbf{M}_0$, and the coefficients are $|W_0^M||W_0^G|^{-1}$ instead of $(n_M^G)^{-1}$; it is easy to see that these are just two ways to write the same thing.

Proof. The second formula (i.e., the spectral side) is just (a) of theorem 7.1 of [A3], because ϕ is cuspidal at infinity.

We show the first formula. We have to compute the value at ϕ of the invariant distributions I_M^G of [A2]. As ϕ_∞ is cuspidal, we see by using the splitting formula (proposition 9.1 of [A2]) as in [A6] §3 that it is enough to compute the I_M^G at infinity, that is, to prove the analog of theorem 5.1 of [A6]. Moreover, by corollary 9.2 of [A2] applied to the set of places $S = \{\infty\}$, and thanks to the cuspidality of ϕ_∞, we see that the term corresponding to $\mathbf{M}$ is non-zero only if $\mathbf{A}_M = \mathbf{A}_{M_\mathbb{R}}$.[3]

So we may assume that $\mathbf{A}_M = \mathbf{A}_{M_\mathbb{R}}$. We want to show that $I_M^G(.,\phi_\infty) = 0$ if $\mathbf{M}$ is not cuspidal and that, for every cuspidal Levi subset $\mathbf{M}$ of $\mathbf{G}$ and for every $\gamma\in\mathbf{M}^0(\mathbb{R})$,

$$I_M^G(\gamma,\phi_\infty) = |D^M(\gamma)|^{1/2}\Phi_M^G(\gamma,\phi_\infty), \qquad (*)$$

where, if $\gamma\theta = (\sigma\theta)u$ is the Jordan decomposition of $\gamma\theta$, then

$$D^M(\gamma) = \det((1-\mathrm{Ad}(\sigma)\circ\theta), \mathrm{Lie}(\mathbf{M}^0)/\mathrm{Lie}(\mathbf{M}^0_{\sigma\theta})).$$

(Note that, if $\mathbf{M}$ is cuspidal, then $\mathbf{A}_M = \mathbf{A}_{M_\mathbb{R}}$.) This implies in particular that $I_M^G(\gamma,\phi_\infty) = 0$ if γ is not θ-semisimple. The rest of the proof of theorem 6.1 of [A6] applies without any changes to the case of nonconnected groups.

The case where $\mathbf{M}$ is not cuspidal is treated in lemma 8.2.6. In the rest of this proof, we assume that $\mathbf{M}$ is cuspidal.

For connected groups, the analog of formula $(*)$ for a semisimple regular γ is theorem 6.4 of [A5] (cf. formula (4.1) of [A6]). Arthur shows in section 5 of [A6] that the analog of $(*)$ for any γ is a consequence of this case.

---

[3] I thank Robert Kottwitz for patiently explaining to me this subtlety of the trace formula.

We first show that formula $(*)$ for a θ-semisimple θ-regular γ implies formula $(*)$ for any γ, by adapting the reasoning in section 5 of [A6]. The reasoning in the second half of page 277 of [A6] applies to the case considered here and shows that it is enough to prove $(*)$ for a $\gamma \in \mathbf{M}^0(\mathbb{R})$ such that $\mathbf{M}^0_{\gamma\theta} = \mathbf{G}^0_{\gamma\theta}$. Lemma 8.2.5 below is the analog of lemma 5.3 of [A6]. Once this lemma is known, the rest of the reasoning of [A6] applies. This is because Arthur reduces to the semisimple regular case by using the results on orbital integrals at unipotent elements of [A6] pp. 275–277, and we can apply the same results here, because these orbitals integrals are taken on the connected group $\mathbf{M}^0_{\sigma\theta}$ (where, as before, $\gamma\theta = (\sigma\theta)u$ is the Jordan decomposition of $\gamma\theta$).

It remains to show formula $(*)$ for a θ-semisimple θ-regular γ. The article [A5] is written in the setting of connected groups, but it is easy to check that, now the invariant trace formula for nonconnected groups is known, all the article up to, and including, corollary 6.3, applies to the general (not necessarily connected) case. We can write statements analogous to theorem 6.4 and lemma 6.6 of [A5] by making the following changes: take a θ-discrete representation π_∞ of $\mathbf{G}^0(\mathbb{R})$ (instead of a discrete series representation of $\mathbf{G}(\mathbb{R})$), and replace the character of π_∞ by the twisted character.

The proof of lemma 6.6 of [A5] applies to the nonconnected case, if we replace Π_{temp} by the set of isomorphism classes of θ-stable tempered representations, Π_{disc} by $\Pi_{\theta\text{-disc}}$ and "regular" by "θ-regular".

The proof of theorem 6.4 of [A5] proceeds by induction on $\mathbf{M}$, starting from the case $\mathbf{M} = \mathbf{G}$, and uses lemma 6.6 of [A5] and three properties of the characters of discrete series representations: the differential equations that they satisfy, the conditions at the boundary of the set of regular elements, and the growth properties. For the twisted characters of θ-discrete representations, there are of course similar differential equations; the bound that we need (in the third property) is proved by Clozel in theorem 5.1 of [Cl2]; as for the boundary conditions, they follow from theorem 7.2 of [Cl2] (called theorem 8.1.3 in this book) and from the case of connected groups. Once these results are known, the reduction to the case $\mathbf{M} = \mathbf{G}$ is the same as in [A5]. But the case $\mathbf{M} = \mathbf{G}$ is exactly lemma 8.1.10. □

Lemma 8.2.5 *Write* $\Phi'_M = |D^M(.)|^{-1/2} I^G_M$.

Let $\mathbf{M}$ *be a cuspidal Levi subset of* $\mathbf{G}$, *and let* $\gamma \in \mathbf{M}^0(\mathbb{R})$ *be such that* $\mathbf{G}^0_{\gamma\theta} = \mathbf{M}^0_{\gamma\theta}$. *Let* $\gamma\theta = (\sigma\theta)u$ *be the Jordan decomposition of* $\gamma\theta$. *Then there exist stable cuspidal functions* $f_1, \dots, f_n$ *on* $\mathbf{M}^0_{\sigma\theta}(\mathbb{R})$ *and a neighborhood* U *of* 1 *in* $\mathbf{M}^0_{\sigma\theta}(\mathbb{R})$, *invariant by* $\mathbf{M}^0_{\sigma\theta}(\mathbb{R})$*-conjugacy, such that, for every* $\mu \in U$,

$$\Phi'_M(\mu\sigma, \phi_\infty) = \sum_{i=1}^{n} \Phi_{M^0_{\sigma\theta}}(\mu, f_i). \qquad (**)$$

Proof. If σ is not θ-elliptic in $\mathbf{M}(\mathbb{R})$, then, by lemma 8.1.9, there exists a proper Levi subset $\mathbf{M}_1$ of $\mathbf{M}_{\mathbb{R}}$ such that $\sigma\theta \in \mathbf{M}_1(\mathbb{R})$ and $\mathbf{M}^0_{\mathbb{R},\sigma\theta} \subset \mathbf{M}^0_1$. If $\mu \in \mathbf{M}^0_{\sigma\theta}(\mathbb{R})$ is small enough, then $\mathbf{M}^0_{(\sigma\theta)\mu}$ is also included in $\mathbf{M}^0_1$. Applying the descent property (corollary 8.3 of [A2]) and using the cuspidality of ϕ_∞, we see that $\Phi'_M(\mu\sigma, \phi_\infty) = 0$, so that we can take $f_i = 0$.

We may therefore assume that σ is θ-elliptic in $\mathbf{M}^0(\mathbb{R})$. We may also assume that $\mathbf{M}$ is standard. Let $\mathbf{T}_H$ be a maximal torus in $\mathbf{M}_{H,\mathbb{R}}$ that is anisotropic modulo $\mathbf{A}_{M_H}$. Then σ is θ-conjugate to an element σ' such that $\sigma'\theta(\sigma') \in \mathbf{T}_H(\mathbb{R})$. As I_M^G is invariant by θ-conjugacy, we may assume that $h := \sigma\theta(\sigma) \in \mathbf{T}_H(\mathbb{R})$. As $\mathbf{G}^0_{\sigma\theta}$ is an inner form of $\mathbf{H}_h$ over $\mathbb{R}$, the maximal torus $\mathbf{T}_H$ of $\mathbf{H}_h$ transfers to a maximal torus $\mathbf{T}$ of $\mathbf{G}^0_{\sigma\theta}$; of course, $h \in \mathbf{T}(\mathbb{R})$. Let U be an invariant neighborhood of 1 in $\mathbf{M}^0_{\sigma\theta}(\mathbb{R})$ small enough so that $\mu\sigma$ is θ-regular if $\mu \in U \cap \mathbf{T}(\mathbb{R})$ is regular in $\mathbf{M}^0_{\sigma\theta}$. Then, by formula $(*)$ in the proof of proposition 8.2.3 above for a θ-regular element (the proof of formula $(*)$ in this case does not depend on the lemma), for every $\mu \in U \cap \mathbf{T}(\mathbb{R})$ that is regular in $\mathbf{M}^0_{\sigma\theta}$

$$\Phi'_M(\mu\sigma, \phi_\infty) = \Phi_M^G(\mu\sigma, \phi_\infty) = (-1)^{\dim(\mathbf{A}_M/\mathbf{A}_G)}\Phi_M^G(\mu\sigma, \Theta_{\pi_\infty^\vee}),$$

and we know that this is equal to

$$\pm|D^H_{M_H}(\mathcal{N}(\mu\sigma))|^{1/2}\Theta_{\Pi_H^\vee}(\mathcal{N}(\mu\sigma))$$

(where the sign depends on the choice of normalized intertwining operator on π_∞). By the proof of lemma 5.3 of [A6] and lemma 4.1 of [A6], there exists $f_1, \dots, f_n$ stable cuspidal on $\mathbf{M}^0_{\sigma\theta}(\mathbb{R})$ such that $(**)$ is satisfied for every $\mu \in U \cap \mathbf{T}(\mathbb{R})$ that is regular in $\mathbf{M}^0_{\sigma\theta}$.

It remains to show that, for this choice of $f_1, \dots, f_n$ and maybe after making U smaller, formula $(**)$ is true for every $\mu \in U$. But the end of the proof of lemma 5.3 of [A6] applies without any changes to the nonconnected case. □

Lemma 8.2.6 *Let* $\mathbf{M}$ *be a Levi subset of* $\mathbf{G}$. *Assume that* $\mathbf{A}_M = \mathbf{A}_{M_\mathbb{R}}$ *and that* $\mathbf{M}$ *is not cuspidal. Then* $I_M^G(., \phi_\infty) = 0$.

Proof. We may assume that $\mathbf{M}$ is standard. We first show that $I_M^G(\gamma, \phi_\infty) = 0$ if γ is θ-regular in $\mathbf{M}^0$. Let $\gamma \in \mathbf{M}^0(\mathbb{R})$ be θ-regular in $\mathbf{M}^0$. We may assume that $\gamma\theta(\gamma) \in \mathbf{M}_H(\mathbb{R})$. The centralizer $\mathbf{T}_H$ of $\gamma\theta(\gamma)$ in $\mathbf{M}_H$ is a maximal torus of $\mathbf{M}_{H,\mathbb{R}}$. By the assumption on $\mathbf{M}$, the torus $\mathbf{T}_H/\mathbf{A}_{M_\mathbb{R}}$ is not anisotropic, so there exists a Levi subgroup $\mathbf{M}_{1,H} \neq \mathbf{M}_{H,\mathbb{R}}$ of $\mathbf{M}_{H,\mathbb{R}}$ such that $\mathbf{T}_H \subset \mathbf{M}_{1,H}$ (e.g., the centralizer of the $\mathbb{R}$-split part of $\mathbf{T}_H$). Let $\mathbf{M}_1$ be the corresponding Levi subset of $\mathbf{G}_\mathbb{R}$. Then $\gamma \in \mathbf{M}_1^0(\mathbb{R})$ and $\mathbf{M}^0_{1,\gamma\theta} = \mathbf{M}^0_{\gamma\theta}$. By the descent formula (theorem 8.3 of [A2]) and the cuspidality of ϕ_∞, $I_M^G(\gamma, \phi_\infty) = 0$.

We now show the statement of the lemma. By formula (2.2) of [A2], it is enough to prove that, for every Levi subset $\mathbf{M}'$ of $\mathbf{G}$ containing $\mathbf{M}$ and every $\gamma \in \mathbf{M}^0(\mathbb{R})$ such that $\mathbf{M}^0_{\gamma\theta} = \mathbf{G}^0_{\gamma\theta}$, $I^G_{M'}(\gamma, \phi_\infty) = 0$. If $\mathbf{M}' \neq \mathbf{M}$, this follows from the descent formula (theorem 8.3 of [A2]) and the cuspidality of ϕ_∞. It remains to show that $I_M^G(\gamma, \phi_\infty) = 0$, if $\gamma \in \mathbf{M}^0(\mathbb{R})$ is such that $\mathbf{M}^0_{\gamma\theta} = \mathbf{G}^0_{\gamma\theta}$. Let $\gamma\theta = (\sigma\theta)u$ be the Jordan decomposition of $\gamma\theta$. By (2.3) of [A2], there exist $f \in C_c^\infty(\mathbf{M}^0(\mathbb{R}))$ and an open neighborhood U of 1 in $\mathbf{M}^0_{\sigma\theta}(\mathbb{R})$ such that, for every $\mu \in U$, $I_M^G(\mu\sigma, \phi_\infty) = O_{\mu\sigma\theta}(f)$. Hence, by the beginning of the proof, $O_{\mu\sigma\theta}(f) = I_M^G(\mu\sigma, \phi_\infty) = 0$ if $\mu \in U$ is such that $\mu\sigma$ is θ-regular. This implies that $O_{\mu\sigma\theta}(f) = 0$ for every $\mu \in U$. On the other hand, after replacing γ by a θ-conjugate, we may assume that $u \in U$. So $I_M^G(\gamma, \phi_\infty) = 0$. □

8.3 STABILIZATION OF THE INVARIANT TRACE FORMULA

In this section, we stabilize the invariant trace formula of proposition 8.2.3 if $\mathbf{H}$ is one of the quasi-split unitary groups of section 2.1. Actually, there is nothing to stabilize; the invariant trace formula is already stable in this case, and we simply show this.

We use the notation and assumptions of sections 8.2 and 5.4.

Proposition 8.3.1 *Assume that* $\mathbf{H}$ *is one of the quasi-split unitary groups of section 2.1 and that* E *is the imaginary quadratic extension of* $\mathbb{Q}$ *that was used to define* $\mathbf{H}$. *Let* $f=\bigotimes_v f_v \in C_c^\infty(\mathbf{H}(\mathbb{A}),\xi_H^{-1})$ *and* $\phi=\bigotimes_v \phi_v \in C_c^\infty(\mathbf{G}^0(\mathbb{A}),\xi^{-1})$ *(where* ξ_H *is the restriction of* ξ *to* $\mathbf{A}_H(\mathbb{R})^0$*). Assume that, for every finite place* v *of* $\mathbb{Q}$*, the functions* f_v *and* ϕ_v *are associated in the sense of [La3] 3.2, that the function* ϕ_∞ *is of the type considered in proposition 8.2.3 and that*

$$f_\infty=\frac{1}{|\Pi_V|}\sum_{\pi_H\in\Pi_V} f_{\pi_H}$$

(cf. corollary 8.1.11). Then there exists a constant $C\in\mathbb{R}^\times$ *(depending only on* $\mathbf{H}$ *and the choice of normalized intertwining operators on* θ*-stable automorphic representations of* $\mathbf{G}^0(\mathbb{A})$*), such that, for every Levi subset* $\mathbf{M}$ *of* $\mathbf{G}$,

$$T^G_{M,\mathrm{geom}}(\phi)=C\frac{d(\mathbf{G})}{\tau(\mathbf{H})}ST^H_{M_H}(f).$$

In particular,

$$T^G(\phi)=C\frac{d(\mathbf{G})}{\tau(\mathbf{H})}ST^H(f).$$

Remark 8.3.2 After maybe choosing different normalized intertwining operators on the θ-stable automorphic representations of $\mathbf{G}^0(\mathbb{A})$, we may assume that C is positive.

Proof. To see that the equalities for the $T^G_{M,\mathrm{geom}}$ imply the equality for T^G, it is enough to notice that the obvious map from the set of $\mathbf{G}^0(\mathbb{Q})$-conjugacy classes of Levi subsets of $\mathbf{M}$ to the set $\mathbf{H}(\mathbb{Q})$-conjugacy classes of Levi subgroups of $\mathbf{H}$ is a bijection, and that $n^H_{M_H}=n^G_M$ if $\mathbf{M}_H$ corresponds to $\mathbf{M}$.

Let $\mathbf{M}$ be a standard cuspidal Levi subset of $\mathbf{G}$. As the morphism $\mathbf{H}^1(K,\mathbf{M}^0)\longrightarrow\mathbf{H}^1(K,\mathbf{G}^0)$ is injective (see the proof of lemma 6.3.4) and $\mathbf{H}^1(K,\mathbf{G}^0)=\{1\}$ for every field K, the assumption on $\mathbf{G}^0$ implies that $d(\mathbf{M}_H,\mathbf{M}^0)=1$, where $d(\mathbf{M}_H,\mathbf{M}^0)$ is defined in [La3] 1.9.3. By lemma 8.1.2 and the fact that the descent formula (corollary 8.3 of [A2]) works just as well for twisted orbital integrals, the proof of lemma 6.3.4 applies in the case considered here and shows that the functions ϕ_M and f_{M_H} are associated at every finite place. Using this fact and lemma 8.3.3, we may apply the stabilization process of chapter 4 of [La3], on the group $\mathbf{M}\rtimes\langle\theta\rangle$, to $T^G_{M,\mathrm{geom}}(\phi)$. As the set of places $\{\infty\}$ is $(\mathbf{M},\mathbf{M}_H)$-essential (by lemma A.2.1 of [CL], whose proof adapts immediately to the case of unitary similitude groups), we get

$$T^G_{M,\mathrm{geom}}(\phi)=C\tau(\mathbf{M}^0)\tau(\mathbf{M}_H)^{-1}d(\mathbf{M})k(\mathbf{M}_H)^{-1}k(\mathbf{H})ST^H_{M_H}(f),$$

with $C \in \mathbb{R}^\times$ (the factor $2^{-\dim(\mathfrak{a}_G)}$ of [La3] 4.3.2 does not appear here because we are taking functions in $C_c^\infty(\mathbf{G}^0(\mathbb{A}), \xi^{-1})$ and not in $C_c^\infty(\mathbf{G}^0(\mathbb{A}))$; and the factor $J_Z(\theta)$ does not appear because, following Arthur, we consider the action of these functions on $L^2(\mathbf{G}^0(\mathbb{Q}) \setminus \mathbf{G}^0(\mathbb{A}), \xi)$ and not on $L^2(\mathbf{G}^0(\mathbb{Q})\mathbf{A}_{G^0}(\mathbb{R})^0\backslash\mathbf{G}^0(\mathbb{A}))$).

To finish the proof, it is enough to check that

$$\frac{\tau(\mathbf{M}^0)}{\tau(\mathbf{M}_H)}\frac{d(\mathbf{M})}{k(\mathbf{M}_H)}k(\mathbf{H}) = d(\mathbf{G})\frac{\tau(\mathbf{G}^0)}{\tau(\mathbf{H})}.$$

By (ii) of lemma 2.3.3, $\tau(\mathbf{G}^0) = \tau(\mathbf{M}^0) = 1$. So the equality above follows from remarks 5.4.3 and 8.1.8. □

In the following lemma, we consider the situation of the beginning of 8.2, so that $\mathbf{H}$ is a cuspidal connected reductive group over $\mathbb{Q}$, E is an imaginary quadratic extension of $\mathbb{Q}$, and $\mathbf{G}^0 = R_{E/\mathbb{Q}}\mathbf{H}_E$. Fix a θ-stable Borel subgroup of $\mathbf{G}^0$ (or, equivalently, a Borel subgroup of $\mathbf{H}$).

Lemma 8.3.3 *Use the notation of [La3] 2.7. Let ϕ_∞ be as in proposition 8.2.3, $\mathbf{M}$ be a standard cuspidal Levi subset of $\mathbf{G}$, and $\gamma \in \mathbf{M}^0(\mathbb{R})$ be θ-semisimple. Set*

$$f_\infty = \frac{1}{|\Pi_V|}\sum_{\pi_H \in \Pi_V} f_{\pi_H}$$

(cf. corollary 8.1.11). Then there exists a constant $C \in \mathbb{R}^\times$ (that is independent of $\mathbf{M}$ and positive for a good choice of normalized intertwining operators) such that

$$\begin{aligned}\sum_{[x]\in\mathfrak{D}(I_\gamma,\mathbf{M}^0;\mathbb{R})} e(\delta_x)\Phi_M^G(\delta_x,\phi_\infty) &= k(\mathbf{M})^{-1}k(\mathbf{G})d(\mathbf{M})S\Phi_M^G(\gamma,\phi_\infty)\\ &= Cd(\mathbf{M})k(\mathbf{M}_H)^{-1}k(\mathbf{H})S\Phi_{M_H}^H(\mathcal{N}\gamma, f_\infty),\end{aligned}$$

and, if $\kappa \in \mathfrak{K}(I_\gamma, \mathbf{M}^0; \mathbb{R}) - \{1\}$,

$$\sum_{[x]\in\mathfrak{D}(I_\gamma,\mathbf{M}^0;\mathbb{R})} e(\delta_x)\langle\kappa,\dot{x}\rangle\Phi_M^G(\delta_x,\phi_\infty) = 0.$$

Proof. Once we notice that $\Phi_M^G(\gamma, \Theta_{\pi^\vee})$ is invariant under stable θ-conjugacy, the proof is exactly the same as in theorem A.1.1 of [CL]. To show the second line of the first equality, use the definitions, theorem 8.1.3, remark 8.1.6, lemma 8.1.2, and the fact that, if $\gamma \in \mathbf{G}^0(\mathbb{R})$ and $h \in \mathcal{N}\gamma$, then $\mathbf{H}_h$ is an inner form of $\mathbf{G}^0_{\gamma\theta}$. □

We finish this section by recalling a few results on the transfer and the fundamental lemma for base change. Assume that we are in the situation of example 8.1.1, with K a local field of characteristic 0. If two functions $f \in C_c^\infty(\mathbf{G}^0(K))$ and $h \in C_c^\infty(\mathbf{H}(K))$ are associated in the sense of [La3] 3.2, we also say that h is a transfer of f to $\mathbf{H}$. Labesse proved the following result.

Theorem 8.3.4 *([La3] theorem 3.3.1 and proposition 3.5.2) Let $f \in C_c^\infty(\mathbf{G}^0(K))$. Then there exists a transfer of f to $\mathbf{H}$.*

Labesse has also proved a result about inverse transfer. We say that an element $\gamma_H \in \mathbf{H}(K)$ *is a norm* if there exists $\gamma \in \mathbf{G}^0(K)$ such that $\gamma_H \in \mathcal{N}\gamma$. Assume that K is non-archimedean.

Proposition 8.3.5 *([La1] propositions 3.3.2, 3.5.3) Let $h \in C_c^\infty(\mathbf{H}(K))$ be such that $SO_{\gamma_H}(h) = 0$ for every semisimple $\gamma_H \in \mathbf{H}(K)$ that is not a norm. Then there exists $f \in C_c^\infty(\mathbf{G}^0(K))$ such that h is a transfer of f.*

So, in order to determine which functions on $\mathbf{H}(K)$ are transfers of functions on $\mathbf{G}^0(K)$, we need to describe the set of norms on $\mathbf{H}(K)$. To do this, we use the results of 2.5 of [La1]. In the next lemma, assume that $\mathbf{H} = \mathbf{G}(\mathbf{U}^*(n_1) \times \cdots \times \mathbf{U}^*(n_r))$ with $n_1, \ldots, n_r \in \mathbb{N}^*$ (notation as in section 2.2) and that $\mathbf{G}^0 = R_{E/\mathbb{Q}}\mathbf{H}_E$, where E is the quadratic extension of $\mathbb{Q}$ used to define $\mathbf{H}$. Take $K = \mathbb{Q}_p$, where p is a prime number.

Lemma 8.3.6 *Let $D_H = \mathbf{H}/\mathbf{H}^{\mathrm{der}}$. Then a semisimple element of $\mathbf{H}(\mathbb{Q}_p)$ is a norm if and only if its image in $D_H(\mathbb{Q}_p)$ is a norm. If p splits and is unramified in E, or if $\mathbf{H} = \mathbf{GU}^*(n)$ with n odd, or if p is unramified in E and (at least) one of the n_i is odd, then every semisimple element of $\mathbf{H}(\mathbb{Q}_p)$ is a norm.*[4]

Proof. Notice that, if p splits and is unramified in E, then $\mathbf{G}^0(\mathbb{Q}_p) \simeq \mathbf{H}(\mathbb{Q}_p) \times \mathbf{H}(\mathbb{Q}_p)$, and the naive norm map $\mathbf{G}^0(\mathbb{Q}_p) \longrightarrow \mathbf{G}^0(\mathbb{Q}_p)$, $g \longmapsto g\theta(g)$, is actually a surjection from $\mathbf{G}^0(\mathbb{Q}_p)$ to $\mathbf{H}(\mathbb{Q}_p)$; there is a similar statement for D_H. So the lemma is trivial in that case.

Hence, for the rest of the proof, we assume that there is only one place $\wp$ of E above p, that is, that p is inert or ramified in E (this is just to avoid a discussion of cases; the results of Labesse apply of course just as well in the general case). As $\mathbf{H}^{\mathrm{der}}$ is simply connected, $D_H(\mathbb{Q}_p) = \mathrm{H}^0_{ab}(\mathbb{Q}_p, \mathbf{H})$, in the notation of [La3] 1.6. Similarly, for every Levi subgroup $\mathbf{M}_H$ of $\mathbf{H}_{\mathbb{Q}_p}$, if we set $D_{M_H} = \mathbf{M}_H/\mathbf{M}_H^{\mathrm{der}}$, then $D_{M_H}(\mathbb{Q}_p) = \mathrm{H}^0_{ab}(\mathbb{Q}_p, \mathbf{M}_H)$. Let γ be a semisimple element of $\mathbf{H}(\mathbb{Q}_p)$, and let $\mathbf{M}_H$ be a Levi subgroup of $\mathbf{H}_{\mathbb{Q}_p}$ such that $\gamma \in \mathbf{M}_H(\mathbb{Q}_p)$ and γ is elliptic in $\mathbf{M}_H$. By proposition 2.5.3 of [La3], γ is a norm if and only if its image in $D_{M_H}(\mathbb{Q}_p)$ is a norm. So, to prove the first statement of the lemma, it is enough to show that an element of $D_{M_H}(\mathbb{Q}_p)$ is a norm if and only if its image by the canonical map $D_{M_H}(\mathbb{Q}_p) \longrightarrow D_H(\mathbb{Q}_p)$ is a norm. As $\mathbf{H}$ does not split over $\mathbb{Q}_p$, $\mathbf{M}_H$ is $\mathbf{H}(\mathbb{Q}_p)$-conjugate to a Levi subgroup of $\mathbf{H}$ (defined over $\mathbb{Q}$), so we may assume that $\mathbf{M}_H$ is a standard Levi subgroup of $\mathbf{H}$ and is defined over $\mathbb{Q}$. By section 2.2, there exist $s, m_1, \ldots, m_r \in \mathbb{N}$ such that $n_i - m_i$ is even for every i and $\mathbf{M}_H$ is isomorphic to the direct product of $\mathbf{G}(\mathbf{U}^*(m_1) \times \cdots \times \mathbf{U}^*(m_r))$ and of s groups obtained by Weil restriction of scalars from general linear groups over E. The derived group of $\mathbf{H}$ is $\mathbf{SU}^*(n_1) \times \cdots \times \mathbf{SU}^*(n_r)$, so the map $\mathbf{G}(\mathbf{U}^*(n_1) \times \ldots \mathbf{U}^*(n_r)) \longrightarrow \mathbb{G}_m \times (R_{E/\mathbb{Q}}\mathbb{G}_m)^r$, $(g_1, \ldots, g_r) \longmapsto (c(g_1), \det(g_1), \ldots, \det(g_r))$ induces an isomorphism

$$D_H \xrightarrow{\sim} \{(\lambda, z_1, \ldots, z_r) \in \mathbb{G}_m \times (R_{E/\mathbb{Q}}\mathbb{G}_m)^r | \forall i, z_i \overline{z}_i = \lambda^{n_i}\}.$$

[4]Note that the first assertion of this lemma is also a consequence of lemma 4.2.1 of [Ha] (and of proposition 2.5.3 of [La3]).

Similarly, there is an isomorphism

$$D_{M_H} \xrightarrow{\sim} D_{M_H,l} \times D_{M_H,h},$$

where $D_{M_H,l} = (R_{E/\mathbb{Q}}\mathbb{G}_m)^s$ and

$$D_{M_H,h} = \{(\lambda, z_1, \ldots, z_r) \in \mathbb{G}_m \times (R_{E/\mathbb{Q}}\mathbb{G}_m)^r | \forall i, z_i\overline{z}_i = \lambda^{m_i} \text{ if } m_i > 0 \text{ and } z_i = 1 \text{ if } m_i = 0\}.$$

The canonical map $D_{M_H} \longrightarrow D_H$ sends the factor $D_{M_H,l}$ to 1 and is induced on the factor $D_{M_H,h}$ by the map $(\lambda, z_1, \ldots, z_r) \longmapsto (\lambda, \lambda^{(n_1-m_1)/2}z_1, \ldots, \lambda^{(n_r-m_r)/2}z_r)$. As every element in $D_{M_H,l}(\mathbb{Q}_p)$ is obviously a norm, it is now clear that an element of $D_{M_H}(\mathbb{Q}_p)$ is a norm if and only if its image in $D_H(\mathbb{Q}_p)$ is a norm.

Assume that $\mathbf{H} = \mathbf{GU}^*(n)$ with n odd, and write $n = 2m+1$, $m \in \mathbb{N}$. Then it is easy to check that the map $\mathbf{H} \longrightarrow R_{E/\mathbb{Q}}\mathbb{G}_m$, $g \longmapsto \det(g)c(g)^{-m}$, induces an isomorphism $D_H \xrightarrow{\sim} R_{E/\mathbb{Q}}\mathbb{G}_m$. So every element of $D_H(\mathbb{Q}_p)$ is a norm, and consequently every semi-simple element of $\mathbf{H}(\mathbb{Q}_p)$ is a norm.

Assume that $\mathbf{H} = \mathbf{G}(\mathbf{U}^*(n_1) \times \cdots \times \mathbf{U}^*(n_r))$ with $n_1, \ldots, n_r \in \mathbb{N}^*$, that n_1 is odd and that p is inert and unramified in E. Write $n_1 = 2m_1 + 1$, $m_1 \in \mathbb{N}$. Then the map $\mathbb{G}_m \times (R_{E/\mathbb{Q}}\mathbb{G}_m)^r$, $(\lambda, z_1, \ldots, z_r) \longmapsto (z_1\lambda^{-m_1}, z_2(z_1^{-1}\lambda^{m_1})^{n_2}, z_3(z_1^{-1}\lambda^{m_1})^{n_3}, \ldots, z_r(z_1^{-1}\lambda^{m_1})^{n_r})$ induces an isomorphism

$$D_H \simeq R_{E/\mathbb{Q}}\mathbb{G}_m \times \mathbf{U}(1)^{r-1}$$

(cf the description of D_H given above). It is obvious that every element of $(R_{E/\mathbb{Q}}\mathbb{G}_m)(\mathbb{Q}_p) = E_p^\times$ is a norm, so, to finish the proof of the lemma, it is enough to show that every element of $\mathbf{U}(1)(\mathbb{Q}_p)$ is a norm. Let $z \in \mathbf{U}(1)(\mathbb{Q}_p)$. Then z is an element of $E_p^\times$ such that $z\overline{z} = 1$, and we want to show that there exists $y \in E_p^\times$ such that $z = y\overline{y}^{-1}$. Write $z = ap^k$, with $a \in \mathcal{O}_{E_p}^\times$ and $k \in \mathbb{Z}$. Then $z\overline{z} = a\overline{a}p^{2k} = 1$, so $k = 0$ and $a\overline{a} = 1$, and we want to show that there exists $b \in \mathcal{O}_{E_p}^\times$ such that $a = b\overline{b}^{-1}$. By Hensel's lemma, it is enough to check the analog of this for the reduction modulo p of a. As p is inert and unramified in E, $\mathcal{O}_{E_p}/(p) = \mathbb{F}_{p^2}$. Let $u : \mathbb{F}_{p^2}^\times \longrightarrow \mathbb{F}_{p^2}^\times$ be the group morphism that sends b to $b\overline{b}^{-1}$. Then $\varphi(b) = b^{1-p}$ for every b, so the image of φ is of cardinality $p^2 - 1/(p-1) = p+1$. But this image is contained in $\mathbf{U}(1)(\mathbb{F}_p)$, and $\mathbf{U}(1)(\mathbb{F}_p) = \{a \in \mathbb{F}_{p^2}^\times | a^{p+1} = 1\}$ is of cardinality $p+1$, so $\varphi(\mathbb{F}_{p^2}^\times) = \mathbf{U}(1)(\mathbb{F}_p)$. This finishes the proof. □

The above lemma (together with the result of Labesse about inverse transfer, that is, proposition 8.3.5, and the fact that the group of norms in a torus contains an open neighborhood of 1) has the following immediate consequence.

Lemma 8.3.7 *If p splits and is unramified in E, or if $\mathbf{H} = \mathbf{GU}^*(n)$ with n odd, or if p is unramified in E and one of the n_i is odd, then every function in $C_c^\infty(\mathbf{H}(\mathbb{Q}_p))$ is a transfer of a function in $C_c^\infty(\mathbf{G}^0(\mathbb{Q}_p))$. In general, the set of functions in $C_c^\infty(\mathbf{H}(\mathbb{Q}_p))$ that are a transfer of a function in $C_c^\infty(\mathbf{G}^0(\mathbb{Q}_p))$ is a subalgebra of $C_c^\infty(\mathbf{H}(\mathbb{Q}_p))$, and it contains all the functions with support a small enough neighborhood of 1.*

Transfer is explicit if we are in an unramified situation. Assume that K is non-archimedean, that the group $\mathbf{H}$ is unramified over K and that the extension E/K is unramified. Let K_G and K_H be hyperspecial maximal compact subgroups of $\mathbf{G}^0(K)$ and $\mathbf{H}(K)$ such that $\mathrm{K}_H = \mathbf{H}(K) \cap \mathrm{K}_G$ and $\theta(\mathrm{K}_G) = \mathrm{K}_G$. The L-morphism $\xi : {}^L\mathbf{H} \longrightarrow {}^L\mathbf{G}^0$ defined in example 8.1.1 induces a morphism of algebras $b : \mathcal{H}(\mathbf{G}^0(\mathrm{K}), \mathrm{K}_G) \longrightarrow \mathcal{H}(\mathbf{H}(K), \mathrm{K}_H)$, called the base change morphism. The following theorem, known under the name "fundamental lemma for base change," is due to Kottwitz (for the unit element of $\mathcal{H}(\mathbf{G}^0(\mathrm{K}), \mathrm{K}_G)$), Clozel and Labesse (for the other elements).

Theorem 8.3.8 *([K6], [Cl3], [La1], [La3] 3.7) Let $f \in \mathcal{H}(\mathbf{G}^0(K), \mathrm{K}_G)$. Then $b(f)$ is a transfer of f to $\mathbf{H}$.*

Let us write down explicit formulas for the base change morphism in the case of unitary groups. Let $\mathbf{H} = \mathbf{G}(\mathbf{U}^*(n_1) \times \cdots \times \mathbf{U}^*(n_r))$, E be the imaginary quadratic extension of $\mathbb{Q}$ used to define $\mathbf{H}$ and p be a prime number that is unramified in E. The groups $\mathbf{G}^0$ and $\mathbf{H}$ have obvious $\mathbb{Z}_p$-models (cf. remark 2.1.1), and we take $\mathrm{K}_G = \mathbf{G}^0(\mathbb{Z}_p)$ and $\mathrm{K}_H = \mathbf{H}(\mathbb{Z}_p)$. Use the notation of chapter 4.

If p is inert in E, the base change morphism is calculated in section 4.2 (with $L = E_p$ and $\mathbf{G} = \mathbf{H}$).

Assume that p splits in E. Then $\mathbf{G}^0_{\mathbb{Q}_p} \simeq \mathbf{H}_{\mathbb{Q}_p} \times \mathbf{H}_{\mathbb{Q}_p}$, and, for every $g = (g_1, g_2) \in \mathbf{G}^0(\mathbb{Q}_p) = \mathbf{H}(\mathbb{Q}_p) \times \mathbf{H}(\mathbb{Q}_p)$, $g_1 g_2 \in \mathcal{N} g$. To simplify notation, we assume that $r = 1$. Then there is an isomorphism (defined in section 4.2)

$$\mathcal{H}(\mathbf{H}(\mathbb{Q}_p), \mathrm{K}_H) \simeq \mathbb{C}[X^{\pm 1}] \otimes \mathbb{C}[X_1^{\pm 1}, \dots, X_n^{\pm 1}]^{\mathfrak{S}_n}.$$

So there is an obvious isomorphism

$$\mathcal{H}(\mathbf{G}^0(\mathbb{Q}_p), \mathrm{K}_G) \simeq \mathbb{C}[Z_1^{\pm 1}] \otimes \mathbb{C}[Z_{1,1}^{\pm 1}, \dots, Z_{n,1}^{\pm 1}]^{\mathfrak{S}_n} \otimes \mathbb{C}[Z_2^{\pm 1}] \otimes \mathbb{C}[Z_{1,2}^{\pm 1}, \dots, Z_{n,2}^{\pm 1}]^{\mathfrak{S}_n},$$

and the base change morphism is induced by

$$Z_j \longmapsto X, \qquad Z_{i,j} \longmapsto X_i.$$

In particular, the base change morphism is surjective if p splits in E. If p is inert in E, then the image of the base change morphism is given in remark 4.2.2; in particular, the base change morphism is surjective if and only if one of the n_i is odd.

The following lemma will be useful in the applications of the next section. We assume again that $\mathbf{H}$ is any connected unramified group on a non-archimedean local field K and that the extension E/K is unramified, and we choose hyperspecial maximal compact subgroups K_H and K_G of $\mathbf{H}(K)$ and $\mathbf{G}^0(K)$ as before.

Lemma 8.3.9 *Let π be a θ-stable admissible irreducible representation of $\mathbf{G}^0(K)$, and A_π be a normalized intertwining operator on π. Let ε be a root of unity such that A_π acts on π^{K_G} by multiplication by ε (such an ε exists because A_π stabilizes π^{K_G} and* $\dim \pi^{\mathrm{K}_G} \leq 1$*). Then, for every $f \in \mathcal{H}(\mathbf{G}^0(K), \mathrm{K}_G)$,*

$$\mathrm{Tr}(\pi(f) A_\pi) = \varepsilon\, \mathrm{Tr}(\pi(f)).$$

8.4 APPLICATIONS

The notation here is slightly different from that used in section 8.3. Let $\mathbf{H} = \mathbf{G}(\mathbf{U}(p_1, q_1) \times \cdots \times \mathbf{U}(p_r, q_r))$ (this group is defined in 2.1), $\mathbf{H}^*$ be a quasi-split inner form of $\mathbf{H}$ (so $\mathbf{H}^* = \mathbf{G}(\mathbf{U}^*(n_1) \times \cdots \times \mathbf{U}^*(n_r))$, where $n_i = p_i + q_i$), E be the imaginary quadratic extension of $\mathbb{Q}$ that was used in the definition of $\mathbf{H}$ and $\mathbf{G}^0 = R_{E/\mathbb{Q}}\mathbf{H}^*_E$. Of course, $\mathbf{G}^0 \simeq R_{E/\mathbb{Q}}\mathbf{H}_E$.

If V is an irreducible algebraic representation of $\mathbf{H}$, let ϕ_V be a twisted pseudo-coefficient of the θ-discrete representation π_V of $\mathbf{G}^0(\mathbb{R})$ associated to $\varphi_{|W_\mathbb{C}}$, where $\varphi : W_\mathbb{R} \longrightarrow {}^L\mathbf{H}$ is a Langlands parameter of the L-packet of the discrete series of $\mathbf{H}(\mathbb{R})$ associated to V.

Let $\mathcal{M}'_\mathbf{G}$ be the set of conjugacy classes of Levi subsets $\mathbf{M}$ of $\mathbf{G}$ such that, for every $i \in \{1, \dots, r\}$, $\mathbf{M}^0 \cap R_{E/\mathbb{Q}}\mathbf{GL}_{n_i,E}$ is equal to $R_{E/\mathbb{Q}}\mathbf{GL}_{n_i,E}$ or to a maximal Levi subgroup of $R_{E/\mathbb{Q}}\mathbf{GL}_{n_i,E}$. Let $\mathbf{M} \in \mathcal{M}'_\mathbf{G}$. Then there exist non-negative integers $n_1^+, n_1^-, \dots, n_r^+, n_r^-$ such that, for every $i \in \{1, \dots, r\}$, $n_i = n_i^+ + n_i^-$, and $\mathbf{M}^0 \cap R_{E/\mathbb{Q}}\mathbf{GL}_{n_i,E} = R_{E/\mathbb{Q}}\mathbf{GL}_{n_i^+,E} \times R_{E/\mathbb{Q}}\mathbf{GL}_{n_i^-,E}$. Let $\mathcal{M}_\mathbf{G}$ be the set of $\mathbf{M} \in \mathcal{M}'_\mathbf{G}$ such that we can choose the n_i^+, n_i^- so that $n_i^- + \cdots + n_r^-$ is even. If $\mathbf{M}$ is in $\mathcal{M}_\mathbf{G}$ and the n_i^+, n_i^- are as above, then we may assume that $n_i^- + \cdots +, n_r^-$ is even; let $(\mathbf{H}_M, s_{H_M}, \eta_{H_M,0})$ be the elliptic endoscopic datum for $\mathbf{H}$ defined by the n_i^+, n_i^- as in proposition 2.3.1, and η_{H_M} be an L-morphism extending $\eta_{H_M,0}$ as in proposition 2.3.2. This defines a bijection between $\mathcal{M}_\mathbf{G}$ and the set $\mathcal{E}_\mathbf{H}$ of 7.2.

Let $\mathbf{M} \in \mathcal{M}_\mathbf{G}$. Let $\xi : {}^L\mathbf{H} = {}^L\mathbf{H}^* \longrightarrow {}^L\mathbf{G}^0$ be the L-morphism defined in example 8.1.1; as $\mathbf{M}^0 = R_{E/\mathbb{Q}}\mathbf{H}_{M,E}$, we get in the same way an L-morphism $\xi_M : {}^L\mathbf{H}_M \longrightarrow {}^L\mathbf{M}^0$. Let η_M be the morphism

$$\begin{array}{rcl} {}^L\mathbf{M}^0 \simeq (\widehat{\mathbf{H}}_M \times \widehat{\mathbf{H}}_M) \rtimes W_\mathbb{Q} & \longrightarrow & {}^L\mathbf{G}^0 \simeq (\widehat{\mathbf{H}} \times \widehat{\mathbf{H}}) \rtimes W_\mathbb{Q} \\ ((h_1, h_2), w) & \longmapsto & ((\eta_{H_M,1}(h_1, w), \eta_{H_M,1}(h_2, w)), w), \end{array}$$

where $\eta_{H_M,1} : {}^L\mathbf{H}_M \longrightarrow \widehat{\mathbf{H}}$ is the first component of η_{H_M}. It is clear that η_M is an L-morphism that makes the following diagram commute:

$$\begin{array}{ccc} {}^L\mathbf{H}_M & \xrightarrow{\eta_{H_M}} & {}^L\mathbf{H} \\ \downarrow{\scriptstyle \xi_M} & & \downarrow{\scriptstyle \xi} \\ {}^L\mathbf{M}^0 & \xrightarrow[\eta_M]{} & {}^L\mathbf{G}^0 \end{array}$$

Note that the embedding $\widehat{\mathbf{M}}^0 \longrightarrow \widehat{\mathbf{G}}^0$ induced by η_M is $W_\mathbb{Q}$-equivariant. Let $\eta_{M,\mathrm{simple}} : {}^L\mathbf{M}^0 \longrightarrow {}^L\mathbf{G}^0$ be the obvious L-morphism extending this embedding (i.e., the one that is equal to identity on $W_\mathbb{Q}$). Write $\eta_M = c_M \eta_{M,\mathrm{simple}}$, where $c_M : W_\mathbb{Q} \longrightarrow Z(\widehat{\mathbf{M}}^0)$ is a 1-cocycle, and let χ_M be the quasi-character of $\mathbf{M}^0(\mathbb{A})$ associated to the class of c_M in $\mathrm{H}^1(W_\mathbb{Q}, Z(\widehat{\mathbf{M}}^0))$. (In general, χ_M can be nontrivial.)

Let S be a set of finite places of $\mathbb{Q}$. Write $\mathbb{A}_S = \prod'_{v \in S} \mathbb{Q}_v$ and $\mathbb{A}^S = \prod'_{v \notin S} \mathbb{Q}_v$. We say that a function $f_S \in C_c^\infty(\mathbf{H}(\mathbb{A}_S))$ satisfies condition (H) if, for every $\mathbf{M} \in \mathcal{M}_G$, there exist a transfer $f_S^{H_M}$ of f_S to $\mathbf{H}_M$ and a function $\phi_{S,M} \in C_c^\infty(\mathbf{M}^0(\mathbb{A}_S))$ such that the functions $\phi_{S,M}$ and $f_S^{H_M}$ are associated at every place in S.

The next lemma gives examples of functions that satisfy condition (H). For every place v of $\mathbb{Q}$, we say that a semisimple element $\gamma \in \mathbf{H}(\mathbb{Q}_v)$ *is a norm* if there exists

$g \in \mathbf{G}^0(\mathbb{Q}_v)$ such that $\gamma \in \mathcal{N}g$ (this condition makes sense because $\mathcal{N}g$ is a stable conjugacy class in $\mathbf{H}^*(\mathbb{Q}_v)$ and $\mathbf{H}$ is an inner form of $\mathbf{H}^*$).

Lemma 8.4.1 *Let v be a finite place of $\mathbb{Q}$.*

(i) Every function in $C_c^\infty(\mathbf{H}(\mathbb{Q}_v))$ with support in a small enough neighborhood of 1 satisfies condition (H).

(ii) Assume that $\mathbf{H}$ is quasi-split over $\mathbb{Q}_v$ (but not necessarily unramified). Then, for every $\phi \in C_c^\infty(\mathbf{G}^0(\mathbb{Q}_v))$ with support in a small enough neighborhood of 1, there exists $f \in C_c^\infty(\mathbf{H}(\mathbb{Q}_v))$ associated to ϕ and satisfying condition (H).

(iii) Assume that v is unramified in E (so $\mathbf{H}_{\mathbb{Q}_v} = \mathbf{H}^*_{\mathbb{Q}_v}$ is unramified). Let $\mathbf{M} \in \mathcal{M}_G$. Then the commutative diagram

$$\begin{array}{ccc} {}^L\mathbf{H}_M & \xrightarrow{\eta_{H_M}} & {}^L\mathbf{H} \\ \downarrow{\scriptstyle \xi_M} & & \downarrow{\scriptstyle \xi} \\ {}^L\mathbf{M}^0 & \xrightarrow[\eta_M]{} & {}^L\mathbf{G}^0 \end{array}$$

gives a commutative diagram

$$\begin{array}{ccc} \mathcal{H}(\mathbf{G}^0(\mathbb{Q}_v), \mathbf{G}^0(\mathbb{Z}_v)) & \longrightarrow & \mathcal{H}(\mathbf{H}(\mathbb{Q}_v), \mathbf{H}(\mathbb{Z}_v)) \\ \downarrow & & \downarrow \\ \chi_{M,v}\mathcal{H}(\mathbf{M}^0(\mathbb{Q}_v), \mathbf{M}^0(\mathbb{Z}_v)) & \longrightarrow & \chi_{\eta_{H_M},v}\mathcal{H}(\mathbf{H}_M(\mathbb{Q}_v), \mathbf{H}_M(\mathbb{Z}_v)) \end{array}$$

(where $\chi_{\eta_{H_M},v}$ is defined as in the last two subsections of 4.2), satisfying the following properties:

- *the upper horizontal arrow is the base change map*
- *the lower horizontal arrow sends a function $\chi_{M,v}\phi_v$, with $\phi_v \in \mathcal{H}(\mathbf{M}^0(\mathbb{Q}_v), \mathbf{M}^0(\mathbb{Z}_v))$, to the function $\chi_{\eta_{H_M},v} f_v$, where*

$$f_v \in \mathcal{H}(\mathbf{H}_M(\mathbb{Q}_v), \mathbf{H}_M(\mathbb{Z}_v))$$

  is the image of ϕ_v by the base change map
- *the left vertical arrow sends a function in $\mathcal{H}(\mathbf{G}^0(\mathbb{Q}_v), \mathbf{G}^0(\mathbb{Z}_v))$ to the product of its constant term at $\mathbf{M}^0$ and of $\chi_{M,v}$*
- *the right vertical arrow is the transfer map defined by η_{H_M} as in section 4.2*

In particular, every function in the image of the base change morphism $\mathcal{H}(\mathbf{G}^0(\mathbb{Q}_v),\mathbf{G}^0(\mathbb{Z}_v)) \longrightarrow \mathcal{H}(\mathbf{H}(\mathbb{Q}_v), \mathbf{H}(\mathbb{Z}_v))$ satisfies condition (H).

(iv) If v is unramified in E and one of the n_i is odd, then every function in $C_c^\infty(\mathbf{H}(\mathbb{Q}_v))$ satisfies condition (H).

Proof. Point (iii) is immediate (because the fundamental lemma is known, cf. section 5.3). Point (iv) is a direct consequence of lemma 8.3.7.

We show (i). For every $\mathbf{M} \in \mathcal{M}_G$, there is a $\mathbf{H}^*(\mathbb{Q})$-conjugacy class of embeddings $\mathbf{H}_M \longrightarrow \mathbf{H}^*$; fix an embedding in this class. Identify $\mathbf{H}_E$ and $\mathbf{H}^*_E$ with

$\mathbb{G}_{m,E} \times \mathbf{GL}_{n_1,E} \times \cdots \times \mathbf{GL}_{n_r,E}$ using the morphism defined in the beginning of section 2.3. Let $\mathbf{M} \in \mathcal{M}_G$. There exists an open neighborhood U_M of 1 in $\mathbf{H}_M(\mathbb{Q}_v)$ such that every semisimple element in U_M is a norm. Choose an open neighborhood V_M of 1 in $\mathbf{H}^*(E \otimes_{\mathbb{Q}} \mathbb{Q}_v) = \mathbf{H}(E \otimes_{\mathbb{Q}} \mathbb{Q}_v)$ such that every semisimple element of $\mathbf{H}_M(\mathbb{Q}_v)$ that is $\mathbf{H}^*(E \otimes_{\mathbb{Q}} \mathbb{Q}_v)$-conjugate to an element of V_M is $\mathbf{H}_M(\mathbb{Q}_v)$-conjugate to an element of U_M (cf. lemma 8.4.2 below).

Let $V = \bigcap_{\mathbf{M} \in \mathcal{M}_G} V_M$ and $U = V \cap \mathbf{H}(\mathbb{Q}_v)$. Then U is an open neighborhood of 1 in $\mathbf{H}(\mathbb{Q}_v)$. Let $f \in C_c^\infty(\mathbf{H}(\mathbb{Q}_v))$ with support contained in U. We show that f satisfies condition (H). For every $\mathbf{M} \in \mathcal{M}_G$, choose a transfer f^{H_M} of f to $\mathbf{H}_M$. To show that there exists a function in $C_c^\infty(\mathbf{M}^0(\mathbb{Q}_v))$ associated to f^{H_M}, it is enough, by proposition 3.3.2 of [La3], to show that, for every semisimple $\gamma \in \mathbf{H}_M(\mathbb{Q}_v)$, $SO_\gamma(f^{H_M}) = 0$ if γ is not a norm. Let $\gamma \in \mathbf{H}_M(\mathbb{Q}_v)$ be semisimple and such that $SO_\gamma(f^{H_M}) \neq 0$. Then, by the definition of the transfer, there exists an image δ of γ in $\mathbf{H}(\mathbb{Q}_v)$ such that $O_\delta(f) \neq 0$. In other words, γ is $\mathbf{H}^*(E \otimes_{\mathbb{Q}} \mathbb{Q}_v)$-conjugate to an element of U. As $U \subset V_M$, this implies that γ is $\mathbf{H}_M(\mathbb{Q}_v)$-conjugate to an element of U_M, hence that γ is conjugate to a norm, that is, that γ is itself a norm.

We show (ii). By (i), it is enough to check that, if U is a neighborhood of 1 in $\mathbf{H}(\mathbb{Q}_v)$, then there exists a neighborhood V of 1 in $\mathbf{G}^0(\mathbb{Q}_v)$ such that every function $\phi \in C_c^\infty(\mathbf{G}^0(\mathbb{Q}_v))$ with support contained in V admits a transfer $f \in C_c^\infty(\mathbf{H}(\mathbb{Q}_v))$ with support contained in U. This follows from the proof of theorem 3.3.1 of [La3]. □

Lemma 8.4.2 *Let F be a local field of characteristic 0, E be a finite extension of F, and $\mathbf{H}$ be a connected reductive group over F. Set $\mathbf{G} = R_{E/F}\mathbf{H}_E$. Let $\mathbf{M}$ be a Levi subgroup of $\mathbf{G}$. Assume that there exists a connected reductive group $\mathbf{H}_M$ on F such that $\mathbf{M} = R_{E/F}\mathbf{H}_{M,E}$ ($\mathbf{H}_M$ is not necessarily a subgroup of $\mathbf{H}$). Let U be a neighborhood of 1 in $\mathbf{H}_M(F)$. Then there exists a neighborhood V of 1 in $\mathbf{G}(F)$ such that, for every semisimple $\gamma \in \mathbf{H}_M(F)$, if γ is $\mathbf{G}(F)$-conjugate to an element of V, then γ is $\mathbf{H}_M(F)$-conjugate to an element of U.*

Proof. Let $(\mathbf{S}_1, \ldots, \mathbf{S}_r)$ be a system of representatives of the set of $\mathbf{H}_M(F)$-conjugacy classes of maximal tori of $\mathbf{H}_M$ (this set is finite because the characteristic of F is 0). For every $i \in \{1, \ldots, r\}$, set $\mathbf{T}_i = R_{E/F}\mathbf{S}_i$ and $U_i = U \cap \mathbf{S}_i(F)$, and choose a neighborhood W_i of 1 in $\mathbf{T}_i(F) = \mathbf{S}_i(E)$ such that $W_i \cap \mathbf{S}_i(F) \subset U_i$. Let $i \in \{1, \ldots, r\}$. Then $\mathbf{T}_i$ is a maximal torus of $\mathbf{M}$, hence of $\mathbf{G}$, so, by lemma 3.1.2 of [La3], there exists a neighborhood V_i of 1 in $\mathbf{G}(F)$ such that, if an element $t \in \mathbf{T}_i(F)$ has a conjugate in V_i, then $t \in W_i$. Set $V = \bigcap_{i=1}^r V_i$.

Let $\gamma \in \mathbf{H}_M(F)$ be semisimple and $\mathbf{G}(F)$-conjugate to an element of V. As γ is semisimple, there exists a maximal torus of $\mathbf{H}_M$ containing γ, so we may assume that there exists $i \in \{1, \ldots, r\}$ such that $\gamma \in \mathbf{S}_i(F)$. In particular, $\gamma \in \mathbf{T}_i(F)$. As γ is $\mathbf{G}(F)$-conjugate to an element of V_i, $\gamma \in W_i$. But $W_i \cap \mathbf{S}_i(F) \subset U_i$, so that $\gamma \in U_i \subset U$. □

We come back to the situation of the beginning of this section. Fix a prime number p that is unramified in E, a neat open compact subgroup $\mathrm{K} = \mathrm{K}^p\mathbf{H}(\mathbb{Z}_p)$

(with $K^p \subset \mathbf{H}(\mathbb{A}_f^p)$) of $\mathbf{H}(\mathbb{A}_f)$, an irreducible algebraic representation V of $\mathbf{H}$ and a function $f^{p,\infty} \in \mathcal{H}(\mathbf{H}(\mathbb{A}_f^p), K^p)$. Assume that $f^{p,\infty}$ satisfies condition (H).

Let $\mathbf{M} \in \mathcal{M}_{\mathbf{G}}$, and define, for every $j \in \mathbb{Z}$, a function $\phi_M^{(j)} = \phi_M^{p,\infty}\phi_{M,p}^{(j)}\phi_{M,\infty} \in C^\infty(\mathbf{M}^0(\mathbb{A}))$, compactly supported modulo $\mathbf{A}_{M^0}(\mathbb{R})^0$, in the following way. Choose $\phi_M^{p,\infty} \in C_c^\infty(\mathbf{M}^0(\mathbb{A}_f^p))$ that is associated at every place to a transfer $(f^{p,\infty})^{H_M}$ of $f^{p,\infty}$ to $\mathbf{H}_M$. The calculations of section 4.2 and (iii) of lemma 8.4.1 show that the function $f_{\mathbf{H}_M,p}^{(j)}$ defined in definition 7.1.6 is the product of $\chi_{\eta_{H_M},p}$ and of a spherical function in the image of the base change map $\mathcal{H}(\mathbf{M}^0(\mathbb{Q}_p), \mathbf{M}^0(\mathbb{Z}_p)) \longrightarrow \mathcal{H}(\mathbf{H}_M(\mathbb{Q}_p), \mathbf{H}_M(\mathbb{Z}_p))$. Take $\phi_{M,p}^{(j)} \in C_c^\infty(\mathbf{M}^0(\mathbb{Q}_p))$ to be $\chi_{M,p}\phi'$, where ϕ' is any spherical function in the inverse image of $\chi_{\eta_{H_M},p}^{-1} f_{\mathbf{H}_M,p}^{(j)}$ by the base change map. To define $\phi_{M,\infty}$, use the notation introduced before and in lemma 7.3.4. Lemma 7.3.4 gives an irreducible algebraic representation V_ω of $\mathbf{H}_M$ for every $\omega \in \Omega_* \simeq \Phi_H(\varphi)$. Take

$$\phi_{M,\infty} = \sum_{\omega \in \Omega_*} \det(\omega)\phi_{V_\omega^*},$$

where $\det(\omega)$ is defined in remark 3.3.2 and $\phi_{V_\omega^*}$ is defined at the beginning of this section.

Let

$$c_M = (-1)^{q(\mathbf{H})}\iota(\mathbf{H}, \mathbf{H}_M)C_M \frac{\tau(\mathbf{H}_M)}{\tau(\mathbf{M}^0)} k(\mathbf{M})^{-1}\langle \mu_H, s_{H_M}\rangle,$$

where μ_H is the cocharacter of $\mathbf{H}_E$ determined by the Shimura datum as in section 2.1 and C_M is the constant of proposition 8.3.1 for $\mathbf{M}$.

Theorem 8.4.3 *For every* $j \in \mathbb{Z}$,

$$\mathrm{Tr}(f^{p,\infty}\Phi_\wp^j, R\Gamma(M^K(\mathbf{H}, \mathcal{X})^*_{\overline{\mathbb{Q}}}, IC^K V_{\overline{\mathbb{Q}}})) = \sum_{\mathbf{M} \in \mathcal{M}_{\mathbf{G}}} c_M T^M(\phi_M^{(j)}),$$

where $(\mathbf{H}, \mathcal{X})$ *is the Shimura datum of 2.1 and* $\Phi_\wp$ *is defined in 7.3.*

Proof. The theorem is an easy consequence of corollary 6.3.2 (and remark 7.2.4), lemma 7.3.4 and proposition 8.3.1 (see also remark 1.3.2 for the choice of p). □

It is possible to deduce from theorem 8.4.3 and proposition 8.2.3 an expression for the logarithm of the L-function (at a good prime number) of the intersection complex $IC^K V$, if K is a small enough open compact subgroup of $\mathbf{G}(\mathbb{A}_f)$.

Remember that we defined in section 2.1 a morphism $\mu_H : \mathbb{G}_{m,E} \longrightarrow \mathbf{H}_E$. The formula for μ_H is

$$\mu_H : \begin{cases} \mathbb{G}_{m,E} & \longrightarrow \quad \mathbf{H}_E = \mathbb{G}_{m,E} \times \mathbf{GL}_{n_1,E} \times \cdots \times \mathbf{GL}_{n_r,E} \\ z & \longmapsto \quad \left(z, \begin{pmatrix} zI_{p_1} & 0 \\ 0 & I_{q_1}\end{pmatrix}, \dots, \begin{pmatrix} zI_{p_r} & 0 \\ 0 & I_{q_r}\end{pmatrix}\right). \end{cases}$$

For every $\mathbf{M} \in \mathcal{M}_G$, let $M_{\mathbf{H}_M}$ be the set of $\mathbf{H}_M(E)$-conjugacy classes of cocharacters $\mu_{H_M} : \mathbb{G}_{m,E} \longrightarrow \mathbf{H}_{M,E}$ such that the cocharacter $\mathbb{G}_{m,E} \xrightarrow{\mu_{H_M}} \mathbf{H}_{M,E} \longrightarrow \mathbf{H}_E$ is

$\mathbf{H}(E)$-conjugate to μ_H. Let $\mathbf{M} \in \mathcal{M}_G$. Write as before $\mathbf{H}_{M,E} = \mathbb{G}_{m,E} \times \mathbf{GL}_{n_1^+,E} \times \mathbf{GL}_{n_1^-,E} \times \cdots \times \mathbf{GL}_{n_r^+,E} \times \mathbf{GL}_{n_r^-,E}$. Then every element μ_{H_M} of $M_{\mathbf{H}_M}$ has a unique representative of the form

$$z \longmapsto \left(z, \begin{pmatrix} zI_{p_1^+} & 0 \\ 0 & I_{q_1^+} \end{pmatrix}, \begin{pmatrix} zI_{p_1^-} & 0 \\ 0 & I_{q_1^-} \end{pmatrix}, \ldots, \begin{pmatrix} zI_{p_r^+} & 0 \\ 0 & I_{q_r^+} \end{pmatrix}, \begin{pmatrix} zI_{p_r^-} & 0 \\ 0 & I_{q_r^-} \end{pmatrix}\right),$$

with $p_i^+ + p_i^- = p_i$. Write $s(\mu_{H_M}) = p_1^- + \cdots + p_r^-$ and $d(\mu_{H_M}) = p_1^+ q_1^+ + p_1^- q_1^- + \cdots + p_r^+ q_r^+ + p_r^- q_r^-$. Let $d = d(\mu_H) = p_1 q_1 + \cdots + p_r q_r$ (d is the dimension of $M^{\mathrm{K}}(\mathbf{H}, \mathcal{X})$, for every open compact subgroup K of $\mathbf{H}(\mathbb{A}_f)$).

Remember that every cocharacter $\mu_{H_M} : \mathbb{G}_{m,E} \longrightarrow \mathbf{H}_{M,E}$ defines a representation $r_{-\mu_{H_M}}$ of ${}^L\mathbf{H}_{M,E}$ (cf. lemma 4.1.1).

We recall the definition of the L-function at a good p of the intersection complex.

Definition 8.4.4 Let p a prime number as in section 1.3, and let $\wp$ be a place of E above p. Set

$$\log L_\wp(s, IC^{\mathrm{K}}V) = \sum_{m \geq 1} \frac{1}{m} (N\wp)^{-ms} \operatorname{Tr}(\Phi_\wp^m, R\Gamma(M^{\mathrm{K}}(\mathbf{G}, \mathcal{X})^*_{\overline{\mathbb{Q}}}, IC^{\mathrm{K}} V_{\overline{\mathbb{Q}}})),$$

where $\Phi_\wp \in W_{E_\wp}$ is a lift of the geometric Frobenius, $N\wp = \#(\mathcal{O}_{E_\wp}/\wp)$ and $s \in \mathbb{C}$ (the series converges for $Re(s) \gg 0$).

Corollary 8.4.5 *Let* K *be a small enough open compact subgroup of* $\mathbf{H}(\mathbb{A}_f)$. *Then there exist functions* $\phi_M \in C^\infty(\mathbf{M}^0(\mathbb{A}))$ *with compact support modulo* $\mathbf{A}_{M^0}(\mathbb{R})^0$, *for every* $\mathbf{M} \in \mathcal{M}_G$, *such that, for every prime number* p *as in section 1.3 (i.e., such that* p *is unramified in* E *and* $\mathrm{K} = \mathrm{K}^p\mathbf{G}(\mathbb{Z}_p)$) *and for every place* $\wp$ *of* E *above* p,

$$\log L_\wp(s, IC^{\mathrm{K}}V) = \sum_{\mathbf{M} \in \mathcal{M}_G} c_M \sum_{t \geq 0} \sum_{\pi_M \in \Pi_{\mathrm{disc}}(\mathbf{M}, t)} a^M_{\mathrm{disc}}(\pi_M)$$

$$\operatorname{Tr}(\pi_M(\phi_M) A_{\pi_M}) \sum_{\mu_{H_M} \in M_{\mathbf{H}_M}} (-1)^{s(\mu_{H_M})} \log L_\wp\left(s - \frac{d}{2}, (\pi_M \otimes \chi_M)_\wp, r_{-\mu_{H_M}}\right),$$

where, for every $\mathbf{M} \in \mathbf{M}_G$ *and* $\pi_M \in \Pi_{\mathrm{disc}}(\mathbf{M}, t)$, $(\pi_M \otimes \chi_M)_\wp$ *is the local component at* $\wp$ *of* $\pi_M \otimes \chi_M$, *seen as a representation of* $\mathbf{H}_M(\mathbb{A}_E)$.

Proof. By lemma 8.4.1, if K is a small enough open compact subgroup of $\mathbf{H}(\mathbb{A}_f)$, then the function $\mathbb{1}_{\mathrm{K}}$ satisfies condition (H). Fix such a K, and assume also that K is neat. Let S be a finite set of prime numbers containing the set of prime numbers that are ramified in E and such that $\mathrm{K} = \mathrm{K}_S\mathrm{K}^S$, with $\mathrm{K}_S \subset \mathbf{H}(\mathbb{A}_S)$ and $\mathrm{K}^S = \prod_{p \notin S} \mathbf{H}(\mathbb{Z}_p)$. For every $\mathbf{M} \in \mathcal{M}_G$, choose a transfer $f_S^{H_M}$ of $\mathbb{1}_{\mathrm{K}_S}$ to $\mathbf{H}_M$ and a function $\phi_{M,S} \in C_c^\infty(\mathbf{M}^0(\mathbb{A}_S))$ associated to $f_S^{H_M}$, and write $\phi_M^S = \chi_{M|\mathbf{M}^0(\mathbb{A}_f^S)} \mathbb{1}_{\mathrm{K}_M^S}$, where $\mathrm{K}_M^S = \prod_{p \notin S} \mathbf{M}^0(\mathbb{Z}_p)$, and $\phi_M = \phi_{M,S}\phi_M^S$.

Let $p \notin S$ and $j \in \mathbb{N}^*$. We want to define, for every $\mu_{H_M} \in M_{\mathbf{H}_M}$, a function $\phi^{(j)}_{\mu_{H_M},p} \in \mathcal{H}(\mathbf{M}^0(\mathbb{Q}_p), \mathbf{M}^0(\mathbb{Z}_p))$. Remember that we fixed a place $\wp$ of E

above p. Let L be the unramified extension of $E_\wp$ of degree j in $\overline{\mathbb{Q}}_p$. For every $\mu_{H_M} \in M_{\mathbf{H}_M}$, let $\phi^{(j)}_{\mu_{H_M},p}$ be the product of $(N\wp)^{j(d-d(\mu_{H_M}))/2}$ and of the image of the function $f_{\mu_{H_M},L}$ in $\mathcal{H}(\mathbf{M}^0(L),\mathbf{M}^0(\mathcal{O}_L))$ (defined by $\mu_{H_M,L}$ as in 4.1) by the morphism

$$\mathcal{H}(\mathbf{H}_M(L),\mathbf{H}_M(\mathcal{O}_L)) \longrightarrow \mathcal{H}(\mathbf{H}_M(E_\wp),\mathbf{H}_M(\mathcal{O}_{E_\wp})) \longrightarrow \mathcal{H}(\mathbf{M}^0(\mathbb{Q}_p),\mathbf{M}^0(\mathbb{Z}_p)),$$

where the first arrow is the base change morphism and the second arrow is

- identity if p is inert in E (so $\mathbf{M}^0(\mathbb{Q}_p) = \mathbf{H}_M(E_\wp)$);
- the morphism $h \longmapsto (h, \mathbb{1}_{\mathbf{H}_M(\mathcal{O}_{E_{\wp'}})})$ if p splits in E and $\wp'$ is the second place of E above p (so $\mathbf{M}^0(\mathbb{Q}_p) = \mathbf{H}_M(E_\wp) \times \mathbf{H}_M(E_{\wp'})$).

Let π_p be an unramified θ-stable representation of $\mathbf{M}^0(\mathbb{Q}_p)$ and $\varphi_{\pi_p} : W_{\mathbb{Q}_p} \longrightarrow {}^L\mathbf{M}^0_{\mathbb{Q}_p}$ be a Langlands parameter of π_p. As π_p is θ-stable, we may assume that φ_{π_p} factors through the image of ${}^L\mathbf{H}_{M,\mathbb{Q}_p} \longrightarrow {}^L\mathbf{M}^0_{\mathbb{Q}_p}$. Let $\varphi_\wp$ be the morphism $W_{E_\wp} \longrightarrow {}^L\mathbf{H}_{M,E_\wp}$ deduced from φ_{π_p}. If p is inert in E, then $\mathbf{H}_M(E_\wp) = \mathbf{M}^0(\mathbb{Q}_p)$, and $\varphi_\wp$ is a Langlands parameter of π_p, seen as a representation of $\mathbf{H}_M(E_\wp)$. If p splits in E and $\wp'$ is the second place of E above p, then $\mathbf{M}^0(\mathbb{Q}_p) = \mathbf{H}_M(E_\wp) \times \mathbf{H}_M(E_{\wp'})$, so $\pi_p = \pi \otimes \pi'$, where π (resp., π') is an unramified representation of $\mathbf{H}_M(E_\wp)$ (resp., $\mathbf{H}_M(E_{\wp'})$). The morphism $\varphi_\wp$ is a Langlands parameter of π. By theorem 4.1.2 and lemma 8.3.9, if A_{π_p} is a normalized intertwining operator on π_p, then

$$\operatorname{Tr}(\pi_p(\phi^{(j)}_{\mu_{H_M},p})A_{\pi_p}) = (N\wp)^{jd/2}\operatorname{Tr}(r_{-\mu_{H_M}} \circ \varphi_\wp(\Phi^j_\wp))\operatorname{Tr}(\pi_p(\mathbb{1}_{\mathbf{M}^0(\mathbb{Z}_p)})A_{\pi_p}).$$

Set

$$\phi^{(j)'}_{M,p} = \sum_{\mu_{H_M} \in M_{\mathbf{H}_M}} (-1)^{s(\mu_{H_M})}\phi^{(j)}_{\mu_{H_M},p}.$$

The calculations of 4.2 and (iii) of lemma 8.4.1 imply that the function $\chi^{-1}_{\eta_{H_M},p} f^{(j)}_{H_M,p} \in \mathcal{H}(\mathbf{H}_M(\mathbb{Q}_p),\mathbf{H}_M(\mathbb{Z}_p))$ is the image by the base change map of the function $\phi^{(j)'}_{M,p}$ (as before, $f^{(j)}_{H_M,p}$ is the function of definition 7.1.6). So we can take $\phi^{(j)}_{M,p} = \chi_{M,p}\phi^{(j)'}_{M,p}$ in theorem 8.4.3, and the corollary follows from this theorem and from proposition 8.2.3. □

Another application of theorem 8.4.3 is the next corollary. In this corollary, E is still an imaginary quadratic extension of $\mathbb{Q}$. Fix $n \in \mathbb{N}^*$, and let θ be the involution $(\lambda, g) \longmapsto (\bar\lambda, \bar\lambda\,{}^t\bar g^{-1})$ of $R_{E/\mathbb{Q}}(\mathbb{G}_{m,E} \times \mathbf{GL}_{n,E})$ (where $(\lambda, g) \longmapsto (\bar\lambda, \bar g)$ is the action of the nontrivial element of $\operatorname{Gal}(E/\mathbb{Q})$). Then θ defines an involution of $\mathbb{C}^\times \times \mathbf{GL}_n(\mathbb{C}) = (R_{E/\mathbb{Q}}(\mathbb{G}_{m,E} \times \mathbf{GL}_{n,E}))(\mathbb{R})$. The morphism θ is the involution induced by the non-trivial element of $\operatorname{Gal}(E/\mathbb{Q})$, if $\mathbb{G}_{m,E} \times \mathbf{GL}_{n,E}$ is identified to $\mathbf{GU}(n)_E$. If $\wp$ is a finite unramified place of E and $\pi_\wp$ is an unramified representation of $\mathbf{GU}(n)(E_\wp)$, let $\log L(s, \pi_\wp)$ (resp., $\log L(s, \pi_\wp, \bigwedge^2)$) be the logarithm of the L-function of $\pi_\wp$ and of the representation $\mathrm{id}_{\mathbb{C}^\times} \otimes \mathrm{st}$ (resp., $\mathrm{id}_{\mathbb{C}^\times} \otimes \bigwedge^2 \mathrm{st}$) of $\widehat{\mathbf{GU}(n)} = \mathbb{C}^\times \times \mathbf{GL}_n(\mathbb{C})$, where st is the standard representation of $\mathbf{GL}_n(\mathbb{C})$.

Corollary 8.4.6 *Let π be a θ-stable cuspidal automorphic representation of $\mathbb{A}_E^\times \times \mathbf{GL}_n(\mathbb{A}_E)$ such that π_∞ is tempered (where ∞ is the unique infinite place of E). Assume that there exists an irreducible algebraic representation V of $\mathbf{GU}(n)$ such that* $\mathrm{ep}(\theta, \pi_\infty \otimes W) \neq 0$, *where W is the θ-stable representation of $\mathbb{C}^\times \times \mathbf{GL}_n(\mathbb{C})$ associated to V (cf. theorem 8.1.5). Let m be the weight of V in the sense of 1.3* (i.e., *the relative integer such that the central subgroup $\mathbb{G}_m$ of $\mathbf{GU}(n)$ acts on V by $x \longmapsto x^m$). Let S be the union of the set of prime numbers that ramify in E and the set of prime numbers under finite places of E where π is ramified. Then there exist a number field K, a positive integer N, and, for every finite place λ of K, a continuous finite-dimensional representation σ_λ of* $\mathrm{Gal}(\overline{\mathbb{Q}}/E)$ *with coefficients in K_λ, such that*

(i) *The representation σ_λ is unramified outside $S \cup \{\ell\}$, where ℓ is the prime number under λ, pure of weight $-m+1-n$ if n is not divisible by* 4 *and mixed with weights between $-m+2(2-n)-1$ and $-m+2(2-n)+1$ if n is divisible by* 4. *If n is divisible by* 4 *and the highest weight of V is regular, then σ_λ is pure of weight $-m+2(2-n)$.*

(ii) *For every place $\wp$ of E above a prime number $p \notin S$, for every finite place $\lambda \not| p$ of K,*

$$\log L_\wp(s, \sigma_\lambda) = N \log L\left(s + \frac{n-1}{2}, \pi_\wp\right)$$

if n is not divisible by 4, *and*

$$\log L_\wp(s, \sigma_\lambda) = N \log L\left(s + (n-2), \pi_\wp, \bigwedge^2\right)$$

if n is divisible by 4 *(where $\pi_\wp$ is the local component at $\wp$ of π, seen as a representation of* $\mathbf{GU}(n)(\mathbb{A}_E)$)*.*

Proof. We can, without changing the properties of π, replace θ by its product with an inner automorphism of $R_{E/\mathbb{Q}}(\mathbb{G}_{m,E} \times \mathbf{GL}_{n,E})$. So we may (and will) assume that $\theta(\lambda, g) = (\bar{\lambda}, \bar{\lambda} J_{p_1,q_1} {}^t\bar{g}^{-1} J_{p_1,q_1}^{-1})$, where $p_1, q_1 \in \mathbb{N}^*$ are such that $p_1 + q_1 = n$ and $J_{p_1,q_1} \in \mathbf{GL}_n(\mathbb{Z})$ is the matrix (defined in 2.1) of the hermitian form that gives the group $\mathbf{GU}(p_1, q_1)$.

Write as before $\mathbf{H} = \mathbf{GU}(p_1, q_1)$ and $\mathbf{G}^0 = R_{E/\mathbb{Q}}\mathbf{H}_E$. Then $\mathbf{G}^0 = R_{E/\mathbb{Q}}\mathbb{G}_{m,E} \times R_{E/\mathbb{Q}}\mathbf{GL}_{n,E}$, and the involution θ of $\mathbf{G}^0$ defined above is equal to the involution induced by the nontrivial element of $\mathrm{Gal}(E/\mathbb{Q})$. Assume that the group $\mathbf{H}$ is quasi-split (but not necessarily unramified) at every finite place of $\mathbb{Q}$. As $\mathrm{ep}(\theta, \pi_\infty \otimes W) \neq 0$ and π_∞ is tempered, remark 8.1.6 and theorem 8.1.5 imply that $\pi_\infty = \pi_{V^*}$, where π_{V^*} is the θ-discrete representation of $\mathbf{G}^0(\mathbb{R})$ associated to V^* as in lemma 8.1.10.

Let $\mathrm{K}_S \subset \mathbf{G}^0(\mathbb{A}_S)$ be an open compact subgroup such that $\mathrm{Tr}(\pi_S(\mathbb{1}_{\mathrm{K}_S}) A_{\pi_S}) \neq 0$ (where A_{π_S} is any intertwining operator on π_S). By lemma 8.4.1, by taking K_S small enough, we may assume that there exists a function $f_S \in C_c^\infty(\mathbf{H}(\mathbb{A}_S))$ associated to $\phi_S := \mathbb{1}_{\mathrm{K}_S}$ and satisfying condition (H). For every $\mathbf{M} \in \mathcal{M}_G$, fix a transfer $f_S^{H_M}$ of f_S to $\mathbf{H}_M$ and a function $\phi_{M,S} \in C_c^\infty(\mathbf{M}^0(\mathbb{A}_S))$ associated to $f_S^{H_M}$. If $p \notin S$, $\mathbf{M} \in \mathcal{M}_G$ and $\phi_p \in \mathcal{H}(\mathbf{G}^0(\mathbb{Q}_p), \mathbf{G}^0(\mathbb{Z}_p))$, let $b(\phi_p) \in \mathcal{H}(\mathbf{H}^0(\mathbb{Q}_p), \mathbf{H}^0(\mathbb{Z}_p))$,

$b(\phi_p)^{H_M} \in C_c^\infty(\mathbf{H}_M(\mathbb{Q}_p))$ and $\phi_{M,p} \in C_c^\infty(\mathbf{M}^0(\mathbb{Q}_p))$ be the functions obtained from ϕ_p by following the arrows of the commutative diagram of point (iii) of lemma 8.4.1. Finally, for every $\mathbf{M} \in \mathcal{M}_G$, define a function $\phi_{M,\infty} \in C^\infty(\mathbf{M}^0(\mathbb{R}))$ from V as in theorem 8.4.3.

Let $\mathrm{K}_{H,S} \subset \mathbf{H}(\mathbb{A}_S)$ be an open compact subgroup small enough for f_S to be bi-invariant under $\mathrm{K}_{H,S}$. Set $\mathrm{K}_H = \mathrm{K}_{H,S} \prod_{p \notin S} \mathbf{H}(\mathbb{Z}_p)$; then K_H is an open compact subgroup of $\mathbf{H}(\mathbb{A}_f)$, and we may assume (by making $\mathrm{K}_{H,S}$ smaller) that K_H is neat. Then the results of section 1.7 apply to $\mathbf{H}$, K_H and every $p \notin S$. In the beginning of section 7.2, we explained how to get a number field K and, for every finite place λ of K, a virtual finite-dimensional λ-adic representation W_λ of $\mathrm{Gal}(\overline{\mathbb{Q}}/E) \times \mathcal{H}(\mathbf{H}(\mathbb{A}_f), \mathrm{K}_H)$ (the cohomology of the complex $IC^{\mathrm{K}_H}V$) such that there is a decomposition

$$W_\lambda = \bigoplus_{\pi_{H,f}} W_\lambda(\pi_{H,f}) \otimes \pi_{H,f}^{\mathrm{K}_H},$$

where the direct sum is taken over the set of isomorphism classes of irreducible admissible representations $\pi_{H,f}$ of $\mathbf{H}(\mathbb{A}_f)$ such that $\pi_{H,f}^{\mathrm{K}_H} \neq 0$, and the $W_\lambda(\pi_{H,f})$ are virtual λ-adic representations of $\mathrm{Gal}(\overline{\mathbb{Q}}/E)$.

Let $\Pi_H(\pi_f)$ be the set of isomorphism classes of irreducible admissible representations $\pi_{H,f}$ of $\mathbf{H}(\mathbb{A}_f)$ such that $\pi_{H,f}^{\mathrm{K}_H} \neq 0$ (so $\pi_{H,f}$ is unramified outside of S), that $W_\lambda(\pi_{H,f}) \neq 0$ and that, for every $p \notin S$, if $\varphi_{\pi_{H,p}} : W_{\mathbb{Q}_p} \longrightarrow {}^L\mathbf{H}_{\mathbb{Q}_p}$ is a Langlands parameter of $\pi_{H,p}$, then the composition of $\varphi_{\pi_{H,p}}$ and of the inclusion ${}^L\mathbf{H}_{\mathbb{Q}_p} \longrightarrow {}^L\mathbf{G}^0_{\mathbb{Q}_p}$ (defined in example 8.1.1) is a Langlands parameter of π_p.

By the multiplicity 1 theorem of Piatetski-Shapiro, $m_{\mathrm{disc}}(\pi) = 1$. So there is a normalized intertwining operator on π such that $m^+_{\mathrm{disc}}(\pi) = 1$ and $m^-_{\mathrm{disc}}(\pi) = 0$; denote this intertwining operator by A_π. Let as before $\mu_H : \mathbb{G}_{m,E} \longrightarrow \mathbf{H}_E$ be the cocharacter defined by the Shimura datum, $r_{-\mu_H}$ be the representation of ${}^L\mathbf{H}_E$ determined by $-\mu_H$ (cf. section 4.1) and $d = p_1 q_1$. Set $\phi^\infty = \phi_S \prod_{p \notin S} \mathbb{1}_{\mathbf{G}^0(\mathbb{Z}_p)}$ and $f^\infty = f_S \prod_{p \notin S} \mathbb{1}_{\mathbf{H}(\mathbb{Z}_p)}$. Let $\wp$ be a finite place of E above a prime number $p \notin S$. Let $\pi_\wp$ be the local component at $\wp$ of π (seen as a representation of $\mathbf{H}(\mathbb{A}_E)$) and $\varphi_\wp : W_{E_\wp} \longrightarrow {}^L\mathbf{H}_{E_\wp}$ be a Langlands parameter of $\pi_\wp$. We are going to show that, for every finite place $\lambda \not| p$ of K and for every $j \in \mathbb{Z}$,

$$c_G(N\wp)^{dj/2} \operatorname{Tr}(r_{-\mu_H} \circ \varphi_\wp(\Phi_\wp^j)) \operatorname{Tr}(\pi_f(\phi^\infty) A_\pi) = \sum_{\pi_{H,f} \in \Pi_H(\pi_f)} \operatorname{Tr}(\pi_{H,f}(f^\infty)) \operatorname{Tr}(\Phi_\wp^j, W_\lambda(\pi_{H,f})), \qquad (*)$$

where $\Phi_\wp \in W_{E_\wp}$ is a lift of the geometric Frobenius. It suffices to show this equality for $j > 0$.

Let λ be a finite place of K such that λ does not divide p. Let $j \in \mathbb{N}^*$. Define a function $\phi_p^{(j)} \in \mathcal{H}(\mathbf{G}^0(\mathbb{Q}_p), \mathbf{G}^0(\mathbb{Z}_p))$ using μ_H and j, as in the proof of corollary 8.4.5. We recall the definition. Let L be an unramified extension of $E_\wp$ of degree j. Then $\phi_p^{(j)}$ is the image of the function $f_{\mu_H,L} \in \mathcal{H}(\mathbf{H}(L), \mathbf{H}(\mathcal{O}_L))$ determined by μ_H as in 4.1 by the morphism

$$\mathcal{H}(\mathbf{H}(L), \mathbf{H}(\mathcal{O}_L)) \longrightarrow \mathcal{H}(\mathbf{H}(E_\wp), \mathbf{H}(\mathcal{O}_{E_\wp})) \longrightarrow \mathcal{H}(\mathbf{G}^0(\mathbb{Q}_p), \mathbf{G}^0(\mathbb{Z}_p)),$$

where the first arrow is the base change morphism and the second arrow is

- identity if p is inert in E (so $\mathbf{G}^0(\mathbb{Q}_p) = \mathbf{H}(E_\wp)$);
- the morphism $h \longmapsto (h, \mathbb{1}_{\mathbf{H}(\mathcal{O}_{E_{\wp'}})})$ if p splits in E and $\wp'$ is the second place of E above p (so $\mathbf{G}^0(\mathbb{Q}_p) = \mathbf{H}(E_\wp) \times \mathbf{H}(E_{\wp'})$).

Then $b(\phi_p^{(j)})$ is the function $f_{H,p}$ defined after theorem 6.2.1. Moreover, by theorem 4.1.2 and lemma 8.3.9, if π_p is the local component at p of π (seen as a representation of $\mathbf{G}^0(\mathbb{A})$) and A_{π_p} is a normalized intertwining operator on π_p, then

$$\mathrm{Tr}(\pi_p(\phi_p^{(j)})A_{\pi_p}) = (N\wp)^{jd/2}\,\mathrm{Tr}(r_{-\mu_H} \circ \varphi_\wp(\Phi_\wp^j))\,\mathrm{Tr}(\pi_p(\mathbb{1}_{\mathbf{G}^0(\mathbb{Z}_p)})A_{\pi_p}).$$

Let $\mathbf{M} \in \mathcal{M}_{\mathbf{G}}$. Let R_M be the set of $\pi_M \in \Pi_{\mathrm{disc}}(\mathbf{M}, t)$, with $t \geq 0$, such that

(i) $\pi_M \otimes \chi_M$ is unramified at every finite place $v \notin S$;
(ii) $a_{\mathrm{disc}}^M(\pi_M) \neq 0$;
(iii) $\mathrm{Tr}(\pi_{M,S}(\phi_{M,S})A_{\pi_M}) \neq 0$ and $\mathrm{Tr}(\pi_{M,\infty}(\phi_{M,\infty})A_{\pi_M}) \neq 0$ (where A_{π_M} is a normalized intertwining operator on π_M);
(iv) if $\mathbf{M} = \mathbf{G}$, then $\pi_M \not\simeq \pi$.

Then R_M is finite.

Let R_H be the set of isomorphism classes of irreducible admissible representations $\pi_{H,f}$ of $\mathbf{H}(\mathbb{A}_f)$ such that

(i) $\pi_{H,f}^{\mathrm{K}_H} \neq 0$;
(ii) $\pi_{H,f} \notin \Pi_H(\pi)$;
(iii) $W_\lambda(\pi_{H,f}) \neq 0$.

Then R_H is also finite.

By the strong multiplicity 1 theorem of Jacquet-Shalika for $\mathbf{G}^0$ (cf. theorem 4.4 of [JS]) and corollary 8.5.3 (cf. also remark 8.5.4), there exists a function $g^{S\cup\{p\}} \in \mathcal{H}(\mathbf{G}^0(\mathbb{A}_f^{S\cup\{p\}}), \mathrm{K}_G^{S\cup\{p\}})$ (where $\mathrm{K}_G^{S\cup\{p\}} = \prod_{v\notin S\cup\{p\}} \mathbf{G}^0(\mathbb{Z}_v)$) such that

- $\mathrm{Tr}(\pi^{S\cup\{p\}}(g^{S\cup\{p\}})A_\pi) = \mathrm{Tr}(\pi^{S\cup\{p\}}(\phi^{S\cup\{p\}})A_\pi) = 1$;
- for every $\pi_{H,f} \in R_H$, $\mathrm{Tr}(\pi_{H,f}^{S\cup\{p\}}(b(g^{S\cup\{p\}}))) = 0$ (where $b(g^{S\cup\{p\}}) \in \mathcal{H}(\mathbf{H}(\mathbb{A}_f^{S\cup\{p\}}), \mathrm{K}_H^{S\cup\{p\}})$ is the base change of $g^{S\cup\{p\}}$);
- for every $\pi_{H,f} \in \Pi_H(\pi)$, $\mathrm{Tr}(\pi_{H,f}^{S\cup\{p\}}(b(g^{S\cup\{p\}}))) = \mathrm{Tr}(\pi_{H,f}{}^{S\cup\{p\}}(f^{S\cup\{p\}}))$ (this actually follows from the first condition and from the fundamental lemma for base change);
- for every $\mathbf{M} \in \mathcal{M}_{\mathbf{G}}$, every $\pi_M \in R_M$ and every normalized intertwining operator A_{π_M} on π_M, $\mathrm{Tr}(\pi_M^{S\cup\{p\}}(g_M^{S\cup\{p\}})A_{\pi_M}) = 0$ (where $g_M^{S\cup\{p\}}$ is the function obtained from $g^{S\cup\{p\}}$ by following the left vertical arrow in the diagram of (iii) of lemma 8.4.1).

Then

$$\mathrm{Tr}(\pi_f(\phi_S g^{S\cup\{p\}}\phi_p^{(j)})A_\pi) = \mathrm{Tr}(\pi_f(\phi^\infty)A_\pi)(N\wp)^{dj/2}\,\mathrm{Tr}(r_{-\mu_H} \circ \varphi_\wp(\Phi_\wp^j)),$$

and, by theorem 8.4.3 (and the fact that $\mathrm{Tr}(\pi_\infty(\phi_{G,\infty})) = 1$),

$$c_G\,\mathrm{Tr}(\pi_f(\phi_S g^{S\cup\{p\}}\phi_p^{(j)})A_\pi) = \sum_{\pi_{H,f}\in\Pi_H(\pi_f)} \mathrm{Tr}(\Phi_\wp^j \times f_S f^{S\cup\{p\}}\mathbb{1}_{\mathbf{H}(\mathbb{Z}_p)}, W_\lambda(\pi_{H,f})\otimes\pi_{H,f}^{\mathrm{K}_H}).$$

This proves equality $(*)$.

Remember that we wanted the group $\mathbf{GU}(p_1, q_1)$ to be quasi-split at every finite place of $\mathbb{Q}$. We use here the calculations of the Galois cohomology of unitary groups of section 2 of [Cl5]. If n is odd, these calculations imply that the group $\mathbf{GU}(p_1, q_1)$ is quasi-split at every finite place of $\mathbb{Q}$ for any p_1 and q_1. Take $p_1 = 1$ and $q_1 = n - 1$. Now assume that n is even. If $n/2$ is odd, take $p_1 = 1$ and $q_1 = n - 1$. If $n/2$ is even, take $p_1 = 2$ and $q_1 = n - 2$. We check that, with these choices, $\mathbf{GU}(p_1, q_1)$ is indeed quasi-split at every finite place of $\mathbb{Q}$. Let D be the discriminant of E. Let q be a prime number. If q does not divide D, then $\mathbf{GU}(p_1, q_1)$ is unramified at q (so, in particular, it is quasi-split). Assume that q divides D. Then the cohomological invariant of $\mathbf{GU}(p_1, q_1)$ at q is 0 if -1 is a norm in $\mathbb{Q}_q$, and $q_1 + n/2 \mod 2$ otherwise. But, by the choice of q_1, $q_1 + n/2$ is always even, so $\mathbf{GU}(p_1, q_1)$ is quasi-split at q.

In the rest of the proof, take p_1 and q_1 as in the discussion above. Note that $d = n - 1$ if n is not divisible by 4, and $d = 2(n-2)$ if n is divisible by 4.

We now apply lemma 7.3.2. As $\mathbf{H}$ splits over E, the representation $r_{-\mu_H}$ of ${}^L\mathbf{H}_E = \widehat{\mathbf{H}} \times W_E$ determined by the cocharacter μ_H of $\mathbf{H}_E$ is trivial on W_E. Let $\mathrm{st}^\vee$ be the contragredient of the standard representation of $\mathbf{GL}_n(\mathbb{C})$ and χ be the character $z \longmapsto z^{-1}$ of $\mathbb{C}^\times$. By lemma 7.3.2, the restriction of $r_{-\mu_H}$ to $\widehat{\mathbf{H}} = \mathbb{C}^\times \times \mathbf{GL}_n(\mathbb{C})$ is $\chi \otimes \mathrm{st}^\vee$ if n is not divisible by 4, and $\chi \otimes \bigwedge^2 \mathrm{st}^\vee$ if n is divisible by 4.

Let $\wp$ be a finite place of E above a prime number $p \notin S$, and $\lambda \nmid p$ be a finite place of K. Fix a Langlands parameter $(z, (z_1, \dots, z_n))$ of $\pi_\wp$ in the maximal torus $\mathbb{C}^\times \times (\mathbb{C}^\times)^n$ of $\widehat{\mathbf{H}} = \mathbb{C}^\times \times \mathbf{GL}_n(\mathbb{C})$. By reasoning as at the beginning of the proof of theorem 7.3.1 (or by applying corollary 8.5.3 and theorem 7.3.1), we see that $\log_{N\wp} |z| \in \frac{1}{2}\mathbb{Z}$. For every $\pi_{H,f} \in \Pi_H(\pi_f)$, let a_i, $i \in I_{\pi_{H,f}}$, be the eigenvalues of $\Phi_\wp$ acting on $W_\lambda(\pi_{H,f})$, $n_i \in \mathbb{Z}$, $i \in I_{\pi_{H,f}}$, be their multiplicities, and

$$b_{\pi_{H,f}} = c_G^{-1} \operatorname{Tr}(\pi_f(\phi^\infty) A_\pi)^{-1} \operatorname{Tr}(\pi_{H,f}(f^\infty))$$

($b_{\pi_{H,f}}$ does not depend on $\wp$). By equality $(*)$, for every $j \in \mathbb{Z}$

$$(N\wp)^{dj/2} z^{-j} \sum_{\substack{J \subset \{1,\dots,n\} \\ |J|=k}} \prod_{l \in J} z_l^{-j} = \sum_{\pi_{H,f} \in \Pi_H(\pi_f)} b_{\pi_{H,f}} \sum_{i \in I_{\pi_{H,f}}} n_i a_i^j, \qquad (**)$$

where $k = 1$ if n is not divisible by 4, and $k = 2$ if n is divisible by 4. So there exists a positive integer N (independent from $\wp$) such that $N_{\pi_{H,f}} := N b_{\pi_{H,f}} \in \mathbb{Z}$, for every $\pi_{H,f} \in \Pi_H(\pi_f)$. Moreover, for every $\pi_{H,f} \in \Pi_H(\pi_f)$ and $i \in I_{\pi_{H,f}}$, the product $N_{\pi_{H,f}} n_i$ is positive. In particular,

$$\sigma_\lambda := \bigoplus_{\pi_{H,f} \in \Pi_H(\pi_f)} N_{\pi_{H,f}} W_\lambda(\pi_{H,f})^\vee$$

is a real representation of $\operatorname{Gal}(\overline{\mathbb{Q}}/E)$ (and not just a virtual representation). Then equality $(**)$ becomes: for every finite place $\wp$ of E above a prime number $p \notin S$, if $\lambda \nmid p$, then, for every $j \in \mathbb{Z}$,

$$N(N\wp)^{-jd/2} \operatorname{Tr}((\mathrm{id}_{\mathbb{C}^\times} \otimes \bigwedge^k \mathrm{st})(\varphi_\wp(\Phi_\wp^j)) = \operatorname{Tr}(\Phi_\wp^j, \sigma_\lambda),$$

where k is as before equal to 1 if n is not divisible by 4, and to 2 if n is divisible by 4. This is point (ii) of the lemma (as $\mathbf{GU}(n)$ is an inner form of $\mathbf{GU}(p_1, q_1)$, we can see $\varphi_\wp$ as the Langlands parameter of the local component at $\wp$ of π, seen as a representation of $\mathbf{GU}(n)(\mathbb{A}_E)$).

It remains to determine the weight of σ_λ. As the algebraic representation V of $\mathbf{GU}(n)$ is pure of weight m in the sense of section 1.3, the complex $IC^{\mathrm{K}}V$ is pure of weight $-m$. Let

$$W_\lambda = \sum_{i=0}^{2d}(-1)^i W_\lambda^i$$

be the decomposition of W_λ according to cohomology degree. For every irreducible admissible representation $\pi_{H,f}$ of $\mathbf{H}(\mathbb{A}_f)$ such that $\pi_{H,f}^{\mathrm{K}_{H,f}} \neq 0$, there is a decomposition

$$W_\lambda(\pi_{H,f}) = \sum_{i=0}^{2d}(-1)^i W_\lambda^i(\pi_{H,f}),$$

and the representation $W_\lambda^i(\pi_{H,f})$ of $\mathrm{Gal}(\overline{\mathbb{Q}}/E)$ is pure of weight $-m+i-2d$.

Remember that $(z,(z_1,\dots,z_n))$ is the Langlands parameter of $\pi_\wp$. Assume first that n is not divisible by 4. Then equality $(**)$ implies that $\log_{N\wp}|z_i| \in \frac{1}{2}\mathbb{Z}$ for every $i \in \{1,\dots,n\}$ (because the a_i and z satisfy the same property). But we know that $-\frac{1+m}{2} < \log_{N\wp}|z_i| < \frac{1-m}{2}$ for every $i \in \{1,\dots,n\}$ (cf. [Cl4] lemma 4.10; note that the conditions on π_∞ imply that π is algebraic regular in the sense of [Cl4], and that Clozel uses a different normalization of the Langlands parameter at $\wp$), so $\log_{N\wp}|z_i| = -\frac{m}{2}$ for every i. This implies that, if $\pi_{H,f} \in \Pi_H(\pi_f)$, then $W^i(\pi_{H,f}) = 0$ for every $i \neq d$. Hence σ_λ is pure of weight $-m-d = -m+1-n$.

Assume now that n is divisible by 4. Equality $(**)$ implies only that $\log_{N\wp}|z_i| \in \frac{1}{4}\mathbb{Z}$. As before, we know that the $\log_{N\wp}|z_i|$ are in $]-\frac{1+m}{2},\frac{1-m}{2}[$, so $\log_p|z_i| \in \{-\frac{1+m}{4},-\frac{m}{4},\frac{1-m}{4}\}$ for every $i \in \{1,\dots,n\}$. Applying $(**)$ again, we see that the only $W^i(\pi_{H,f})$ that can appear in σ_λ are those with $d-1 \leq i \leq d+1$. This proves the bounds on the weights of σ_λ. Assume that the highest weight of V is regular. Then, by lemma 7.3.5, $W^i = 0$ if $i \neq d$, so σ_λ is of weight $-m-d = -m+2(2-n)$. □

To formulate the last two corollaries, we will use the following definition of Clozel.

Definition 8.4.7 (cf. [Cl4] 1.2.3 or [Cl5] 3.1) Let π be a cuspidal automorphic representation of $\mathbf{GL}_n(\mathbb{A})$ (resp., $\mathbf{GL}_n(\mathbb{A}_E)$), where E is an imaginary quadratic extension of $\mathbb{Q}$). Then π is called *algebraic* if there exist a Langlands parameter $\varphi : W_\mathbb{R} \longrightarrow \mathbf{GL}_n(\mathbb{C})$ (resp., $\varphi : W_\mathbb{C} \longrightarrow \mathbf{GL}_n(\mathbb{C})$) of π_∞ and $p_1,\dots,p_n,q_1,\dots,q_n \in \mathbb{Z}$ such that, for every $z \in W_\mathbb{C} = \mathbb{C}^\times$,

$$\varphi(z) = \begin{pmatrix} z^{p_1+\frac{n-1}{2}}\overline{z}^{q_1+\frac{n-1}{2}} & & 0 \\ & \ddots & \\ 0 & & z^{p_n+\frac{n-1}{2}}\overline{z}^{q_n+\frac{n-1}{2}} \end{pmatrix}.$$

We may assume that $p_1 \geq \cdots \geq p_n$. The representation π is called *regular algebraic* if $p_1 > \cdots > p_n$.

If π is regular algebraic, then there is an algebraic representation W of $\mathbf{GL}_n$ (resp., $R_{E/\mathbb{Q}}\mathbf{GL}_{n,E}$) associated to π as in [Cl4] 3.5 and [Cl5] 3.2 : the highest weight of W is $(p_1, p_2 + 1, \ldots, p_n + (n-1))$ (resp., $((p_1, p_2 + 1, \ldots, p_n + (n-1)), (q_n, q_{n-1} + 1, \ldots, q_1 + (n-1))))$. We say that π is *very regular* if the highest weight of W is regular.

We summarize a few results of Clozel about regular algebraic representations in the next lemma.

Lemma 8.4.8 *Let E be an imaginary quadratic extension of $\mathbb{Q}$. Let $\mathbf{G}^0 = \mathbf{GL}_n$ or $R_{E/\mathbb{Q}}\mathbf{GL}_{n,E}$.*

(i) Let π be a cuspidal automorphic representation of $\mathbf{G}^0(\mathbb{A})$. Then the following conditions are equivalent.

(a) π is regular algebraic.

(b) The infinitesimal character of π_∞ is that of an algebraic representation of $\mathbf{G}^0$.

(c) There exist an algebraic representation W of $\mathbf{G}^0$ and a character ε of $\mathbf{G}^0(\mathbb{R})$ of order 2 such that $\mathrm{H}^*(\mathfrak{g}, \mathrm{K}'_\infty; \varepsilon(\pi_\infty \otimes W^*)) \neq 0$, where $\mathfrak{g} = \mathrm{Lie}(\mathbf{G}^0(\mathbb{C}))$ and K'_∞ is the set of fixed points of a Cartan involution of $\mathbf{G}^0(\mathbb{R})$.

Moreover, if π is regular algebraic, then π_∞ is essentially tempered.

(ii) Assume that $\mathbf{G}^0 = R_{E/\mathbb{Q}}\mathbf{GL}_{n,E}$. Let θ be the involution of $\mathbf{G}^0$ defined by $g \longmapsto {}^t\overline{g}^{-1}$. Let π be a θ-stable cuspidal automorphic representation of $\mathbf{G}^0(\mathbb{A})$. Then the following conditions are equivalent.

(a) π is regular algebraic.

(b) There exists a θ-stable algebraic representation W of $\mathbf{G}^0$ such that $\mathrm{ep}(\theta, \pi_\infty \otimes W^*) \neq 0$ (where $\mathrm{ep}(\theta, .)$ is defined before theorem 8.1.5).

Moreover, if π is regular algebraic, then π_∞ is tempered.

Proof

(i) The equivalence of (a) and (b) is obvious from the definition, and (c) implies (b) by Wigner's lemma. The fact that (a) implies (c) is lemma 3.14 of [Cl4] (cf. also proposition 3.5 of [Cl5] for the case $\mathbf{G}^0 = R_{E/\mathbb{Q}}\mathbf{GL}_{n,E}$). The last sentence of (i) is lemma 4.19 of [Cl4].

(ii) It is obvious (by Wigner's lemma and (i)) that (b) implies (a). The fact that (a) implies (b) is proved in proposition 3.5 of [Cl5]. The last sentence is a remark made at the beginning of 3.2 of [Cl5]: if π is regular algebraic, then π_∞ is essentially tempered; as it is also θ-stable, it must be tempered. □

As in the lemma above, denote by θ the automorphism $g \longmapsto {}^t\overline{g}^{-1}$ of $R_{E/\mathbb{Q}}\mathbf{GL}_{n,E}$.

Corollary 8.4.9 *Let π be a θ-stable cuspidal automorphic representation of $\mathbf{GL}_n(\mathbb{A}_E)$ that is regular algebraic. Let S be the union of the set of prime numbers that* *ramify in E and of the set of prime numbers under finite places of E where π is ramified. Then there exist a number field K, a positive integer N, and, for every finite place λ of K, a continuous finite-dimensional representation σ_λ of* $\mathrm{Gal}(\overline{\mathbb{Q}}/E)$ *with coefficients in K_λ, such that:*

(i) The representation σ_λ is unramified outside $S \cup \{\ell\}$, where ℓ is the prime number under λ, pure of weight $1-n$ if n is not divisible by 4 and mixed with weights between $2(2-n)-1$ and $2(2-n)+1$ if n is divisible by 4. If n is divisible by 4 and π is very regular, then σ_λ is pure of weight $2(2-n)$.

(ii) For every place $\wp$ of E above a prime number $p \notin S$, for every finite place $\lambda \not| p$ of K,

$$\log L_\wp(s, \sigma_\lambda) = N \log L\left(s + \frac{n-1}{2}, \pi_\wp\right)$$

if n is not divisible by 4, and

$$\log L_\wp(s, \sigma_\lambda) = N \log L\left(s + (n-2), \pi_\wp, \bigwedge^2\right)$$

if n is divisible by 4 (where $\pi_\wp$ is the local component at $\wp$ of π, seen as a representation of $\mathbf{U}(n)(\mathbb{A}_E)$*).*

(The first L-function is the one associated to the standard representation of $\mathbf{GL}_n(\mathbb{C}) = \widehat{\mathbf{U}(n)}$*, and the second L-function is the one associated to the exterior square of the standard representation.)*

Proof. It is enough to show that there exists a character $\chi : \mathbb{A}_E/E^\times \longrightarrow \mathbb{C}^\times$ such that $\chi \otimes \pi$ satisfies the conditions of corollary 8.4.6, with V of weight 0. This follows from lemma VI.2.10 of [HT]. □

Using the base change theorem of Arthur and Clozel ([AC]), it is possible to deduce from the last corollary results about self-dual automorphic representations of $\mathbf{GL}_n(\mathbb{A})$. Here, we will treat only the case n odd (which is simpler); in the general case, the next corollary would not hold for all quadratic imaginary extensions E.

Corollary 8.4.10 *Assume that n is odd. Let τ be a self-dual cuspidal automorphic representation of* $\mathbf{GL}_n(\mathbb{A})$*, and assume that τ is regular algebraic. Let E be a quadratic imaginary extension of $\mathbb{Q}$. Write S for the union of the set of prime numbers that ramify in E and the set of prime numbers where τ is ramified. Then there exist a number field K, a positive integer N and, for every finite place λ of K, a (continuous finite-dimensional) representation σ_λ of* $\mathrm{Gal}(\overline{\mathbb{Q}}/E)$ *with coefficients in K_λ, such that:*

(i) the representations σ_λ are unramified outside S and pure of weight $1-n$;

(ii) for every finite place $\wp$ of E above a prime number $p \notin S$, for every finite place $\lambda \not| p$ of K, for every $j \in \mathbb{Z}$,

$$\mathrm{Tr}(\sigma_\lambda(\Phi_\wp^j)) = N(N\wp)^{j(n-1)/2}\,\mathrm{Tr}(\varphi_{\tau_p}(\Phi_\wp^j)),$$

where $\varphi_{\tau_p} : W_{\mathbb{Q}_p} \longrightarrow$ $\mathbf{GL}_n(\mathbb{C})$ is a Langlands parameter of τ_p and $\Phi_\wp \in W_{E_\wp}$ is a lift of the geometric Frobenius.

In particular, τ satisfies the Ramanujan-Petersson conjecture at every unramified place.

Proof. Let θ be as before the involution $g \longmapsto {}^t\overline{g}^{-1}$ of $R_{E/\mathbb{Q}}\mathbf{GL}_{n,E}$. If V is an irreducible algebraic representation of $\mathbf{GL}_n$, it defines a θ-discrete representation π_V of $\mathbf{GL}_n(E \otimes_{\mathbb{Q}} \mathbb{R})$ as in lemma 8.1.10.

Let π be the automorphic representation of $\mathbf{GL}_n(\mathbb{A}_E)$ obtained from τ by base change (cf. [AC] theorem III.4.2). Because n is odd, π is necessarily cuspidal (this follows from (b) of the same theorem). [5] By the definition of base change, π is regular algebraic. Let $(p_1, \dots, p_n) \in \mathbb{Z}^n$, with $p_1 \geq \cdots \geq p_n$, be the n-uple of integers associated to τ as in definition 8.4.7. As τ is self-dual, $p_i + p_{n+1-i} = 1 - n$ for every $i \in \{1, \dots, n\}$. For every $i \in \{1, \dots, n\}$, set $a_i = p_i + i - 1$. Then $\sum_{i=1}^n a_i = 0$. Let V be the irreducible algebraic representation of $\mathbf{GL}_n$ with highest weight $(a_1, \dots, a_n)$, and let W^* be the θ-stable algebraic representation of $R_{E/\mathbb{Q}}\mathbf{GL}_{n,E}$ defined by V^* as in remark 8.1.6. (As the notation suggests, W^* is the contragredient of the irreducible algebraic representation W of $R_{E/\mathbb{Q}}\mathbf{GL}_{n,E}$ associated to π as in definition 8.4.7.) By proposition 3.5 of [Cl5], $\mathrm{ep}(\theta, \pi_\infty \otimes W^*) \neq 0$. As π_∞ is tempered, theorem 8.1.5 and remark 8.1.6 imply that $\pi_\infty \simeq \pi_V$, so that π_∞ is θ-discrete.

We may therefore apply corollary 8.4.9 to π. We get a family of representations σ_λ of $\mathrm{Gal}(\overline{\mathbb{Q}}/E)$. Point (i) follows from (i) of corollary 8.4.9. It remains to check the equality in point (ii).

Let $\mathbf{H} = \mathbf{U}(n)$, $\mathbf{H}' = \mathbf{GL}_n$, $\mathbf{G}^0 = R_{E/\mathbb{Q}}\mathbf{H}_E = R_{E/\mathbb{Q}}\mathbf{H}'_E$, and let θ' be the involution $g \longmapsto \overline{g}$ of $\mathbf{G}^0$. As in 2.3, let $\Phi_n \in \mathbf{GL}_n(\mathbb{Z})$ be the matrix with coefficients: $(\Phi_n)_{i,j} = (-1)^{i-1}\delta_{i,n+1-j}$. There is an isomorphism $\widehat{\mathbf{G}}^0 \simeq \mathbf{GL}_n(\mathbb{C}) \times \mathbf{GL}_n(\mathbb{C})$ such that

- the embedding $\widehat{\mathbf{H}}' = \mathbf{GL}_n(\mathbb{C}) \longrightarrow \widehat{\mathbf{G}}^0$ is $g \longmapsto (g, g)$;
- the embedding $\widehat{\mathbf{H}} = \mathbf{GL}_n(\mathbb{C}) \longrightarrow \widehat{\mathbf{G}}^0$ is $g \longmapsto (g, \Phi_n {}^t g^{-1} \Phi_n^{-1})$;
- for every $(g, h) \in \widehat{\mathbf{G}}^0$, $\hat{\theta}(g, h) = (\Phi_n {}^t h^{-1} \Phi_n^{-1}, \Phi_n {}^t g^{-1} \Phi_n^{-1})$ and $\hat{\theta}'(g, h) = (h, g)$.

Let $\mathbf{T}$ be the diagonal torus of $\mathbf{G}^0$. Let $p \notin S$ be a prime number. Denote by $x = ((y_1, \dots, y_n), (z_1, \dots, z_n)) \in \widehat{\mathbf{T}}^{\mathrm{Gal}(\overline{\mathbb{Q}}_p/\mathbb{Q}_p)}$ the Langlands parameter of π_p. As π_p is θ-stable and θ'-stable, we may assume that $\hat{\theta}(x) = \hat{\theta}'(x) = x$, that is, that $y_i = z_i = y_{n+1-i}^{-1}$ for every $i \in \{1, \dots, n\}$. Assume that p is inert in E. Then $\mathbf{H}(E_p) \simeq \mathbf{H}'(E_p)$, and the Langlands parameter of π_p, seen as a representation of $\mathbf{H}(E_p)$ or $\mathbf{H}'(E_p)$, is $(y_1^2, \dots, y_n^2)$; on the other hand, the Langlands parameter of τ_p is $(y_1, \dots, y_n)$, hence the image of Φ_{E_p} by φ_{τ_p} is $(y_1^2, \dots, y_n^2)$. Assume that p splits in E, and let $\wp$ and $\wp'$ be the places of E above p. Then $\mathbf{G}^0(\mathbb{Q}_p) = \mathbf{H}(E_\wp) \times \mathbf{H}(E_{\wp'}) = \mathbf{H}'(E_\wp) \times \mathbf{H}'(E_{\wp'})$. Write $\pi_p = \pi_\wp \otimes \pi_{\wp'} = \pi'_\wp \otimes \pi'_{\wp'}$, where $\pi_\wp$ (resp., $\pi_{\wp'}$, resp., $\pi'_\wp$, resp., $\pi'_{\wp'}$) is an unramified representation of $\mathbf{H}(E_\wp)$ (resp., $\mathbf{H}(E_{\wp'})$, resp., $\mathbf{H}'(E_\wp)$, resp., $\mathbf{H}'(E_{\wp'})$). Then the Langlands parameter of τ_p, $\pi_\wp$, $\pi_{\wp'}$, $\pi'_\wp$ or $\pi'_{\wp'}$ is $(y_1, \dots, y_n)$. These calculations show that point (ii) follows from (ii) of corollary 8.4.9. □

[5] I thank Sug Woo Shin for pointing out this useful fact to me.

8.5 A SIMPLE CASE OF BASE CHANGE

As an application of the techniques in this chapter (and of the knowledge about automorphic representations of general linear groups), it is possible to obtain some weak base change results between general unitary groups and general linear groups. These results are spelled out in this section.

Use the notation at the beginning of section 8.4; in particular, $\mathbf{H} = \mathbf{G}(\mathbf{U}(p_1, q_1) \times \cdots \times \mathbf{U}(p_r, q_r))$, $\mathbf{G}^0 = R_{E/\mathbb{Q}}\mathbf{H}_E$ and $\xi : {}^L\mathbf{H} \longrightarrow {}^L\mathbf{G}^0$ is the "diagonal" L-morphism.

As before, if $\mathbf{L}$ is a connected reductive group over $\mathbb{Q}$, p is a prime number where $\mathbf{L}$ is unramified and $\pi_{L,p}$ is an unramified representation of $\mathbf{L}(\mathbb{Q}_p)$, we will denote by $\varphi_{\pi_{L,p}} : W_{\mathbb{Q}_p} \longrightarrow \widehat{\mathbf{L}} \rtimes W_{\mathbb{Q}_p}$ a Langlands parameter of $\pi_{L,p}$.

Definition 8.5.1 Let $\mathbf{L}$ be a Levi subgroup of $\mathbf{G}^0$ (it does not have to be the identity component of a Levi subset of $\mathbf{G}$). Then there is a $W_{\mathbb{Q}}$-embedding $\widehat{\mathbf{L}} \longrightarrow \widehat{\mathbf{G}}^0$, unique up to $\widehat{\mathbf{G}}^0$-conjugacy; fix such an embedding, and let $\eta_L : {}^L\mathbf{L} \longrightarrow {}^L\mathbf{G}^0$ be the obvious L-morphism extending it (i.e., the L-morphism whose restriction to $W_{\mathbb{Q}}$ is the identity).

As $\mathbf{G}^0$ is isomorphic to $R_{E/\mathbb{Q}}(\mathbb{G}_{m,E} \times \mathbf{GL}_{n_1,E} \times \cdots \times \mathbf{GL}_{n_r,E})$, $\mathbf{L}$ is a direct product of $R_{E/\mathbb{Q}}\mathbb{G}_{m,E}$ and of groups of the type $R_{E/\mathbb{Q}}\mathbf{GL}_{m,E}$, $m \in \mathbb{N}^*$. Choose an isomorphism $\mathbf{L} \simeq R_{E/\mathbb{Q}}(\mathbb{G}_{m,E} \times \mathbf{GL}_{m_1,E} \times \cdots \times \mathbf{GL}_{m_l,E})$, and denote by θ_L the automorphism $(x, g_1, \dots, g_l) \longmapsto (\overline{x}, \overline{x}{}^t\overline{g_1}^{-1}, \dots, \overline{x}{}^t\overline{g_l}^{-1})$ of $\mathbf{L}$; the class of θ_L in the group of outer automorphisms of $\mathbf{L}$ does not depend on the choices. (Note also that θ and θ_{G^0} are equal up to an inner automorphism, so we can take $\theta = \theta_{G^0}$.)

Let π_H be an irreducible admissible representation of $\mathbf{H}(\mathbb{A})$ and π_L be an irreducible admissible representation of $\mathbf{L}(\mathbb{A})$. Let v be a finite place where π_H and π_L are unramified. We say that π_H and π_L *correspond to each other at* v if $\xi \circ \varphi_{\pi_{H,v}}$ and $\eta_L \circ \varphi_{\pi_{L,v}}$ are $\widehat{\mathbf{G}}^0$-conjugate.

Remark 8.5.2 Let v be a finite place of $\mathbb{Q}$ that is unramified in E, $\mathbf{L}$ be a Levi subgroup of $\mathbf{G}^0$, $\pi_{L,v}$ be an unramified representation of $\mathbf{L}(\mathbb{Q}_v)$ and $\pi_{H,v}$ be an unramified representation of $\mathbf{H}(\mathbb{Q}_v)$. Then $\xi \circ \varphi_{\pi_{H,v}}$ and $\eta_L \circ \varphi_{\pi_{L,v}}$ are $\widehat{\mathbf{G}}^0$-conjugate if and only if, for every $\phi_v \in \mathcal{H}(\mathbf{G}^0(\mathbb{Q}_v), \mathbf{G}^0(\mathbb{Z}_v))$,

$$\operatorname{Tr}(\pi_v(b_\xi(\phi_v))) = \operatorname{Tr}(\pi_{L,v}((\phi_v)_L)),$$

where $(\phi_v)_L$ is the constant term of ϕ_v at $\mathbf{L}$ (and b_ξ is, as in 4.2, the base change map $\mathcal{H}(\mathbf{G}^0(\mathbb{Q}_v), \mathbf{G}^0(\mathbb{Z}_v)) \longrightarrow \mathcal{H}(\mathbf{H}(\mathbb{Q}_v), \mathbf{H}(\mathbb{Z}_v)))$.

Corollary 8.5.3 *(i) Let π_H be an irreducible admissible representation of $\mathbf{H}(\mathbb{A})$. Assume that*

- *there exists a neat open compact subgroup K_H of $\mathbf{H}(\mathbb{A}_f)$ such that $\pi_{H,f}^{\mathrm{K}_H} \neq \{0\}$;*
- *there exist an irreducible algebraic representation V of $\mathbf{H}$ and $i \in \mathbb{Z}$ such that $m_{\mathrm{disc}}(\pi_H) \neq 0$ and $\mathrm{H}^i(\mathfrak{h}, \mathrm{K}'_\infty; \pi_{H,\infty} \otimes V) \neq 0$, where $\mathfrak{h} = \mathrm{Lie}(\mathbf{H}(R)) \otimes \mathbb{C}$ and the notation is that of remark 7.2.5 (or lemma 7.3.5).*

(In other words, π_H is a discrete automorphic representation of $\mathbf{H}(\mathbb{A})$ and appears in the intersection cohomology of some Shimura variety associated to $\mathbf{H}$.)

Then there exist a Levi subgroup $\mathbf{L}$ of $\mathbf{G}^0$, a cuspidal automorphic representation π_L of $\mathbf{L}(\mathbb{A})$, and an automorphic character χ_L of $\mathbf{L}(\mathbb{A})$ such that $\pi_L \otimes \chi_L^{-1}$ is θ_L-stable and π_H and π_L correspond to each other at almost every finite place. If $\mathbf{L} = \mathbf{G}^0$, then we can take $\chi_L = 1$, and π_L is regular algebraic.

(ii) *Assume that $\mathbf{H}$ is quasi-split at every finite place of $\mathbb{Q}$. Let π be a θ-stable cuspidal automorphic representation of $\mathbf{G}^0(\mathbb{A})$. Assume that there exists an irreducible algebraic representation V of $\mathbf{H}$ such that $\mathrm{Tr}(\pi_\infty(\phi_V)) \neq 0$, where $\phi_V \in C^\infty(\mathbf{G}^0(\mathbb{R}))$ is associated to V as in lemma 8.1.10. (In other words, π is regular algebraic, cf. lemma 8.4.8.) Let S be the union of the set of finite places that are ramified in E and of the set of places where π is ramified. Then there exists an automorphic representation π_H of $\mathbf{H}(\mathbb{A})$, unramified outside S, such that π and π_H correspond to each other at every finite place $v \notin S$. Moreover,*

- *π_H satifies the conditions of (i) for V and some K_H;*
- *if π'_H satisfies the conditions of (i) and is isomorphic to π_H at almost every finite place, then π'_H is cuspidal (in particular, π_H is cuspidal);*
- *in the notation of 7.2, for every $\underline{e} \in \mathcal{F}_{\mathbf{H}}$, $R_{\underline{e}}(\pi_{H,f}) = \varnothing$ (i.e., π_H "does not come from an endoscopic group of $\mathbf{H}$").*

Remark 8.5.4 Note that, using (ii) of the corollary, we can strengthen (i) a little. We obtain the following statement. If, in (i), $\mathbf{H}$ is quasi-split at every finite place of $\mathbb{Q}$ and $\mathbf{L} = \mathbf{G}^0$, then π_H is unramified at v as soon as $\pi := \pi_L$ is, and π_H and π correspond to each other at every finite place of $\mathbb{Q}$ where π is unramified.

In the rest of this section, we use the following notation. If S is a set of places of $\mathbb{Q}$, then $\mathbb{Z}_S = \prod_{v\in S} \mathbb{Z}_v$ and $\mathbb{Z}^S = \prod_{v\notin S\cup\{\infty\}} \mathbb{Z}_v$.

Proof. We show (i). Let π_H, K_H and V be as in (i). Let W_λ be the virtual λ-adic representation of $\mathcal{H}(\mathbf{H}(\mathbb{A}_f), \mathrm{K}_H) \times \mathrm{Gal}(\overline{\mathbb{Q}}/E)$ defined by K_H and V as in section 7.2. After replacing K_H by a smaller open compact subgroup of $\mathbf{H}(\mathbb{A}_f)$, we may assume that $\mathbb{1}_{\mathrm{K}_H}$ satisfies condition (H) of 8.4 (cf. (i) of lemma 8.4.1) and that $\mathrm{K}_H = \prod_{v\neq\infty} \mathrm{K}_{H,v}$. For every set S of finite places of $\mathbb{Q}$, we will write $\mathrm{K}_{H,S} = \prod_{v\in S} \mathrm{K}_{H,v}$ and $\mathrm{K}_H^S = \prod_{v\notin S\cup\{\infty\}} \mathrm{K}_{H,v}$. Let S be a finite set of finite places of $\mathbb{Q}$ such that, for every $v \notin S \cup \{\infty\}$, $\mathbf{H}$ is unramified at v and $\mathrm{K}_{H,v} = \mathbf{H}(\mathbb{Z}_v)$. Set $f_S = \mathbb{1}_{\mathrm{K}_{H,S}}$.

Let $\Pi'(\pi_H)$ be the set of isomorphism classes of irreducible admissible representations π'_H of $\mathbf{H}(\mathbb{A})$ such that

- π'_H satisfies the conditions of (i) for the same K_H and V as π_H;
- for almost every finite place v of $\mathbb{Q}$ such that $\mathbf{H}$ is unramified at v, $\xi \circ \varphi_{\pi_{H,v}}$ and $\xi \circ \varphi_{\pi'_{H,v}}$ are $\widehat{\mathbf{G}}^0$-conjugate.

(We use the notation $\Pi'(\pi_H)$ to avoid confusion with the set Π_H of the proof of (ii).) Then $\Pi'(\pi_H)$ is finite (in fact, the set of π'_H that satisfy the first condition defining $\Pi'(\pi_H)$ is already finite). So there exist a finite set $T \supset S$ of finite places of $\mathbb{Q}$ and a function $f_{T-S} \in \mathcal{H}(\mathbf{H}(\mathbb{A}_{T-S}), \mathrm{K}_{H,T-S})$ such that

- f_{T-S} is in the image of the base change map $b_\xi : \mathcal{H}(\mathbf{G}^0(\mathbb{A}_{T-S}), \mathbf{G}^0(\mathbb{Z}_{T-S})) \longrightarrow \mathcal{H}(\mathbf{H}(\mathbb{A}_{T-S}), \mathrm{K}_{H,T-S})$ (hence, by (iii) of lemma 8.4.1, it satisfies condition (H)).
- Let π'_H be an irreducible admissible representation of $\mathbf{H}(\mathbb{A})$ that satisfies the conditions of (i) for K_H and V. Then $\mathrm{Tr}(\pi'_{H,T-S}(f_{T-S})) = 1$ if $\pi'_H \in \Pi'(\pi_H)$, and 0 otherwise.
- For every $v \notin T$ finite, for every $\pi'_H \in \Pi'(\pi_H)$, $\xi \circ \varphi_{\pi_{H,v}}$ and $\xi \circ \varphi_{\pi'_{H,v}}$ are $\widehat{\mathbf{G}}^0$-conjugate.

Write $f_T = f_S f_{T-S}$. As in 8.4, f_T determines functions $\phi_{M,T} \in C_c^\infty(\mathbf{M}^0(\mathbb{A}_T))$, for $\mathbf{M} \in \mathcal{M}_{\mathbf{G}}$.

Fix a prime number $p \notin T$. Then, with the notation of section 7.2, for every $m \in \mathbb{Z}$ and every $f^{T\cup\{p\}}$ in the image of the base change map $b_\xi : \mathcal{H}(\mathbf{G}^0(\mathbb{A}_f^{T\cup\{p\}}), \mathbf{G}^0(\mathbb{Z}^{T\cup\{p\}})) \longrightarrow \mathcal{H}(\mathbf{H}(\mathbb{A}_f^{T\cup\{p\}}), \mathrm{K}_H^{T\cup\{p\}})$,

$$\begin{aligned}&\mathrm{Tr}(\Phi_\wp^m f_T f^{T\cup\{p\}}, W_\lambda) =\\ &\mathrm{Tr}(\pi_{H,f}^{T\cup\{p\}}(f^{T\cup\{p\}})) \sum_{\pi'_H\in\Pi'(\pi_H)} \dim((\pi'_{H,S})^{\mathrm{K}_{H,S}}) \,\mathrm{Tr}(\Phi_\wp^m, W_\lambda(\pi'_H)).\end{aligned}$$

Note that the virtual representation $\sum_{\pi'_H\in\Pi'(\pi_H)} \dim((\pi'_{H,S})^{\mathrm{K}_{H,S}}) W_\lambda(\pi'_H)$ of $\mathrm{Gal}(\overline{\mathbb{Q}}/E)$ is not trivial. This is proved as the fact that (2) implies (1) in remark 7.2.5 (see this proof for more details) : Let w be the weight of V in the sense of 1.3. Then, for every $\pi'_H \in \Pi'(\pi_H)$, $W_\lambda(\pi'_H) = \sum_{i\in\mathbb{Z}}(-1)^i W_\lambda^i(\pi'_H)$, with $W_\lambda^i(\pi'_H)$ a true (not virtual) representation of $\mathrm{Gal}(\overline{\mathbb{Q}}/E)$ that is unramified and of weight $-w + i$ at almost every place of E. Hence there can be no cancellation in the sum $\sum_{\pi'_H\in\Pi'(\pi_H)} \dim((\pi'_{H,S})^{\mathrm{K}_{H,S}}) W_\lambda(\pi'_H)$. (And the assumptions on π_H imply that $W_\lambda(\pi_H)$ is not trivial, by remark 7.2.5.) So there exists an integer $m \in \mathbb{Z}$ such that

$$C := \sum_{\pi'_H\in\Pi'(\pi_H)} \dim((\pi'_{H,S})^{\mathrm{K}_{H,S}}) \,\mathrm{Tr}(\Phi_\wp^m, W_\lambda(\pi'_H)) \neq 0.$$

For every $\mathbf{M} \in \mathcal{M}_{\mathbf{G}}$, write $\phi_{M,p} = \phi_{M,p}^{(m)}$, where $\phi_{M,p}^{(m)}$ is as in section 8.4, and define $\phi_{M,\infty}$ as in loc. cit. Then, by theorem 8.4.3 and the calculations above (and (iii) of lemma 8.4.1), for every $\phi^{T\cup\{p\}} \in \mathcal{H}(\mathbf{G}^0(\mathbb{A}_f^{T\cup\{p\}}), \mathbf{G}^0(\mathbb{Z}^{T\cup\{p\}}))$,

$$C\,\mathrm{Tr}(\pi_{H,f}^{T\cup\{p\}}(f^{T\cup\{p\}})) = \sum_{\mathbf{M}\in\mathcal{M}_{\mathbf{G}}} c_M T^M(\phi_M),$$

where $f^{T\cup\{p\}} = b_\xi(\phi^{T\cup\{p\}})$ and $\phi_M = \phi_{M,T}\phi_{M,p}\phi_M^{T\cup\{p\}}\phi_{M,\infty}$, with $\phi_M^{T\cup\{p\}}$ equal to the product of $\chi_{M|\mathbf{G}^0(\mathbb{A}_f^{T\cup\{p\}})}$ and of the constant term of $\phi^{T\cup\{p\}}$ at $\mathbf{M}^0$.

By lemma 8.5.5 below, the right-hand side of this equality, seen as a linear form T over $\mathcal{H}(\mathbf{G}^0(\mathbb{A}_f^{T\cup\{p\}}), \mathbf{G}^0(\mathbb{Z}^{T\cup\{p\}}))$, is a finite linear combination of linear maps of the form $\phi^{T\cup\{p\}} \longmapsto \mathrm{Tr}(\pi_L((\phi^{T\cup\{p\}})_L))$, with $\mathbf{L}$, π_L and $(\phi^{T\cup\{p\}})_L$ as in this lemma. By the strong multiplicity 1 theorem of Jacquet and Shalika (cf. theorem 4.4 of [JS]), there exist a Levi subgroup $\mathbf{L}$ of $\mathbf{G}^0$, a cuspidal automorphic representation π_L of $\mathbf{L}(\mathbb{A})$, an automorphic character χ_L of $\mathbf{L}(\mathbb{A})$, a scalar $a \in \mathbb{C}^\times$, a finite set $\Sigma \supset T \cup \{p\}$ of finite places of $\mathbb{Q}$, and a function $\phi_{\Sigma - T\cup\{p\}} \in \mathcal{H}(\mathbf{G}^0(\mathbb{A}_{\Sigma-T\cup\{p\}}), \mathbf{G}^0(\mathbb{Z}_{\Sigma-T\cup\{p\}}))$ such that $\pi_L \otimes \chi_L^{-1}$ is θ_L-stable and, for every $\phi^\Sigma \in \mathcal{H}(\mathbf{G}^0(\mathbb{A}_f^\Sigma), \mathbf{G}^0(\mathbb{Z}^\Sigma))$,

$$T(\phi_{\Sigma-T\cup\{p\}}\phi^\Sigma) = a \operatorname{Tr}\left(\pi_L^\Sigma((\phi^\Sigma)_L)\right)$$

(a is nonzero because T is nonzero, and that in turn follows from the fact that $C \neq 0$; the existence of χ_L such that $\pi_L \otimes \chi_L^{-1}$ is θ_L-stable comes from lemma 8.5.5).

Let $D = C\operatorname{Tr}(\pi_{H,\Sigma-T\cup\{p\}}(b_\xi(\phi_{\Sigma-T\{p\}}))$. Then we finally find that, for every $\phi^\Sigma \in \mathcal{H}(\mathbf{G}^0(\mathbb{A}_f^\Sigma), \mathbf{G}^0(\mathbb{Z}^\Sigma))$,

$$D\operatorname{Tr}(\pi_{H,f}^\Sigma(b_\xi(\phi^\Sigma))) = a\operatorname{Tr}(\pi_L^\Sigma((\phi^\Sigma)_L)).$$

In particular, $D = a \neq 0$. This equality implies that π_H and π_L correspond to each other at every finite place $v \notin \Sigma$. Assume that $\mathbf{L} = \mathbf{G}^0$. Then it is obvious from the definition of T and from the definition of $\phi_{G,\infty}$ in section 8.4 that the infinitesimal character of π_L is equal to that of an algebraic representation of $\mathbf{G}^0$. By (i) of lemma 8.4.8, π_L is regular algebraic. This finishes the proof of (i).

We show (ii). Assume that $\mathbf{H}$ is quasi-split at every finite place of v, and let π, V and S be as (ii). Fix an open compact subgroup $\mathrm{K} = \prod_v \mathrm{K}_v$ of $\mathbf{G}^0(\mathbb{A}_f)$ such that $\operatorname{Tr}(\pi_f(\mathbb{1}_\mathrm{K})A_{\pi_f}) \neq 0$, that $\mathrm{K}_v = \mathbf{G}^0(\mathbb{Z}_v)$ for $v \notin S$ and that $\mathbb{1}_\mathrm{K}$ has a transfer to $\mathbf{H}$ satisfying condition (H) (such a K exists by lemma 8.4.1). Set $\mathrm{K}_S = \prod_{v\in S}\mathrm{K}_v$, $\phi_S = \mathbb{1}_{\mathrm{K}_S}$, and let $f_S \in C_c^\infty(\mathbf{H}(\mathbb{A}_S))$ be a transfer of ϕ_S satisfying condition (H). Choose an open compact subgroup $\mathrm{K}_{H,S}$ of $\mathbf{H}(\mathbb{A}_S)$ such that f_S is bi-$\mathrm{K}_{H,S}$-invariant, and set $\mathrm{K}_H = \mathrm{K}_{H,S}\prod_{v\notin S\cup\{\infty\}}\mathbf{H}(\mathbb{Z}_v)$. After making $\mathrm{K}_{H,S}$ smaller, we may assume that K_H is neat. Let W_λ be the λ-adic virtual representation of $\mathcal{H}(\mathbf{H}(\mathbb{A}_f), \mathrm{K}_H) \times \operatorname{Gal}(\overline{\mathbb{Q}}/E)$ defined by K_H and V as in 7.2.

Let $\phi^S \in \mathcal{H}(\mathbf{G}^0(\mathbb{A}_f^S), \mathbf{G}^0(\mathbb{Z}^S))$. By theorem 8.4.3 applied to a prime number $p \notin S$ where ϕ^S is $\mathbb{1}_{\mathbf{G}^0(\mathbb{Z}_p)}$ and to $j = 0$, there exist scalars $c'_M \in \mathbb{R}$, for $\mathbf{M} \in \mathcal{M}_\mathbf{G}$, such that

$$\operatorname{Tr}(f_S b_\xi(\phi^S), W_\lambda) = \sum_{\mathbf{M}\in\mathcal{M}_\mathbf{G}} c'_M T^M(\phi_{S,M}\phi_M^S\phi_{M,\infty}),$$

where $\phi_{S,M} \in C_c^\infty(\mathbf{M}^0(\mathbb{A}_S))$ and $\phi_M^S \in C_c^\infty(\mathbf{M}^0(\mathbb{A}_f^S))$ are the functions associated to ϕ_S and ϕ^S as in the beginning of section 8.4 (so ϕ_M^S is the product of $\chi_{M|\mathbf{M}^0(\mathbb{A}_f^S)}$ and of the constant term of ϕ^S at $\mathbf{M}^0$) and $\phi_{M,\infty} \in C^\infty(\mathbf{M}^0(\mathbb{R}))$ is obtained from V as at the beginning of section 8.4 (so $\phi_{G,\infty}$ is the function ϕ_V that appears in the statement of (ii)). But, as in the proof of proposition 7.1.4, we see that, for any $\mathbf{M} \in \mathcal{M}_\mathbf{G}$, the function $f_{H_M,p}^{(0)}$ of definition 7.1.6 is equal to the product of $\frac{\iota_{\mathbf{H},\mathbf{H}_M}}{\iota(\mathbf{H},\mathbf{H}_M)}$ and of a transfer of $\mathbb{1}_{\mathbf{H}(\mathbb{Z}_p)}$ to $\mathbf{H}_M$. So, for every $\mathbf{M} \in \mathcal{M}_\mathbf{G}$, $c'_M = \frac{\iota_{\mathbf{H},\mathbf{H}_M}}{\iota(\mathbf{H},\mathbf{H}_M)} c_M$; in particular, c'_M does not depend on ϕ^S, and $c'_G \neq 0$ (because all the signs in the definition of $\iota_{\mathbf{H},\mathbf{H}}$ are obviously equal to 1).

Consider the function T that sends ϕ^S to $\sum_{\mathbf{M}\in\mathcal{M}_\mathbf{G}} c'_M T^M(\phi_{S,M}\phi^S_M\phi_{M,\infty})$. It is a linear form on $\mathcal{H}(\mathbf{G}^0(\mathbb{A}^S_f),\mathbf{G}^0(\mathbb{Z}^S))$. By lemma 8.5.5, T is a finite linear combination of characters on $\mathcal{H}(\mathbf{G}^0(\mathbb{A}^S_f),\mathbf{G}^0(\mathbb{Z}^S))$ of the form $\phi^S \longmapsto \mathrm{Tr}(\pi^S_L((\phi^S)_L))$, where $\mathbf{L}$ is a Levi subgroup of $\mathbf{G}^0$ and π_L is a cuspidal automorphic representation of $\mathbf{L}(\mathbb{A})$. By the strong multiplicity 1 theorem of Jacquet and Shalika, these characters are pairwise distinct. Hence, by the choice of ϕ_S, the assumption on π_∞, and the fact that $c'_G \neq 0$, the coefficient of the character $\phi^S \longmapsto \mathrm{Tr}(\pi^S(\phi^S))$ in T is nonzero.

Let R_H be the set of isomorphism classes of irreducible admissible representations of $\mathbf{H}(\mathbb{A})$ that satisfy the conditions of (i) for K_H and V. Then R_H is finite. Define an equivalence relation $\sim$ on R_H in the following way. If $\pi_H, \pi'_H \in R_H$, then $\pi_H \sim \pi'_H$ if and only if, for every finite place $v \notin S$ of $\mathbb{Q}$, $\xi\circ\varphi_{\pi_{H,v}}$ and $\xi\circ\varphi_{\pi'_{H,v}}$ are $\widehat{\mathbf{G}}^0$-conjugate (π_H and π'_H are necessarily unramified at v because we have chosen K_H to be hyperspecial outside S). Let $\Pi_H \in R_H/\!\sim$. Then, if $\phi^S \in \mathcal{H}(\mathbf{G}^0(\mathbb{A}^S),\mathbf{G}^0(\mathbb{Z}^S))$, the value $\mathrm{Tr}(\pi^S_H(b_\xi(\phi^S)))$ is the same for every $\pi_H \in \Pi_H$; denote it by $\mathrm{Tr}(\Pi^S_H(b_\xi(\phi^S)))$. Let

$$c(\Pi_H) = \sum_{\pi_H\in\Pi_H} \dim(W_\lambda(\pi_H))\,\mathrm{Tr}(\pi_{H,S}(f_S))$$

(where the $W_\lambda(\pi_H)$ are as in section 7.2).

By the definition of the $W_\lambda(\pi_H)$, $W_\lambda = \sum_{\pi_H\in R_H} W_\lambda(\pi_H)\otimes\pi^{\mathrm{K}_H}_{H,f}$. Hence, for every $\phi^S \in \mathcal{H}(\mathbf{G}^0(\mathbb{A}^S),\mathbf{G}^0(\mathbb{Z}^S))$,

$$T(\phi^S) = \mathrm{Tr}(f_S b_\xi(\phi^S), W_\lambda) = \sum_{\Pi_H\in R_H/\sim} c(\Pi_H)\,\mathrm{Tr}(\Pi^S_H(b_\xi(\phi^S))).$$

As the characters of $\mathcal{H}(\mathbf{G}^0(\mathbb{A}^S),\mathbf{G}^0(\mathbb{Z}^S))$ are linearly independant, there exists $\Pi_H \in R_H/\sim$ such that, for every $\phi^S \in \mathcal{H}(\mathbf{G}^0(\mathbb{A}^S),\mathbf{G}^0(\mathbb{Z}^S))$, $\mathrm{Tr}(\Pi^S_H(b_\xi(\phi^S))) = \mathrm{Tr}(\pi^S(\phi^S))$. Let π_H be any element of Π_H. It is unramified outside S and corresponds to π at every finite $v \notin S$.

It remains to prove the last three properties on π_H. The first one is true because $\pi_H \in R_H$ by construction. The second and third ones are easy consequences of (i) and of the strong multiplicity 1 theorem for $\mathbf{G}^0$. □

Lemma 8.5.5 *Let S be a set of finite places of $\mathbb{Q}$ that contains all the finite places that ramify in E and let V be an irreducible algebraic representation of $\mathbf{H}$. Fix functions $\phi_{M,S} \in C_c^\infty(\mathbf{M}^0(\mathbb{A}_S))$, $\mathbf{M} \in \mathcal{M}_\mathbf{G}$, and let $\phi_{M,\infty}$, $\mathbf{M} \in \mathcal{M}_\mathbf{G}$, be the functions associated to V as in 8.4. Consider the linear form $T : \mathcal{H}(\mathbf{G}^0(\mathbb{A}^S_f),\mathbf{G}^0(\mathbb{Z}^S)) \longrightarrow \mathbb{C}$ that sends ϕ^S to*

$$\sum_{\mathbf{M}\in\mathcal{M}_\mathbf{G}} a_M T^M(\phi_{M,S}\phi^S_M),$$

where the ϕ^S_M are obtained from ϕ^S as in 8.4 and the a_M are complex constants.

Then T is a finite linear combination of linear maps of the form $\phi^S \longmapsto \mathrm{Tr}(\pi_L^S((\phi^S)_L))$, where

(a) $\mathbf{L}$ *is a Levi subgroup of* $\mathbf{G}^0$;

(b) π_L *is a cuspidal automorphic representation of* $\mathbf{L}(\mathbb{A})$ *such that there exists an automorphic character* χ_L *of* $\mathbf{L}(\mathbb{A})$ *such that* $\pi_L \otimes \chi_L^{-1}$ *is* θ_L*-stable; if* $\mathbf{L} = \mathbf{G}^0$, *then we can take* $\chi_L = 1$;

(c) $(\phi^S)_L$ *is the constant term of* ϕ^S *at* $\mathbf{L}$.

Proof. If we did not care about condition (b) on π_L, the lemma would be an easy consequence of proposition 8.2.3, (iii) of lemma 8.4.1, and lemma 8.3.9. As we do, we must know more precisely what kind of automorphic representations appear on the spectral side of $T^M(\phi_M)$. This is the object of lemma 8.5.6 below. Once we know this lemma (for all $\mathbf{M} \in \mathcal{M}_{\mathbf{G}}$), the proof is straightforward. □

Lemma 8.5.6[6] *Let V be an irreducible algebraic representation of* $\mathbf{H}$, *and let* $\phi_V \in C^\infty(\mathbf{G}^0(\mathbb{R}))$ *be the function associated to V as in 8.4. Let $t \geq 0$, and let* $\pi \in \Pi_{\mathrm{disc}}(\mathbf{G}, t)$ *such that* $a_{disc}(\pi)\,\mathrm{Tr}(\pi_\infty(\phi_V)A_{\pi_\infty}) \neq 0$ *(the notation is that of section 8.2). Choose a Levi subgroup* $\mathbf{L}$ *of* $\mathbf{G}^0$, *a parabolic subgroup* $\mathbf{Q}$ *of* $\mathbf{G}^0$ *with Levi subgroup* $\mathbf{L}$ *and a cuspidal automorphic representation* π_L *of* $\mathbf{L}(\mathbb{A})$ *such that π is a subquotient of the parabolic induction* $Ind_{\mathbf{Q}}^{\mathbf{G}^0}\pi_L$.

Then there exists an automorphic character χ_L *such that* $\pi_L \otimes \chi_L$ *is* θ_L*-stable.*

Proof. Remember that, in this section, we take θ on $\mathbf{G}^0 = R_{E/\mathbb{Q}}(\mathbb{G}_{m,E} \times \mathbf{GL}_{n_1,E} \times \cdots \times \mathbf{GL}_{n_r,E})$ to be $(x, g_1, \dots, g_r) \longmapsto (\overline{x}, \overline{x}\,{}^t\overline{g}_1^{-1}, \dots, \overline{x}\,{}^t\overline{g}_1^{-1})$ (θ is determined only up to inner automorphism, and this is a possible choice of θ). Let $\mathbf{T}_H$ be the diagonal torus of $\mathbf{H}$. Then $\mathbf{T}_H(\mathbb{C})$ is the diagonal torus of $\mathbf{H}(\mathbb{C}) \simeq \mathbb{C}^\times \times \mathbf{GL}_{n_1}(\mathbb{C}) \times \cdots \times \mathbf{GL}_{n_r}(\mathbb{C})$. Let $\mathbf{T} = R_{E/\mathbb{Q}}\mathbf{T}_{H,E}$, let θ' be the automorphism of $\mathbf{H}(\mathbb{C})$ defined by the same formula as θ, and choose an isomorphism $\mathbf{G}^0(\mathbb{C}) \simeq \mathbf{H}(\mathbb{C}) \times \mathbf{H}(\mathbb{C})$ such that θ corresponds to the automorphism $(h_1, h_2) \longmapsto (\theta'(h_2), \theta'(h_1))$ of $\mathbf{H}(\mathbb{C}) \times \mathbf{H}(\mathbb{C})$ and that $\mathbf{T}(\mathbb{C})$ is sent to $\mathbf{T}_H(\mathbb{C}) \times \mathbf{T}_H(\mathbb{C})$. Let $\mathfrak{t}_H = \mathrm{Lie}(\mathbf{T}_H)(\mathbb{C})$ and $\mathfrak{t} = \mathrm{Lie}(\mathbf{T})(\mathbb{C})$. Then $\mathfrak{t} = \mathfrak{t}_H \oplus \mathfrak{t}_H$, and θ acts on $\mathfrak{t}$ by $(t_1, t_2) \longmapsto (\iota(t_2), \iota(t_1))$, where ι is the involution $(t, (t_{i,j})_{1\leq i\leq r, 1\leq j\leq n_i}) \longmapsto (t, (t - t_{i,j})_{1\leq i\leq r, 1\leq j\leq n_i})$ of $\mathfrak{t}_H = \mathbb{C} \oplus \mathbb{C}^{n_1} \oplus \cdots \oplus \mathbb{C}^{n_r}$. Let $\lambda \in \mathfrak{t}_H^*$ be a representative of the infinitesimal character of V, seen as a representation of $\mathbf{H}(\mathbb{R})$. Then λ is regular, and $(\lambda, \iota(\lambda)) \in \mathfrak{t}^* = \mathfrak{t}_H^* \oplus \mathfrak{t}_H^*$ represents the infinitesimal character of the θ-stable representation W of $\mathbf{G}^0(\mathbb{R})$ associated to V (as in theorem 8.1.5). By the definition of ϕ_V in theorem 8.1.5, the assumption that $\mathrm{Tr}(\pi_\infty(\phi_V)A_{\pi_\infty}) \neq 0$ implies that $\mathrm{ep}(\theta, \pi_\infty \otimes W) \neq 0$, and, by Wigner's lemma, this implies that the infinitesimal character of π_∞ is $(-\lambda, -\iota(\lambda))$. In particular, the infinitesimal character of π_∞ is regular.

Let $\mathbf{L}$ be a Levi subgroup of $\mathbf{G}^0$. We may assume that $\mathbf{L}$ is standard (in particular, it is stable by θ, and $\theta_{|\mathbf{L}} = \theta_L$), and we write

$$\mathbf{L} = R_{E/\mathbb{Q}}\left(\mathbb{G}_{m,E} \times \prod_{j=1}^{r}\prod_{k=1}^{l_j} \mathbf{GL}_{m_{jk},E}\right),$$

[6] Most of this lemma (and of its proof) was worked out during conversations with Sug Woo Shin.

where the m_{jk} are positive integers such that $n_j = m_{j,1} + \cdots + m_{j,l_j}$ for every $j \in \{1, \ldots, r\}$. We use the notation of [A3], in particular of Section 4 of that article. Then $\mathfrak{a}_L = \mathbb{R} \oplus \bigoplus_{j=1}^{r} \bigoplus_{k=1}^{l_j} \mathbb{R}$, and θ acts by $(t, (t_{ik})) \longmapsto (t, (t - t_{ik}))$. In particular, $\mathfrak{a}_{G^0} = \mathbb{R} \oplus \bigoplus_{j=1}^{r} \mathbb{R}$, and θ acts by $(t, (t_j)) \longmapsto (t, (t - t_j))$, so $\mathfrak{a}_G = \mathfrak{a}_{G^0}^{\theta=1} = \mathbb{R}$. The group $W^{G^0}(\mathfrak{a}_L)$ is equal to the group of linear automorphisms of $\mathfrak{a}_L$ that are induced by an element of $\mathrm{Nor}_{\mathbf{G}^0(\mathbb{Q})}(\mathbf{L})$, so it is embedded in an obvious way in $\mathfrak{S}_{l_1} \times \cdots \times \mathfrak{S}_{l_r}$ (but this embedding is far from being an equality in general; for example, $W^{G^0}(\mathfrak{a}_L)$ contains the factor $\mathfrak{S}_{l_j}$ if and only if $m_{j,1} = \cdots = m_{j,l_j} = n_j/l_j$). The set $W^G(\mathfrak{a}_L)$ is equal to $W^{G^0}(\mathfrak{a}_L)\theta_L$.

It will be useful to determine the subset $W^G(\mathfrak{a}_L)_{\mathrm{reg}}$ of regular elements of $W^G(\mathfrak{a}_L)$. Remember ([A3] p. 517) that an element s of $W^G(\mathfrak{a}_L)$ is in $W^G(\mathfrak{a}_L)_{\mathrm{reg}}$ if and only if $\det(s-1)_{\mathfrak{a}_L/\mathfrak{a}_G} \neq 0$. By the above calculations, $\mathfrak{a}_L/\mathfrak{a}_G = \bigoplus_{j=1}^{r} \bigoplus_{k=1}^{l_j} \mathbb{R}$, and θ_L acts by multiplication by -1. Let $s \in W^G(\mathfrak{a}_L)$, and write $s = \sigma\theta_L$, with $\sigma = (\sigma_1, \ldots, \sigma_r) \in W^{G^0}(\mathfrak{a}_L) \subset \mathfrak{S}_{l_1} \times \cdots \times \mathfrak{S}_{l_r}$. Then $\sigma\theta_L$ acts on $\mathfrak{a}_L/\mathfrak{a}_G$ by $(\lambda_{j,k})_{1 \leq j \leq r, 1 \leq k \leq l_j} \longmapsto (-\lambda_{j,\sigma_j^{-1}(k)})_{1 \leq j \leq r, 1 \leq k \leq l_j}$, so it is regular if and only if σ, acting in the obvious way on $\mathbb{R}^{l_1} \times \cdots \times \mathbb{R}^{l_r}$, does not have -1 as an eigenvalue. This is equivalent to the fact that, for every $j \in \{1, \ldots, r\}$, there are only cycles of odd length in the decomposition of σ_j as a product of cycles with disjoint supports.

We will need another fact. Let $s = (\sigma_1, \ldots, \sigma_r)\theta_L \in W^G(\mathfrak{a}_L)$ as before (we do not assume that s is regular). Let $\pi_{L,\infty}$ be an irreducible admissible representation of $\mathbf{L}(\mathbb{R})$. As $\mathbf{L}$ is standard, $\mathbf{T}$ is a maximal torus of $\mathbf{L}$. Assume that the infinitesimal character of $\pi_{L,\infty}$ has a representative of the form $(\mu, \iota(\mu)) \in \mathfrak{t}^*$, with $\mu \in \mathfrak{t}_H^*$ regular, and that $\pi_{L,\infty}$ is s-stable. Then, for every $j \in \{1, \ldots, r\}$, there are only cycles of length ≤ 2 in the decomposition of σ_j as a product of cycles with disjoint supports (in other words, $\sigma_j^2 = 1$ for every $j \in \{1, \ldots, r\}$). To prove this, we assume (to simplify notation) that $r = 1$; the general case is similar, because it is possible to reason independently on each factor $\mathbf{GL}_{n_j,E}$. Write $n = n_1$, $\sigma = \sigma_{n_1} \in \mathfrak{S}_n$, and $\mathbf{L} = R_{E/\mathbb{Q}}(\mathbb{G}_{m,E} \times \mathbf{GL}_{m_1,E} \times \cdots \times \mathbf{GL}_{m_l,E})$, with $n = m_1 + \cdots + m_l$. For every $j \in \{1, \ldots, l\}$, set $I_j = \{m_1 + \cdots + m_{j-1} + 1, \ldots, m_1 + \cdots + m_j\} \subset \{1, \ldots, n\}$. Let $(\mu, \iota(\mu))$, with $\mu \in \mathfrak{t}_H^*$, be a representative of the infinitesimal character of $\pi_{L,\infty}$. Write $\mu = (\mu_0, \ldots, \mu_n) \in \mathfrak{t}_H^* = \mathbb{C} \oplus \mathbb{C}^n$. As $\pi_{L,\infty}$ is $\sigma\theta_L$-stable, its infinitesimal character is $\sigma\theta_L$-stable. This means that there exists $\tau \in \mathfrak{S}_n$ such that

(a) for every $j \in \{1, \ldots, l\}$, $\tau(I_j) = I_{\sigma(j)}$ (i.e., τ, seen as an element of the Weyl group of $\mathbf{T}(\mathbb{Q})$ in $\mathbf{G}^0(\mathbb{Q})$, normalizes $\mathbf{L}$ and induces σ on $\mathfrak{a}_L$);

(b) $(-\mu_1, \ldots, -\mu_n) = (\mu_{\tau(1)}, \ldots, \mu_{\tau(n)})$ (i.e., $(\mu, \iota(\mu))$ is stable by $\tau\theta$, where τ is again seen as an element of the Weyl group of $\mathbf{T}(\mathbb{Q})$ in $\mathbf{G}^0(\mathbb{Q})$).

Assume that, in the decomposition of σ as a product of cycles with disjoint supports, there is a cycle of length ≥ 3. Then, by (a), there is also a cycle of length ≥ 3 in the decomposition of τ as a product of cycles with disjoint supports. But, by (b), this contradicts the fact that μ is regular.

We now come back to the proof of the lemma. By the definition of the coefficients $a_{\mathrm{disc}} = a_{\mathrm{disc}}^G$ in section 4 of [A3], the fact that $a_{\mathrm{disc}}(\pi) \neq 0$ implies that there exists a

Levi subgroup $\mathbf{L}$ of $\mathbf{G}^0$ (that we may assume to be standard), a discrete automorphic representation π_L of $\mathbf{L}(\mathbb{A})$, and an element s of $W^G(\mathfrak{a}_L)_{\mathrm{reg}}$ such that π_L is s-stable and π is a subquotient of the parabolic induction of π_L (where, for example, we use the standard parabolic subgroup of $\mathbf{G}^0$ having $\mathbf{L}$ as Levi subgroup). In particular, the infinitesimal character of $\pi_{L,\infty}$ is represented by $(-\lambda, -\iota(\lambda))$ (and it is regular). By the two facts proved above, s is equal to θ_L. That is, π_L is a θ_L-stable discrete automorphic representation of $\mathbf{L}^0(\mathbb{A})$. Hence, if we know that the lemma is true for discrete automorphic representations (and for any Levi subgroup of $\mathbf{G}^0$), then we know that it is true in general. So we may assume that π is discrete.

From now on, we assume that the automorphic representation π of $\mathbf{G}^0(\mathbb{A})$ is discrete (and of course θ-stable). Let $\mathbf{L}$ be a standard Levi subgroup of $\mathbf{G}^0$ and π_L be a cuspidal automorphic representation of $\mathbf{L}(\mathbb{A})$ such that π is a subquotient of the parabolic induction of $\mathbf{L}$ (as before, use the standard parabolic subgroup of $\mathbf{G}^0$ with Levi subgroup $\mathbf{L}$). We want to show that π_L is θ_L-stable after we twist it by an automorphic character. As π is θ-stable, there exists $s \in W^G(\mathfrak{a}_L)$ such that π_L is s-stable (note that s does not have to be regular now). We also know that π is discrete; so, by a result of Moeglin and Waldspurger (the main theorem of [MW]), there exist $m_1, \dots, m_r, l_1, \dots, l_r \in \mathbb{N}$, an automorphic character χ of $\mathbb{A}_E^\times$ and cuspidal automorphic representations τ_j of $\mathbf{GL}_{m_j}(\mathbb{A}_E)$, for $1 \leq j \leq r$, such that

- $n_j = l_j m_j$ for every $j \in \{1, \dots, r\}$;
- $\mathbf{L} = R_{E/\mathbb{Q}}(\mathbb{G}_{m,E} \times (\mathbf{GL}_{m_1,E})^{l_1} \times \cdots \times (\mathbf{GL}_{m_r,E})^{l_r})$;
- $\pi_L = \chi \otimes \pi_1 \otimes \cdots \otimes \pi_r$, where π_j is the cuspidal automorphic representation $\tau_j|\det|^{(l_j-1)/2} \otimes \tau_j|\det|^{(l_j-3)/2} \otimes \cdots \otimes \tau_j|\det|^{(1-l_j)/2}$ of $\mathbf{GL}_{m_j}(\mathbb{A}_E)^{l_j}$.

Write $s = \sigma\theta_L$, with $\sigma = (\sigma_1, \dots, \sigma_r) \in \mathfrak{S}_{l_1} \times \cdots \times \mathfrak{S}_{l_r} = W^{G^0}(\mathfrak{a}_L)$. As π is a subquotient of the parabolic induction of π_L, the infinitesimal character of $\pi_{L,\infty}$ is represented by $(-\lambda, -\iota(\lambda))$, and so, by the fact proved above, $\sigma^2 = 1$. For every $j \in \{1, \dots, r\}$, denote by θ_{m_j} the automorphism $g \longmapsto {}^t\overline{g}^{-1}$ of $\mathbf{GL}_{m_j,E}$. Then the fact that π_L is $\sigma\theta_L$-stable means that, for every $j \in \{1, \dots, r\}$ and for every $k \in \{1, \dots, l_j\}$,

$$\tau_j|\det|^{(l_j-2k+1)/2} \simeq (\tau_j|\det|^{(l_j-2\sigma_j(k)+1)/2}) \circ \theta_{m_j} = (\tau_j \circ \theta_{m_j})|\det|^{(2\sigma_j(k)-1-l_j)/2},$$

that is,

$$\tau_j \circ \theta_{m_j} \simeq \tau_j|\det|^{l_j+1-k-\sigma_j(k)}.$$

In particular, we see by taking the absolute values of the central characters in the equality above that, for every $j \in \{1, \dots, r\}$, the function $k \longmapsto k + \sigma_j(k)$ is constant on $\{1, \dots, l_j\}$. But, if $j \in \{1, \dots, r\}$, then $\sum_{k=1}^{l_j}(k + \sigma_j(k)) = l_j(l_j + 1)$, so $k + \sigma_j(k) = l_j + 1$ for every $k \in \{1, \dots, l_j\}$. This show that $\tau_j \simeq \tau_j \circ \theta_{m_j}$ for every $j \in \{1, \dots, r\}$, and it easily implies that π_L is θ_L-stable after we twist it by an automorphic character. □

Chapter Nine

The twisted fundamental lemma

The goal of this chapter is to show that, for the special kind of twisted endoscopic transfer that appears in the stabilization of the fixed point formula and for the groups considered in this book (and some others), the twisted fundamental lemma for the whole Hecke algebra follows from the twisted fundamental lemma for the unit element. No attempt has been made to prove a general result, and the method is absolutely not original: it is simply an adaptation of the method used by Hales in the untwisted case ([H2]), and this was inspired by the method used by Clozel in the case of base change ([Cl3]) and by the simplification suggested by the referee of the article [Cl3].

The definitions and facts about twisted groups recalled in section 8.1 will be used freely in this chapter.

9.1 NOTATION

We will consider the following situation. Let F be a local non-archimedean field of characteristic 0. Fix an algebraic closure $\overline{F}$ of F, let $\Gamma_F = \mathrm{Gal}(\overline{F}/F)$, and denote by F^{ur} the maximal unramified extension of F in $\overline{F}$. Fix a uniformizer ϖ_F of F. Let $\mathbf{G}$ be a connected reductive unramified group over F. Assume that $\mathbf{G}$ is defined over $\mathcal{O}_F$ and that $\mathbf{G}(\mathcal{O}_F)$ is a hyperspecial maximal compact subgroup of $\mathbf{G}(F)$. For such a group $\mathbf{G}$, write $\mathcal{H}_G = \mathcal{H}(\mathbf{G}(F), \mathbf{G}(\mathcal{O}_F)) := C_c^\infty(\mathbf{G}(\mathcal{O}_F)\backslash\mathbf{G}(F)/\mathbf{G}(\mathcal{O}_F))$. Let $(\mathbf{H}, s, \eta_0)$ be an endoscopic triple for $\mathbf{G}$ (in the sense of [K4] 7.4). Assume that

- $\mathbf{H}$ is unramified over F, $\mathbf{H}$ is defined over $\mathcal{O}_F$, and $\mathbf{H}(\mathcal{O}_F)$ is a hyperspecial maximal compact subgroup of $\mathbf{H}(F)$;
- there exists an L-morphism $\eta : {}^L\mathbf{H} \longrightarrow {}^L\mathbf{G}$ extending η_0 and unramified, that is, coming by inflation from an L-morphism $\widehat{\mathbf{H}} \rtimes \mathrm{Gal}(K/F) \longrightarrow \widehat{\mathbf{G}} \rtimes \mathrm{Gal}(K/F)$, where K is a finite unramified extension of F.

Choose a generator σ of $W_{F^{ur}/F}$. Let E/F be a finite unramified extension of F in $\overline{F}$, and let $d \in \mathbb{N}^*$ be the degree of E/F. Write $R = R_{E/F}\mathbf{G}_E$. Let θ be the automorphism of R induced by the image of σ in $\mathrm{Gal}(E/F)$.

Kottwitz explained in [K9] pp. 179–180 how to get from this twisted endoscopic data for $(R, \theta, 1)$ (in the sense of [KS] 2.1). We recall his construction. There is an obvious isomorphism $\widehat{R} = \widehat{\mathbf{G}}^d$, and the actions of $\widehat{\theta}$ and σ are given by the formulas

$$\widehat{\theta}(g_1, \dots, g_d) = (g_2, \dots, g_d, g_1),$$

$$\sigma(g_1, \dots, g_d) = (\sigma(g_2), \dots, \sigma(g_d), \sigma(g_1)).$$

In particular, the diagonal embedding $\widehat{\mathbf{G}} \longrightarrow \widehat{R}$ is W_F-equivariant, hence extends in an obvious way to an L-morphism ${}^L\mathbf{G} \longrightarrow {}^LR$. Let $\xi' : {}^L\mathbf{H} \longrightarrow {}^LR$ be the composition of the morphism $\eta : {}^L\mathbf{H} \longrightarrow {}^L\mathbf{G}$ and of this L-morphism. As F is local, we may assume that $s \in Z(\widehat{\mathbf{H}})^{\Gamma_F}$. Let $t_1, \dots, t_d \in Z(\widehat{\mathbf{H}})^{\Gamma_F}$ be such that $s = t_1 \dots t_d$. Set $t = (t_1, \dots, t_d) \in \widehat{R}$. Let $\xi : {}^L\mathbf{H} \longrightarrow {}^LR$ be the morphism such that

- $\xi_{|\widehat{\mathbf{H}}}$ is the composition of $\eta_0 : \widehat{\mathbf{H}} \longrightarrow \widehat{\mathbf{G}}$ and of the diagonal embedding $\widehat{\mathbf{G}} \longrightarrow \widehat{R}$;
- for every $w \in W_F$ that is a pre-image of $\sigma \in W_{F^{ur}/F}$, $\xi(1, w) = (t, 1)\xi'(1, w)$.

Then $(\mathbf{H}, {}^L\mathbf{H}, t, \xi)$ are twisted endoscopic data for $(R, \theta, 1)$. Kottwitz shows ([K9] p. 180) that the equivalence class of these twisted endoscopic data does not depend on the choice of $t_1, \dots, t_d$.

The morphism ξ induces a morphism

$$\mathcal{H}_R \longrightarrow \mathcal{H}_H,$$

that will be denoted by b_ξ (in 4.2, this morphism is explicitly calculated for the unitary groups of section 2.1).

Let Δ_ξ be the transfer factors defined by ξ, normalized as in [Wa3] 4.6. The twisted fundamental lemma for a function $f \in \mathcal{H}_R$ is the following statement. For every $\gamma_H \in \mathbf{H}(F)$ semisimple and strongly $\mathbf{G}$-regular,

$$\Lambda(\gamma_H, f) := SO_{\gamma_H}(b_\xi(f)) - \sum_\delta \Delta_\xi(\gamma_H, \delta) O_{\delta\theta}(f) = 0,$$

where the sum is taken over the set of θ-conjugacy classes of θ-semisimple $\delta \in R(F)$. (Remember that a semisimple $\gamma_H \in \mathbf{H}(F)$ is called *strongly* **G**-*regular* if it has an image in $\mathbf{G}(F)$ whose centralizer is a torus.)

Remark 9.1.1 There is an obvious variant of this statement where E is replaced by a finite product of finite unramified extensions of F such that $\mathrm{Aut}_F(E)$ is a cyclic group.

9.2 LOCAL DATA

Notation is as in section 9.1. Fix a Borel subgroup $\mathbf{B}_G$ (resp., $\mathbf{B}_H$) of $\mathbf{G}$ (resp., $\mathbf{H}$) defined over $\mathcal{O}_F$ and a maximal torus $\mathbf{T}_G \subset \mathbf{B}_G$ (resp., $\mathbf{T}_H \subset \mathbf{B}_H$) defined over $\mathcal{O}_F$. Let $\mathbf{T}_R = R_{E/F}\mathbf{T}_{G,E}$ and $\mathbf{B}_R = R_{E/F}\mathbf{B}_{G,E}$. Denote by I_R (resp., I_H) the Iwahori subgroup of $R(F)$ (resp., $\mathbf{H}(F)$) defined by the Borel subgroup $\mathbf{B}_R$ (resp., $\mathbf{B}_H$).

Let $\Pi(R)$ (resp., $\Pi(\mathbf{H})$) be the set of equivalence classes of irreducible θ-stable representations π of $R(F)$ (resp., of irreducible representations π_H of $\mathbf{H}(F)$) such that $\pi^{I_R} \neq \{0\}$ (resp., $\pi_H^{I_H} \neq \{0\}$). For every $\pi \in \Pi(R)$, fix a normalized intertwining operator A_π on π (remember that an *intertwining operator* on π is an $R(F)$-equivariant isomorphism $\pi \xrightarrow{\sim} \pi \circ \theta$, and that an intertwining operator A_π is called *normalized* if $A_\pi^d = 1$). If π is unramified, choose the intertwining operator that fixes the elements of the subspace $\pi^{R(\mathcal{O}_F)}$.

The definition of local data used here is the obvious adaptation of the definition of Hales ([H2] 4.1).

Definition 9.2.1 *Local data* for R and $(\mathbf{H}, {}^L\mathbf{H}, s, \xi)$ are the data of a set I and of two families of complex numbers, $(a_i^R(\pi))_{i\in I,\pi\in\Pi(R)}$ and $(a_i^H(\pi_H))_{i\in I,\pi_H\in\Pi(\mathbf{H})}$, such that, for every $i \in I$, the numbers $a_i^R(\pi)$ (resp., $a_i^H(\pi_H)$) are zero for almost every $\pi \in \Pi(R)$ (resp., $\pi_H \in \Pi(\mathbf{H})$) and that, for every $f \in \mathcal{H}_R$, the following conditions are equivalent:

(a) *for every* $i \in I$, $\sum_{\pi\in\Pi(R)} a_i^R(\pi)\,\mathrm{Tr}(\pi(f)A_\pi) = \sum_{\pi_H\in\Pi(\mathbf{H})} a_i^H(\pi_H) \times \mathrm{Tr}(\pi_H(b_\xi(f)))$;

(b) *for every* $\gamma_H \in H(F)$ *that is semisimple, elliptic, and strongly* $\mathbf{G}$*-regular,* $\Lambda(\gamma_H, f) = 0$.

Proposition 9.2.2 *Assume that* $\mathbf{G}$ *is adjoint, that the endoscopic triple* $(\mathbf{H}, s, \eta_0)$ *for* $\mathbf{G}$ *is elliptic, that the center of* $\mathbf{H}$ *is connected, and that there exist local data for* R *and* $(\mathbf{H}, {}^L\mathbf{H}, s, \xi)$.

Then, for every $f \in \mathcal{H}_R$ *and for every* $\gamma_H \in \mathbf{H}(F)$ *semisimple elliptic and strongly* $\mathbf{G}$*-regular,* $\Lambda(\gamma_H, f) = 0$.

Remark 9.2.3 If $\mathbf{G}$ is adjoint and $(\mathbf{H}, s, \eta_0)$ is elliptic, then the morphism $\xi : {}^L\mathbf{H} \longrightarrow {}^L R$ comes by inflation from a morphism $\widehat{\mathbf{H}} \rtimes \mathrm{Gal}(K/F) \longrightarrow \widehat{R} \rtimes \mathrm{Gal}(K/F)$, where K is a finite unramified extension of F. Let us prove this. By the definition of t, $\xi(1 \rtimes \sigma^d) = \xi'(s \rtimes \sigma^d)$ (remember that ξ' is the composition of $\eta : {}^L\mathbf{H} \longrightarrow {}^L\mathbf{G}$ and the "diagonal embedding" ${}^L\mathbf{G} \longrightarrow {}^L R$). We know that $s \in Z(\widehat{\mathbf{H}})^{\Gamma_F} Z(\widehat{\mathbf{G}})$. As $\mathbf{G}$ is adjoint, $Z(\widehat{\mathbf{G}})$ is finite; because the endoscopic triple $(\mathbf{H}, s, \eta_0)$ is elliptic, $Z(\widehat{\mathbf{H}})^{\Gamma_F}$ is also finite. The finite subgroup $Z(\widehat{\mathbf{G}})Z(\widehat{\mathbf{H}})^{\Gamma_F}$ of $Z(\widehat{\mathbf{H}})$ is invariant by σ^d, so there exists $k \in \mathbb{N}^*$ such that the restriction of σ^{dk} to this subgroup is trivial. Let $s' = s\sigma^d(s)\ldots\sigma^{d(k-1)}(s)$. Then s' is fixed by σ^d, and $(s \rtimes \sigma^d)^k = s' \rtimes \sigma^{dk}$. As s' is in the finite group $Z(\widehat{\mathbf{G}})Z(\widehat{\mathbf{H}})^{\Gamma_F}$, there exists $l \in \mathbb{N}^*$ such that $s'^l = 1$. Then

$$\xi(1 \rtimes \sigma^{dkl}) = \xi'((s \rtimes \sigma^d)^{kl}) = \xi'((s' \rtimes \sigma^{dk})^l) = \xi'(1 \rtimes \sigma^{dkl}).$$

By the assumption on η, there exists $r \in \mathbb{N}^*$ such that $\eta(1 \rtimes \sigma^r) = 1 \rtimes \sigma^r$. So we get finally : $\xi(1 \rtimes \sigma^{dklr}) = 1 \rtimes \sigma^{dklr}$.

Before proving the proposition, we show a few lemmas.

Remember that an element of $\mathbf{H}(F)$ or $\mathbf{G}(F)$ is called *strongly compact* if it belongs to a compact subgroup, and *compact* if its image in the adjoint group of $\mathbf{H}$ (resp., $\mathbf{G}$) is strongly compact (cf. [H1] §2). Every semisimple elliptic element of $\mathbf{G}(F)$ or $\mathbf{H}(F)$ is compact, and an element that is stably conjugate to a compact element is also compact (this follows easily from the characterization of compact elements in [H1] §2). If the center of $\mathbf{G}$ is anisotropic (e.g., if $\mathbf{G}$ is adjoint), then an element of $\mathbf{G}(F)$ is compact if and only if it is strongly compact.

Lemma 9.2.4 *Assume that the centers of* $\mathbf{G}$ *and* $\mathbf{H}$ *are anisotropic. Let* $A : \mathcal{H}_R \longrightarrow \mathbb{C}$ *be a linear form. Assume that for every* $f \in \mathcal{H}_R$*, if* $\Lambda_{\gamma_H}(f) = 0$ *for every* $\gamma_H \in \mathbf{H}(F)$ *semisimple elliptic and strongly* $\mathbf{G}$*-regular, then* $A(f) = 0$.

Then A *is a (finite) linear combination of linear forms* $\gamma_H \longmapsto \Lambda(\gamma_H, f)$*, with* $\gamma_H \in \mathbf{H}(F)$ *semisimple elliptic and strongly* $\mathbf{G}$*-regular.*

Proof. Let $U_H \subset \mathbf{H}(F)$ be the set of compact elements of $\mathbf{H}(F)$ and U_R be the set of θ-semisimple elements of $R(F)$ whose norm contains a compact element of $\mathbf{G}(F)$. Then U_H is compact modulo $\mathbf{H}(F)$-conjugacy and U_R is compact modulo θ-conjugacy (these notions are defined before theorem 2.8 of [Cl3]). By the twisted version of the Howe conjecture (theorem 2.8 of [Cl3]), the vector space of distributions on $\mathcal{H}_R$ generated by the $f \longmapsto O_{\delta\theta}(f)$, $\delta \in U_R$ and the $f \longmapsto O_{\gamma_H}(b_\xi(f))$, $\gamma_H \in U_F$, is finite-dimensional. If $\gamma_H \in \mathbf{H}(F)$ is elliptic semisimple, then $\gamma_H \in U_H$, and every image of γ_H in $R(F)$ is in U_R. In particular, the vector space generated by the distributions $\gamma_H \longmapsto \Lambda(\gamma_H, f)$, for $\gamma_H \in \mathbf{H}(F)$ elliptic semisimple and strongly $\mathbf{G}$-regular, is finite-dimensional. The lemma follows from this. □

Let $\mathbf{S}_H$ be the maximal split subtorus of $\mathbf{T}_H$, $\mathbf{S}_R$ be the maximal split subtorus of $\mathbf{T}_R$, and $\Omega_H = \Omega(\mathbf{S}_H(F), \mathbf{H}(F))$, $\Omega_R = \Omega(\mathbf{S}_R(F), R(F))$ be the relative Weyl groups. Identify $\mathcal{H}_H$ (resp., $\mathcal{H}_R$) to $\mathbb{C}[\widehat{\mathbf{S}}_H/\Omega_H]$ (resp., $\mathbb{C}[\widehat{\mathbf{S}}_R/\Omega_R]$) by the Satake isomorphism. If $z \in \widehat{\mathbf{S}}_H$ (resp., $\widehat{\mathbf{S}}_R$) and $f \in \mathcal{H}_H$ (resp., $\mathcal{H}_R$), write $f(z)$ for $f(z\Omega_H)$ (resp., $f(z\Omega_R)$).

We recall the definition of the morphism $b_\xi : \mathcal{H}_R \longrightarrow \mathcal{H}_H$ induced by ξ (cf. [Bo] 6 and 7). The group Ω_H (resp., Ω_R) is naturally isomorphic to the subgroup of Γ_F-fixed points of the Weyl group $\Omega(\widehat{\mathbf{T}}_H, \widehat{\mathbf{H}})$ (resp., $\Omega(\widehat{\mathbf{T}}_R, \widehat{R})$). Let N_H (resp., N_R) be the inverse image of Ω_H (resp., Ω_R) in $\mathrm{Nor}_{\widehat{\mathbf{H}}}(\widehat{\mathbf{T}}_H)$ (resp., $\mathrm{Nor}_{\widehat{R}}(\widehat{\mathbf{T}}_R)$), let $Y_H = \widehat{\mathbf{S}}_H$ (resp., $Y_R = \widehat{\mathbf{S}}_R$), and let $(\widehat{\mathbf{H}} \rtimes \sigma)_{\mathrm{ss}}$ (resp., $(\widehat{R} \rtimes \sigma)_{\mathrm{ss}}$) be the set of semisimple elements of $\widehat{\mathbf{H}} \rtimes \sigma \subset \widehat{\mathbf{H}} \rtimes W_{F^{\mathrm{ur}}/F}$ (resp., $\widehat{R} \rtimes \sigma \subset \widehat{R} \rtimes W_{F^{\mathrm{ur}}/F}$) (remember that σ is a fixed generator of $W_{F^{\mathrm{ur}}/F}$). As $X_*(\mathbf{S}_H) = X_*(\mathbf{T}_H)^{\Gamma_F}$ (resp., $X_*(\mathbf{S}_R) = X_*(\mathbf{T}_R)^{\Gamma_F}$), the group Ω_H (resp., Ω_R) acts naturally on Y_H (resp., Y_R). Moreover,

- the restriction to $(\widehat{\mathbf{T}}_H^{\Gamma_F})^0$ (resp., $(\widehat{\mathbf{T}}_R^{\Gamma_F})^0$) of the morphism $\nu : \widehat{\mathbf{T}}_H \longrightarrow Y_H$ (resp., $\nu : \widehat{\mathbf{T}}_R \longrightarrow Y_R$) dual of the inclusion $\mathbf{S}_H \subset \mathbf{T}_H$ (resp., $\mathbf{S}_R \subset \mathbf{T}_R$) is an isogeny ([Bo] 6.3);
- the map $\widehat{\mathbf{T}}_H \rtimes \sigma \longrightarrow Y_H$ (resp., $\widehat{\mathbf{T}}_R \rtimes \sigma \longrightarrow Y_R$) that sends $t \rtimes \sigma$ to $\nu(t)$ induces a bijection
$$(\widehat{\mathbf{T}}_H \rtimes \sigma)/\operatorname{Int} N_H \xrightarrow{\sim} Y_H/\Omega_H$$
$$(\text{resp.,} \qquad (\widehat{\mathbf{T}}_R \rtimes \sigma)/\operatorname{Int} N_R \xrightarrow{\sim} Y_R/\Omega_R)$$
([Bo], lemma 6.4);
- the inclusion induces a bijection
$$(\widehat{\mathbf{T}}_H \rtimes \sigma)/\operatorname{Int} N_H \xrightarrow{\sim} (\widehat{\mathbf{H}} \rtimes \sigma)_{\mathrm{ss}}/\operatorname{Int} \widehat{\mathbf{H}}$$
$$(\text{resp.,} \qquad (\widehat{\mathbf{T}}_R \rtimes \sigma)/\operatorname{Int} N_R \xrightarrow{\sim} (\widehat{R} \rtimes \sigma)_{\mathrm{ss}}/\operatorname{Int} \widehat{R})$$
([Bo], lemma 6.5).

In particular, we get bijections $\varphi_H : (\widehat{\mathbf{H}} \rtimes \sigma)_{ss}/\operatorname{Int}\widehat{\mathbf{H}} \xrightarrow{\sim} Y_H/\Omega_H$ and $\varphi_R : (\widehat{R} \rtimes \sigma)_{ss}/\operatorname{Int}\widehat{R} \xrightarrow{\sim} Y_R/\Omega_R$. The morphism $\xi : {}^L\mathbf{H} \longrightarrow {}^L R$ is unramified, hence it induces a morphism $(\widehat{\mathbf{H}} \rtimes \sigma)_{ss}/\operatorname{Int}\widehat{\mathbf{H}} \longrightarrow (\widehat{R} \rtimes \sigma)_{ss}/\operatorname{Int}\widehat{R}$, and this gives a morphism $b_\xi^* : Y_H/\Omega_H \longrightarrow Y_R/\Omega_R$. The morphism $b_\xi : \mathbb{C}[Y_R/\Omega_R] \longrightarrow \mathbb{C}[Y_H/\Omega_H]$ is the dual of b_ξ^*.

Let Y_H^u (resp., Y_R^u) be the maximal compact subgroup of Y_H (resp., Y_R).

Lemma 9.2.5 *The morphism $b_\xi^* : Y_H/\Omega_H \longrightarrow Y_R/\Omega_R$ sends Y_H^u/Ω_H to Y_R^u/Ω_R and $(Y_H - Y_H^u)/\Omega_H$ to $(Y_R - Y_R^u)/\Omega_R$.*

Proof. Let K be an unramified extension of F such that $\mathbf{H}$ and $\mathbf{G}$ split over K; write $r = [K : F]$. For every $g \rtimes \sigma \in \widehat{\mathbf{H}} \rtimes \sigma$ or $\widehat{R} \rtimes \sigma$, write $N(g \rtimes \sigma) = g\sigma(g)\ldots\sigma^{r-1}(g)$.

Let G' be the set of complex points of a linear algebraic group over $\mathbb{C}$. Copying the definition of [H1] §2, say that an element $g \in G'$ is *strongly compact* if there exists a compact subgroup of G' containing g. It is easy to see that this is equivalent to the fact that there exists a faithful representation $\rho : G' \longrightarrow \mathbf{GL}_m(\mathbb{C})$ such that the eigenvalues of $\rho(g)$ all have module 1. So a morphism of algebraic groups over $\mathbb{C}$ sends strongly compact elements to strongly compact elements.

Let $g \in \widehat{\mathbf{H}}$ be such that $g \rtimes \sigma$ is semisimple. We show that $\varphi_H(g \rtimes \sigma) \in Y_H^u/\Omega_H$ if and only if $N(g \rtimes \sigma)$ is strongly compact. After replacing $g \rtimes \sigma$ by a $\widehat{\mathbf{H}}$-conjugate, we may assume that $g \in \widehat{\mathbf{T}}_H$. Assume that $N(g \rtimes \sigma)$ is strongly compact. Then $\nu(N(g \rtimes \sigma)) = \nu(g)^r \in Y_H$ is strongly compact and this implies that $\nu(g)$ is strongly compact, that is, $\nu(g) \in Y_H^u$. Assume now that $\varphi_H(g \rtimes \sigma) = \nu(g)\Omega_H \in Y_H^u/\Omega_H$, that is, $\nu(g) \in Y_H^u$. Then $\nu(N(g \rtimes \sigma)) = \nu(g)^r \in Y_H^u$. Moreover, $\widehat{\mathbf{T}}_H$ is commutative, so $N(g \rtimes \sigma) \in \widehat{\mathbf{T}}_H^{\Gamma_F}$. As the restriction of ν to $\widehat{\mathbf{T}}_H^{\Gamma_F}$ is finite, it is easy to see that the fact that $\nu(N(g \rtimes \sigma)) \in Y_H^u$ implies that $N(g \rtimes \sigma)$ is strongly compact.

Of course, there is a similar statement for R. Hence, to finish the proof, it is enough to show that, for every $g \rtimes \sigma \in \widehat{\mathbf{H}} \rtimes \sigma$, $N(g \rtimes \sigma)$ is strongly compact if and only if $N(\xi(g \rtimes \sigma))$ is strongly compact. Let $\xi_0 : \widehat{\mathbf{H}} \longrightarrow \widehat{R}$ be the morphism induced by ξ. Write $t' \rtimes \sigma = \xi(1 \rtimes \sigma)$. Then, for every $g \in \widehat{\mathbf{H}}$, $N(\xi(g \rtimes \sigma)) = \xi_0(N(g \rtimes \sigma))N(t' \rtimes \sigma)$. By remark 9.2.3, we may assume (after replacing K by a bigger unramified extension of F) that $\xi(1 \rtimes \sigma^r) = 1 \rtimes \sigma^r$, that is, $N(t' \rtimes \sigma) = 1$. Then the statement of the lemma follows from the injectivity of ξ_0. □

Lemma 9.2.6 *For every $\delta \in R(F)$ that is θ-regular θ-semisimple and θ-elliptic, for every $\gamma_H \in \mathbf{H}(F)$ that is regular semisimple and elliptic, the distributions $f \longmapsto O_{\delta\theta}(f)$ and $f \longmapsto O_{\gamma_H}(b_\xi(f))$ on $\mathcal{H}_R$ are tempered.*

Proof. Remember that a distribution on $\mathcal{H}_R$ is called tempered if it extends continuously to the Schwartz space of rapidly decreasing bi-$R(\mathcal{O}_R)$-invariant functions on $R(F)$ (defined, for example, in section 5 of [Cl3]). For the first distribution of the lemma, this is proved in lemma 5.2 of [Cl3]. Moreover, the distribution $f_H \longmapsto O_{\gamma_H}(f_H)$ on $\mathcal{H}_H$ is tempered (this is a particular case of lemma 5.2 of [Cl3]). So, to prove that the second distribution of the lemma is tempered, it is enough to show that the morphism $b_\xi : \mathcal{H}_R \longrightarrow \mathcal{H}_H$ extends to the Schwartz spaces. To show this last statement, it is enough to show, by the proof of lemma 5.1 of [Cl3], that b_ξ^* sends Y_H^u/Ω_H to Y_R^u/Ω_R. This follows from lemma 9.2.5 above. □

Call a θ-semisimple element of $R(F)$ *θ-compact* if its norm is compact. Let $\mathbb{1}_c$ be the characteristic function of the set of semisimple compact elements of $\mathbf{H}(F)$, and $\mathbb{1}_{\theta\text{-}c}$ be the characteristic function of the set of θ-semisimple θ-compact elements of $R(F)$. If π_H is an irreducible admissible representation of $\mathbf{H}(F)$, π is a θ-stable irreducible admissible representation of $R(F)$ and A_π is a normalized intertwining operator on π, define the compact trace of π_H and the twisted θ-compact trace of π by the formulas

$$\mathrm{Tr}_c(\pi_H(f_H)) := \mathrm{Tr}(\pi_H(\mathbb{1}_c f_H)), \qquad f_H \in C_c^\infty(\mathbf{H}(F)),$$

$$\mathrm{Tr}_{\theta\text{-}c}(\pi(f)A_\pi) := \mathrm{Tr}(\pi(\mathbb{1}_{\theta\text{-}c} f)A_\pi), \qquad f \in C_c^\infty(R(F)).$$

Lemma 9.2.7 *Let π be a θ-stable irreducible admissible representation of $R(F)$, and let A_π be a normalized intertwining operator on π. Assume that the distribution $f \longmapsto \mathrm{Tr}_{\theta\text{-}c}(\pi(f)A_\pi)$ on $\mathcal{H}_R$ is not identically zero. Then $\pi \in \Pi(R)$.*

Proof. By the corollary to proposition 2.4 of [Cl3], there exists a θ-stable parabolic subgroup $\mathbf{P} \supset \mathbf{B}_R$ of R such that $\pi_{\mathbf{N}_P}$ is unramified, where $\mathbf{N}_P$ is the unipotent radical of $\mathbf{P}$ and $\pi_{\mathbf{N}_P}$ is the unnormalized Jacquet module (the module of $\mathbf{N}_P(F)$-coinvariants of π). So, if $\mathbf{N}$ is the unipotent radical of $\mathbf{B}_R$, then $\pi_{\mathbf{N}}$ is unramified. By proposition 2.4 of [Cas], this implies that $\pi^{I_R} \neq \{0\}$. □

Lemma 9.2.8 *Assume that the centers of $\mathbf{G}$ and $\mathbf{H}$ are anisotropic and that the center of $\mathbf{H}$ is connected. Let $\delta \in R(F)$ be θ-regular θ-semisimple and θ-elliptic, and $\gamma_H \in \mathbf{H}(F)$ be regular semisimple and elliptic. Then the distribution $f \longmapsto O_{\delta\theta}(f)$ on $\mathcal{H}_R$ is a linear combination of distributions $f \longmapsto \mathrm{Tr}_{\theta\text{-}c}(\pi(f)A_\pi)$, with $\pi \in \Pi(R)$, and the distribution $f_H \longmapsto SO_{\gamma_H}(f_H)$ on $\mathcal{H}_H$ is a linear combination of distributions $f_H \longmapsto \mathrm{Tr}_c(\pi_H(f_H))$, with $\pi_H \in \Pi(\mathbf{H})$ coming from an element of $\Pi(\mathbf{H}_{\mathrm{ad}})$.*

Proof. We show the first assertion of the lemma. Let $f \in \mathcal{H}_R$ be such that $\mathrm{Tr}_{\theta\text{-}c}(\pi(f)A_\pi) = 0$ for every $\pi \in \Pi(R)$; let us show that $O_{\delta\theta}(f) = 0$. As δ is θ-elliptic, hence θ-compact, $O_{\delta\theta}(f) = O_{\delta\theta}(\mathbb{1}_{\theta\text{-}c} f)$. But $\mathrm{Tr}(\pi(\mathbb{1}_{\theta\text{-}c} f)A_\pi) = \mathrm{Tr}_{\theta\text{-}c}(\pi(f)A_\pi) = 0$ for every $\pi \in \Pi(R)$, so, by the main theorem of [KRo] and lemma 9.2.7, $O_{\delta\theta}(\mathbb{1}_{\theta\text{-}c} f) = 0$.

On the other hand, by the twisted version of the Howe conjecture (theorem 2.8 of [Cl3]), the space generated by the distributions (on $\mathcal{H}_R$) $f \longmapsto \mathrm{Tr}_{\theta\text{-}c}(\pi(f)A_\pi)$, $\pi \in \Pi(R)$, is finite-dimensional. This implies the first assertion.

We show the second assertion of the lemma. As $Z(\mathbf{H})$ is anisotropic and connected, lemma 9.4.4 implies that, for every $f_H \in \mathcal{H}_H$, $SO_{\gamma_H}(f_H) = SO_{\gamma'_H}(f'_H)$, where γ'_H is the image of γ_H in $\mathbf{H}_{\mathrm{adj}}(F)$ and f'_H is the image of f_H in $\mathcal{H}_{H_{\mathrm{adj}}}$

(defined in lemma 9.4.4). So the second assertion of the lemma follows from the first, applied to the group $\mathbf{H}_{\mathrm{adj}}$ (with $\theta = 1$). □

Identify the group of unramified characters of $\mathbf{T}_R(F)$ to Y_R in the usual way (cf. [Bo] 9.5). For every $z \in Y_R$, let ψ_z be the unramified character of $\mathbf{T}_R(F)$ corresponding to z and denote by $I(z)$ the representation of $R(F)$ obtained by (normalized) parabolic induction from ψ_z:

$$I(z) = \mathrm{Ind}_{\mathbf{B}_R}^{R}(\delta_{B_R}^{1/2} \otimes \psi_z),$$

where, if $\mathbf{N}$ is the unipotent radical of $\mathbf{B}_R$, $\delta_{B_R}(t) = |\det(\mathrm{Ad}(t), \mathrm{Lie}(\mathbf{N}))|_F$ for every $t \in \mathbf{T}_G(F)$ ($\delta_{B_R}^{1/2} \otimes \psi_z$ is seen as a character on $\mathbf{B}_R(F)$ via the projection $\mathbf{B}_R(F) \longrightarrow \mathbf{T}_R(F)$). If $\widehat{\theta}(z) = z$, then $\psi_z = \psi_z \circ \theta$, so $I(z)$ is θ-stable. In that case, let $A_{I(z)}$ be the linear endomorphism of the space of $I(z)$ that sends a function f to the function $x \longmapsto f(\theta(x))$ (remember that the space of $I(z)$ is a space of functions $R(F) \longrightarrow \mathbb{C}$); then $A_{I(z)}$ is a normalized intertwining operator on $I(z)$. We will use similar notation for $\mathbf{H}$ (without the intertwining operators, of course).

Lemma 9.2.9 *Assume that* $\mathbf{G}$ *is adjoint and that the center of* $\mathbf{H}$ *is anisotropic (the second assumption is true, for example, if* $\mathbf{G}$ *is adjoint and the endoscopic triple* $(\mathbf{H}, s, \eta_0)$ *is elliptic). Let* $\pi \in \Pi(R)$. *Then there exist a* θ*-stable* $z' \in Y_R - Y_R^u$, *a* θ*-stable subquotient* π' *of* $I(z')$ *and a normalized intertwining operator* $A_{\pi'}$ *on* π' *such that, for every* $f \in \mathcal{H}_R$, $\mathrm{Tr}_{\theta\text{-}c}(\pi(f)A_\pi) = \mathrm{Tr}_{\theta\text{-}c}(\pi'(f)A_{\pi'})$.

Similarly, if $\pi_H \in \Pi(\mathbf{H})$ *comes from a representation in* $\Pi(\mathbf{H}_{\mathrm{ad}})$, *then there exist* $z'_H \in Y_H - Y_H^u$ *and a subquotient* π'_H *of* $I(z'_H)$ *such that, for every* $f_H \in \mathcal{H}_H$, $\mathrm{Tr}_c(\pi_H(f_H)) = \mathrm{Tr}_c(\pi'_H(f_H))$.

Proof. By proposition 2.6 of [Cas], there exists $z_0 \in Y_R$ such that π is a subrepresentation of $I(z_0)$. By examining the proof of this proposition, we see that $\theta(z_0) \in \Omega_R z_0$. By lemma 4.7 of [Cl3], there exists a θ-stable z in $\Omega_R z_0$. Then π is a subquotient of $I(z)$ (because $I(z_0)$ and $I(z)$ have the same composition factors). If $z \notin Y_R^u$, this finishes the proof of the first statement (take $z' = z$ and $\pi' = \pi$). Assume that $z \in Y_R^u$. As $\mathbf{G}$ is adjoint, by a result of Keys (cf. [Ke], in particular the end of section 3), the representation $I(z)$ is irreducible, hence $\pi = I(z)$. Let $z' \in Y_R - Y_R^u$ be θ-stable. The unramified characters χ_z and $\chi_{z'}$ corresponding to z and z' are equal on the set of θ-compact elements of $\mathbf{T}_R(F)$. Hence, by theorem 3 of [vD], $\mathrm{Tr}_{\theta\text{-}c}(\pi(f)A_\pi) = \mathrm{Tr}_{\theta\text{-}c}(I(z')(f)A_{I(z')})$ for every $f \in C_c^\infty(R(F))$. This finishes the proof of the first statement (take $\pi' = I(z')$).

The same reasoning (without the twist by θ) applies to π_H, or rather to the representation of $\mathbf{H}_{\mathrm{ad}}(F)$ inducing π_H; note that, as the center of $\mathbf{H}$ is anisotropic, $Y_H = Y_{H_{\mathrm{ad}}}$. We need the fact that π_H comes from a representation in $\Pi(\mathbf{H}_{\mathrm{ad}})$ to apply Keys's result. □

In the following lemma, $\mathbf{N}$ is the unipotent radical of $\mathbf{B}_R$ and, for every representation π of $R(F)$, π_N is the $\mathbf{T}_R(F)$-module of $\mathbf{N}(F)$-coinvariants of π.

Lemma 9.2.10 *Let* π *be a* θ*-stable admissible representation of* $R(F)$ *of finite length, and let* A_π *be an intertwining operator on* π. *The semisimplification of*

$\delta_B^{-1/2} \otimes \pi_{\mathbf{N}}$ is a sum of characters of $\mathbf{T}_R(F)$; let $z_1, \dots, z_n$ be the points of Y_R corresponding to the θ-stable unramified characters that appear in that way. Then the distribution $f \longmapsto \mathrm{Tr}(\pi(f)A_\pi)$ on $\mathcal{H}_R$ is a linear combination of distributions $f \longmapsto f(z_i)$, $1 \le i \le n$. Moreover, if π is a subquotient of $I(z)$, with $z \in Y_R$ θ-stable, then the z_i are all in $\Omega_R z$.

Of course, there is a similar result for $\mathbf{H}$.

Proof. As π and its semisimplification have the same character, we may assume that π is irreducible. We may also assume that π is unramified (otherwise the result is trivial). By proposition 2.6 of [Cas], there exists $z \in Y_R$ such that π is a subquotient of $I(z)$. Reasoning as in the proof of lemma 9.2.9 above, we may assume that z is θ-stable. By corollary 2.2 of [Cas], $I(z)^{R(\mathcal{O}_F)}$ is 1-dimensional; hence $I(z)^{R(\mathcal{O}_F)} = \pi^{R(\mathcal{O}_F)}$. By the explicit description of a basis of $I(z)^{R(\mathcal{O}_F)}$ in [Car] 3.7 and the definition of the Satake transform (see, for example, [Car] 4.2), for every $f \in \mathcal{H}_R$, $\mathrm{Tr}(f, I(z)^{R(\mathcal{O}_F)}) = \mathrm{Tr}(f, \pi^{R(\mathcal{O}_F)}) = f(z)$. As $\pi^{R(\mathcal{O}_F)}$ is 1-dimensional and stable by A_π, the restriction of A_π to this subspace is multiplication by a scalar. So the distribution $f \longmapsto \mathrm{Tr}(\pi(f)A_\pi)$ is equal to a scalar multiple of the distribution $z \longmapsto f(z)$. By theorem 3.5 of [Car], the z_i are all in $\Omega_R z$. This finishes the proof of the lemma. □

For every $\lambda \in X^*(Y_R)$, set

$$f_\lambda = \sum_{\omega \in \Omega_R} \lambda^\omega \in \mathbb{C}[Y_R]^{\Omega_R} \simeq \mathbb{C}[Y_R/\Omega_R] \simeq \mathcal{H}_R.$$

Lemma 9.2.11 *There exists a nonempty open cone C in $X^*(Y_R) \otimes_{\mathbb{Z}} \mathbb{R}$ such that*

(a) for every θ-stable $z \in Y_R$, for every θ-stable subquotient π of $I(z)$, and for every intertwining operator A_π on π, the restriction to $C \cap X^*(Y_R)$ of the function $\lambda \longmapsto \mathrm{Tr}_{\theta\text{-}c}(\pi(f_\lambda)A_\pi)$ on $X^*(Y_R)$ is a linear combination of the functions $\lambda \longmapsto \lambda(\omega z)$, $\omega \in \Omega_R$.

Assume moreover that there exists an admissible embedding $\mathbf{T}_H \longrightarrow \mathbf{G}$ with image $\mathbf{T}_G$[1] *and that the center of $\mathbf{G}$ is connected (both assumptions are automatic if $\mathbf{G}$ is adjoint, cf. lemma 9.4.6).*

Then there exists a nonempty open cone C in $X^*(Y_R)\otimes_{\mathbb{Z}}\mathbb{R}$ that satisfies condition (a) above and also the following condition:

(b) for every $z_H \in Y_H$, for every subquotient π_H of $I(z_H)$, the restriction to $C \cap X^*(Y_R)$ of the function $\lambda \longmapsto \mathrm{Tr}_c(\pi_H(b_\xi(f_\lambda)))$ on $X^*(Y_R)$ is a linear combination of the functions $\lambda \longmapsto \lambda(\omega b_\xi^*(z_H))$, with $\omega \in \Omega_R$.

Proof. We will need some new notation. If $\mathbf{P} \supset \mathbf{B}_R$ is a parabolic subgroup of R, let $\mathbf{N}_P$ be the unipotent radical of $\mathbf{P}$, $\mathbf{M}_P$ be the Levi subgroup of $\mathbf{P}$ that contains $\mathbf{T}_R$, $\Omega_{M_P} = \Omega(\mathbf{S}_R(F), \mathbf{M}_P(F))$ be the relative Weyl group of $\mathbf{M}_P$, δ_P be the function

[1] It would be enough to assume this over an unramified extension K/F such that the base change morphism $\mathcal{H}(R(K), R(\mathcal{O}_K)) \longrightarrow \mathcal{H}(R(F), R(\mathcal{O}_F))$ is surjective.

$\gamma \longmapsto |\det(\mathrm{Ad}(\gamma), \mathrm{Lie}(\mathbf{N}_P))|_F$ on $\mathbf{P}(F)$, $\mathfrak{a}_{M_P} = \mathrm{Hom}(X^*(\mathbf{A}_{M_P}), \mathbb{R})$, and $a_P = \dim(\mathfrak{a}_{M_P})$. Assume that $\mathbf{P}$ is θ-stable. Let $\mathbf{P}_0$ and $\mathbf{M}_0$ be the parabolic subgroup and the Levi subgroup of $\mathbf{G}$ corresponding to $\mathbf{P}$ and $\mathbf{M}_P$ (cf. example 8.1.1). Denote by $H_{M_0} : \mathbf{M}_0(F) \longrightarrow \mathfrak{a}_{M_0} := \mathrm{Hom}(X^*(\mathbf{A}_{M_0}), \mathbb{R})$ the Harish-Chandra morphism (cf. [A1] p. 917), $\hat{\tau}_{P_0}^G : \mathfrak{a}_T := \mathrm{Hom}(X^*(\mathbf{T}_G), \mathbb{R}) \longrightarrow \{0, 1\}$ the characteristic function of the obtuse Weyl chamber defined by $\mathbf{P}_0$ (cf. [A1] p. 936) and $\hat{\chi}_{N_0} = \hat{\tau}_{P_0}^G \circ H_{M_0}$ (there is a canonical injective morphism $\mathfrak{a}_{M_0} \subset \mathfrak{a}_T$). Define a function $\hat{\chi}_{N_P,\theta}$ on $\mathbf{M}_P(F)$ by $\hat{\chi}_{N_P,\theta}(m) = \hat{\chi}_{N_0}(\mathcal{N}m)$ if $m \in \mathbf{M}_P(F)$ is θ-semisimple, and $\hat{\chi}_{N_P,\theta} = 0$ otherwise. If π is a θ-stable admissible representation of $R(F)$ of finite length and A_π is an intertwining operator on π, denote by $\pi_{\mathbf{N}_P}$ the Jacquet module of π (the module of $\mathbf{N}_P$-coinvariants of π) and by A_π the intertwining operator on $\pi_{\mathbf{N}_P}$ induced by A_π. If $f \in \mathcal{H}_R$, denote by $f^{(P)} \in \mathcal{H}_{\mathbf{M}_P}$ the constant term of f at $\mathbf{P}$.

Let π be a θ-stable admissible representation of $R(F)$ of finite length and A_π be an intertwining operator on π. The corollary to proposition 2.4 of [Cl3] says that, for every $f \in \mathcal{H}_R$,

$$\mathrm{Tr}_{\theta\text{-}c}(\pi(f)A_\pi) = \sum_{\mathbf{P}} (-1)^{a_P - a_G}\, \mathrm{Tr}((\delta_P^{-1/2} \otimes \pi_{\mathbf{N}_P})(\hat{\chi}_{N_P,\theta} f^{(P)})A_\pi), \qquad (*)$$

where the sum is taken over the set of θ-stable parabolic subgroups $\mathbf{P}$ of R that contain $\mathbf{B}_R$.

Let $N : \mathfrak{a}_{\mathbf{T}_R} \longrightarrow \mathfrak{a}_{\mathbf{T}_G}, \lambda \longmapsto \lambda + \theta(\lambda) + \cdots + \theta^{d-1}(\lambda)$, and identify $X^*(Y_R) \otimes_{\mathbb{Z}} \mathbb{R}$ to $\mathfrak{a}_{\mathbf{T}_R}$. Let $\lambda \in X^*(Y_R)$ and let $\mathbf{P} \supset \mathbf{B}_R$ be a θ-stable parabolic subgroup of R. Then $f_\lambda^{(P)} = \sum_{\omega \in \Omega_R} \lambda^\omega \in \mathbb{C}[Y_R]^{\Omega_{M_P}} \simeq \mathcal{H}_{M_P}$, and it follows easily from the definitions that, for every $\omega \in \Omega_R$,

$$\hat{\chi}_{N_P,\theta} \sum_{\omega' \in \Omega_{M_P}} \lambda^{\omega'\omega} = \hat{\tau}_{P_0}^G(N(\lambda^\omega)) \sum_{\omega' \in \Omega_{M_P}} \lambda^{\omega'\omega}.$$

From the definition of the functions $\hat{\tau}_{P_0}^G$, it is clear that there exists a finite union $D \subset \mathfrak{a}_{\mathbf{T}_G}$ of hyperplanes (containing the origin) such that, for every parabolic subgroup $\mathbf{P}_0$ of $\mathbf{G}$, $\hat{\tau}_{P_0}^G$ is constant on the connected components of $\mathfrak{a}_T - D$ (take for D the union of the kernels of the fundamental weights of $\mathbf{T}_G$ in $\mathbf{B}_G$). Then $D' := N^{-1}(D) \subset \mathfrak{a}_{T_R}$ is a finite union of hyperplanes and, for every θ-stable parabolic subgroup $\mathbf{P} \supset \mathbf{B}_R$ of R, the function $\hat{\tau}_{P_0}^G \circ N$ is constant on the connected components of $\mathfrak{a}_{T_R} - D'$. After replacing D' by $\bigcup_{\omega \in \Omega_R} \omega(D')$, we may assume that, for every connected component C of $\mathfrak{a}_{T_R} - D'$, for all $\lambda, \lambda' \in C$, for every θ-stable parabolic subgroup $\mathbf{P} \supset \mathbf{B}_R$ of R, and for every $\omega \in \Omega_R$,

$$\hat{\tau}_{P_0}^G \circ N(\lambda^\omega) = \hat{\tau}_{P_0}^G \circ N((\lambda')^\omega).$$

Let C be a connected component of $\mathfrak{a}_{T_R} - D'$. The calculations above show that there exist subsets Ω'_{M_P} of Ω_R, indexed by the set of θ-stable parabolic subgroups $\mathbf{P}$ of R containing $\mathbf{B}_R$, such that, for every $\lambda \in C$, for every $\mathbf{P}$,

$$\hat{\chi}_{N_P,\theta} f_\lambda^{(P)} = \sum_{\omega \in \Omega'_{M_P}} \lambda^\omega.$$

Let $z \in Y_R$ be θ-stable, let π be a θ-stable subquotient of $I(z)$, and let A_π be an intertwining operator on π. For every θ-stable parabolic subgroup $\mathbf{P} \supset \mathbf{B}_R$ of R,

$$\delta_{B_R}^{-1/2} \otimes \pi_{\mathbf{N}_{B_R}} = \delta_{B_R \cap M_P}^{-1/2} \otimes (\delta_P^{-1/2} \otimes \pi_{\mathbf{N}_P})_{\mathbf{N}_{B_R \cap M_P}}.$$

By formula $(*)$, the calculation of the functions $\hat{\chi}_{N_P,\theta} f_\lambda^{(P)}$ above and lemma 9.2.10 (applied to the representations $\delta_P^{-1/2} \otimes \pi_{\mathbf{N}_P}$), the restriction to $C \cap X^*(Y_R)$ of the function $\lambda \longmapsto \mathrm{Tr}_{\theta\text{-}c}(\pi(f_\lambda)A_\pi)$ is a linear combination of functions $\lambda \longmapsto \lambda(\omega z)$, with $\omega \in \Omega_R$. Hence C is a cone satisfying condition (a) of the lemma.

We now show the second statement of the lemma. After replacing the embeddings of $\widehat{\mathbf{T}}_H$ and $\widehat{\mathbf{T}}_G$ in $\widehat{\mathbf{H}}$ and $\widehat{\mathbf{G}}$ by conjugates, we may assume that ξ_0 induces a Γ_F-equivariant isomorphism $\widehat{\mathbf{T}}_H \xrightarrow{\sim} \widehat{\mathbf{T}}_G$. Use this isomorphism to identify $\mathbf{T}_H$ and $\mathbf{T}_G$. By the definition of ξ, the restriction to $\widehat{\mathbf{T}}_H$ of $\xi_0 : \widehat{\mathbf{H}} \longrightarrow \widehat{\mathbf{G}}$ induces a Γ_F-equivariant morphism $\widehat{\mathbf{T}}_H \longrightarrow \widehat{\mathbf{T}}_R$. Let $t' \rtimes \sigma = \xi(1 \rtimes \sigma)$. Then t' centralizes the image (by the diagonal embedding) of $\widehat{\mathbf{T}}_G$ in $\widehat{R} = \widehat{\mathbf{G}}^d$; as $\widehat{\mathbf{G}}^{\mathrm{der}}$ is simply connected, $t' \in \widehat{\mathbf{T}}_R$. The isomorphism $\mathbf{T}_H \simeq \mathbf{T}_G$ fixed above induces an isomorphism $\mathfrak{a}_{\mathbf{T}_G} \simeq \mathfrak{a}_{\mathbf{T}_H}$, and we can see the morphism $N : \mathfrak{a}_{\mathbf{T}_R} \longrightarrow \mathfrak{a}_{\mathbf{T}_G}$, $\lambda \longmapsto \lambda + \theta(\lambda) + \cdots + \theta^{d-1}(\lambda)$ defined above as a morphism $\mathfrak{a}_{\mathbf{T}_R} \longrightarrow \mathfrak{a}_{\mathbf{T}_H}$. We may identify $\mathfrak{a}_{\mathbf{T}_R}$ and $\mathfrak{a}_{\mathbf{T}_H}$ to $X^*(Y_R) \otimes_{\mathbb{Z}} \mathbb{R}$ and $X^*(Y_H) \otimes_{\mathbb{Z}} \mathbb{R}$, and then N sends $X^*(Y_R)$ to $X^*(Y_H)$. It is easy to see that $b_\xi : \mathbb{C}[Y_R/\Omega_R] \longrightarrow \mathbb{C}[Y_H/\Omega_H]$ sends f_λ to $|\Omega_H|^{-1} \sum_{\omega \in \Omega_R} \lambda(t') f_{N(\lambda^\omega)}$, for every $\lambda \in X^*(Y_R)$, where $\lambda(t')$ denotes the value of λ at the image of t' by the obvious morphism $\widehat{\mathbf{T}}_R \longrightarrow Y_R = \widehat{\mathbf{S}}_R$. Let $D_H \subset \mathfrak{a}_{\mathbf{T}_H}$ be the union of the kernels of the fundamental weights of $\mathbf{T}_H$ in $\mathbf{B}_H$, and let D'_H be the union of the $\omega(N^{-1}(D_H))$, for $\omega \in \Omega_R$. Then, for every connected component C of $\mathfrak{a}_{\mathbf{T}_R} - D'_H$, for all $\lambda, \lambda' \in C$, for every parabolic subgroup $\mathbf{P}_H \supset \mathbf{B}_H$ of $\mathbf{H}$ and for every $\omega \in \Omega_R$,

$$\hat{\tau}_{P_H}^H \circ N(\lambda^\omega) = \hat{\tau}_{P_H}^H \circ N((\lambda')^\omega).$$

By the untwisted version of the reasoning above (applied to the calculation of compact traces of representations of $\Pi(\mathbf{H})$), such a connected component C satisfies condition (b). Hence a connected component of $\mathfrak{a}_{\mathbf{T}_R} - (D' \cup D'_H)$ satisfies conditions (a) and (b). □

The next lemma will be used in section 9.3. It is a vanishing result similar to proposition 3.7.2 of [La3].

Lemma 9.2.12 *Assume that there exists an admissible embedding $\mathbf{T}_H \longrightarrow \mathbf{G}$ with image $\mathbf{T}_G$ and that the center of $\mathbf{G}$ is connected. Let $\gamma_H \in \mathbf{H}(F)$ be semisimple elliptic and strongly $\mathbf{G}$-regular. Assume that, for every θ-semisimple $\delta \in R(F)$, no element of $\mathcal{N}\delta$ is an image of γ_H in $\mathbf{G}(F)$. Then, for every $f \in \mathcal{H}_R$, $O_{\gamma_H}(b_\xi(f)) = 0$.*

As the condition on γ_H is stable by stable conjugacy, the lemma implies that, under the same hypothesis on γ_H, $SO_{\gamma_H}(b_\xi(f)) = 0$ for every $f \in \mathcal{H}_R$.

Proof. The proof is an adaptation of the proof of proposition 3.7.2 of [La3]. We first reformulate the condition on γ_H. By proposition 2.5.3 of [La3], a semisimple elliptic element of $\mathbf{G}(F)$ is a norm if and only if its image in $\mathbf{H}^0_{\mathrm{ab}}(F, \mathbf{G})$ is a norm. Using the proof of proposition 1.7.3 of [La3], we get canonical isomorphisms $\mathrm{H}^0_{\mathrm{ab}}(F, \mathbf{G}) = \mathrm{H}^1(\Gamma_F, Z(\widehat{\mathbf{G}}))^D$ and $\mathrm{H}^0_{\mathrm{ab}}(F, \mathbf{H}) = \mathrm{H}^1(\Gamma_F, Z(\widehat{\mathbf{H}}))^D$ (where ^{D} means Pontryagin dual). As there is a canonical Γ_F-equivariant embedding $Z(\widehat{\mathbf{G}}) \subset Z(\widehat{\mathbf{H}})$, we get a canonical morphism $\mathbf{H}^0_{\mathrm{ab}}(F, \mathbf{H}) \longrightarrow \mathbf{H}^0_{\mathrm{ab}}(F, \mathbf{G})$. The condition of the lemma on γ_H is equivalent to the following condition: the image of γ_H in $\mathbf{H}^0_{\mathrm{ab}}(F, \mathbf{G})$ is not a norm.

Assume that γ_H satisfies this condition. Then there exists a character χ of $\mathbf{H}^0_{\mathrm{ab}}(F, \mathbf{G})$ that is trivial on the norms and such that $\chi_H(\gamma_H) \neq 1$, where χ_H is the character of $\mathbf{H}(F)$ obtained by composing χ and the morphism $\mathbf{H}(F) \longrightarrow \mathbf{H}^0_{\mathrm{ab}}(F, \mathbf{H}) \longrightarrow \mathbf{H}^0_{\mathrm{ab}}(F, \mathbf{G})$. By the proof of lemma 3.7.1 of [La3], χ induces a character χ_{T_H} of $\mathbf{T}_H(F)$. Let us show that, for every $f \in \mathcal{H}_R$, $b_\xi(f) = \chi_H b_\xi(f)$ (this finishes the proof of the lemma, because $\chi_H(\gamma'_H) \neq 1$ for every $\gamma'_H \in \mathbf{H}(F)$ that is conjugate to γ_H). To do this, we imitate the proof of lemma 3.7.1 of [La3] and we show that $\mathrm{Tr}(\pi(b_\xi(f))) = \mathrm{Tr}(\pi(\chi_H b_\xi(f)))$ for every $f \in \mathcal{H}_R$ and every unramified representation π of $\mathbf{H}(F)$. Identify $\mathbf{T}_H$ to $\mathbf{T}_G$ by an admissible embedding, and let $N : X_*(\mathbf{S}_R) \longrightarrow X_*(\mathbf{S}_H)$ be the norm morphism as in the proof of lemma 9.2.11 above. By the proof of the second part of lemma 9.2.11, every function in $\mathcal{H}_H = \mathbb{C}[\widehat{\mathbf{S}}_H/\Omega_H] = \mathbb{C}[\widehat{\mathbf{S}}_H]^{\Omega_H}$ that is in the image of b_ξ is a linear combination of elements $N(\mu)$, with $\mu \in X_*(\mathbf{S}_R)$. If $z \in \widehat{\mathbf{S}}_H$ and χ_z is the unramified character of $\mathbf{T}_H(F)$ corresponding to z, let π_z be the unramified representation of $\mathbf{H}(F)$ obtained from χ_z (π_z is the unique unramified subquotient of $I(z)$, see, e.g., [Car] p. 152). Finally, let $z_0 \in \widehat{\mathbf{S}}_H$ be the element corresponding to the unramified character χ_{T_H} of $\mathbf{T}_H(F)$. As χ_{T_H} is trivial on the norms, $N(\mu)(z_0) = 1$ for every $\mu \in X_*(\mathbf{S}_R)$.

Let $f \in \mathcal{H}_R$. By lemma 9.2.10, for every $z \in \widehat{\mathbf{S}}_H$, $\mathrm{Tr}(\pi_z(b_\xi(f)))$ is a linear combination of the $b_\xi(f)(\omega z)$, with $\omega \in \Omega_H$. Hence, by the discussion above, for every $z \in \widehat{\mathbf{S}}_H$, $\mathrm{Tr}(\pi_z(b_\xi(f))) = \mathrm{Tr}(\pi_{zz_0}(b_\xi(f)))$, but $\mathrm{Tr}(\pi_{zz_0}(b_\xi(f))) = \mathrm{Tr}(\pi_z(\chi_H b_\xi(f)))$, so $\mathrm{Tr}(\pi_z(b_\xi(f))) = \mathrm{Tr}(\pi_z(\chi_H b_\xi(f)))$. This implies that $b_\xi(f) = \chi_H b_\xi(f)$. □

Proof of proposition 9.2.2. Note that, by lemma 9.4.6, there exists an admissible embedding $\mathbf{T}_H \longrightarrow \mathbf{G}$ with image $\mathbf{T}_G$.

Let $(a_i^R(\pi))_{i \in I, \pi \in \Pi(R)}$ and $(a_i^H(\pi_H))_{i \in I, \pi_H \in \Pi(\mathbf{H})}$ be the local data. By the definition of local data, it is enough to show that, for every $i \in I$ and every $f \in \mathcal{H}_R$,

$$\sum_{\pi \in \Pi(R)} a_i^R(\pi)\,\mathrm{Tr}(\pi(f) A_\pi) = \sum_{\pi_H \in \Pi(\mathbf{H})} a_i^H(\pi_H)\,\mathrm{Tr}(\pi_H(b_\xi(f))).$$

Fix $i \in I$, and let A be the distribution on $\mathcal{H}_R$ defined by

$$A(f) = \sum_{\pi \in \Pi(R)} a_i^R(\pi)\,\mathrm{Tr}(\pi(f) A_\pi) - \sum_{\pi_H \in \Pi(\mathbf{H})} a_i^H(\pi_H)\,\mathrm{Tr}(\pi_H(b_\xi(f))).$$

We want to show that $A = 0$. The distribution A is a sum of characters of $\mathcal{H}_R$. In other words, there exist $z_1, \dots, z_n \in Y_R$ such that A is a linear combination of the distributions $z_i \longmapsto f(z_i)$; we may assume that z_i and z_j are not Ω_R-conjugate if

$i \neq j$. Write

$$A(f) = \sum_{i=1}^{n} c_i f(z_i),$$

with $c_1, \dots, c_n \in \mathbb{C}$. By the definition of local data and lemma 9.2.4, A is a linear combination of distributions $f \longmapsto \Lambda(\gamma_H, f)$, with $\gamma_H \in \mathbf{H}(F)$ elliptic semi-simple and strongly $\mathbf{G}$-regular. By lemma 9.2.6, the distribution A is tempered. By lemma 5.5 of [Cl3], we may assume that $z_1, \dots, z_n \in Y_R^u$. On the other hand, by lemma 9.2.8, the distribution A is a linear combination of distributions $f \longmapsto \mathrm{Tr}_{\theta\text{-}c}(\pi(f)A_\pi)$ and $f \longmapsto \mathrm{Tr}_c(\pi_H(b_\xi(f)))$, with $\pi \in \Pi(R)$ and $\pi_H \in \Pi(\mathbf{H})$ coming from an element of $\Pi(\mathbf{H}_{adj})$. By lemma 9.2.11, there exist a nonempty open cone C of $X^*(Y_R) \otimes_{\mathbb{Z}} \mathbb{R}$ and $y_1, \dots, y_m \in Y_R$ such that the restriction to $C \cap X^*(Y_R)$ of the function $\lambda \longmapsto A(f_\lambda)$ on $X^*(Y_R)$ is a linear combination of the functions $\lambda \longmapsto \lambda(y_i)$, $1 \leq i \leq m$. By the explicit description of the y_i given in lemma 9.2.11 and lemmas 9.2.9 and 9.2.5, we may assume that $y_1, \dots, y_m \in Y_R - Y_R^u$. Let $d_1, \dots, d_m \in \mathbb{C}$ be such that

$$A(f_\lambda) = \sum_{i=1}^{m} d_i \lambda(y_i)$$

for every $\lambda \in C \cap X^*(Y_R)$. Consider the characters φ and φ' of $X^*(Y_R)$ defined by $\varphi(\lambda) = \sum_{i=1}^{n} \sum_{\omega \in \Omega_R} c_i \lambda(\omega z_i)$ and $\varphi'(\lambda) = \sum_{i=1}^{m} d_i \lambda(y_i)$. Then $\varphi(\lambda) = \varphi'(\lambda) = A(f_\lambda)$ if $\lambda \in C \cap X^*(Y_R)$. As $C \cap X^*(Y_R)$ generates the group $X^*(Y_R)$, the characters φ and φ' are equal. But the family $(\lambda \longmapsto \lambda(z))_{z \in Y_R}$ of characters of $X^*(Y_R)$ is free and $\{\omega z_i, 1 \leq i \leq n, \omega \in \Omega_R\} \cap \{y_1, \dots, y_m\} = \varnothing$ (because the first set is included in Y_R^u and the second set is included in $Y_R - Y_R^u$), so $\varphi = \varphi' = 0$. By the linear independance of the characters $\lambda \longmapsto \lambda(\omega z_i)$, this implies that $c_1 = \cdots = c_n = 0$, hence, finally, that $A = 0$.

9.3 CONSTRUCTION OF LOCAL DATA

The goal of this section is to construct local data. The method is global and uses the trace formula. The first thing to do is to show that there exists a global situation that gives back the situation of section 9.1 at one place.

Lemma 9.3.1 *Let F, E, $\mathbf{G}$, $(\mathbf{H}, s, \eta_0)$ and η be as in section 9.1. Assume that there exists a finite unramified extension K of E such that the groups $\mathbf{G}$ and $\mathbf{H}$ split over K and that the morphisms η and ξ come from morphisms $\widehat{\mathbf{H}} \rtimes \mathrm{Gal}(K/F) \longrightarrow \widehat{\mathbf{G}} \rtimes \mathrm{Gal}(K/F)$ and $\widehat{\mathbf{H}} \rtimes \mathrm{Gal}(K/F) \longrightarrow \widehat{R} \rtimes \mathrm{Gal}(K/F)$ (if $\mathbf{G}$ is adjoint, then such a K exists by remark 9.2.3). Then, for every $r \in \mathbb{N}^*$, there exist a number field k_F, finite Galois extensions $k_K/k_E/k_F$, a finite set S_0 of finite places of k_F, an element v_0 of S_0, connected reductive groups $\underline{\mathbf{G}}$ and $\underline{\mathbf{H}}$ over k_F, and L-morphisms $\underline{\eta} : {}^L\underline{\mathbf{H}} \longrightarrow {}^L\underline{\mathbf{G}}$ and $\underline{\xi} : {}^L\underline{\mathbf{H}} \longrightarrow {}^L\underline{R}$, where $\underline{R} = R_{k_E/k_F}\underline{\mathbf{G}}_{k_E}$, such that:*

(i) The groups $\underline{\mathbf{H}}$ and $\underline{\mathbf{G}}$ are quasi-split over k_F and split over k_K.

(ii) The set S_0 has r elements. Let $v \in S_0$. Then the place v is inert in k_K, and there are isomorphisms $k_{F,v} \simeq F$, $k_{E,v} \simeq E$, $k_{K,v} \simeq K$, $\underline{\mathbf{G}}_v \simeq \mathbf{G}$, $\underline{\mathbf{H}}_v \simeq \mathbf{H}$.

The group $\underline{\mathbf{G}}$ *has elliptic maximal tori* $\mathbf{T}$ *that stay elliptic over* $k_{F,v}$. *Moreover, the obvious morphism* $\mathrm{Gal}(k_{K,v}/k_{F,v}) \longrightarrow \mathrm{Gal}(k_K/k_F)$ *is an isomorphism (in particular, the extension* k_E/k_F *is cyclic).*

(iii) k_F *is totally imaginary.*

(iv) $(\underline{\mathbf{H}}, {}^L\underline{\mathbf{H}}, s, \eta)$ *are endoscopic data for* $(\underline{\mathbf{G}}, 1, 1)$, *and* $(\underline{\mathbf{H}}, {}^L\underline{\mathbf{H}}, t, \xi)$ *are endoscopic data for* $(\underline{R}, \underline{\theta}, 1)$, *where* $\underline{\theta}$ *is the automorphism of* $\underline{R}$ *defined by the generator of* $\mathrm{Gal}(k_E/k_F)$ *given by the isomorphism* $\mathrm{Gal}(E/F) \simeq \mathrm{Gal}(k_{E,v_0}/k_{F,v_0}) \xrightarrow{\sim} \mathrm{Gal}(k_E/k_F)$ *of (ii).*

(v) For every $v \in S_0$, $\underline{\eta}_v$ *corresponds to* $\eta : {}^L\mathbf{H} \longrightarrow {}^L\mathbf{G}$ *and* $\underline{\xi}_v$ *corresponds to* $\xi : {}^L\mathbf{H} \longrightarrow {}^L R$ *(by the isomorphisms of (ii)). For every infinite place* v *of* k_F, *the morphism* $\underline{\eta}_v : \widehat{\underline{\mathbf{H}}} \times W_{\mathbb{C}} = \widehat{\mathbf{H}} \times W_{\mathbb{C}} \longrightarrow \widehat{\underline{\mathbf{G}}} \times W_{\mathbb{C}} = \widehat{\mathbf{G}} \times W_{\mathbb{C}}$ *is equal to* $\eta_0 \times \mathrm{id}_{W_{\mathbb{C}}}$.

Moreover:

(vi) There exist infinitely many places of k_F *that split totally in* k_K.

(vii) For every finite set S *of places of* k_F *such that* $S_0 \not\subset S$, *the group* $\underline{\mathbf{H}}(k_F)$ *is dense in* $\prod_{v\in S} \underline{\mathbf{H}}(k_{F,v})$. *The same statement is true if* $\underline{\mathbf{H}}$ *is replaced by* $\underline{\mathbf{G}}$, $\underline{R}$ *or by a torus of* $\underline{\mathbf{H}}$, $\underline{\mathbf{G}}$ *or* $\underline{R}$.

Proof. If $r = 1$, the existence of k_F, k_E, k_K, $\underline{\mathbf{G}}$, $\underline{\mathbf{H}}$, and $S_0 = \{v_0\}$ satisfying (i), (ii), and (iii) is a consequence of the proof of proposition 11.1 of [Wa1]. As in [Cl3] p. 293, we pass from the case $r = 1$ to the general case by replacing k_F by an extension of degree r where v_0 splits totally (this extension is necessarily linearly disjoint from k_K, because v_0 is inert in k_K). By the last sentence of (ii), η gives an L-morphism $\widehat{\underline{\mathbf{H}}} \rtimes \mathrm{Gal}(k_K/k_F) \longrightarrow \widehat{\underline{\mathbf{G}}} \rtimes \mathrm{Gal}(k_K/k_F)$ and ξ gives an L-morphism $\widehat{\underline{\mathbf{H}}} \rtimes \mathrm{Gal}(k_K/k_F) \longrightarrow \widehat{\underline{R}} \rtimes \mathrm{Gal}(k_K/k_F)$. Take as $\underline{\eta}$ and $\underline{\xi}$ the L-morphisms ${}^L\underline{\mathbf{H}} \longrightarrow {}^L\underline{\mathbf{G}}$ and ${}^L\underline{\mathbf{H}} \longrightarrow {}^L\underline{R}$ that make the obvious diagrams commute. Then (iv) and (v) are clear. Point (vi) follows from the Čebotarev density theorem (cf. [Ne], chapter VII, theorem 13.4, and in particular corollary 13.6). As all the places of S_0 are inert in k_K, (vii) follows from (b) of lemma 1 of [KRo]. □

The main result of this section is the next proposition.

Proposition 9.3.2 *Assume that* $\mathbf{G}$ *is adjoint and that the endoscopic triple* $(\mathbf{H}, s, \eta_0)$ *is elliptic. Let* k_F, k_E, *etc., be as in lemma 9.3.1 above, with* $r = 3$. *Assume that, for almost every place* v *of* k_F, *the fundamental lemma for the unit element of the Hecke algebra is known for* $(\underline{R}_v, \underline{\theta}_v, 1)$ *and* $(\underline{\mathbf{H}}_v, {}^L\underline{\mathbf{H}}_v, t, \underline{\xi}_v)$ *(at almost every place* v *of* F, *the local situation is as in remark 9.1.1).*

Then there exist local data for R *and* $(\mathbf{H}, {}^L\mathbf{H}, t, \xi)$.

The proposition is proved at the end of this section, after a few lemmas.

We will need a simple form of the trace formula, due originally to Deligne and Kazhdan (see the article [He] of Henniart, sections 4.8 and 4.9, for the untwisted case, and lemma I.2.5 of the book [AC] of Arthur and Clozel for the twisted case). The next lemma is the obvious generalization (to groups that are not necessarily $\mathbf{GL}_n$) of lemma I.2.5 of [AC], and the proof of this lemma applies without any change (in condition (3′) on page 14 of [AC], the assumption that the functions ϕ_{w_i} are all coefficients of the *same* supercuspidal representation is not necessary).

Lemma 9.3.3 *Let F be a number field, E/F be a cyclic extension of degree d and $\mathbf{G}$ be a connected adjoint group over F. Set $R = R_{E/F}\mathbf{G}_E$, fix a generator of* Gal(E/F)*, and let θ be the automorphism of R induced by this generator. Let $\phi \in C_c^\infty(R(F))$. Denote by $r(\phi)$ the endomorphism of $L^2 := L^2(R(F) \setminus R(\mathbb{A}_F))$ obtained by making ϕ act by right convolution, and let I_θ be the endomorphism $f \longmapsto f \circ \theta^{-1}$ of L^2. Assume that*

- *(0) $\phi = \bigotimes_v \phi_v$, where the tensor product is taken over the set of places v of F and $\phi_v \in C_c^\infty(R(F_v))$ for every v; moreover, at almost every finite place v where R is unramified, ϕ_v is the characteristic function of a hyperspecial maximal compact subgroup of $R(F_v)$.*
- *(1) There exists a finite place v of F that splits totally in E and such that, on $R(F_v) \simeq \mathbf{G}(F_v)^d$, $\phi_v = \phi_1 \otimes \cdots \otimes \phi_d$, where the $\phi_i \in C_c^\infty(\mathbf{G}(F_v))$ are supercuspidal functions (in the sense of [He] 4.8).*
- *(2) There exists a finite place v of F such that the support of ϕ_v is contained in the set of θ-elliptic θ-semisimple and strongly θ-regular elements of $R(F_v)$.*

Then $r(\phi)I_\theta$ sends L^2 to the subspace of cuspidal functions (in particular, the endomorphism $r(\phi)I_\theta$ has a trace), and

$$\mathrm{Tr}(r(\phi)I_\theta) = \sum_\delta \mathrm{vol}(\mathbf{G}_{\delta\theta}(F) \setminus \mathbf{G}_{\delta\theta}(\mathbb{A}_F))O_{\delta\theta}(\phi),$$

where the sum is taken over the set of θ-conjugacy classes of θ-elliptic θ-semisimple and strongly θ-regular $\delta \in R(F)$.

Lemma 9.3.4 *([H2] lemma 5.1) Let F, $\mathbf{G}$ and $\mathbf{H}$ be as in 9.1 (in particular, F is a non-archimedean local field, $\mathbf{G}$ is an unramified group over F and $\mathbf{H}$ is an unramified endoscopic group of $\mathbf{G}$). Assume that the centers of $\mathbf{G}$ and $\mathbf{H}$ are anisotropic. Let $\mathbf{T}$ be an unramified elliptic maximal torus of $\mathbf{H}$; assume that $\mathbf{T}$ is defined over $\mathcal{O}_F$ and that $\mathbf{T}(F) \subset \mathbf{H}(\mathcal{O}_F)$. Let $j : \mathbf{T} \longrightarrow \mathbf{G}$ be an admissible embedding defined over $\mathcal{O}_F$. Set $N = \mathrm{Nor}_{\mathbf{G}(F)}(j(\mathbf{T}(F)))$, and make N act on $\mathbf{T}(F)$ via j.*

Then there exist functions $f \in C_c^\infty(\mathbf{G}(F))$ and $f_H \in C_c^\infty(\mathbf{H}(F))$ such that :

- *f and f_H are supercuspidal (in the sense of [He] 4.8; in particular, a linear combination of coefficients of supercuspidal representations is a supercuspidal function);*
- *the function $\gamma \longmapsto O_\gamma(f)$ (resp., $\gamma_H \longmapsto SO_{\gamma_H}(f_H)$) on $\mathbf{G}(F)$ (resp., $\mathbf{H}(F)$) is not identically zero, and its support is contained in the set of semisimple strongly regular (resp., strongly $\mathbf{G}$-regular) elements that are conjugate to an element of $j(\mathbf{T}(F))$ (resp., $\mathbf{T}(F)$);*
- *the function $\gamma_H \longmapsto O_{\gamma_H}(f_H)$ on $\mathbf{T}(F)$ is invariant under the action of N.*

The next two lemmas will be useful to construct transfers (and inverse transfers) of certain functions. The first lemma is a particular case of a theorem of Vignéras (theorem A of [Vi]).

Lemma 9.3.5 *Let F be a non-archimedean local field and $\mathbf{G}$ be a connected reductive group over F. Denote by $\mathbf{G}(F)_{ss\text{-}reg}$ the set of semisimple strongly regular elements of $\mathbf{G}(F)$. Let $\Gamma : \mathbf{G}(F)_{ss\text{-}reg} \longrightarrow \mathbb{C}$ be a function that is invariant by $\mathbf{G}(F)$-conjugacy and such that, for every $\gamma \in \mathbf{G}(F)_{ss\text{-}reg}$, the restriction of Γ to $\mathbf{G}_\gamma(F) \cap \mathbf{G}(F)_{ss\text{-}reg}$ is locally constant with compact support. Then there exists $f \in C_c^\infty(\mathbf{G}(F)_{ss\text{-}reg})$ such that, for every $\gamma \in \mathbf{G}(F)_{ss\text{-}reg}$, $\Gamma(\gamma) = O_\gamma(f)$.*

We will need a twisted variant of this lemma and some consequences of it. In the next lemma, F is still a non-archimedean local field and $\mathbf{G}$ a connected reductive group over F. We also assume that F is of characteristic 0. Let E be a finite étale F-algebra such that $\mathrm{Aut}_F(E)$ is cyclic. Set $R = R_{E/F}\mathbf{G}_E$, fix a generator of $\mathrm{Aut}_F(E)$, and denote by θ the automorphism of R induced by this generator (so the situation is that of example 8.1.1, except that E does not have to be a field). Use the definitions of 8.1. Let $\delta \in R(F)$ be θ-semisimple and strongly θ-regular, and write $\mathbf{T} = R_{\delta\theta}$. As in [La3] 1.8, set

$$\mathfrak{D}(\mathbf{T}, R; F) = \mathrm{Ker}(\mathbf{H}^1(F, \mathbf{T}) \longrightarrow \mathbf{H}^1(F, R)).$$

As F is local and non-archimedean, the pointed set $\mathfrak{D}(\mathbf{T}, \mathbf{G}; F)$ is canonically isomorphic to an abelian group (cf. [La3] lemma 1.8.3), so we will see $\mathfrak{D}(\mathbf{T}, R; F)$ as an abelian group. If $\delta' \in R(F)$ is stably θ-conjugate to δ, then it defines an element $\mathrm{inv}(\delta, \delta')$ of $\mathfrak{D}(\mathbf{T}, R; F)$. The map $\delta' \longmapsto \mathrm{inv}(\delta, \delta')$ induces a bijection from the set of θ-conjugacy classes in the stable θ-conjugacy class of δ to the set $\mathfrak{D}(\mathbf{T}, R; F)$ (cf. [La3] 2.3). Remember also (cf. [A4] §1) that $\mathbf{T}$ is a torus of R and that, if $(\mathbf{T}(F)\delta)_{\mathrm{reg}}$ is the set of strongly θ-regular elements of $\mathbf{T}(F)\delta$, then the map $u : (\mathbf{T}(F)\delta)_{\mathrm{reg}} \times \mathbf{T}(F) \setminus R(F) \longrightarrow R(F)$, $(g, x) \longmapsto x^{-1}g\theta(x)$, is finite on its image and open.

Now forget about δ and fix a maximal torus $\mathbf{T}$ of $\mathbf{G}$ (that can also be seen as a torus of R via the obvious embedding $\mathbf{G} \subset R$). Let Ω be the set of θ-semisimple strongly θ-regular $\delta \in R(F)$ such that there exists $x \in R(\overline{F})$ with $R_{\delta\theta} = x\mathbf{T}x^{-1}$. If κ is a character of $\mathfrak{D}(\mathbf{T}, R; F)$, $f \in C_c^\infty(\Omega)$ and $\delta \in \Omega$ is such that $R_{\delta\theta} = \mathbf{T}$, set

$$O^\kappa_{\delta\theta}(f) = \sum_{\delta'} \langle \mathrm{inv}(\delta, \delta'), \kappa \rangle O_{\delta'\theta}(f),$$

where the sum is taken over the set of θ-conjugacy classes in the stable θ-conjugacy class of δ. Then $O^\kappa_{\delta\theta}(f)$ is a (twisted) κ-orbital integral of f. (This definition is a particular case of [La3] 2.7).

Lemma 9.3.6 *Let $\mathbf{T}$ and Ω be as above. Let $\Gamma : \Omega \longrightarrow \mathbb{C}$ be a function that is invariant by θ-conjugacy and such that, for every $\delta \in \Omega$, the restriction of Γ to $R_{\delta\theta}(F)\delta \cap \Omega$ is locally constant with compact support. Then there exists $f \in C_c^\infty(\Omega)$ such that, for every $\delta \in \Omega$, $\Gamma(\delta) = O_{\delta\theta}(f)$.*

Let κ be a character of $\mathfrak{D}(\mathbf{T}, R; F)$. Assume that, for every $\delta \in \Omega$ such that $R_{\delta\theta} = \mathbf{T}$, for every $\delta' \in \Omega$ that is stably θ-conjugate to δ, $\Gamma(\delta) = \langle \mathrm{inv}(\delta, \delta'), \kappa \rangle \Gamma(\delta')$. Then there exists $g \in C_c^\infty(\Omega)$ such that

(a) for every $\delta \in \Omega$ such that $R_{\delta\theta} = \mathbf{T}$, $\Gamma(\delta) = O^\kappa_{\delta\theta}(g)$;

(b) *for every character* κ' *of* $\mathfrak{D}(\mathbf{T}, R; F)$ *such that* $\kappa' \neq \kappa$ *and for every* $\delta \in \Omega$ *such that* $R_{\delta\theta} = \mathbf{T}$, $O^{\kappa'}_{\delta\theta}(g) = 0$.

Moreover, for every character κ *of* $\mathfrak{D}(\mathbf{T}, R; F)$*, for every* $\delta \in \Omega$*, and every open neighborhood* Ω' *of* δ*, there exists a function* Γ *satisfying the conditions above and such that* $\Gamma(\delta) \neq 0$ *and that* $\Gamma(\delta') = 0$ *if* δ' *is not stably* θ*-conjugate to an element of* Ω'.

Proof. Let $\mathbf{T} = \mathbf{T}_1, \dots, \mathbf{T}_n$ be a system of representatives of the set of $R(F)$-conjugacy classes of tori of R (defined over F) that are equal to a $x\mathbf{T}x^{-1}$, with $x \in R(\overline{F})$. For every $i \in \{1, \dots, n\}$, denote by Ω_i the set of $\delta \in \Omega$ such that $R_{\delta\theta}$ is $R(F)$-conjugate to $\mathbf{T}_i$. The Ω_i are pairwise disjoint open subsets of Ω, and $\Omega = \bigcup_{i=1}^n \Omega_i$. Let $i \in \{1, \dots, n\}$. If $\delta \in \Omega_i$ is such that $R_{\delta\theta} = \mathbf{T}_i$, let u_δ be the function $(\mathbf{T}_i(F)\delta)_{\text{reg}} \times \mathbf{T}_i(F) \setminus R(F) \longrightarrow R(F), (x, y) \longmapsto y^{-1}xy$. Then Ω_i is the union of the images of the u_δ, this images are open, and two of these images are either equal or disjoint. So there exists a finite family $(\delta_{ij})_{j \in J_i}$ of elements of Ω_i such that $R_{\delta_{ij}\theta} = \mathbf{T}_i$ for every j and that $\Omega_i = \coprod_{j \in J_i} \operatorname{Im}(u_{\delta_{ij}})$. Write $u_{ij} = u_{\delta_{ij}}$, $\Omega_{ij} = \operatorname{Im}(u_{ij})$ and $\Gamma_{ij} = \mathbb{1}_{\Omega_{ij}}\Gamma$. Then the functions Γ_{ij} are invariant by θ-conjugacy, and $\Gamma = \sum_{i,j} \Gamma_{ij}$.

For every $i, i' \in \{1, \dots, n\}$, let $A(i, i')$ be the (finite) set of isomorphisms $\mathbf{T}_i \xrightarrow{\sim} \mathbf{T}_{i'}$ (over F) of the form $\operatorname{Int}(x)$, with $x \in R(\overline{F})$. If $i = i'$, write $A(i) = A(i, i')$. Let $i \in \{1, \dots, n\}$. For every $j \in J_i$, the support of $\Gamma_{ij} \circ u_{ij}$ is contained in a set of the form $\omega \times \mathbf{T}_i(F) \setminus R(F)$, with ω a compact subset of $(\mathbf{T}_i(F)\delta_{ij})_{\text{reg}}$. So it is easy to see that there exist open compact subsets ω_{ik}, $k \in K_i$, of $\mathbf{T}_i(F)$, and functions $\Gamma_{ijk} \in C_c^\infty(\Omega_{ij})$, invariant by θ-conjugacy, such that

(1) for every $j \in J_i$ and $k \in K_i$, $\omega_{ik}\delta_{ij} \subset (\mathbf{T}_i(F)\delta_{ij})_{\text{reg}}$, and the support of the function Γ_{ijk} is contained in $u_{ij}(\omega_{ik}\delta_{ij} \times \mathbf{T}_i(F) \setminus R(F))$;

(2) for every $k \in K_i$, the images of the ω_{ik} by the elements of $A(i)$ are pairwise disjoint;

(3) for every $j \in J_i$, $\Gamma_{ij} = \sum_{k \in K_i} \Gamma_{ijk}$.

Let $i \in \{1, \dots, n\}$, $j \in J_i$ and $k \in K_i$. By point (2) above, the restriction of u_{ij} to $\omega_{ik}\delta_{ij} \times \mathbf{T}_i(F) \setminus R(F)$ is injective. Let U_i be an open compact subset of volume 1 of $\mathbf{T}_i(F) \setminus \mathbf{G}(F)$. Denote by f_{ijk} the product of Γ_{ijk} and the characteristic function of $u_{ij}(\omega_{ik}\delta_{ij} \times U_i)$. Then $f_{ijk} \in C_c^\infty(\Omega)$ and, for every $\delta \in \Omega$, $O_{\delta\theta}(f_{ijk}) = \Gamma_{ijk}(\delta)$. So the function $f := \sum_{ijk} f_{ijk}$ satisfies the condition of the first statement of the lemma.

Let κ be a character of $\mathfrak{D}(\mathbf{T}, R; F)$. Assume that Γ satisfies the condition of the second statement of the lemma. If κ' is a character of $\mathfrak{D}(\mathbf{T}, R; F)$ and $\delta \in \Omega$ is such that $R_{\delta\theta} = \mathbf{T}$, then

$$O^{\kappa'}_{\delta\theta}(f) = \sum_{\delta'} \langle \operatorname{inv}(\delta, \delta'), \kappa' \rangle O_{\delta'\theta}(f) = \sum_{\delta'} \langle \operatorname{inv}(\delta, \delta'), \kappa' \rangle \Gamma(\delta')$$

$$= \Gamma(\delta) \sum_{\delta'} \frac{\langle \operatorname{inv}(\delta, \delta'), \kappa' \rangle}{\langle \operatorname{inv}(\delta, \delta'), \kappa \rangle},$$

where the sum is taken over the set of θ-conjugacy classes in the stable θ-conjugacy class of δ. So we can take $g = |\mathfrak{D}(\mathbf{T}, R; F)|^{-1} f$.

We show the last statement of the lemma. Let κ be a character of $\mathfrak{D}(\mathbf{T}, R; F)$. Choose (arbitrarily) an element j_0 of J_1, and write $\delta_1 = \delta_{1,j_0}$. For every $i \in \{1, \dots, n\}$, let J_i' be the set of $j \in J_i$ such that δ_{ij} is stably θ-conjugate to an element of $\mathbf{T}_1(F)\delta_1$; after translating (on the left) δ_{ij} by an element of $\mathbf{T}_i(F)$, we may assume that δ_{ij} is stably θ-conjugate to δ_1 for every $j \in J_i'$. For every $i \in \{1, \dots, n\}$ and $j \in J_i'$, choose $x_{ij} \in R(\overline{F})$ such that $\delta_{ij} = x_{ij}\delta_1\theta(x_{ij})^{-1}$, and let a_{ij} be the element of $A(1, i)$ induced by $\mathrm{Int}(x_{ij})$. Let $\omega \subset \mathbf{T}_1(F)$ be an open compact subset such that

- $1 \in \omega$;
- for every $i \in \{1, \dots, n\}$, the images of ω by the elements of $A(1, i)$ are pairwise disjoint;
- for every $i \in \{1, \dots, n\}$, $j \in J_i'$ and $a \in A(1, i)$, $a(\omega)\delta_{ij} \subset (\mathbf{T}_i(F)\delta_{ij})_{\mathrm{reg}}$, and the function $x \longmapsto \langle \mathrm{inv}(x\delta_1, a(x)\delta_{ij}), \kappa\rangle$ is constant on ω.

For every $i \in \{1, \dots, n\}$ and $j \in J_i'$, let Γ_{ij} be the product of the characteristic function of $u_{ij}(a_{ij}(\omega)\delta_{ij} \times \mathbf{T}_i(F) \setminus R(F))$ and of $\langle \mathrm{inv}(y\delta_1, a_{ij}(y)\delta_{ij}), \kappa\rangle^{-1}$, where y is any element of ω. Set $\Gamma = \sum_{i,j} \Gamma_{ij}$. Let $\delta \in \Omega$ be such that $R_{\delta\theta} = \mathbf{T}$. Then $\Gamma(\delta) = |A(1)|$ if δ is θ-conjugate to an element of $\omega\delta_1$, and $\Gamma(\delta) = 0$ otherwise (in particular, Γ is not identically zero). Let $\delta \in R(F)$ be stably θ-conjugate to δ_1. There exists a unique pair (i, j), with $i \in \{1, \dots, n\}$ and $j \in J_i'$, such that δ is θ-conjugate to an element of $\mathbf{T}_i(F)\delta_{ij}$. If δ is not stably θ-conjugate to an element of $\omega\delta_1$, then δ is not θ-conjugate to an element of $a_{ij}(\omega)\delta_{ij}$, and $\Gamma(\delta) = 0$. Otherwise, $\Gamma(\delta) = \langle \mathrm{inv}(\delta_1, \delta), \kappa\rangle^{-1}|A(1)| = \langle \mathrm{inv}(\delta_1, \delta), \kappa\rangle^{-1}\Gamma(\delta_1)$.

Let $\delta \in \Omega$. After changing the order of the $\mathbf{T}_i$ and choosing other representatives for the δ_{1j}, we may assume that the fixed δ_1 is δ. As it is always possible to replace ω by a smaller open compact (containing 1), this proves the last statement of the lemma. □

The two lemmas above have the following consequence.

Lemma 9.3.7 *Assume that F, E, $\mathbf{G}$, R, and θ are as in lemma 9.3.6. Let $(\mathbf{H}, s, \eta_0)$ be an endoscopic triple for $\mathbf{G}$. Assume that this local situation comes from a global situation as in lemma 9.3.1. In particular, $\mathbf{H}$ is the first element of endoscopic data $(\mathbf{H}, {}^L\mathbf{H}, t, \xi)$ for $(R, \theta, 1)$. Let Δ_ξ be the transfer factors defined by ξ (with any normalization). Then*

(i) Every function $f \in C_c^\infty(R(F))$ with support in the set of θ-semisimple strongly θ-regular elements admits a transfer to $\mathbf{H}$.

(ii) Let $\mathbf{T}_H$ be a maximal torus of $\mathbf{H}$. Choose an admissible embedding $j : \mathbf{T}_H \longrightarrow \mathbf{G}$, and make $N := \mathrm{Nor}_{\mathbf{G}(F)}(j(\mathbf{T}_H(F)))$ act on $\mathbf{T}_H(F)$ via j. Let $f_H \in C_c^\infty(\mathbf{H}(F))$ be a function with support in the set of strongly regular elements that are stably conjugate to an element of $\mathbf{T}_H(F)$. Assume that the function $\mathbf{T}_H(F) \longrightarrow \mathbb{C}$, $\gamma_H \longmapsto O_{\gamma_H}(f_H)$, is invariant under the action of N. Then there exists $f \in C_c^\infty(R(F))$ such that f_H is a transfer of f to $\mathbf{H}$.

The notion of transfer (or of "matching functions") in that case is defined in [KS] 5.5.

Proof. To prove (i), define a function Γ_H on the set of semisimple strongly **G**-regular elements of $\mathbf{H}(F)$ by $\Gamma_H(\gamma_H) = \sum_\delta \Delta_\xi(\gamma_H, \delta) O_{\delta\theta}(f)$, where the sum is taken over the set of θ-conjugacy classes of $R(F)$, and apply lemma 9.3.5 to Γ_H. To show (ii), construct a function Γ on the set of θ-semisimple strongly θ-regular elements of $R(F)$ in the following way : If there does not exist any $\gamma_H \in \mathbf{H}(F)$ such that $\Delta_\xi(\gamma_H, \delta) \neq 0$, set $\Gamma(\delta) = 0$. If there exists $\gamma_H \in \mathbf{H}(F)$ such that $\Delta_\xi(\gamma_H, \delta) \neq 0$, set $\Gamma(\delta) = \Delta_\xi(\gamma_H, \delta)^{-1} SO_{\gamma_H}(f_H)$. The function Γ is welldefined by the assumption on f_H (and lemma 5.1.B of [KS]). So (ii) follows from theorem 5.1.D of [KS] and from lemma 9.3.6. □

Remark 9.3.8 We need to be able to compare the groups of endoscopic characters of [KS] and [La3]. In the situation of the lemma above, but with F global or local (and allowed to be archimedean), if $\mathbf{T}_R$ is a θ-stable maximal torus of R coming from a torus $\mathbf{T}$ of $\mathbf{G}$, Labesse defined groups $\mathfrak{K}(\mathbf{T}, R; F)_1 \subset \mathfrak{K}(\mathbf{T}, R; F)$ ([La3] 1.8) and Kottwitz and Shelstad defined groups $\mathfrak{K}(\mathbf{T}_R, \theta, R)_1 \subset \mathfrak{K}(\mathbf{T}_R, \theta, R)$ ([KS] 6.4; Kottwitz and Shelstad assume that F is a number field, but it is possible to write the same definitions if F is local, erasing of course the quotient by Ker^1 in the definition of $\mathfrak{K}_1$). As we are interested in endoscopic data for the triple $(R, \theta, 1)$ (whose third element, which could in general be an element of $\mathrm{H}^1(W_F, Z(\widehat{\mathbf{G}}))$, is trivial here), we must use the group $\mathfrak{K}(\mathbf{T}_R, \theta, \mathbf{T})_1$ (cf. [KS] 7.1 and 7.2) to parametrize this data. Labesse showed that the groups $\mathfrak{K}(\mathbf{T}, R, F)$ and $\mathfrak{K}(\mathbf{T}_R, \theta, R)$ are canonically isomorphic (cf. the end of [La3] 2.6). Using the techniques of [La3] 1.7, it is easy to see that this isomorphism identifies $\mathfrak{K}(\mathbf{T}, R; F)_1$ and $\mathfrak{K}(\mathbf{T}_R, \theta, R)_1$.

The next lemma explains what happens if $E = F^d$.

Lemma 9.3.9 *Let F be a local or global field, $\mathbf{G}$ be a connected reductive group over F and $d \in \mathbb{N}^*$. Set $R = \mathbf{G}^d$, and let θ be the automorphism of R that sends $(g_1, \dots, g_d)$ to $(g_2, \dots, g_d, g_1)$. Then*

- *(i) The set of equivalence classes of endoscopic data for $(R, \theta, 1)$ is canonically isomorphic to the set of equivalence classes of endoscopic data for $(\mathbf{G}, 1, 1)$.*
- *(ii) Assume that the field F is local. Let $\phi \in C_c^\infty(R(F))$. Assume that $\phi = \phi_1 \otimes \cdots \otimes \phi_d$, with $\phi_1, \dots, \phi_d \in C_c^\infty(\mathbf{G}(F))$. Then, for every $\gamma = (\gamma_1, \dots, \gamma_d) \in R(F)$,*

$$O_{\gamma\theta}(\phi) = O_{\gamma_1 \dots \gamma_d}(\phi_1 * \cdots * \phi_d)$$

(provided, of course, that the measures are normalized in compatible ways).

Proof. Point (ii) is a particular case of [AC] I.5. Point (i) is almost obvious. We explain how the isomorphism is constructed. The dual group of R is $\widehat{R} = \widehat{\mathbf{G}}^d$, with the diagonal action of $\mathrm{Gal}(\overline{F}/F)$, so the diagonal embedding $\widehat{\mathbf{G}} \longrightarrow \widehat{R}$ extends in an obvious way to an L-morphism $\eta : {}^L\mathbf{G} \longrightarrow {}^LR$. If $(\mathbf{H}, \mathcal{H}, s, \xi)$ are endoscopic data for $(\mathbf{G}, 1, 1)$, they define endoscopic data $(\mathbf{H}, \mathcal{H}, \eta(s), \eta \circ \xi)$ for $(R, \theta, 1)$. Conversely, let $(\mathbf{H}, \mathcal{H}, s, \xi)$ be endoscopic data for $(R, \theta, 1)$. Write

$\xi(h \rtimes w) = (\xi_1(h \rtimes w), \ldots, \xi_d(h \rtimes w)) \rtimes w$, with $h \rtimes w \in \mathcal{H} \simeq \widehat{\mathbf{H}} \rtimes W_F$, and $s = (s_1, \ldots, s_d)$. Let $\xi_G : \mathcal{H} \longrightarrow {}^L\mathbf{G}$, $h \rtimes w \longmapsto \xi_1(h \rtimes w) \rtimes w$. Then $(\mathbf{H}, \mathcal{H}, s_1 \ldots s_d, \xi_G)$ are endoscopic data for $(\mathbf{G}, 1, 1)$. □

The next lemma is the analog of a statement proved in [H2], pp. 20–22. It is proved in exactly the same way, using the twisted version of the Paley-Wiener theorem (cf. the article [DeM] of Delorme and Mezo) instead of the untwisted version. (The statement on the support of functions in E is shown by using the control over the support of the functions given by theorem 3 of [DeM].)

Lemma 9.3.10 *Let* $\mathbf{G}$ *be a connected reductive group over* $\mathbb{C}$, $(\mathbf{H}, s, \eta_0)$ *be an endoscopic triple for* $\mathbf{G}$ *and* $d \in \mathbb{N}^*$. *Set* $R = \mathbf{G}^d$, *and denote by* θ *the automorphism of* R *defined by* $\theta(g_1, \ldots, g_d) = (g_2, \ldots, g_d, g_1)$. *Let* $\eta = \eta_0 \times \mathrm{id}_{W_\mathbb{C}} : {}^L\mathbf{H} \longrightarrow {}^L\mathbf{G}$ *be the obvious extension of* η_0, *and* ξ *be the composition of* η *and of the obvious ("diagonal") embedding* ${}^L\mathbf{G} \longrightarrow {}^LR$. *Fix maximal compact subgroups* K_G *and* K_H *of* $\mathbf{G}(\mathbb{C})$ *and* $\mathbf{H}(\mathbb{C})$, *write* $\mathrm{K}_R = \mathrm{K}_G^d$ *and denote by* $C_c^\infty(\mathbf{G}(\mathbb{C}), \mathrm{K}_G)$ *(resp.,* $C_c^\infty(\mathbf{H}(\mathbb{C}), \mathrm{K}_H)$, $C_c^\infty(R(\mathbb{C}), \mathrm{K}_R)$*) the space of* C^∞ *functions with compact support and* K_G*-finite (resp.,* K_H*-finite,* K_R*-finite) on* $\mathbf{G}(\mathbb{C})$ *(resp.,* $\mathbf{H}(\mathbb{C})$, $R(\mathbb{C})$*). Let* $\Pi(\mathbf{H})$ *(resp.,* $\Pi_\theta(R)$*) be the set of isomorphism classes of irreducible unitary representations of* $\mathbf{H}(\mathbb{C})$ *(resp. of* θ*-stable irreducible unitary representations of* $R(\mathbb{C})$*), and* $\Pi_{\mathrm{temp}}(\mathbf{H})$ *(resp.* $\Pi_{\theta\text{-temp}}(R)$*) be the subset of tempered representations. For every* $\pi \in \Pi_\theta(R)$, *choose a normalized intertwining operator* A_π *on* π. *For every* $\pi \in \Pi_{\theta\text{-temp}}(R)$, *let* $\Pi_H(\pi)$ *be the set of* $\pi_H \in \Pi_{\mathrm{temp}}(\pi)$ *whose functorial transfer to* R *is* π *(so* π_H *is in* $\Pi_H(\pi)$ *if and only if there exists a Langlands parameter* $\varphi_H : W_\mathbb{C} \longrightarrow {}^L\mathbf{H}$ *of* π_H *such that* $\xi \circ \varphi_H$ *is a Langlands parameter of* π*). Let* $N : C_c^\infty(R(\mathbb{C}), \mathrm{K}_R) = C_c^\infty(\mathbf{G}(\mathbb{C}), \mathrm{K}_G)^{\otimes d} \longrightarrow C_c^\infty(\mathbf{G}(\mathbb{C}), \mathrm{K}_G)$ *be the morphism of* $\mathbb{C}$*-algebras such that, for every* $f_1, \ldots, f_d \in C_c^\infty(\mathbf{G}(\mathbb{C}), \mathrm{K}_G)$, $N(f_1 \otimes \cdots \otimes f_d) = f_1 * \cdots * f_d$.

Then there exist a vector space $E \subset C_c^\infty(R(\mathbb{C}), \mathrm{K}_R)$ *and a compact subset* C *of* $\mathbf{H}(\mathbb{C})$ *such that*

(i) There exists $f \in E$ *and a transfer* $f^H \in C_c^\infty(\mathbf{H}(\mathbb{C}), \mathrm{K}_H)$ *of* $N(f)$ *to* $\mathbf{H}$ *such that the stable orbital integrals of* f^H *are not identically zero on the set of regular semisimple elliptic elements of* $\mathbf{H}(\mathbb{C})$.

(ii) For every $f \in E$ *and every transfer* f^H *of* $N(f)$, $SO_{\gamma_H}(f^H) = 0$ *if* γ_H *is not conjugate to an element of* C.

(iii) Let $(a(\pi))_{\pi \in \Pi_\theta(R)}$ *and* $(b(\pi_H))_{\pi_H \in \Pi(\mathbf{H})}$ *be families of complex numbers such that, for every* $f \in E$ *and every transfer* $f^H \in C_c^\infty(\mathbf{H}(\mathbb{C}), \mathrm{K}_H)$ *of* $N(f)$ *to* $\mathbf{H}$, *the sums* $A(f) := \sum_{\pi \in \Pi(R)} a(\pi)\,\mathrm{Tr}(\pi(f)A_\pi)$ *and* $A_H(f^H) := \sum_{\pi_H \in \Pi(\mathbf{H})} b(\pi_H)\,\mathrm{Tr}(\pi_H(f^H))$ *are absolutely convergent. Then the following conditions are equivalent:*

(A) for every $f \in E$ *and every transfer* $f^H \in C_c^\infty(\mathbf{H}(\mathbb{C}), \mathrm{K}_H)$ *of* $N(f)$ *to* $\mathbf{H}$, $A(f) = A_H(f^H)$;

(B) for every $f \in E$, *for every transfer* $f^H \in C_c^\infty(\mathbf{H}(\mathbb{C}), \mathrm{K}_H)$ *of* $N(f)$ *to* $\mathbf{H}$ *and for every* $\pi \in \Pi_{\theta\text{-temp}}(R)$, $a(\pi)\,\mathrm{Tr}(\pi(f)A_\pi) = \sum_{\pi_H \in \Pi_H(\pi)} b(\pi_H) \times \mathrm{Tr}(\pi_H(f^H))$.

The next lemma is proved in [Cl3], p. 292.

Lemma 9.3.11 *Notation is as in section 9.1. Let $f \in \mathcal{H}_R$. If there exists a dense subset D of the set of elliptic semisimple strongly* **G***-regular elements of* $\mathbf{H}(F)$ *such that* $\Lambda(\gamma_H, f) = 0$ *for every* $\gamma_H \in D$*, then* $\Lambda(\gamma_H, f) = 0$ *for every elliptic semisimple strongly* **G***-regular* $\gamma_H \in \mathbf{H}(F)$.

Lemma 9.3.12 *Let F be a number field, E be a cyclic extension of F and* **G** *be a connected reductive group over F. Set $R = R_{E/F}\mathbf{G}_E$, choose a generator of* $\mathrm{Gal}(E/F)$ *and denote by θ the automorphism of R induced by this generator. Let K be a finite extension of E such that* **G** *splits over K. Assume that the center of* **G** *is connected and that there exists a finite place v of F, inert in K, such that the morphism* $\mathrm{Gal}(K_v/F_v) \longrightarrow \mathrm{Gal}(K/F)$ *is an isomorphism. Then localization induces an injective map from the set of equivalence classes of endoscopic data for* $(R, \theta, 1)$ *to the set of equivalence classes of endoscopic data for* $(R_v, \theta_v, 1)$.

Proof. Let $(\mathbf{H}, \mathcal{H}, s, \xi)$, $(\mathbf{H}', \mathcal{H}', s', \xi')$ be endoscopic data for $(R, \theta, 1)$ whose localizations are equivalent (as endoscopic data for $(R_v, \theta_v, 1)$). By lemma 9.3.9, the endoscopic data for $(R_K, \theta_K, 1)$ defined by $(\mathbf{H}, \mathcal{H}, s, \xi)$ and $(\mathbf{H}', \mathcal{H}', s', \xi')$ are equivalent to endoscopic data coming from endoscopic data for $(\mathbf{G}_K, 1, 1)$. As the derived group of $\widehat{\mathbf{G}}$ is simply connected (because the center of **G** is connected) and $\mathbf{G}_K$ is split, if $(\mathbf{G}', \mathcal{G}', s_G, \xi_G)$ are endoscopic data for $(\mathbf{G}_K, 1, 1)$, then $\mathbf{G}'$ is split, so $\mathcal{G}' \simeq \widehat{\mathbf{G}'} \times W_K$, and we may assume that ξ_G is the product of an embedding $\widehat{\mathbf{G}'} \longrightarrow \widehat{R}$ and of the identity on W_K. Hence, after replacing $(\mathbf{H}, \mathcal{H}, s, \xi)$ and $(\mathbf{H}', \mathcal{H}', s', \xi')$ by equivalent data, we may assume that ξ and ξ' come from L-morphisms $\widehat{\mathbf{H}} \rtimes \mathrm{Gal}(K/F) \longrightarrow \widehat{R} \rtimes \mathrm{Gal}(K/F)$ and $\widehat{\mathbf{H}}' \rtimes \mathrm{Gal}(K/F) \longrightarrow \widehat{R} \rtimes \mathrm{Gal}(K/F)$. As the data $(\mathbf{H}, \mathcal{H}, s, \xi)$ and $(\mathbf{H}', \mathcal{H}', s', \xi')$ are equivalent at v, we may identify $\widehat{\mathbf{H}}$ and $\widehat{\mathbf{H}}'$ and assume that $s = s'$. As $\mathrm{Gal}(K_v/F_v) \xrightarrow{\sim} \mathrm{Gal}(K/F)$ (and $\mathrm{Gal}(\overline{F}/K)$ acts trivially on $\widehat{\mathbf{H}}$ and $\widehat{\mathbf{H}}'$), the isomorphism $\widehat{\mathbf{H}} = \widehat{\mathbf{H}}'$ extends to an isomorphism $\mathcal{H} \simeq \mathcal{H}'$ that identifies ξ and ξ'. So the data $(\mathbf{H}, \mathcal{H}, s, \xi)$ and $(\mathbf{H}', \mathcal{H}', s', \xi')$ are equivalent, and the lemma is proved. □

Proof of proposition 9.3.2. Write $S_0 = \{v_0, v_1, v_2\}$. Identify k_{F,v_0}, k_{E,v_0}, etc, to F, E, etc., We will prove the proposition by applying the twisted trace formula on $\underline{R}$ to functions whose local component at v_0 is a function in $\mathcal{H}_R$.

Let $\mathbf{T}_G$ be an elliptic maximal torus of $\underline{\mathbf{G}}$ such that $\mathbf{T}_{G,k_{F,v_1}}$ is also elliptic. Fix an admissible embedding $\mathbf{T}_H \longrightarrow \underline{\mathbf{G}}$ with image $\mathbf{T}_G$, where $\mathbf{T}_H$ is an elliptic maximal torus of $\underline{\mathbf{H}}$, and let $\mathbf{T}_R = R_{k_E/k_F}\mathbf{T}_{G,k_E}$. Let κ be the element of $\mathfrak{K}(\mathbf{T}_R, \theta, k_F)_1 = \mathfrak{K}(\mathbf{T}_G, \underline{R}; k_F)_1$ (cf. remark 9.3.8) associated to the endoscopic data $(\underline{\mathbf{H}}, {}^L\underline{\mathbf{H}}, t, \xi)$ by the map of [KS] 7.2. Write κ_{v_1} for the image of κ by the localization map $\mathfrak{K}(\mathbf{T}_H, \underline{R}; k_F) \longrightarrow \mathfrak{K}(\mathbf{T}_{H,v_1}, \underline{R}_{v_1}; k_{F,v_1})$ (cf. [La3] p. 43). Choose a function $\phi_{v_1} \in C_c^\infty(\underline{R}(k_{F,v_1}))$ that satisfies the conditions of lemma 9.3.6 (such that the support of ϕ_{v_1} is contained in the union of the stable θ-conjugates of $\mathbf{T}_R(k_{F,v_1})$, that the κ_{v_1}-orbital integrals of ϕ_{v_1} are not all zero and that the κ'_{v_1}-orbital integrals of ϕ_{v_1} are all zero if $\kappa'_{v_1} \neq \kappa_{v_1}$). Let $f_{v_1}^H$ be a transfer of ϕ_{v_1} to $\underline{\mathbf{H}}_{v_1}$. Let $f_{v_2}^H \in C_c^\infty(\underline{\mathbf{H}}(k_{F,v_2}))$ be a function with support in the set of semisimple strongly

$\mathbf{G}$-regular elements, whose orbital integrals are constant on stable conjugacy classes and whose stable orbital integrals are not all zero (such a function exists by lemma 9.3.6, applied with $\theta = 1$ and $\kappa = 1$). Fix a function $\phi_{v_2} \in C_c^\infty(\underline{R}(k_{F,v_2}))$ such that $f_{v_2}^H$ is a transfer of ϕ_{v_2} (such a function exists by lemma 9.3.7).

Let v_3 and v_4 be finite places of k_F where all the data are unramified (i.e., where the situation is as in remark 9.1.1); assume moreover that v_3 splits totally in k_E (this is possible by (vi) of lemma 9.3.1). Let $f_{v_3} \in C_c^\infty(\underline{\mathbf{G}}(k_{F,v_3}))$ be as in lemma 9.3.4. Write $\phi_{v_3} = f_{v_3} \otimes \cdots \otimes f_{v_3} \in C_c^\infty(\underline{R}(k_{F,v_3}))$ (where we identified $\underline{R}(k_{F,v_3})$ to $\underline{\mathbf{G}}(k_{F,v_3})^d$), and choose a transfer $f_{v_3}^H$ of ϕ_{v_3} (such a transfer exists by lemma 9.3.7). Let $f_{v_4}^H \in C_c^\infty(\underline{\mathbf{H}}(k_{F,v_4}))$ be as in lemma 9.3.4. Choose a function $\phi_{v_4} \in C_c^\infty(\underline{R}(k_{F,v_4}))$ such that $f_{v_4}^H$ is a transfer of ϕ_{v_4} (such a function exists by lemma 9.3.7).

Let S_∞ be the set of infinite places of k_F (by (iii) of lemma 9.3.1, they are all complex). Write $\mathbf{H}_\infty = \prod_{v \in S_\infty} \underline{\mathbf{H}}(k_{F,v})$ and $R_\infty = \prod_{v \in S_\infty} \underline{R}(k_{F,v})$. Let E be a subspace of $C_c^\infty(R_\infty)$ and C_∞ be a compact subset of $\mathbf{H}_\infty$ that satisfy the conditions of lemma 9.3.10. Let $\phi_{0,\infty} \in E$ and $f_{0,\infty}^H$ be a transfer of $\phi_{0,\infty}$ such that the stable orbital integrals of $f_{0,\infty}^H$ on elliptic elements of $\mathbf{H}_\infty$ are not all zero.

Let D_1 be the set of semisimple strongly $\mathbf{G}$-regular elliptic elements of $\mathbf{H}(F)$ coming from a $\gamma_H \in \underline{\mathbf{H}}(k_F)$ such that

- there exist $\delta \in \underline{R}(k_F)$ and an image γ of γ_H in $\underline{\mathbf{G}}(k_F)$ such that $\gamma \in \mathcal{N}\delta$;
- for every $v \in \{v_1, v_2, v_3, v_4\}$, $SO_{\gamma_H}(f_v^H) \neq 0$;
- $SO_{\gamma_H}(f_{0,\infty}^H) \neq 0$.

Let D_2 be the set of semisimple strongly $\mathbf{G}$-regular elliptic elements of $\mathbf{H}(F)$ that have no image in $\mathbf{G}(F)$ that is a norm. By (vii) of lemma 9.3.1, $D := D_1 \cup D_2$ is dense in the set of semisimple strongly $\mathbf{G}$-regular elliptic elements of $\mathbf{H}(F)$. By lemma 9.3.11, we may replace the set of semi-simple strongly $\mathbf{G}$-regular elliptic elements of $\mathbf{H}(F)$ by D in the definition of local data. By lemma 9.2.12, we may even replace D by D_1 in this definition.

Let $\gamma_H \in D_1$ (we use the same notation for the element of $\mathbf{H}(F)$ and for the element of $\underline{\mathbf{H}}(k_F)$ that induces it). Let S be a finite set of finite places of k_F such that $\{v_0, v_1, v_2, v_3, v_4\} \subset S$ and that, for every finite place $v \notin S$ of k_F, all the data are unramified at v, $\gamma_H \in \underline{\mathbf{H}}(\mathcal{O}_{k_{F,v}})$ and the fundamental lemma for the unit of the Hecke algebra is known for $(\underline{R}_v, \underline{\theta}_v, 1)$ and $(\underline{\mathbf{H}}_v, {}^L\underline{\mathbf{H}}_v, t, \underline{\xi}_v)$. For every $v \in S \backslash \{v_0, v_1, v_2, v_3, v_4\}$, choose associated functions f_v^H and ϕ_v such that $SO_{\gamma_H}(f_v^H) \neq 0$ (this is possible by the end of lemma 9.3.6). Let $C_0 \ni \gamma_H$ be a compact subset of $\mathbf{H}(F)$ that meets all the conjugacy classes of semisimple elliptic elements of $\mathbf{H}(F)$ (such a C_0 exists because the center of $\mathbf{H}$ is anisotropic). By proposition 8.2 of [K7], there is only a finite number of conjugacy classes of semi-simple elements γ'_H of $\underline{\mathbf{H}}(k_F)$ such that $\gamma'_H \in C_0$, $\gamma'_H \in C_\infty$, $SO_{\gamma'_H}(f_v^H) \neq 0$ for every $v \in S - \{v_0\}$ and $\gamma'_H \in \underline{\mathbf{H}}(\mathcal{O}_{k_{F,v}})$ for every finite place $v \notin S$. By the end of lemma 9.3.6, after adding a place in S and fixing well-chosen functions at that place, we may assume that γ_H is, up to conjugacy, the only semisimple element of $\underline{\mathbf{H}}(k_F)$ that satisfies the list of properties given above. For every finite place $v \notin S$ of k_F, take $\phi_v = \mathbb{1}_{\underline{R}(\mathcal{O}_{k_{F,v}})}$ and $f_v^H = \mathbb{1}_{\underline{\mathbf{H}}(\mathcal{O}_{k_{F,v}})}$.

Let $\phi_{v_0} \in \mathcal{H}_R$ and $f_{v_0}^H = b_\xi(\phi_{v_0})$. Fix $\phi_\infty \in E$ and a transfer f_∞^H of ϕ_∞, and set $\phi = \phi_\infty \otimes \bigotimes_{v\neq\infty} \phi_v$ and $f^H = f_\infty^H \otimes \bigotimes_{v\neq\infty} f_v^H$. Then lemma 9.3.3 applies to f and ϕ, thanks to the choice of the functions at v_3 and v_4. As in sections 5.4 and 8.2, let $T^{\underline{R}\rtimes\underline{\theta}}$ and $T^{\underline{\mathbf{H}}}$ be the distributions of the $\underline{\theta}$-twisted invariant trace formula on $\underline{R}$ and of the invariant trace formula on $\underline{\mathbf{H}}$.

By lemma 9.3.3, $T^{\underline{\mathbf{H}}}(f)$ is equal to the strongly regular elliptic part of the trace formula for $\underline{\mathbf{H}}$, so we may use the stabilization of [L3]. By the choice of $f_{v_2}^H$, the only endoscopic group of $\underline{\mathbf{H}}$ that appears is $\underline{\mathbf{H}}$ itself; so we need neither the transfer hypothesis nor the fundamental lemma to stabilize $T^{\underline{\mathbf{H}}}(f)$. We get

$$T^{\underline{\mathbf{H}}}(f^H) = ST^{\underline{\mathbf{H}}}_{**}(f^H),$$

where $ST^{\underline{\mathbf{H}}}_{**}$ is the distribution denoted by ST_e^{**} in [KS] 7.4 (the strongly $\underline{\mathbf{G}}$-regular elliptic part of the stable trace formula for $\underline{\mathbf{H}}$). Moreover, by the choice of f^H,

$$ST^{\underline{\mathbf{H}}}_{**}(f^H) = a_{\phi_\infty} SO_{\gamma_H}(b_\xi(\phi_{v_0})),$$

where a_{ϕ_∞} is the product of $SO_{\gamma_H}(f_\infty)$ and a nonzero scalar that does not depend on ϕ_∞ and ϕ_{v_0}.

Similarly, by lemma 9.3.3, $T^{\underline{R}\rtimes\underline{\theta}}(\phi)$ is equal to the strongly $\underline{\theta}$-regular $\underline{\theta}$-elliptic part of the trace formula for $\underline{R} \rtimes \underline{\theta}$, se we may apply the stabilization of chapters 6 and 7 of [KS]. By the choice of ϕ_{v_1}, the only endoscopic data of $(\underline{R}, \underline{\theta}, 1)$ that appear are $(\underline{\mathbf{H}}, \mathcal{H}, t, \underline{\xi})$. (Equality (7.4.1) of [KS] expresses $T^{\underline{R}\rtimes\underline{\theta}}(\phi)$ as a sum over elliptic endoscopic data of $(\underline{R}, \underline{\theta}, 1)$. The proofs of lemma 7.3.C and theorem 5.1.D of [KS] show that the global κ-orbital integrals that appear in this sum are products of local κ-orbital integrals. By the choice of ϕ_{v_1}, these products of local κ-orbital integrals are zero for the endoscopic data that are not equivalent to $(\underline{\mathbf{H}}, \mathcal{H}, t, \underline{\xi})$ at the place v_1. By lemma 9.3.12, $(\underline{\mathbf{H}}, \mathcal{H}, t, \underline{\xi})$ are the only endoscopic data satisfying this condition.) By equality (7.4.1) and the proof of lemma 7.3.C of [KS], we get

$$T^{\underline{R}\rtimes\underline{\theta}}(\phi) = b_{\phi_\infty} \sum_{\delta} \Delta_\xi(\gamma_H, \delta) O_{\delta\underline{\theta}}(\phi_{v_0}),$$

where the sum is taken over the set of θ-conjugacy classes of $R(F)$ and b_{ϕ_∞} is the product of $SO_{\gamma_H}(f_\infty^H)$ and of a nonzero scalar that does not depend on ϕ_{v_0} and ϕ_∞.

Hence $\Lambda(\gamma_H, \phi_{v_0}) = 0$ if and only if, for every $\phi_\infty \in E$, $a_{\phi_\infty} T^{\underline{R}\rtimes\underline{\theta}}(\phi) - b_{\phi_\infty} T^{\underline{\mathbf{H}}}(f^H) = 0$ (the "only if" part comes from the fact that $a_{\phi_\infty} b_{\phi_\infty} \neq 0$ for at least one choice of ϕ_∞). By lemmas 9.3.3 and 9.3.10, this last condition is equivalent to a family of identities of the form

$$\mathrm{Tr}(\pi_\infty(\phi_\infty)A_{\pi_\infty}) \sum_{\pi_0\in\Pi(R)} a(\pi_0)\,\mathrm{Tr}(\pi_0(\phi_{v_0}A_{\pi_0}))$$

$$= \sum_{\pi_{H,\infty}\in\Pi_{H_\infty}(\pi_\infty)} \mathrm{Tr}(\pi_{H,\infty})(f_\infty^H) \sum_{\pi_{H,0}\in\Pi(\mathbf{H})} b(\pi_{H,\infty}, \pi_{H,0})\,\mathrm{Tr}(\pi_{H,0}(b_\xi(\phi_{v_0}))),$$

for $\pi_\infty \in \Pi_{\theta\text{-temp}}(R_\infty)$ and $\phi_\infty \in E$, where the notation is as in lemma 9.3.10. By Harish-Chandra's finiteness theorem (cf. [BJ] 4.3(i)), the sums that appear in these equalities have only a finite number of nonzero terms.

Finally, we have shown that the identity $\Lambda(\gamma_H, \phi_{v_0}) = 0$ is equivalent to a family of identities like those that appear in the definition of local data. To obtain local data for R and $(\mathbf{H}, {}^L\mathbf{H}, t, \xi)$, we simply have to repeat this process for all the elements of D_1. □

9.4 TECHNICAL LEMMAS

We use the notation of section 9.1.

Let $\gamma_H \in \mathbf{H}(F)$ be semisimple and $\gamma \in \mathbf{G}(F)$ be an image of γ_H. Let $\mathbf{M}_H$ be a Levi subgroup of $\mathbf{H}$ such that $\gamma_H \in \mathbf{M}_H(F)$ and $\mathbf{M}_{H,\gamma_H} = \mathbf{H}_{\gamma_H}$. Langlands and Shelstad ([LS2] §1, see also section 7 of [K13]) associated to such a $\mathbf{M}_H$ a Levi subgroup $\mathbf{M}$ of $\mathbf{G}$ such that $\gamma \in \mathbf{M}(F)$ and $\mathbf{M}_\gamma = \mathbf{G}_\gamma$, an endoscopic triple $(\mathbf{M}_H, s_M, \eta_{M,0})$ for $\mathbf{M}$ and an L-morphism $\eta_M : {}^L\mathbf{M}_H \longrightarrow {}^L\mathbf{M}$ extending $\eta_{M,0}$ and such that there is a commutative diagram

$$\begin{array}{ccc} {}^L\mathbf{M}_H & \xrightarrow{\eta_M} & {}^L\mathbf{M} \\ \downarrow & & \downarrow \\ {}^L\mathbf{H} & \xrightarrow{\eta} & {}^L\mathbf{G} \end{array}$$

where the left (resp., right) vertical map is in the canonical $\widehat{\mathbf{H}}$-conjugacy (resp., $\widehat{\mathbf{G}}$-conjugacy) class of L-morphisms ${}^L\mathbf{M}_H \longrightarrow {}^L\mathbf{H}$ (resp., ${}^L\mathbf{M} \longrightarrow {}^L\mathbf{G}$). (If $\mathbf{G}$ and $\mathbf{H}$ are as in section 2.3, this construction is made more explicit in section 4.3.) Write $\mathbf{M}_R = R_{E/F}\mathbf{M}$ (a θ-stable Levi subgroup of R), and let θ_M be the restriction of θ to $\mathbf{M}_R$. As in section 9.1, associate to $(\mathbf{M}_H, s_M, \eta_M)$ endoscopic data $(\mathbf{M}_H, {}^L\mathbf{M}_H, t_M, \xi_M)$ for $(\mathbf{M}_R, \theta_M, 1)$.

The next lemma is a generalization of the beginning of [H1] 12, and can be proved in exactly the same way (because there is a descent formula for twisted orbital integrals, cf., for example, corollary 8.3 of [A2]). Note that we also need to use lemma 4.2.1 of [H2] (and the remarks below it).

Lemma 9.4.1 *Assume that, for every proper Levi subgroup $\mathbf{M}_H$ of $\mathbf{H}$, the twisted fundamental lemma is known for $\mathbf{M}_R$ and $(\mathbf{M}_H, {}^L\mathbf{M}_H, t_M, \xi_M)$ and for all the functions of $\mathcal{H}_{M_R}$. Then, for every $f \in \mathcal{H}_R$ and every $\gamma_H \in \mathbf{H}(F)$ that is semisimple and not elliptic,*

$$\Lambda(\gamma_H, f) = 0.$$

Lemma 9.4.2 *Let χ be a character of $R(F)$ such that $\chi = \chi \circ \theta$. Then χ is constant on the θ-semisimple stable θ-conjugacy classes of $R(F)$.*

Proof. If $\theta = 1$ (i.e., if $E = F$), this is lemma 3.2 of [H2]. In the general case, the result follows from the case $\theta = 1$ and from lemma 2.4.3 of [La3]. □

Notation is still as in 9.1. Let $\mathbf{B}_G$ be a Borel subgroup of $\mathbf{G}$ and $\mathbf{T}_G$ be a Levi subgroup of $\mathbf{B}_G$. Assume that the center of $\mathbf{G}$ is connected and that there exist a maximal torus $\mathbf{T}_H$ of $\mathbf{H}$ and an admissible embedding $\mathbf{T}_H \longrightarrow \mathbf{G}$ with image $\mathbf{T}_G$ (if $\mathbf{G}$ is adjoint, this is always the case by lemma 9.4.6). Use the same notation as in lemma 9.2.11 (in particular, $\mathbf{T}_R = R_{E/F}\mathbf{T}_{G,E}$).

Let $Z(R)_\theta$ be the group of θ-coinvariants of the center $Z(R)$ of R. There is a canonical injective morphism $N : Z(R)_\theta \longrightarrow Z(\mathbf{H})$ (cf. [KS] 5.1, p. 53). Let Z be a subtorus of $Z(R)_\theta$; denote by Z_R the inverse image of Z in $Z(R)$ and by Z_H the image of Z in $Z(\mathbf{H})$. Let χ_H be an unramified character of $Z_H(F)$. Write $\chi_R = (\lambda_C^{-1}(\chi_H \circ N))_{|Z_R(F)}$, where $\lambda_C : Z(R)(F) \longrightarrow \mathbb{C}^\times$ is the character defined in [KS] 5.1 p. 53; then χ_R is also unramified, by (i) of lemma 9.4.3 below.

Let $\mathcal{H}_{R,\chi_R}$ (resp., $\mathcal{H}_{H,\chi_H}$) be the algebra of functions $f : R(F) \longrightarrow \mathbb{C}$ (resp., $f : \mathbf{H}(F) \longrightarrow \mathbb{C}$) that are right and left invariant by $R(\mathcal{O}_F)$ (resp., $\mathbf{H}(\mathcal{O}_F)$), have compact support modulo $Z_R(F)$ (resp., $Z_H(F)$), and such that, for every $(z,x) \in Z_R(F) \times R(F)$ (resp., $Z_H(F) \times \mathbf{H}(F)$), $f(zx) = \chi_R^{-1}(z)f(x)$ (resp., $f(zx) = \chi_H^{-1}(z)f(x)$). The product is the convolution product that sends (f,g) to

$$f * g : x \longmapsto \int_{Z_R(F)\backslash R(F)} f(xy^{-1})g(y)dy$$

$$\text{(resp.,} \quad f * g : x \longmapsto \int_{Z_H(F)\backslash H(F)} f(xy^{-1})g(y)dy).$$

There is a surjective morphism $\nu_R : \mathcal{H}_R \longrightarrow \mathcal{H}_{R,\chi_R}$ (resp., $\nu_H : \mathcal{H}_H \longrightarrow \mathcal{H}_{H,\chi_H}$) that sends f to $x \longmapsto \int_{Z_R(F)} \chi_R^{-1}(z)f(zx)\mathrm{d}z$ (resp., $x \longmapsto \int_{Z_H(F)} \chi_H^{-1}(z)f(zx)\mathrm{d}z$). If $\gamma_H \in \mathbf{H}(F)$ and $f \in \mathcal{H}_{H,\chi_H}$, then we can define $O_{\gamma_H}(f)$ by the usual formula (the integral converges). If $\delta \in R(F)$ and $f \in \mathcal{H}_{R,\chi_R}$, set

$$O_{\delta\theta}(f) = \int_{Z_R(F)R_{\delta\theta}(F)\backslash R(F)} f(x^{-1}\delta\theta(x))dx$$

($f(x^{-1}\delta\theta(x))$ depends only on the class of x in $Z_R(F)R_{\delta\theta}(F) \setminus R(F)$, because χ_R is trivial on elements of the form $z^{-1}\theta(z)$, $z \in Z_R(F)$). We define κ-orbital integrals and stable orbital integrals as before for functions of $\mathcal{H}_{R,\chi_R}$ and $\mathcal{H}_{H,\chi_H}$.

Lemma 9.4.3

(i) As in the proof of lemma 9.2.11, write $\xi(1 \rtimes \sigma) = t' \rtimes \sigma$; we may assume that $t' \in \widehat{\mathbf{T}}_R$. As $Z(R)$ is an unramified torus, there is a canonical surjective morphism $Z(R)(F) \longrightarrow X_*(Z(R)_d)$ with kernel the maximal compact subgroup of $Z(R)(F)$, where $Z(R)_d$ is the maximal split subtorus of $Z(R)$ (cf. [Bo] 9.5). Define a character λ'_C on $Z(R)(F)$ in the following way. If $z \in Z(R)(F)$, and if $\mu \in X_*(Z(R)_d) = X^*(\widehat{Z(R)_d})$ is the image of z by the above morphism, $\lambda'_C(z)$ is the value of μ at the image of $t'^{-1} \in \widehat{\mathbf{T}}_R$ by the canonical morphism $\widehat{\mathbf{T}}_R \longrightarrow \widehat{Z(R)} \longrightarrow \widehat{Z(R)}_d$.
Then $\lambda_C = \lambda'_C$ (in particular, λ_C is unramified).

(ii) *There exists a morphism* $b_{\xi,\chi_H} : \mathcal{H}_{R,\chi_R} \longrightarrow \mathcal{H}_{H,\chi_H}$ *that makes the following diagram commute*

$$\begin{array}{ccc} \mathcal{H}_R & \xrightarrow{\nu_R} & \mathcal{H}_{R,\chi_R} \\ {\scriptstyle b_\xi}\downarrow & & \downarrow{\scriptstyle b_{\xi,\chi_H}} \\ \mathcal{H}_H & \xrightarrow[\nu_H]{} & \mathcal{H}_{H,\chi_H} \end{array}.$$

Let $\gamma_H \in \mathbf{H}(F)$ *be semisimple and strongly* $\mathbf{G}$*-regular. Use the morphism* $b_{\xi,\chi_H} : \mathcal{H}_{R,\chi_R} \longrightarrow \mathcal{H}_{H,\chi_H}$ *to define a linear form* $\Lambda_{\chi_H}(\gamma_H, .)$ *on* $\mathcal{H}_{R,\chi_R}$ *that is the analog of the linear form* $\Lambda(\gamma_H, .)$ *on* $\mathcal{H}_R$ *of 9.1 (use the same formula). Then the following conditions are equivalent:*

(a) *for every* $z \in Z_H(F)$*, for every* $f \in \mathcal{H}_R$*,* $\Lambda(z\gamma_H, f) = 0$*;*
(b) *for every* $z \in Z_H(F)$*, for every* $f \in \mathcal{H}_{R,\chi_R}$*,* $\Lambda_{\chi_H}(z\gamma_H, f) = 0$.

Proof. Point (i) follows from the definitions of λ_C ([KS] 5.1) and of the transfer factor Δ_{III} ([KS] 4.4).

We show (ii). Let $z \in Z(R)(F)$. For every function $f : R(F) \longrightarrow \mathbb{C}$, let $R_z f$ be the function $x \longmapsto f(zx)$. Denote by λ_z the image of z by the canonical map $Z(R)(F) \subset \mathbf{T}_R(F) \longrightarrow X_*(Y_R)$. Then, for every $\lambda \in X^*(Y_R)$, $R_z f_\lambda = f_{\lambda+\lambda_z}$. Moreover, it is easy to see that, for every $\lambda \in X^*(Y_R)$, $\nu_R^{-1}(\nu_R(f_\lambda))$ is generated by the functions $\chi_R(z)^{-1} R_z f_\lambda$, $z \in Z_R(F)$. There are similar statements for $\mathbf{H}$ instead of R.

To show the existence of the morphism $b_{\xi,\chi_H} : \mathcal{H}_{R,\chi_R} \longrightarrow \mathcal{H}_{H,\chi_H}$, it is enough to show that, for every $\lambda \in X^*(Y_R)$, all the elements of $\nu_R^{-1}(\nu_R(f_\lambda))$ have the same image by $\nu_H \circ b_\xi$. Let $\lambda \in X^*(Y_R)$. Let $z \in Z_R(F)$; denote by z_H the image of z in $Z_H(F)$. It is enough to show that $b_\xi(\chi_R^{-1}(z) R_z f_\lambda) = \chi_H(z_H)^{-1} R_{z_H} b_\xi(f_\lambda)$. By the explicit calculation of $b_\xi f_\lambda$ in the proof of lemma 9.2.11, $b_\xi(R_z f_\lambda) = \lambda_z(t') R_{z_H} b_\xi(f_\lambda)$; hence the equality that we are trying to prove follows from (i) and from the definition of χ_R.

Let $f \in \mathcal{H}_R$ and $\delta \in R(F)$. It is easy to see that

$$O_{\delta\theta}(\nu_R(f)) = \int_{Z(F)} \chi_R^{-1}(z) O_{z\delta\theta}(f) dz = \int_{Z(F)} \chi_R^{-1}(z) O_{\delta\theta}(R_z f) dz$$

($O_{z\delta\theta}(f)$ depends only on the image of z in $Z(F)$, because the function $\delta \longmapsto O_{\delta\theta}(f)$ is invariant by θ-conjugacy). Similarly, for every $f \in \mathcal{H}_H$ and $\gamma_H \in \mathbf{H}(F)$,

$$\begin{aligned} O_{\gamma_H}(\nu_H(f)) &= \int_{Z_H(F)} \chi_H^{-1}(z_H) O_{z_H\gamma_H}(f) dz_H \\ &= \int_{Z_H(F)} \chi_H^{-1}(z_H) O_{\gamma_H}(R_{z_H} f) dz_H. \end{aligned}$$

Remember ([KS] 5.1) that λ_C is such that, for every semisimple strongly regular $\gamma_H \in \mathbf{H}(F)$, every θ-semisimple strongly θ-regular $\delta \in R(F)$ and every $z \in Z(R)(F)$,

$$\Delta_\xi(z_H\gamma_H, z\delta) = \lambda_C^{-1}(z) \Delta_\xi(\gamma_H, \delta),$$

where z_H is the image of z in $Z(\mathbf{H})(F)$. By this fact and the above formulas for the orbital integrals, it is clear that (a) implies (b).

Let $\gamma_H \in \mathbf{H}(F)$ be semisimple strongly $\mathbf{G}$-regular. Assume that (b) is satisfied for γ_H; we want to show (a). Let $\lambda \in X^*(Y_R)$. Denote by ${}^0Z$ (resp., ${}^0Z_R, {}^0Z_H$) the maximal compact subgroup of $Z(F)$ (resp., $Z_R(F)$, $Z_H(F)$). The function f_λ is obviously invariant by translation by ${}^0Z_R$, and, moreover, all the unramified θ-stable characters of $R(F)$ are constant on its support (because f_λ is a linear combination of characteristic functions of sets $R(\mathcal{O}_F)\mu(\varpi_F)R(\mathcal{O}_F)$, where ϖ_F is a uniformizer of F and $\mu \in X^*(Y_R) = X_*(\mathbf{T}_R)^{\Gamma_F}$ is such that $\lambda\mu^{-1}$ is a cocharacter of R^{der}). Hence, for $z \in Z_R(F)$, $R_z f_\lambda$ depends only on the image of z in ${}^0Z_R\backslash Z_R(F)$ and, for every θ-semisimple stable θ-conjugacy class C of $R(F)$, there exists a unique $z \in {}^0Z_R\backslash Z_R(F)$ such that, for every $z' \in {}^0Z_R\backslash Z_R(F) - \{z\}$ and every $\delta \in C$, $O_{\delta\theta}(R_{z'} f_\lambda) = 0$ (use lemma 9.4.2). There are similar results for $\mathbf{H}$ and $b_\xi(f_\lambda)$.

Let C be the set of $\delta \in R(F)$ such that $\Delta_\xi(\gamma_H, \delta) \neq 0$. Then C is either the empty set or a θ-semisimple θ-regular stable θ-conjugacy class. So, by the reasoning above and the formulas for the orbital integrals of $\nu_R(f_\lambda)$ and $\nu_H(b_\xi(f_\lambda))$, there exists $z \in {}^0Z\backslash Z(F)$ such that $O_{\delta\theta}(\nu_R(f_\lambda)) = \chi_R(z)^{-1} O_{\delta\theta}(R_z f_\lambda)$, for every $\delta \in C$, and that $SO_{\gamma_H}(\nu_H(b_\xi(f_\lambda))) = \chi_H(z_H)^{-1} SO_{\gamma_H}(R_{z_H} b_\xi(f_\lambda))$, where $z_H \in {}^0Z_H\backslash Z_H(F)$ is the image of z. If $z \neq 1$, then $\Lambda(\gamma_H, f_\lambda) = 0$, because all the orbital integrals that appear in this expression are zero. If $z = 1$, then $\Lambda(\gamma_H, f_\lambda) = 0$ by condition (b) and the properties of λ_C. □

We still denote by F a non-archimedean local field (of characteristic 0) and by E a finite unramified extension of F. Let $\mathbf{G}$ be a connected unramified group over F, defined over $\mathcal{O}_F$ and such that $\mathbf{G}(\mathcal{O}_F)$ is a hyperspecial maximal compact subgroup of $\mathbf{G}(F)$. Set $R = R_{E/F}\mathbf{G}_E$, and let θ be the automorphism of R induced by a chosen generator of $\mathrm{Gal}(E/F)$. Let Z_G be a subtorus of $Z(\mathbf{G})$ defined over $\mathcal{O}_F$, and let $Z_R = R_{E/F} Z_{G,E}$. Let $\mathbf{G}' = \mathbf{G}/Z_G$, $R' = R/Z_R = R_{E/F}\mathbf{G}'_E$, $u : R \longrightarrow R'$ be the obvious morphism, $\mathcal{H}' = \mathcal{H}_{R'}$, and $\mathcal{H}$ be the convolution algebra of functions $R(F) \longrightarrow \mathbb{C}$ that are bi-invariant by $R(\mathcal{O}_F)$, invariant by $Z_R(F)$, and with compact support modulo $Z_R(F)$ (with the notation of lemma 9.4.3, $\mathcal{H} = \mathcal{H}_{R,1}$). As Z_R is connected, we see as in [Cl3] 6.1 (p. 284) that Lang's theorem (cf. for example, theorem 4.4.17 of [Sp]) and Hensel's lemma imply that $u : R(\mathcal{O}_F) \longrightarrow R'(\mathcal{O}_F)$ is surjective. So u induces a morphism of algebras $\varphi : \mathcal{H} \longrightarrow \mathcal{H}'$ (for every $f \in \mathcal{H}$ and every $x \in R'(F)$, $\varphi(f)(x)$ is equal to 0 if $x \notin u(R(F))$ and to $f(u^{-1}(x))$ if $x \in u(R(F))$). For every δ in $R(F)$ or $R'(F)$, denote by $C(\delta)$ (resp., $C_{st}(\delta)$) the θ-conjugacy (resp., stable θ-conjugacy) class of δ.

Lemma 9.4.4 *Assume that θ acts trivially on $\mathrm{H}^1(F, Z_R)$.*

Let $\delta \in R(F)$ be θ-semisimple; write $\delta' = u(\delta)$. Then $u(C_{st}(\delta)) = C_{st}(\delta')$. So there exists a (necessarily finite) family $(\delta_i)_{i\in I}$ of elements of $R(F)$ that are stably θ-conjugate to δ, such that $C(\delta') = \coprod_{i\in I} u(C(\delta_i))$. Moreover, for every $f \in \mathcal{H}$,

$$O_{\delta'\theta}(\varphi(f)) = \sum_{i\in I} O_{\delta_i\theta}(f).$$

(As always, we use the Haar measures on $R(F)$ and $R'(F)$ such that the volumes of $R(\mathcal{O}_F)$ and $R'(\mathcal{O}_F)$ are equal to 1.)

Proof. It is clear that $u(C_{st}(\delta)) \subset C_{st}(\delta')$. We show the other inclusion. Let $\gamma' \in R'(F)$ be stably θ-conjugate to δ'. As $u(R(F)) = \mathrm{Ker}(R'(F) \longrightarrow \mathrm{H}^1(F, Z_R))$ is the intersection of kernels of θ-stable characters of $R'(F)$, lemma 9.4.2 implies that there exists $\gamma \in R(F)$ such that $\gamma' = u(\gamma)$. It is easy to see that γ and δ are stably θ-conjugate.

Fix a family $(\delta_i)_{i\in I}$ as in the statement of the lemma, and write $\mathrm{K} = R(\mathcal{O}_F)$, $\mathrm{K}' = R'(\mathcal{O}_F)$. We show the equality of orbital integrals. Let $f \in \mathcal{H}$. We may assume that $f = \mathbb{1}_A$, where A is a subset of $R(F)$ such that $A = Z_R(F)\mathrm{K}A\mathrm{K}$. Then $\varphi(f) = \mathbb{1}_{u(A)}$, so

$$O_{\delta'\theta}(\varphi(f)) = \sum_{\gamma'} \mathrm{vol}(u(A) \cap R'_{\gamma'\theta}(F))^{-1},$$

where the sum is taken over a set of representatives γ' of the K'-θ-conjugacy classes of elements of $u(A)$ that are θ-conjugate to δ' (in $R'(F)$). There are similar formulas for the twisted orbital integrals of f at the δ_i; for every $i \in I$,

$$O_{\delta_i\theta}(f) = \sum_{\gamma} \mathrm{vol}((A \cap R_{\gamma\theta}(F))Z_R(F)/Z_R(F))^{-1},$$

where the sum is taken over a set of representatives γ of the K-θ-conjugacy classes of elements of A that are θ-conjugate to δ_i (in $R(F)$). To show the formula of the lemma, it is therefore enough to notice that, for every $\gamma \in R(F)$, u induces an isomorphism $(A \cap R_{\gamma\theta}(F))Z_R(F)/Z_R(F) \xrightarrow{\sim} u(A) \cap R'_{u(\gamma)\theta}(F)$. □

We use again the notation of 9.1. Assume that $\eta(1 \rtimes \sigma) \in \widehat{\mathbf{G}}^{\mathrm{der}} \rtimes \sigma$ and that $s \in \widehat{\mathbf{G}}^{\mathrm{der}}$. Then $(\mathbf{H}, {}^L\mathbf{H}, s, \eta)$ defines in an obvious way endoscopic data $(\mathbf{H}', {}^L\mathbf{H}', s, \eta')$ for $\mathbf{G}' := \mathbf{G}/Z(\mathbf{G})^0$ (because $\widehat{\mathbf{G}}' = \widehat{\mathbf{G}}^{\mathrm{der}}$). As in section 9.1, we get from these endoscopic data $(\mathbf{H}', {}^L\mathbf{H}', t', \xi')$ for (R', θ), where $R' = R_{E/F}\mathbf{G}'_E$.

Lemma 9.4.5 *Assume that θ acts trivially on $\mathrm{H}^1(F, Z(R)^0)$ and that* **G** *and* **H** *satisfy the conditions of lemma 9.4.3.*

Then the fundamental lemma is true for (R, θ) and $(\mathbf{H}, {}^L\mathbf{H}, t, \xi)$ if and only if it is true for (R', θ) and $(\mathbf{H}', {}^L\mathbf{H}', t', \xi')$.

Proof. Let $Z_R = Z(R)^0$, and let Z_H be the image of $Z(\mathbf{G})^0$ in $Z(\mathbf{H})$. Then $\mathbf{H}' = \mathbf{H}/Z_H$. It is easy to check that, if $\gamma_H \in \mathbf{H}(F)$ is semisimple and strongly $\mathbf{G}$-regular, if $\delta \in R(F)$ is θ-semisimple and strongly θ-regular, and if (γ_H, δ) is sent to $(\gamma'_H, \delta') \in \mathbf{H}'(F) \times R'(F)$ by the obvious projection, then $\Delta_\xi(\gamma_H, \delta) = \Delta_{\xi'}(\gamma'_H, \delta')$. Apply lemma 9.4.3 with $\chi_H = 1$ and $\chi_R = 1$ (this is possible because the character λ_C that appears in this lemma is trivial, thank to the assumption that $\eta(1 \rtimes \sigma) \in \widehat{\mathbf{G}}^{\mathrm{der}} \rtimes \sigma$). This lemma shows that we may replace the Hecke algebras of R and **H** by the Hecke algebras of $Z_R(F)$-invariant or $Z_H(F)$-invariant functions. To finish the proof, apply lemma 9.4.4. □

The next lemma and its proof were communicated to me by Robert Kottwitz. (Any mistakes that I may have inserted are my sole responsibility.)

Lemma 9.4.6 *Let F be a non-archimedean local field of characteristic 0, $\mathbf{G}$ be an adjoint quasi-split group over F, and $(\mathbf{H}, s, \eta_0)$ be an endoscopic triple for $\mathbf{G}$. Fix a Borel subgroup $\mathbf{B}$ (resp., $\mathbf{B}_H$) of $\mathbf{G}$ (resp., $\mathbf{H}$) and a Levi subgroup $\mathbf{T}_G$ (resp., $\mathbf{T}_H$) of $\mathbf{B}$ (resp., $\mathbf{B}_H$). Then there exists an admissible embedding $\mathbf{T}_H \longrightarrow \mathbf{G}$ with image $\mathbf{T}_G$.*

Proof. Write $\Gamma = \mathrm{Gal}(\overline{F}/F)$. Choose embeddings $\widehat{\mathbf{T}}_G \subset \widehat{\mathbf{B}} \subset \widehat{\mathbf{G}}$ and $\widehat{\mathbf{T}}_H \subset \widehat{\mathbf{B}}_H \subset \widehat{\mathbf{H}}$ that are preserved by the action of Γ on $\widehat{\mathbf{G}}$ and $\widehat{\mathbf{H}}$.

As F is local, we may assume that $s \in Z(\widehat{\mathbf{H}})^\Gamma$. By the definition of an endoscopic triple, for every $\tau \in \Gamma$, there exists $g_\tau \in \widehat{\mathbf{G}}$ such that, for every $h \in \widehat{\mathbf{H}}$,

$$g_\tau \tau(\eta_0(h)) g_\tau^{-1} = \eta_0(\tau(h)). \qquad (*)$$

In particular, the $\widehat{\mathbf{G}}$-conjugacy class of $\eta_0(s)$ is fixed by the action of Γ on $\widehat{\mathbf{G}}$. By lemma 4.8 of [Cl3], $\eta_0(s)$ is $\widehat{\mathbf{G}}$-conjugate to an element of $\widehat{\mathbf{T}}_G^\Gamma$. Replacing η_0 by a $\widehat{\mathbf{G}}$-conjugate, we may assume that $\eta_0(s) \in \widehat{\mathbf{T}}_G^\Gamma$. Then

$$\widehat{\mathbf{T}}_G \subset \mathrm{Cent}_{\widehat{\mathbf{G}}}(\eta_0(s)) = \mathrm{Cent}_{\widehat{\mathbf{G}}}(\eta_0(s))^0 = \widehat{\mathbf{H}}$$

($\mathrm{Cent}_{\widehat{\mathbf{G}}}(\eta_0(s))$ is connected because $\widehat{\mathbf{G}}$ is semisimple and simply connected), so by further conjugating η_0 by an element in $\eta_0(\widehat{\mathbf{H}})$ (which does not change $\eta_0(s)$, since $s \in Z(\widehat{\mathbf{H}})$), we may also assume that $\eta_0(\widehat{\mathbf{T}}_H) = \widehat{\mathbf{T}}_G$ and $\eta_0(\widehat{\mathbf{B}}_H) = \widehat{\mathbf{B}} \cap \eta_0(\widehat{\mathbf{H}})$.

Since $\eta_0(s)$ is fixed by Γ, for every $\tau \in \Gamma$, $g_\tau \eta_0(s) g_\tau^{-1} = \eta_0(s)$, so that $g_\tau \in \mathrm{Cent}_{\widehat{\mathbf{G}}}(\eta_0(s)) = \eta_0(\widehat{\mathbf{H}})$. Moreover $(*)$, together with the fact that Γ preserves $(\widehat{\mathbf{B}}, \widehat{\mathbf{T}}_G)$ and $(\widehat{\mathbf{B}}_H, \widehat{\mathbf{T}}_H)$, implies that $h_\tau := \eta_0^{-1}(g_\tau)$ conjugates $(\widehat{\mathbf{B}}_H, \widehat{\mathbf{T}}_H)$ into itself. Therefore $h_\tau \in \widehat{\mathbf{T}}_H$, and $(*)$ now shows that η_0 induces a Γ-equivariant isomorphism $\widehat{\mathbf{T}}_H \xrightarrow{\sim} \widehat{\mathbf{T}}_G$. Dual to this is an admissible isomorphism $\mathbf{T}_H \xrightarrow{\sim} \mathbf{T}_G$. □

Let F be a non-archimedean local field of characteristic 0. Let $n, n_1, \dots, n_r \in \mathbb{N}^*$. Set $\mathbf{PGL}_n = \mathbf{GL}_n / Z(\mathbf{GL}_n)$. For every quadratic extension E of F, set $\mathbf{PGU}(n, E) = \mathbf{GU}(n, E)/Z(\mathbf{GU}(n, E))$, where $\mathbf{GU}(n, E)$ is the unitary group defined by the extension E/F and by the hermitian form with matrix

$$J_n := \begin{pmatrix} 0 & & 1 \\ & \cdot^{\cdot^{\cdot}} & \\ 1 & & 0 \end{pmatrix} \in \mathbf{GL}_n(\mathbb{Z}).$$

More generally, set $\mathbf{P}(\mathbf{U}(n_1, E) \times \cdots \times \mathbf{U}(n_r, E)) = (\mathbf{GU}(n_1, E) \times \cdots \times \mathbf{GU}(n_r, E))/Z$, where $Z = R_{E/\mathbb{Q}}\mathbb{G}_m$, embedded diagonally. Set $\mathbf{PGSO}_n = \mathbf{GSO}(J_n)/Z(\mathbf{GSO}(J_n))$, where $\mathbf{GSO}(J_n) = \mathbf{GO}(J_n)^0$, and $\mathbf{PGSp}_{2n} = \mathbf{GSp}(J_n')/Z(\mathbf{GSp}(J_n'))$, where

$$J_n' = \begin{pmatrix} 0 & J_n \\ -J_n & 0 \end{pmatrix} \in \mathbf{GL}_{2n}(\mathbb{Z}).$$

If $Y^1, \dots, Y^r \in \{\mathbf{GSO}, \mathbf{GSp}\}$, we denote by $\mathbf{P}(Y^1_{n_1} \times \cdots \times Y^r_{n_r})$ the quotient of $Y^1_{n_1} \times \cdots \times Y^r_{n_r}$ by $\mathbb{G}_m$ embedded diagonally.

Lemma 9.4.7 *Let* $\mathbf{G}$ *be a simple adjoint unramified group over* F.

(i) If $\mathbf{G}$ *is of type* A*, then there exist a finite unramified extension* K *of* F*, a quadratic unramified extension* E *of* K*, and a positive integer* n *such that* $\mathbf{G} = R_{K/F}\mathbf{PGL}_n$ *or* $\mathbf{G} = R_{K/F}\mathbf{PGU}(n, E)$*. If* $\mathbf{G} = R_{K/F}\mathbf{PGL}_n$*, then* $\mathbf{G}$ *has no nontrivial elliptic endoscopic groups. If* $\mathbf{G} = R_{K/F}\mathbf{PGU}(n, E)$*, then the elliptic endoscopic groups of* $\mathbf{G}$ *are the* $R_{K/F}\mathbf{P}(\mathbf{GU}(n_1, E) \times \mathbf{GU}(n_2, E))$*, with* $n_1, n_2 \in \mathbb{N}$ *such that* $n = n_1 + n_2$ *and* n_2 *is even.*

(ii) If $\mathbf{G}$ *is of type* B*, then there exist a finite unramified extension* K *of* F *and a positive integer* n *such that* $\mathbf{G} = R_{K/F}\mathbf{PGSO}_{2n+1}$*. The elliptic endoscopic groups of* $\mathbf{G}$ *are the* $R_{K/F}\mathbf{P}(\mathbf{GSO}_{2n_1+1} \times \mathbf{GSO}_{2n_2+1})$*, with* $n_1, n_2 \in \mathbb{N}$ *such that* $n = n_1 + n_2$.

(iii) If $\mathbf{G}$ *is of type* C*, then there exist a finite unramified extension* K *of* F *and a positive integer* n *such that* $\mathbf{G} = R_{K/F}\mathbf{PGSp}_{2n}$*. The elliptic endoscopic groups of* $\mathbf{G}$ *are the* $\mathbf{P}(\mathbf{GSO}_{2n_1} \times \mathbf{GSp}_{2n_2})$*, with* $n_1, n_2 \in \mathbb{N}$ *such that* $n = n_1 + n_2$ *and* $n_1 \neq 1$.

In particular, if $\mathbf{G}$ is adjoint of type A, B or C, then the hypothesis of proposition 9.2.2 on the center of $\mathbf{H}$ (i.e., that this center be connected) is satisfied.

Proof. Let K be the smallest extension of F on which $\mathbf{G}$ splits, and fix a generator σ of $\mathrm{Gal}(K/F)$. Then $\mathbf{G}_K \simeq (\mathbf{G}')^r$, where $r \in \mathbb{N}^*$ and $\mathbf{G}'$ is a split adjoint simple group over K. Let θ be the automorphism (over K) of $(\mathbf{G}')^r$ induced by σ. If $\mathbf{G}$ is of type B or C, then $\mathbf{G}'$ is also of type B or C, so $\mathbf{G}'$ is equal to $\mathbf{PGSO}_n$ or $\mathbf{PGSp}_{2n}$, and $\mathbf{G}'$ has no nontrivial outer automorphisms (cf. [Di] IV.6 and IV.7), so we may assume that θ acts by permuting the factors of $(\mathbf{G}')^r$. As K is the smallest extension on which $\mathbf{G}$ splits, θ has to be an n-cycle. So $\mathbf{G} \simeq R_{K/F}\mathbf{G}'$. To compute the elliptic endosocopic triples for $\mathbf{G}$, we may assume that $K = F$. Then $\mathbf{G}$ is split and has a connected center, so its endoscopic groups are also split (cf. definition 1.8.1 of [Ng]). From this observation, it is easy to see that the elliptic endoscopic groups of $\mathbf{G}$ are the ones given in the statement of the lemma.

Assume that $\mathbf{G}$ is of type A. Then there exists $n \in \mathbb{N}^*$ such that $\mathbf{G}' = \mathbf{PGL}_{n,K}$, and $\mathrm{Out}(\mathbf{G}')$ is isomorphic to $\mathbb{Z}/2\mathbb{Z}$ (cf. [Di] IV.6). We may assume that $\theta \in (\mathbb{Z}/2\mathbb{Z})^r \rtimes \mathfrak{S}_r$, where $(\mathbb{Z}/2\mathbb{Z})^r$ acts on $(\mathbf{G}')^r$ via the isomorphism $\mathbb{Z}/2\mathbb{Z} \simeq \mathrm{Out}(\mathbf{G}')$ (and a splitting of $\mathrm{Aut}(\mathbf{G}') \longrightarrow \mathrm{Out}(\mathbf{G}')$) and $\mathfrak{S}_r$ acts on $(\mathbf{G}')^r$ by permuting the factors. Write $\theta = \epsilon \rtimes \tau$, with $\varepsilon \in (\mathbb{Z}/2\mathbb{Z})^r$ and $\tau \in \mathfrak{S}_r$. As in the first case, τ has to be an n-cycle. After conjugating τ by an element of $(\mathbb{Z}/2\mathbb{Z})^r \rtimes \mathfrak{S}_r$, we may assume that $\varepsilon \in \{(1, \dots, 1), (-1, 1, \dots, 1)\}$ (because $\varepsilon_1 \rtimes \tau$ and $\varepsilon_2 \rtimes \tau$ are conjugate if and only if there exists $\eta \in (\mathbb{Z}/2\mathbb{Z})^r$ such that $\varepsilon_1\varepsilon_2 = \eta\tau(\eta)$, and the image of the morphism $(\mathbb{Z}/2\mathbb{Z})^r \longrightarrow (\mathbb{Z}/2\mathbb{Z})^r$, $\eta \longmapsto \eta\tau(\eta)$ is $\{(e_1, \dots, e_r) \in (\mathbb{Z}/2\mathbb{Z})^r | e_1 \dots e_r = 1\}$). If $\theta = (1, \dots, 1) \rtimes \tau$, then $\mathbf{G} \simeq R_{K/F}\mathbf{PGL}_n$, and it is not hard to see that $\mathbf{G}$ has no nontrivial elliptic endoscopic triples. Assume that $\theta = (-1, 1, \dots, 1) \rtimes \tau$. Then θ is of order $2r$, so $[K : F] = 2r$ and $\mathbf{G} = R_{K'/F}\mathbf{PGU}(n, K)$, where K' is the subfield of K fixed by $\theta^r (= (-1, \dots, -1) \rtimes 1)$. The calculation of the elliptic endoscopic triples of $\mathbf{G}$ is done just as in proposition 2.3.1 (with the obvious changes). □

9.5 RESULTS

Proposition 9.5.1 *Let $X \in \{A, B, C\}$. Let F be a non-archimedean local field and $\mathbf{G}$ be an adjoint unramified group over F, of type X. Assume that there exists $N \in \mathbb{N}^*$ such that, for all F', E', $\mathbf{G}'$, R', and $(\mathbf{H}', {}^L\mathbf{H}', t', \xi')$ as in section 9.1, the twisted fundamental lemma is true for the unit of the Hecke algebra if $\mathbf{G}'$ is adjoint of type X, $\dim(\mathbf{G}') \leq \dim(\mathbf{G})$, and the residual characteristic of F' does not divide N.*

Then, for every finite unramified extension E of F and for all twisted endoscopic data $(\mathbf{H}, {}^L\mathbf{H}, t, \xi)$ for $R := R_{E/F}\mathbf{G}_E$ as in section 9.1, the twisted fundamental lemma is true for R and $(\mathbf{H}, {}^L\mathbf{H}, t, \xi)$ and for all the functions in the Hecke algebra.

Proof. By lemma 9.4.1, lemma 9.4.6, lemma 9.4.7, lemma 9.3.1, proposition 9.3.2, and proposition 9.2.2, the twisted fundamental lemma for $\mathbf{G}$ follows from the twisted fundamental lemma for all proper Levi subgroups of $\mathbf{G}$ (if $\mathbf{G}$ has no elliptic maximal torus, then lemma 9.4.1 is enough to see this). But, by the classification of adjoint unramified groups of type X given in lemma 9.4.7, every proper Levi subgroup of $\mathbf{G}$ is isomorphic to a group $\mathbf{G}_0 \times \mathbf{G}_1 \times \cdots \times \mathbf{G}_r$, with $\mathbf{G}_1, \ldots, \mathbf{G}_r$ of the form $R_{K/F}\mathbf{GL}_m$, where K is a finite unramified extension of F and $m \in \mathbb{N}^*$, and $\mathbf{G}_0$ adjoint unramified of type X and such that $\dim(\mathbf{G}_0) < \dim(\mathbf{G})$. If $1 \leq i \leq r$, $\mathbf{G}_i$ has no nontrivial elliptic endoscopic groups, so the twisted fundamental lemma for $\mathbf{G}_i$ follows from descent (lemma 9.4.1) and from the fundamental lemma for stable base change, which has been proved in the case of general linear groups by Arthur and Clozel ([AC], chapter I, proposition 3.1). Hence, to prove the proposition, it suffices to reason by induction on the dimension of $\mathbf{G}$. □

Corollary 9.5.2 *We use the notation of section 9.1. If $F = \mathbb{Q}_p$, $\mathbf{G}$ is one of the unitary groups $\mathbf{G}(\mathbf{U}^*(n_1) \times \cdots \times \mathbf{U}^*(n_r))$ of 2.1 and the morphism η is the morphism η_{simple} of 4.2, then the twisted fundamental lemma is true.*

Proof. As the center of $\mathbf{G}$ is connected, the corollary follows from proposition 9.5.1 and from lemma 9.4.5, so it is enough to check that the hypotheses of this lemma are satisfied. The endoscopic triple $(\mathbf{H}, s, \eta_0)$ satisfies the hypotheses of lemma 9.4.3, by the explicit description of the endoscopic triples of $\mathbf{G}$ given in proposition 2.3.1. It is obvious that $s \in \widehat{\mathbf{G}}^{\mathrm{der}}$ and $\eta(1 \rtimes \sigma) \in \widehat{\mathbf{G}}^{\mathrm{der}} \rtimes \sigma$. Finally, the center of $\mathbf{G}$ is an induced torus, so its first Galois cohomology group on any extension of F is trivial. □

The result that we really need in this book is formula $(*)$ of section 5.3. We recall this formula. Notation is still as in section 9.1, with E a field. Let Δ_η be the transfer factors for the morphism $\eta : {}^L\mathbf{H} \longrightarrow {}^L\mathbf{G}$, with the normalization given by the $\mathcal{O}_F$-structures on $\mathbf{H}$ and $\mathbf{G}$ (cf. [H1] II 7 or [Wa3] 4.6). If $\delta \in R(F)$ is θ-semisimple and $\gamma \in \mathcal{N}\delta$, Kottwitz defined in [K9] §7 p. 180 an element $\alpha_p(\gamma, \delta)$ of $X^*(Z(\mathbf{G}_\gamma)^{\Gamma_F})$ (remember that $\mathbf{G}_\gamma = \mathrm{Cent}_{\mathbf{G}}(\gamma)^0$). The result that we want to prove is the following. For every $\gamma_H \in \mathbf{H}(F)$ semisimple, for every $f \in \mathcal{H}_R$, let γ be an image of γ_H in $\mathbf{G}(F)$ (such a γ exists because $\mathbf{G}$ is quasi-split). Then

$$SO_{\gamma_H}(b_\xi(f)) = \sum_{\delta} \langle \alpha_p(\gamma, \delta), s \rangle \Delta_\eta(\gamma_H, \gamma) e(R_{\delta\theta}) O_{\delta\theta}(f), \qquad (*)$$

where the sum is taken over the set of θ-semisimple θ-conjugacy classes δ of $R(F)$ such that $\gamma \in \mathcal{N}\delta$, $R_{\delta\theta}$ is the connected component of 1 of the centralizer of $\delta\theta$ in R and $e(R_{\delta\theta})$ is the sign defined by Kottwitz in [K2].

Corollary 9.5.3 *Assume that $F = \mathbb{Q}_p$ and that $\mathbf{G}$ is one of the unitary groups $\mathbf{G}(\mathbf{U}^*(n_1) \times \cdots \times \mathbf{U}^*(n_r))$ of 2.1. Then formula $(*)$ above is true.*

Proof. If γ_H is strongly regular, then formula $(*)$ follows from corollary 9.5.2 and from corollary A.2.10 of the appendix.

The reduction from the general case to the case where γ_H is strongly regular is done in section A.3 of the appendix (see in particular proposition A.3.14). □

Remark 9.5.4 The last two corollaries are also true (with the same proof) for any group $\mathbf{G}$ with connected center and such that all its endoscopic triples satisfy the conditions of lemma 9.4.5. Examples of such groups are the symplectic groups of [M3] (cf. proposition 2.1.1 of [M3]).

Appendix

Comparison of two versions of twisted transfer factors

Robert Kottwitz

In order to stabilize the Lefschetz formula for Shimura varieties over finite fields, one needs to use twisted transfer factors for cyclic base change. Now these twisted transfer factors can be expressed in terms of standard transfer factors, the ratio between the two being given by a Galois cohomological factor involving an invariant denoted by $\operatorname{inv}(\gamma, \delta)$ in [KS]. However, in the stabilization of the Lefschetz formula it is more natural to use a different invariant $\alpha(\gamma, \delta)$. The purpose of this appendix is to relate the invariants $\operatorname{inv}(\gamma, \delta)$ and $\alpha(\gamma, \delta)$ (see theorem A.2.5), and then to justify the use made in [K9] of transfer factors

$$\Delta_0(\gamma_H, \delta) = \Delta_0(\gamma_H, \gamma)\langle \alpha(\gamma, \delta), s\rangle^{-1},$$

first in the case when γ_H is strongly G-regular semisimple (see corollary A.2.10) and then in the more general case in which γ_H is assumed only to be (G, H)-regular (see proposition A.3.14, where, however, the derived group of G is assumed to be simply connected).

I would like to thank Sophie Morel for her very helpful comments on a first version of this appendix.

A.1 COMPARISON OF $\Delta_0(\gamma_H, \delta)$ AND $\Delta_0(\gamma_H, \gamma)$

In the case of cyclic base change the twisted transfer factors $\Delta_0(\gamma_H, \delta)$ of [KS] are closely related to the standard transfer factors $\Delta_0(\gamma_H, \gamma)$ of [LS1]. This fact, first observed by Shelstad [Sh2] in the case of base change for $\mathbb{C}/\mathbb{R}$, was one of several guiding principles used to arrive at the general twisted transfer factors defined in [KS]. Thus there is nothing really new in this section. After reviewing some basic notions, we prove proposition A.1.10, which gives the precise relationship between $\Delta_0(\gamma_H, \delta)$ and $\Delta_0(\gamma_H, \gamma)$.

A.1.1 Set-up

We consider a finite cyclic extension E/F of local fields of characteristic zero. We put $d := [E : F]$ and choose a generator σ of $\operatorname{Gal}(E/F)$. In addition we choose an algebraic closure $\overline{F}$ of F that contains E. We write Γ for the absolute Galois group $\operatorname{Gal}(\overline{F}/F)$ and W_F for the absolute Weil group of F. There is then a canonical homomorphism $W_F \to \Gamma$ that will go unnamed.

We also consider a quasi-split connected reductive group G over F. Put $R_G :=$ $\mathrm{Res}_{E/F}(G_E)$, where G_E is the E-group obtained from G by extension of scalars and $\mathrm{Res}_{E/F}$ denotes Weil's restriction of scalars. As usual there is a natural automorphism θ of R_G inducing σ on $G(E)$ via the canonical identification $R_G(F) =$ $G(E)$.

A.1.2 Description of $\hat{R}_G$

For any Γ-group A (i.e., a group A equipped with an action of Γ) we obtain by restriction a Γ_E-group A_E (with Γ_E denoting the subgroup $\mathrm{Gal}(\overline{F}/E)$ of Γ), and we write $I(A)$ for the Γ-group obtained from A_E by induction from Γ_E to Γ.

Then $I(A)$ has the following description in terms of A. Let J denote the set of embeddings of E in $\overline{F}$ over F, with j_0 denoting the inclusion $E \subset \overline{F}$. The group Γ acts on the left of J by $\tau j := \tau \circ j$ (for $\tau \in \Gamma$, $j \in J$), and the group $\mathrm{Gal}(E/F)$ acts on the right of J by $j\sigma^i := j \circ \sigma^i$. An element $x \in I(A)$ is then a map $j \mapsto x_j$ from J to A. An element $\tau \in \Gamma$ acts on $x \in I(A)$ by

$$(\tau x)_j := \tau(x_{\tau^{-1}j}).$$

There is a *right* action of $\mathrm{Gal}(E/F)$ on the Γ-group $I(A)$ given by

$$(x\sigma^i)_j := x_{j\sigma^{-i}}.$$

We have $\hat{R}_G = I(\hat{G})$ as Γ-group. Bearing in mind that for any automorphisms θ_1, θ_2 of a connected reductive group one has the rule $\widehat{\theta_1\theta_2} = \hat{\theta}_2\hat{\theta}_1$, we see that the natural left action of $\mathrm{Gal}(E/F)$ on R_G is converted into a *right* action of $\mathrm{Gal}(E/F)$ on $\hat{R}_G$, and hence that the automorphism $\hat{\theta}$ of $\hat{R}_G$ is given by

$$(\hat{\theta}x)_j = x_{j\sigma^{-1}}.$$

There is an obvious embedding

$$A \hookrightarrow I(A)$$

of Γ-groups, sending $a \in A$ to the constant map $J \to A$ with value a, and this map identifies A with the group of fixed points of $\mathrm{Gal}(E/F)$ on $I(A)$. In particular we get

$$i : \hat{G} \simeq (\hat{R}_G)^{\hat{\theta}} \hookrightarrow \hat{R}_G,$$

which we extend to an embedding

$$i : {}^LG \to {}^LR_G$$

by mapping $g\tau$ to $i(g)\tau$ (for $g \in \hat{G}$, $\tau \in W_F$). Note that $i({}^LG)$ is the group of fixed points of the automorphism ${}^L\theta$ of ${}^LR_G$ defined by

$${}^L\theta(x\tau) := \hat{\theta}(x)\tau$$

for $x \in \hat{R}_G$, $\tau \in W_F$.

A.1.3 Endoscopic groups and twisted endoscopic groups

Let (H, s, η) be an endoscopic datum for G. Thus $s \in Z(\hat{H})^\Gamma$ and $\eta : {}^L H \to {}^L G$ is an L-homomorphism that restricts to an isomorphism $\hat{H} \to (\hat{G}_{\eta(s)})^\circ$. When the derived group of G is not simply connected, we should actually allow for a z-extension of H, as in [LS1] and [KS], but since this wrinkle does not perturb the arguments below in any nontrivial way, we prefer to ignore it.

Following Shelstad [Sh2], we now explain how to regard H as a twisted endoscopic group for (R_G, θ). Let $\mathcal{Z}$ denote the centralizer of $i\eta(\hat{H})$ in $\hat{R}_G$. Since the centralizer of $\eta(\hat{H})$ in $\hat{G}$ is $\eta(Z(\hat{H}))$, we see that $\mathcal{Z}$ is the subgroup of $\hat{R}_G$ consisting of all maps $J \to \eta(Z(\hat{H}))$. Thus, as a group, $\mathcal{Z}$ can be identified with $I(Z(\hat{H}))$. Since $Z(\hat{H})$ is a Γ-group, so too is $I(Z(\hat{H})) = \mathcal{Z}$, but the embedding $\mathcal{Z} \hookrightarrow \hat{R}_G$ is *not* Γ-equivariant. The subgroup $\mathcal{Z}$ is however stable under $\hat{\theta}$.

Using $s \in Z(\hat{H})^\Gamma$, we now define an element $\tilde{s} \in \mathcal{Z}$ by the rule

$$\tilde{s}_j := \begin{cases} s & \text{if } j = j_0, \\ 1 & \text{if } j \neq j_0. \end{cases} \qquad \text{(A.1.3.1)}$$

Thus $\tilde{s}$ maps to s under the norm map $\mathcal{Z} \to Z(\hat{H})$ (given by $x \mapsto \prod_{j \in J} x_j$). It is easy to see that the composition

$$\hat{H} \xrightarrow{\eta} \hat{G} \xrightarrow{i} \hat{R}_G$$

identifies $\hat{H}$ with the identity component of the $\hat{\theta}$-centralizer of $\tilde{s}$ in $\hat{R}_G$.

A.1.4 Allowed embeddings

We now have part of what is needed to view H as a twisted endoscopic group for (R_G, θ), but in addition to $\tilde{s}$ we need suitable $\tilde{\eta} : \mathcal{H} \to {}^L R_G$. In the situation of interest in the next section, we may even take $\mathcal{H} = {}^L H$, so this is the only case we will discuss further.

When $\mathcal{H} = {}^L H$, in order to get a twisted endoscopic datum $(H, \tilde{s}, \tilde{\eta})$ for (R_G, θ), we need for $\tilde{\eta} : {}^L H \to {}^L R_G$ to be one of Shelstad's *allowed embeddings* [Sh2], which is to say that $\tilde{\eta}$, $i\eta$ must have the same restriction to $\hat{H}$, and that $\tilde{\eta}({}^L H)$ must be contained in the group of fixed points of the automorphism $\mathrm{Int}(\tilde{s}) \circ {}^L\theta$ of ${}^L R_G$.

In subsection A.2.6 we will see that, when E/F is an unramified extension of p-adic fields and σ is the Frobenius automorphism, there exists a canonical allowed embedding $\tilde{\eta}$ determined by $\tilde{s}$. In this section, however, we work with an arbitrary allowed embedding.

We are going to use $\tilde{\eta}$ to produce a 1-cocycle of W_F in $\mathcal{Z} \xrightarrow{1-\hat{\theta}} \mathcal{Z}$. For this we need to compare (as in [KS]) $\tilde{\eta}$ to the L-homomorphism

$$i\eta : {}^L H \to {}^L G \to {}^L R_G.$$

Since $\tilde{\eta}$ and $i\eta$ agree on $\hat{H}$, there is a unique 1-cocycle a of W_F in $\mathcal{Z}$ such that

$$\tilde{\eta}(\tau) = a_\tau i\eta(\tau)$$

for all $\tau \in W_F$. The pair $(a^{-1}, \tilde{s})$ is a 1-cocycle of W_F in $\mathcal{Z} \xrightarrow{1-\hat{\theta}} \mathcal{Z}$. Here one must not forget that the Γ-action on $\mathcal{Z}$ comes from viewing it as $I(Z(\hat{H}))$. In fact the map $\tilde{\eta} \mapsto a$ sets up a bijection between allowed embeddings $\tilde{\eta}$ and 1-cocycles a of W_F in $\mathcal{Z}$ such that $(a^{-1}, \tilde{s})$ is a 1-cocycle in $\mathcal{Z} \xrightarrow{1-\hat{\theta}} \mathcal{Z}$.

A.1.5 Canonical twisted and standard transfer factors

We now choose an F-splitting [LS1], p. 224 for our quasi-split group G. This choice determines canonical transfer factors $\Delta_0(\gamma_H, \gamma)$ (see [LS1], p. 248).

Our F-splitting of G can also be viewed as a σ-invariant E-splitting, and therefore gives rise to an (F, θ)-splitting [KS], p. 61 of R_G, which then determines canonical twisted transfer factors $\Delta_0(\gamma_H, \delta)$ (see [KS], p. 62). Our goal is to express $\Delta_0(\gamma_H, \delta)$ as the product of $\Delta_0(\gamma_H, \gamma)$ and a simple Galois cohomological factor involving an invariant $\mathrm{inv}(\gamma, \delta)$ that we are now going to discuss.

It may be useful to recall (though it will play no role in this appendix) that when G is unramified, and we fix an $\mathcal{O}$-structure on G for which $G(\mathcal{O})$ is a hyperspecial maximal compact subgroup of $G(F)$, there is an obvious notion of $\mathcal{O}$-splitting, namely, an F-splitting that is defined over $\mathcal{O}$ and reduces modulo the maximal ideal in $\mathcal{O}$ to a splitting for the special fiber of G. When such an $\mathcal{O}$-splitting is used, and H is also unramified, the transfer factors $\Delta_0(\gamma_H, \gamma)$ so obtained are the ones needed for the fundamental lemma for the spherical Hecke algebra on G obtained from $G(\mathcal{O})$. In the case that E/F is unramified, the same is true for the twisted fundamental lemma for the spherical Hecke algebra for $G(E)$ obtained from $G(\mathcal{O}_E)$.

A.1.6 Definition of the invariant inv(γ, δ)

We consider a maximal F-torus T_H of H and an admissible isomorphism $T_H \simeq T$ between T_H and a maximal F-torus T of G. We consider γ_H in $T_H(F)$ whose image γ in $T(F)$ is strongly G-regular. The standard transfer factor $\Delta_0(\gamma_H, \gamma)$ is then defined. We also consider $\delta \in R_G(F) = G(E)$ whose abstract norm ([KS], 3.2) is the stable conjugacy class of γ. The twisted transfer factor $\Delta_0(\gamma_H, \delta)$ is then defined.

The position of δ relative to γ is measured by

$$\mathrm{inv}(\gamma, \delta) \in H^1(F, R_T \xrightarrow{1-\theta} R_T),$$

whose definition ([KS], p. 63) we now recall. Our assumption that the abstract norm of δ is γ does *not* imply that δ is stably θ-conjugate to an element in the F-points of the θ-stable maximal F-torus R_T of R_G. It does, however, imply that there exists $g \in R_G(\overline{F})$ such that $g(N\delta)g^{-1} = \gamma$, where $N : R_G \to R_G$ is the F-morphism $x \mapsto x\theta(x)\theta^2(x) \cdots \theta^{d-1}(x)$, and γ is viewed as an element of $R_G(F) = G(E)$ via the obvious inclusion $G(F) \subset G(E)$. Put $\delta' := g\delta\theta(g)^{-1}$ and define a 1-cocycle t of Γ by $t_\tau := g\tau(g)^{-1}$ (for $\tau \in \Gamma$). Note that the strong regularity of γ implies that its centralizer in R_G is R_T.

Lemma A.1.7 *The pair (t^{-1}, δ') is a 1-cocycle of Γ in $R_T \xrightarrow{1-\theta} R_T$.*

Proof. We first check that $\delta' \in R_T(\overline{F})$. Observe that $N\delta' = \gamma$. Therefore

$$\gamma = \theta(\gamma) = \theta(N\delta') = (\delta')^{-1}(N\delta')\delta' = (\delta')^{-1}\gamma\delta',$$

which shows that δ' centralizes γ and hence lies in R_T. We note for later use that the θ-centralizer of δ' in R_G is T, viewed as the subtorus of θ-fixed points in R_T.

Next, a short calculation using the definitions of δ' and t_τ shows that

$$(\delta')^{-1}t_\tau\tau(\delta') = \theta(t_\tau). \tag{A.1.7.1}$$

To see that $t_\tau \in R_T(\overline{F})$, we begin by noting that (A.1.7.1) says that t_τ θ-conjugates $\tau(\delta')$ into δ'. Now

$$N(\tau(\delta')) = \tau(N\delta') = \tau(\gamma) = \gamma = N(\delta'),$$

showing that the two elements $\tau(\delta')$ and δ' in R_T have the same image under the norm homomorphism $N : R_T \to R_T$, and hence that there exists $u \in R_T(\overline{F})$ that θ-conjugates δ' into $\tau(\delta')$. Thus $t_\tau u$ lies in the θ-centralizer (namely $T = R_T^\theta$) of δ', which implies that t_τ lies in $R_T(\overline{F})$.

The 1-cocycle condition for (t^{-1}, δ') is none other than (A.1.7.1), and the proof is complete. □

Definition A.1.8 We define $\mathrm{inv}(\gamma, \delta)$ to be the class in $H^1(F, R_T \xrightarrow{1-\theta} R_T)$ of the 1-cocycle (t^{-1}, δ').

A.1.9 Main proposition

The last thing to do before stating proposition A.1.10 is to relate $\mathcal{Z}$ to $\hat{R}_T$. This is very easy. Since T_H is a maximal torus in H, there is a canonical Γ-equivariant embedding $Z(\hat{H}) \hookrightarrow \hat{T}_H$. Our admissible isomorphism $T_H \simeq T$ yields $\hat{T}_H \simeq \hat{T}$, so that we end up with a Γ-equivariant embedding $Z(\hat{H}) \hookrightarrow \hat{T}$, to which we may apply our restriction-induction functor I, obtaining a Γ-equivariant embedding

$$k : \mathcal{Z} \hookrightarrow \hat{R}_T,$$

which is compatible with the $\hat{\theta}$-actions as well. We then obtain an induced homomorphism

$$H^1(W_F, \mathcal{Z} \xrightarrow{1-\hat{\theta}} \mathcal{Z}) \to H^1(W_F, \hat{R}_T \xrightarrow{1-\hat{\theta}} \hat{R}_T). \tag{A.1.9.1}$$

Near the end of subsection A.1.4 we used $\tilde{s}$,$\tilde{\eta}$ to produce a 1-cocycle $(a^{-1}, \tilde{s})$ of W_F in $\mathcal{Z} \xrightarrow{1-\hat{\theta}} \mathcal{Z}$, to which we may apply the homomorphism k, obtaining a 1-cocycle in $\hat{R}_T \xrightarrow{1-\hat{\theta}} \hat{R}_T$, which, since k is injective, we may as well continue to denote simply by $(a^{-1}, \tilde{s})$. Recall from appendix A of [KS] that there is a $\mathbb{C}^\times$-valued pairing $\langle\cdot,\cdot\rangle$ between $H^1(F, R_T \xrightarrow{1-\theta} R_T)$ and $H^1(W_F, \hat{R}_T \xrightarrow{1-\hat{\theta}} \hat{R}_T)$. Thus it makes sense to form the complex number

$$\langle \mathrm{inv}(\gamma, \delta), (a^{-1}, \tilde{s})\rangle.$$

Proposition A.1.10 *There is an equality*

$$\Delta_0(\gamma_H, \delta) = \Delta_0(\gamma_H, \gamma)\langle \mathrm{inv}(\gamma, \delta), (a^{-1}, \tilde{s})\rangle.$$

Proof. Since the restricted root system ([KS], 1.3) of R_T can be identified with the root system of T, we may use the same a-data and χ-data for T and R_T. When this is done, one has

$$\Delta_I(\gamma_H, \delta) = \Delta_I(\gamma_H, \gamma),$$
$$\Delta_{II}(\gamma_H, \delta) = \Delta_{II}(\gamma_H, \gamma),$$
$$\Delta_{IV}(\gamma_H, \delta) = \Delta_{IV}(\gamma_H, \gamma).$$

It remains only to prove that

$$\Delta_{III}(\gamma_H, \delta) = \Delta_{III}(\gamma_H, \gamma)\langle \mathrm{inv}(\gamma, \delta), (a^{-1}, \tilde{s})\rangle.$$

To do so we must recall how Δ_{III} is defined. We use (see [LS1]) the chosen χ-data to obtain embeddings

$$\xi_1 : {}^L T \hookrightarrow {}^L G,$$
$$\xi_2 : {}^L T \hookrightarrow {}^L H.$$

Replacing ξ_1 by a conjugate under $\hat{G}$, we may assume that $\eta\xi_2$ and ξ_1 agree on $\hat{T}$, and then there exists a unique 1-cocycle b of W_F in $\hat{T}$ such that

$$(\eta\xi_2)(\tau) = \xi_1(b_\tau \tau)$$

for all $\tau \in W_F$. We then have (see [LS1], p. 246)

$$\Delta_{III}(\gamma_H, \gamma) = \langle \gamma, b\rangle,$$

where $\langle\cdot,\cdot\rangle$ now denotes the Langlands pairing between $T(F)$ and $H^1(W_F, \hat{T})$.

Similarly we have two embeddings $i\xi_1, \tilde{\eta}\xi_2 : {}^L T \hookrightarrow {}^L R_G$ that agree on $\hat{T}$, and therefore there exists a unique 1-cocycle c of W_F in $\hat{R}_T$ (which arises here because it is the centralizer in $\hat{R}_G$ of $(i\xi_1)(\hat{T})$) such that

$$(\tilde{\eta}\xi_2)(\tau) = c_\tau\big((i\xi_1)(\tau)\big)$$

for all $\tau \in W_F$. Then $(c^{-1}, \tilde{s})$ is a 1-cocycle of W_F in $\hat{R}_T \xrightarrow{1-\hat{\theta}} \hat{R}_T$, and (see pp. 40, 63 of [KS])

$$\Delta_{III}(\gamma_H, \delta) = \langle \mathrm{inv}(\gamma, \delta), (c^{-1}, \tilde{s})\rangle.$$

It is clear from the definitions that the 1-cocycles a, b, c satisfy the equality $c = ab$, in which we use $k : \mathcal{Z} \hookrightarrow \hat{R}_T$ and $\hat{T} = (\hat{R}_T)^{\hat{\theta}} \hookrightarrow \hat{R}_T$ to view a, b as 1-cocycles in $\hat{R}_T$. Therefore

$$(c^{-1}, \tilde{s}) = (a^{-1}, \tilde{s})(b^{-1}, 1),$$

which shows that

$$\Delta_{III}(\gamma_H, \delta) = \langle \mathrm{inv}(\gamma, \delta), (a^{-1}, \tilde{s})\rangle \, \langle \mathrm{inv}(\gamma, \delta), (b, 1)\rangle^{-1}.$$

It remains only to observe that $\langle \mathrm{inv}(\gamma, \delta), (b, 1)\rangle^{-1} = \langle \gamma, b\rangle$, a consequence of the first part of lemma A.1.12, to be proved next. Here we use the obvious fact that the image of $\mathrm{inv}(\gamma, \delta)$ under $H^1(F, R_T \xrightarrow{1-\theta} R_T) \to T(F)$ is γ. □

A.1.11 Compatibility properties for the pairing $\langle\cdot,\cdot\rangle$

In this subsection we consider a homomorphism $f : T \to U$ of F-tori. We follow all the conventions of appendix A in [KS] concerning $H^1(F, T \to U)$ and $H^1(W_F, \hat{U} \to \hat{T})$. We denote by K the kernel of f and by C the cokernel of f. Of course C is necessarily a torus, and we now assume that K is also a torus. Dual to the exact sequence

$$1 \to K \to T \to U \to C \to 1$$

is the exact sequence

$$1 \to \hat{C} \to \hat{U} \to \hat{T} \to \hat{K} \to 1,$$

which we use to identify $\hat{C}$ with $\ker \hat{f}$ and $\hat{K}$ with $\operatorname{cok} \hat{f}$. From [KS], p. 119 we obtain two long exact sequences, the relevant portions of which are

$$H^1(F, K) \xrightarrow{i'} H^1(F, T \to U) \xrightarrow{j'} C(F),$$
$$H^1(W_F, \hat{C}) \xrightarrow{\hat{i}'} H^1(W_F, \hat{U} \to \hat{T}) \xrightarrow{\hat{j}'} \hat{K}^\Gamma.$$

The following lemma concerns the compatibility of these two exact sequences with the pairing [KS] between $H^1(F, T \to U)$ and $H^1(W_F, \hat{U} \to \hat{T})$.

Lemma A.1.12 *The pairing $\langle\cdot,\cdot\rangle$ satisfies the following two compatibilities.*

1. *Let $x \in H^1(F, T \to U)$ and $c \in H^1(W_F, \hat{C})$. Then*
$$\langle x, \hat{i}'c\rangle = \langle j'x, c\rangle^{-1},$$
where the pairing on the right is the Langlands pairing between $C(F)$ and $H^1(W_F, \hat{C})$.
2. *Let $k \in H^1(F, K)$ and $\hat{x} \in H^1(W_F, \hat{U} \to \hat{T})$. Then*
$$\langle i'k, \hat{x}\rangle = \langle k, \hat{j}'\hat{x}\rangle,$$
where the pairing on the right is the Tate-Nakayama pairing between the groups $H^1(F, K)$ and $\hat{K}^\Gamma$.

Proof. Using the fact that the pairing in [KS] is functorial in $T \to U$ (apply this functoriality to $(K \to 1) \to (T \to U)$ and $(T \to U) \to (1 \to C)$), we reduce the lemma to the case in which one of T,U is trivial, which can then be handled using the compatibilities (A.3.13) and (A.3.14) of [KS]. □

A.2 RELATION BETWEEN inv (γ, δ) AND $\alpha(\gamma, \delta)$

We retain all the assumptions and notation of the previous section. In particular, we have the invariant $\mathrm{inv}(\gamma, \delta) \in H^1(F, R_T \xrightarrow{1-\theta} R_T)$. Throughout this section we assume that E/F is an unramified extension of p-adic fields, and that σ is the Frobenius automorphism of E/F. In this situation there is another invariant measuring the position of δ relative to γ. This invariant arose naturally in [K9] in the course of stabilizing the Lefschetz formula for Shimura varieties over finite fields. This second invariant, denoted $\alpha(\gamma, \delta)$, lies in the group $B(T)$ introduced in [K5] and studied further in [K12].

The goal of this section is to compare $\mathrm{inv}(\gamma, \delta)$ and $\alpha(\gamma, \delta)$ and then to rewrite the ratio of $\Delta_0(\gamma_H, \delta)$ and $\Delta_0(\gamma_H, \gamma)$ in terms of $\alpha(\gamma, \delta)$ rather than $\mathrm{inv}(\gamma, \delta)$. Since the two invariants lie in different groups, the reader may be wondering what it means to compare them. Note, however, that $H^1(F, T)$ injects naturally into both $H^1(F, R_T \xrightarrow{1-\theta} R_T)$ and $B(T)$, which suggests that we need a group A and a commutative diagram of the form

$$\begin{array}{ccc} H^1(F,T) & \longrightarrow & H^1(F, R_T \xrightarrow{1-\theta} R_T) \\ \downarrow & & \downarrow \\ B(T) & \longrightarrow & A \end{array}$$

in which the two new arrows are injective. It should seem plausible that A ought to be a group $B(R_T \xrightarrow{1-\theta} R_T)$ bearing the same relation to $H^1(F, R_T \xrightarrow{1-\theta} R_T)$ as $B(T)$ does to $H^1(F, T)$.

Such a group has already been introduced and studied in sections 9–13 of [K12]. The rest of this section will lean heavily on those sections of [K12], whose *raison d'être* is precisely this application to twisted transfer factors.

This section begins with a review of the relevant material from [K12], and then recalls the definition of $\alpha(\gamma, \delta)$. Next comes a theorem comparing $\mathrm{inv}(\gamma, \delta)$ and $\alpha(\gamma, \delta)$. The two invariants do *not* become equal in $B(R_T \xrightarrow{1-\theta} R_T)$; the relation between them is more subtle, as we will see in theorem A.2.5. Finally, we express the ratio of twisted to standard transfer factors in terms of $\alpha(\gamma, \delta)$.

A.2.1 Review of $B(T \to U)$

Let L denote the completion of the maximal unramified extension F^{un} of F in $\overline{F}$. We use σ to denote the Frobenius automorphism of L/F. We are already using σ to denote the Frobenius automorphism of E/F, but since $E \subset F^{\mathrm{un}} \subset L$ and σ on L restricts to σ on E, this abuse of notation should lead to no confusion.

In this subsection $f : T \to U$ will denote any homomorphism of F-tori. We then have the group ([K12], 12.2)

$$B(T \to U) := H^1(\langle\sigma\rangle, T(L) \to U(L)).$$

Elements of $B(T \to U)$ can be represented by simplified 1-cocycles ([K12], 12.1) (t, u), where $t \in T(L)$, $u \in U(L)$ satisfy the cocycle condition $f(t) = u^{-1}\sigma(u)$. Simplified 1-coboundaries are pairs $(t^{-1}\sigma(t), f(t))$ with $t \in T(L)$.

In [K12], 11.2 a canonical isomorphism

$$\mathrm{Hom}_{\mathrm{cont}}(B(T \to U), \mathbb{C}^\times) \simeq H^1(W_F, \hat{U} \to \hat{T})$$

is constructed; here we implicitly used the canonical isomorphism ([K12], 12.2) between $B(T \to U)$ and $\mathbf{B}(T \to U)$, but as we have no further use for $\mathbf{B}(T \to U)$, we will not review its definition. In particular we have a $\mathbb{C}^\times$-valued pairing between

$B(T \to U)$ and $H^1(W_F, \hat{U} \to \hat{T})$. Moreover there is a natural injection ([K12], 9.4)

$$H^1(F, T \to U) \to B(T \to U). \tag{A.2.1.1}$$

Our pairing restricts to one between $H^1(F, T \to U)$ and $H^1(W_F, \hat{U} \to \hat{T})$, and this restricted pairing agrees ([K12], 11.1) with the one in appendix A of [KS].

Now we come to the material in [K12], §13, which concerns the case in which our homomorphism of tori is of the very special form $R_T \xrightarrow{1-\theta} R_T$ for some F-torus T. In this case it is shown that the exact sequence ([K12], (13.3.2))

$$1 \to B(T) \to B(R_T \xrightarrow{1-\theta} R_T) \to T(F) \to 1$$

has a canonical splitting, so that there is a canonical direct product decomposition

$$B(R_T \xrightarrow{1-\theta} R_T) = B(T) \times T(F). \tag{A.2.1.2}$$

Similarly it is shown that the exact sequence ([K12], (13.3.8))

$$1 \to H^1(W_F, \hat{T}) \to H^1(W_F, \hat{R}_T \xrightarrow{1-\hat{\theta}} \hat{R}_T) \to \hat{T}^\Gamma \to 1$$

has a canonical splitting, so that there is a canonical direct product decomposition

$$H^1(W_F, \hat{R}_T \xrightarrow{1-\hat{\theta}} \hat{R}_T) = \hat{T}^\Gamma \times H^1(W_F, \hat{T}). \tag{A.2.1.3}$$

Let $x \in B(R_T \xrightarrow{1-\theta} R_T)$ and $\hat{x} \in H^1(W_F, \hat{R}_T \xrightarrow{1-\hat{\theta}} \hat{R}_T)$. As we have seen, we may then pair x with $\hat{x}$, obtaining $\langle x, \hat{x}\rangle \in \mathbb{C}^\times$. Using (A.2.1.2) and (A.2.1.3), we decompose x as $(x_1, x_2) \in B(T) \times T(F)$, and $\hat{x}$ as $(\hat{x}_1, \hat{x}_2) \in \hat{T}^\Gamma \times H^1(W_F, \hat{T})$. We also have the pairing $\langle x_1, \hat{x}_1\rangle$ coming from the canonical isomorphisms $B(T) = X_*(T)_\Gamma = X^*(\hat{T}^\Gamma)$ of [K5, K12], as well as the Langlands pairing $\langle x_2, \hat{x}_2\rangle$.

Lemma A.2.2 *There is an equality*

$$\langle x, \hat{x}\rangle = \langle x_1, \hat{x}_1\rangle\langle x_2, \hat{x}_2\rangle^{-1}.$$

Proof. This follows from [K12], proposition 13.4 together with the obvious analog of lemma A.1.12 with $H^1(F, T \to U)$ replaced by $B(T \to U)$. □

A.2.3 Review of $\alpha(\gamma, \delta)$

Our assumptions on γ, δ are the same as in A.1.6. However the group R_G will play no role in the definition of $\alpha(\gamma, \delta)$, so we prefer to view δ as an element of $G(E)$ such that $N\delta = \delta\sigma(\delta)\cdots\sigma^{d-1}(\delta)$ is conjugate in $G(\overline{F})$ to our strongly regular element $\gamma \in T(F)$. Then, since $H^1(F^{\mathrm{un}}, T)$ is trivial, there exists $c \in G(F^{\mathrm{un}}) \subset G(L)$ such that

$$c\gamma c^{-1} = N\delta. \tag{A.2.3.1}$$

Now define $b \in G(F^{\mathrm{un}})$ by $b := c^{-1}\delta\sigma(c)$. Applying σ to (A.2.3.1), we find that b centralizes γ, hence lies in $T(F^{\mathrm{un}}) \subset T(L)$. Making a different choice of c replaces b by a σ-conjugate under $T(F^{\mathrm{un}})$. Thus it makes sense to define $\alpha(\gamma, \delta) \in B(T)$ as the σ-conjugacy class of b.

A.2.4 Precise relation between inv(γ,δ) and $\alpha(\gamma,\delta)$

Now that we have reviewed $\alpha(\gamma,\delta)$, we can prove one of the main results of this appendix. We denote by $\mathrm{inv}^B(\gamma,\delta)$ the image of $\mathrm{inv}(\gamma,\delta)$ under the canonical injection (A.2.1.1)

$$H^1(F, R_T \xrightarrow{1-\theta} R_T) \hookrightarrow B(R_T \xrightarrow{1-\theta} R_T).$$

Theorem A.2.5 *Under the canonical isomorphism*

$$B(R_T \xrightarrow{1-\theta} R_T) = B(T) \times T(F),$$

the element $\mathrm{inv}^B(\gamma,\delta)$ *goes over to the pair* $(\alpha(\gamma,\delta)^{-1},\gamma)$.

Proof. As usual when working with cocycles, one has to make various choices. In this proof, unless the choices are made carefully, $\mathrm{inv}^B(\gamma,\delta)$ will differ from $(\alpha(\gamma,\delta)^{-1},\gamma)$ by a complicated 1-cocycle in $R_T \xrightarrow{1-\theta} R_T$ that one would then have to recognize as a 1-coboundary. We will take care that this does not happen.

We have already discussed R_G, $\mathrm{inv}(\gamma,\delta)$, and $\alpha(\gamma,\delta)$. In particular we have chosen $c \in G(F^{\mathrm{un}})$ such that $c\gamma c^{-1} = N\delta$ and used it to form the element $b = c^{-1}\delta\sigma(c) \in T(L)$ representing $\alpha(\gamma,\delta) \in B(T)$. In order to define $\mathrm{inv}(\gamma,\delta)$ we need to choose an element $g \in R_G(\overline{F})$ such that $g(N\delta)g^{-1} = \gamma$. The best choice for g is by no means the most obvious one. The one we choose lies in $R_G(F^{\mathrm{un}})$ and is given by a certain function $J \to G(F^{\mathrm{un}})$.

Recall that J is the set of F-embeddings of E in $\overline{F}$, and that $j_0 \in J$ is the inclusion $E \subset \overline{F}$. We now identify J with $\mathbb{Z}/d\mathbb{Z}$, with $i \in \mathbb{Z}/d\mathbb{Z}$ corresponding to the embedding $e \mapsto \sigma^i e$ of E in $\overline{F}$. Thus $R_G(F^{\mathrm{un}})$ becomes identified with the set of functions $i \mapsto x_i$ from $\mathbb{Z}/d\mathbb{Z}$ to $G(F^{\mathrm{un}})$, and the same is true with L in place of F^{un}. The Galois action of σ on $x \in R_G(F^{\mathrm{un}})$ is then given by

$$(\sigma x)_i = \sigma(x_{i-1}),$$

while the effect on x of the automorphism θ of R_G is given by

$$(\theta x)_i = x_{i+1}.$$

For $i = 0, 1, \dots, d-1$ we put $g_i := c^{-1}\delta\sigma(\delta)\sigma^2(\delta)\cdots\sigma^{i-1}(\delta)$. In particular, $g_0 = c^{-1}$. Then $i \mapsto g_i$ is the desired element $g \in R_G(F^{\mathrm{un}})$ satisfying $g(N\delta)g^{-1} = \gamma$. We leave this computation to the reader, remarking only that γ corresponds to the element $i \mapsto \gamma$ in $R_G(F^{\mathrm{un}})$, while δ corresponds to $i \mapsto \sigma^i(\delta)$, so that $N\delta$ corresponds to $i \mapsto \sigma^i(\delta)\sigma^{i+1}(\delta)\cdots\sigma^{i+d-1}(\delta)$.

Since our chosen g lies in $R_G(F^{\mathrm{un}})$, the 1-cocycle $t_\tau = g\tau(g)^{-1} \in R_T(F^{\mathrm{un}})$ is unramified, which is to say that t_τ depends only on the restriction of τ to F^{un}. Thus we get a well-defined element $u \in R_T(F^{\mathrm{un}})$ by putting $u := t_\tau^{-1}$ for any $\tau \in \Gamma$ such that τ restricts to σ on F^{un}. It is then clear from the definitions that $\mathrm{inv}^B(\gamma,\delta)$ is represented by the simplified 1-cocycle $(u,\delta') \in R_T(L) \times R_T(L)$. Here, as before, $\delta' = g\delta\theta(g)^{-1}$.

The element (u,δ') can be written as the product of two simplified 1-cocycles (u',t'), (u'',t'') in $R_T(L) \times R_T(L)$. Of course elements in $R_T(L)$ are given by functions $\mathbb{Z}/d\mathbb{Z} \to T(L)$. We take u' to be the constant function with value b^{-1}. We take t' to be the identity. We take u'' to be the function given by

$$u''_i = \begin{cases} \gamma & \text{if } i = 0 \text{ in } \mathbb{Z}/d\mathbb{Z}, \\ 1 & \text{otherwise.} \end{cases}$$

Finally, we take t'' to be the function given by

$$t''_i = \begin{cases} \gamma & \text{if } i = -1 \text{ in } \mathbb{Z}/d\mathbb{Z}, \\ 1 & \text{otherwise.} \end{cases}$$

It is straightforward to verify that $u = u'u''$ and $\delta' = t't''$. Since $t' = 1$ and u' is fixed by θ, it is clear that (u', t') is a 1-cocycle. So too is (u'', t''), since its product with (u', t') is a 1-cocycle.

Since b represents $\alpha(\gamma, \delta) \in B(T)$, and since the image of b^{-1} under $T = R_T^\theta \hookrightarrow R_T$ is u', it is clear that $(u', t') = (b^{-1}, 1)$ represents the image of $\alpha(\gamma, \delta)^{-1}$ under the canonical injection $B(T) \hookrightarrow B(R_T \xrightarrow{1-\theta} R_T)$.

It remains only to verify that (u'', t'') represents the image of γ under the canonical splitting of the natural surjection

$$B(R_T \xrightarrow{1-\theta} R_T) \twoheadrightarrow T(F).$$

Since this surjection sends (u'', t'') to $t''_0 t''_1 \cdots t''_{d-1} = \gamma$, we just need to check that the class of (u'', t'') lies in the subgroup of $B(R_T \xrightarrow{1-\theta} R_T)$ complementary to $B(T)$ that is described in [K12], p. 326. This is clear, since (u'', t'') has the form $(\sigma(x), x)$ for $x = t''$, and every value of $i \mapsto t''_i$ lies in $T(L)^{\langle\sigma\rangle} = T(F)$. □

A.2.6 More about allowed embeddings

As mentioned before, now that we are taking E/F to be an unramified extension of p-adic fields, and σ to be the Frobenius automorphism of E/F, there is a canonical choice of allowed embedding $\tilde{\eta} : {}^L H \to {}^L R_G$ determined by $\tilde{s}$. As we have seen, giving $\tilde{\eta}$ is the same as giving a 1-cocycle a of W_F in $\mathcal{Z}$ such that $(a^{-1}, \tilde{s})$ is a 1-cocycle of W_F in $\mathcal{Z} \xrightarrow{1-\hat{\theta}} \mathcal{Z}$.

Before describing the canonical choice for the 1-cocycle a, we need to recall the exact sequence

$$1 \to I \to W_F \to \langle\sigma\rangle \to 1,$$

where I denotes the inertia subgroup of Γ. By an *unramified* 1-cocycle of W_F in $\mathcal{Z}$ we mean one which is inflated from a 1-cocycle of $\langle\sigma\rangle$ in $\mathcal{Z}^I$. Note that giving a 1-cocycle of $\langle\sigma\rangle$ in $\mathcal{Z}^I$ is the same as giving an element in $\mathcal{Z}^I$, namely, the value of the 1-cocycle on the canonical generator σ of $\langle\sigma\rangle$.

Lemma A.2.7 *The element $\tilde{s}$ satisfies the following properties.*

(1) $\tilde{s}_j \in Z(\hat{H})^\Gamma$ *for all* $j \in J$.
(2) $\tilde{s} \in \mathcal{Z}^I$.
(3) $\hat{\theta}(\tilde{s}) = \sigma(\tilde{s})$.

Proof. (1) Recall that $\tilde{s}_{j_0} = s$ and that $\tilde{s}_j = 1$ for $j \neq j_0$. Since $s \in Z(\hat{H})^\Gamma$, we conclude that (1) is true.

(2) Since E/F is unramified, the inertia group I acts trivially on J. Therefore for $\tau \in I$ we have

$$(\tau\tilde{s})_j = \tau(\tilde{s}_j) = \tilde{s}_j,$$

showing that τ fixes $\tilde{s}$.

(3) Again using that all values of $\tilde{s}$ are fixed by Γ, we compute that

$$(\sigma(\tilde{s}))_j = \sigma(\tilde{s}_{\sigma^{-1}j}) = \tilde{s}_{\sigma^{-1}j} = \tilde{s}_{j\sigma^{-1}} = (\hat{\theta}(\tilde{s}))_j,$$

showing that $\sigma(\tilde{s}) = \hat{\theta}(\tilde{s})$. □

Corollary A.2.8 *Let a be the unramified 1-cocycle of W_F in $\mathcal{Z}$ sending σ to $\tilde{s}$. Then $(a^{-1}, \tilde{s})$ is a 1-cocycle of W_F in $\mathcal{Z} \xrightarrow{1-\hat{\theta}} \mathcal{Z}$.*

Proof. It follows from the second part of the lemma that a is a valid unramified 1-cocycle, and it follows from the third part of the lemma that $(a^{-1}, \tilde{s})$ satisfies the 1-cocycle condition. □

Combining this corollary with theorem A.2.5, we obtain

Theorem A.2.9 *There is an equality*

$$\langle \mathrm{inv}(\gamma, \delta), (a^{-1}, \tilde{s})\rangle = \langle \alpha(\gamma, \delta), s\rangle^{-1},$$

the pairing on the right being the usual one between $B(T)$ and $\hat{T}^\Gamma$.

Proof. Using simplified 1-cocycles of W_F in $\mathcal{Z} \xrightarrow{1-\hat{\theta}} \mathcal{Z}$, the 1-cocycle $(a^{-1}, \tilde{s})$ becomes $(\tilde{s}^{-1}, \tilde{s})$, which is of the form (d^{-1}, d) for $d = \tilde{s}$. Moreover, $\tilde{s}_j \in Z(\hat{H})^\Gamma = (Z(\hat{H})^I)^{\langle\sigma\rangle}$ for all $j \in J$. It follows from the discussion in [K12], pp. 327, 328, 331 that $(\tilde{s}^{-1}, \tilde{s})$ represents a class lying in the canonical subgroup of $H^1(W_F, \hat{R}_T \xrightarrow{1-\hat{\theta}} \hat{R}_T)$ complementary to $H^1(W_F, \hat{T})$. It then follows from theorem A.2.5, lemma A.2.2 and the previous corollary that

$$\langle \mathrm{inv}(\gamma, \delta), (a^{-1}, \tilde{s})\rangle = \langle \alpha(\gamma, \delta)^{-1}, s\rangle\langle\gamma, 1\rangle^{-1} = \langle \alpha(\gamma, \delta), s\rangle^{-1}.$$

We used that the image of $(a^{-1}, \tilde{s})$ under $H^1(W_F, \mathcal{Z} \xrightarrow{1-\hat{\theta}} \mathcal{Z}) \to Z(\hat{H})^\Gamma$ is s, which boils down to the fact that the product of the d values of $\tilde{s}$ is equal to s. □

Corollary A.2.10 *When we use the allowed embedding $\tilde{\eta}$ determined by the special 1-cocycle $(a^{-1}, \tilde{s})$ described above, the twisted transfer factor $\Delta_0(\gamma_H, \delta)$ is related to the standard transfer factor $\Delta_0(\gamma_H, \gamma)$ by the equality*

$$\Delta_0(\gamma_H, \delta) = \Delta_0(\gamma_H, \gamma)\langle\alpha(\gamma, \delta), s\rangle^{-1}.$$

Proof. Use the previous theorem together with proposition A.1.10. □

Corollary A.2.10 justifies the use of $\langle\alpha(\gamma_0; \delta), s\rangle\Delta_p(\gamma_H, \gamma_0)$ as twisted transfer factors in [K9], (7.2), at least for strongly G-regular γ_H. Under the additional assumption that the derived group of G is simply connected, the next section will treat all (G, H)-regular γ_H. That $\langle \alpha(\gamma_0; \delta), s \rangle$ (rather than its inverse) appears in [K9] is not a mistake; it is due to the fact that the normalization of transfer factors, both standard and twisted, used in [K9] is opposite (see [K9], p. 178) to the one used in [LS1], [KS]. However, there are some minor mistakes in the last two lines of page 179 of [K9]: each of the five times that η appears it should be replaced by $\tilde{\eta}$, and the symbols $= t \rtimes \sigma$ near the end of the next to last line should all be deleted.

A.3 MATCHING FOR (G, H)-REGULAR ELEMENTS

In this section G, $F \subset E \subset L$, σ, H are as in section A.2. However, we will now consider transfer factors and matching of orbital integrals for all (G, H)-regular semisimple $\gamma_H \in H(F)$. For simplicity we assume that the derived group of G is simply connected, as this ensures the connectedness of the centralizer G_γ of any semisimple γ in G.

A.3.1 Image of the stable norm map

We begin by recalling two facts about the stable norm map, which we will use to prove a lemma needed later when we prove vanishing of certain stable orbital integrals for non-norms.

Let D denote the quotient of G by its derived group (which we have assumed to be simply connected).

Proposition A.3.2 (Labesse) *Let γ be an elliptic semisimple element in $G(F)$. Then γ is a stable norm from $G(E)$ if and only if the image of γ in $D(F)$ is a norm from $D(E)$.*

Proof. This is a special case of proposition 2.5.3 in [La3]. Of course the implication $\Longrightarrow$ is obvious and is true even when γ is not elliptic. □

Proposition A.3.3 (Haines) *Let M be a Levi subgroup of G and let γ be a semisimple element in $M(F)$ such that $G_\gamma \subset M$. Then γ is a stable norm from $G(E)$ if and only if it is a stable norm from $M(E)$.*

Proof. This is part of lemma 4.2.1 in [Ha]. □

These two results have the following easy consequence.

Lemma A.3.4 *Let γ be a semisimple element in $G(F)$ that is not a stable norm from $G(E)$. Then there exists a neighborhood V of γ in $G(F)$ such that no semisimple element in V is a stable norm from $G(E)$.*

Proof. Let A be the split component of the center of G_γ. The centralizer M of A in G is then a Levi subgroup of G containing G_γ. Note that γ is elliptic in $M(F)$.

The property of having a simply connected derived group is inherited by M, and we write D_M for the quotient of M by its derived group.

Since γ is not a stable norm from $G(E)$, it is certainly not a stable norm from $M(E)$. By Labesse's result the image $\overline{\gamma}$ of γ in $D_M(F)$ is not a norm from $D_M(E)$. Since the image of the norm homomorphism $D_M(E) \to D_M(F)$ is an open subgroup of $D_M(F)$, there is an open neighborhood of $\overline{\gamma}$ in $D_M(F)$ consisting entirely of non-norms. Certainly any semisimple element of $M(F)$ in the preimage V_1 of this neighborhood is not a stable norm from $M(E)$.

Consider the regular function $m \mapsto \det(1 - \mathrm{Ad}(m); \mathrm{Lie}(G)/\mathrm{Lie}(M))$ on M. Let M' be the Zariski open subset of M where this regular function does not vanish. Equivalently, M' is the set of points $m \in M$ whose centralizer in $\mathrm{Lie}(G)$ is contained in $\mathrm{Lie}(M)$, or, in other words, whose connected centralizer in G is contained in M. In particular, γ belongs to $M'(F)$, so that $M'(F)$ is another open neighborhood of γ. Applying Haines's result, we see that no semisimple element in the open neighborhood $V_2 := V_1 \cap M'(F)$ of γ in $M(F)$ is a stable norm from $G(E)$.

Finally, consider the morphism $G \times M' \to G$ sending (g, m') to $gm'g^{-1}$. It is a submersion, so the image V of $G(F) \times V_2$ provides the desired open neighborhood V of γ in $G(F)$. □

A.3.5 Review of $\alpha(\gamma, \delta)$ in the general case

Let γ be a semisimple element in $G(F)$ and $I := G_\gamma$, a connected reductive F-group. Suppose that γ is the stable norm of some θ-semisimple $\delta \in G(E)$, and let J denote the θ-centralizer $\{x \in R_G : x^{-1}\delta\theta(x) = \delta\}$ of δ, another connected reductive F-group.

There exists $c \in G(L)$ such that

$$c\gamma c^{-1} = N\delta, \tag{A.3.5.1}$$

where, as before, $N\delta = \delta\sigma(\delta)\cdots\sigma^{d-1}(\delta) \in G(E)$. Now define $b \in G(L)$ by $b := c^{-1}\delta\sigma(c)$. Applying σ to (A.3.5.1), we find that b centralizes γ, hence lies in $I(L)$. Making a different choice of c replaces b by a σ-conjugate under $I(L)$. Thus it makes sense to define $\alpha(\gamma, \delta) \in B(I)$ as the σ-conjugacy class of b.

Lemma A.3.6 *The element $\alpha(\gamma, \delta)$ is basic in $B(I)$.*

Proof. We are free to compute $\alpha(\gamma, \delta)$ using any c satisfying (A.3.5.1), and therefore we may assume that $c \in G(F^{\mathrm{un}})$. Thus there exists a positive integer r such that c is fixed by σ^{dr}. Inside the semidirect product $I(L) \rtimes \langle\sigma\rangle$ we then have $(b\sigma)^{dr} = \gamma^r\sigma^{dr}$, and, since γ is central in I, it follows that b is basic [K5] in $I(L)$. □

Since b is basic, we may use it (see [K5], [K12]) to twist the Frobenius action on $I(L)$, obtaining an inner twist I' of I such that $I'(L) = I(L)$ and with the Frobenius actions $\sigma_{I'}, \sigma_I$ on $I'(L), I(L)$, respectively, being related by $\sigma_{I'}(x) = b\sigma_I(x)b^{-1}$ for all $x \in I'(L) = I(L)$.

Recall that we are writing elements $x \in R_G$ as functions $i \mapsto x_i$ from $\mathbb{Z}/d\mathbb{Z}$ to G. There is a homomorphism $p : R_G \to G$ given by $p(x) := x_0$, but it is only defined over E (not over F). The centralizer $G_{N\delta}$ of $N\delta \in G(E)$ is also defined over E, and p restricts to an E-isomorphism $p_J : J \to G_{N\delta}$. Since $c\gamma c^{-1} = N\delta$, the inner automorphism Int(c) induces an L-isomorphism $I \to G_{N\delta}$. Therefore $x \mapsto c^{-1} p_J(x) c$ induces an L-isomorphism $\psi : J \to I$.

Lemma A.3.7 *The L-isomorphism $\psi : J \to I$ is an F-isomorphism $J \to I'$. In other words, when we use b to twist the Frobenius action of σ on I, we obtain J.*

Proof. Let $x \in J(L)$. We must show that $\psi(\sigma(x)) = b\sigma(\psi(x))b^{-1}$. The left side works out to $c^{-1}\sigma(x)_0 c = c^{-1}\sigma(x_{-1})c$, while the right side works out to $(c^{-1}\delta\sigma(c))\sigma(c^{-1}x_0 c)(c^{-1}\delta\sigma(c))^{-1} = c^{-1}\delta\sigma(x_0)\delta^{-1}c$, so we just need to observe that $\delta\sigma(x_0)\delta^{-1} = \sigma(x_{-1})$, a consequence of the fact that $\sigma(x)$ θ-centralizes δ (apply p to the equality $\delta\theta(\sigma(x))\delta^{-1} = \sigma(x)$). □

A.3.8 Comparison of $\alpha(\gamma, \delta)$ with $\alpha(\gamma_t, \delta_t)$

Our next task is to compare $\alpha(\gamma, \delta)$ with $\alpha(\gamma', \delta')$ for suitable (γ', δ') near (γ, δ) with γ' regular in G. This will be needed in order to understand the behavior of twisted transfer factors near (γ, δ). As usual in Harish-Chandra's method of semisimple descent, we obtain suitable (γ', δ') in the following way.

We retain all the notation of the previous subsection. Choose an elliptic maximal torus T in I. Since T is elliptic, it automatically transfers to the inner form J of I. Let us now see more concretely how this comes about.

From lemma A.3.7 we know that the L-isomorphism $\psi : J \to I$ is an F-isomorphism $J \to I'$. Let $i \in I(L)$. Then the L-isomorphism $\psi^{-1} \circ \mathrm{Int}(i) : I \to J$ serves to transfer T from I to J if and only if its restriction to T is defined over F. This happens if and only if $b\sigma(iti^{-1})b^{-1} = i\sigma(t)i^{-1}$ for all $t \in T(L)$, or, equivalently, if and only if $i^{-1}b\sigma(i) \in T(L)$. Here we used that $T(L)$ is Zariski dense in $T(\bar{L})$.

Since T is elliptic in I, the image of the map $B(T) \to B(I)$ is the set $B(I)_b$ of basic elements in $B(I)$ ([K5], proposition 5.3). Therefore the fiber over $\alpha(\gamma, \delta)$ is nonempty, which means we may choose $i \in I(L)$ such that $b_T := i^{-1}b\sigma(i) \in T(L)$. As above we then obtain an F-embedding $k : T \hookrightarrow J$ (given by the restriction to T of $\psi^{-1} \circ \mathrm{Int}(i) : I \to J$). A standard twisting argument identifies the fiber over $\alpha(\gamma, \delta)$ with $\ker[H^1(F, T) \to H^1(F, J)]$, the set that indexes the $J(F)$-conjugacy classes of embeddings $k' : T \hookrightarrow J$ that are stably conjugate to k. Therefore by varying the choice of i, we obtain all the different ways $k' : T \hookrightarrow J$ of transferring T to J. We will work with our particular i, b_T, k, but of course everything we do will also apply to the other choices we could have made.

Now we are in a position to compute $\alpha(\gamma', \delta')$ for certain suitably regular (γ', δ') near (γ, δ). Let $t \in T(F)$ and put $\delta_t := k(t)\delta \in R_G(F) = G(E)$. Using that $k(t)$ θ-centralizes δ, we see that the stable norm of δ_t is represented by $\gamma_t := t^d\gamma \in T(F)$. Let U denote the Zariski open subset of T consisting of those $t \in T$ such that γ_t is

G-regular. For $t \in U(F)$ the centralizer of γ_t in G is T, and therefore $\alpha(\gamma_t, \delta_t)$ lies in $B(T)$.

Proposition A.3.9 *For $t \in U(F)$ the element $\alpha(\gamma_t, \delta_t) \in B(T)$ is represented by $tb_T \in T(L)$.*

Proof. Recall that $\psi : J \to I$ is given by $\mathrm{Int}(c^{-1}) \circ p_J$. Therefore $k^{-1} : k(T) \to T$ is given by $\mathrm{Int}(i^{-1}c^{-1}) \circ p_J$. Now $N\delta_t = p(k(t^d))N\delta$. Applying $\mathrm{Int}(i^{-1}c^{-1})$ to both sides of this equality (and bearing in mind that i centralizes γ), we find that $\mathrm{Int}(i^{-1}c^{-1})(N\delta_t) = t^d\gamma$. Therefore $\alpha(\gamma_t, \delta_t)$ is represented by $i^{-1}c^{-1}\delta_t\sigma(c)\sigma(i)$. The identification $R_G(F) = G(E)$ is induced by $p : R_G \to G$, so that in $G(E)$ we have the equality $\delta_t = p_J(k(t))\delta$. Therefore our representative for $\alpha(\gamma_t, \delta_t)$ can be rewritten as $\big(\mathrm{Int}(i^{-1}c^{-1})(p_J(k(t)))\big)i^{-1}c^{-1}\delta\sigma(c)\sigma(i)$, which simplifies to tb_T, as desired. □

Let K denote the kernel of the homomorphism $T(F) \to B(T)$ that sends $t \in T(F)$ to the σ-conjugacy class of t in $T(L)$. It follows easily from [K12, §7], that K is an open subgroup of $T(F)$. The previous proposition then has the immediate corollary:

Corollary A.3.10 *For all $t \in U(F) \cap K$ the element $\alpha(\gamma_t, \delta_t) \in B(T)$ maps to $\alpha(\gamma, \delta)$ under the map $B(T) \to B(I)$ induced by $T \subset I$.*

This corollary is exactly what will be needed in the descent argument to come, through the intermediary of proposition A.3.12.

A.3.11 Twisted transfer factors for (G, H)-regular γ_H

Consider a (G, H)-regular semisimple element γ_H in $H(F)$. The centralizer I_H of γ_H in H is connected ([K7], lemma 3.2). Choose an elliptic maximal torus T_H in I_H. In particular T_H is a maximal torus in H containing γ_H. Choose an admissible embedding $T_H \hookrightarrow G$. We write γ, T for the images under this embedding of γ_H, T_H, respectively. Then from [K7] the centralizer I of γ in G is an inner twist of I_H. Of course T is an elliptic maximal torus in I, and our chosen isomorphism $T_H \cong T$ exhibits T as the transfer of T_H to the inner twist I of I_H.

We need a twisted transfer factor $\Delta_0(\gamma_H, \delta)$ for any θ-semisimple $\delta \in G(E)$ whose stable norm is γ. These were not defined in [KS], but in the current context, that of cyclic base change for unramified E/F, with σ being the Frobenius automorphism and the derived group of G being simply connected, they were defined in [K9] by the formula

$$\Delta_0(\gamma_H, \delta) := \Delta_0(\gamma_H, \gamma)\langle\alpha(\gamma, \delta), s\rangle^{-1}, \tag{A.3.11.1}$$

with $\Delta_0(\gamma_H, \gamma)$ defined as in [LS2], 2.4. (See the comment following corollary A.2.10 concerning the opposite normalization of transfer factors used in [K9].) The pairing occurring in this formula is between $B(I)_b$ and $Z(\hat{I})^\Gamma$, and comes from the canonical isomorphism ([K5], [K12]) $B(I)_b \simeq X^*(Z(\hat{I})^\Gamma)$. In forming this pairing, we view $s \in Z(\hat{H})^\Gamma$ as an element of $Z(\hat{I})^\Gamma$ via

$$Z(\hat{H})^\Gamma \subset Z(\hat{I}_H)^\Gamma = Z(\hat{I})^\Gamma.$$

By corollary A.2.10 this definition of $\Delta_0(\gamma_H, \delta)$ agrees with the one in [KS] when γ_H is G-regular.

We now apply the work we did in the previous subsection to (γ, δ). With notation as in that subsection we can now formulate

Proposition A.3.12 *There is an open neighborhood of* 1 *in* $T(F)$ *such that*

$$\Delta_0(t^d\gamma_H, k(t)\delta) = \Delta_0(\gamma_H, \delta)$$

for all t *in this neighborhood for which* $t^d\gamma_H$ *is* G*-regular. In writing* $t^d\gamma_H$ *we are viewing* t *as an element in* $T_H(F)$ *via our chosen isomorphism* $T_H \simeq T$.

Proof. This follows from corollary A.3.10 and the fact that $t \mapsto \Delta_0(t^d\gamma_H, t^d\gamma)$ is defined and constant near $t = 1$ (see [LS2], 2.4, where γ_H, γ are said to be equisingular). □

A.3.13 Matching of orbital integrals for (G, H)-regular γ_H

We continue with γ_H, γ, I_H, I as in the previous subsection. For $f^H \in C_c^\infty(H(F))$ we consider the stable orbital integral

$$SO_{\gamma_H}(f^H) = \sum_{\gamma'_H} e(I_{\gamma'_H})O_{\gamma'_H}(f^H),$$

where the sum is taken over conjugacy classes of $\gamma'_H \in H(F)$ that are stably conjugate to γ_H, and $I_{\gamma'_H}$ denotes the (connected) centralizer of γ'_H in H. For $f \in C_c^\infty(G(E))$ we consider the endoscopic linear combination of twisted orbital integrals

$$TO_{\gamma_H}(f) = \sum_{\delta} e(J_\delta)\Delta_0(\gamma_H, \delta)TO_\delta(f)$$

determined by γ_H. Thus the sum is taken over twisted conjugacy classes of $\delta \in G(E)$ whose stable norm is γ, and J_δ is the twisted centralizer of δ. When γ is not a stable norm from $G(E)$, we have $TO_{\gamma_H}(f) = 0$, since the sum occurring in its definition is then empty.

Proposition A.3.14 *Suppose that*

$$SO_{\gamma_H}(f^H) = TO_{\gamma_H}(f)$$

for all G*-regular semisimple* γ_H *in* $H(F)$*. Then the same equality holds for all* (G, H)*-regular semisimple* γ_H *in* $H(F)$.

Proof. We will just sketch the proof since it is essentially the same as that of proposition 2 in [K8], p. 640, as well as those of lemma 2.4.A in [LS2] and proposition 7.2 in [Cl3].

Fix a (G, H)-regular semisimple $\gamma_H \in H(F)$. Introduce T_H, T as in the previous subsection. Assume for the moment that γ is a stable norm. Looking at the degree 0 part of the germs about 1 of the functions $t \mapsto SO_{t^d\gamma_H}(f^H)$ and $t \mapsto TO_{t^d\gamma_H}(f)$ on $T(F)$, we conclude from proposition A.3.12 that

$$m\sum_{\gamma'_H}(-1)^{q(I_{\gamma'_H})}O_{\gamma'_H}(f^H) = m\sum_{\delta}(-1)^{q(J_\delta)}\Delta_0(\gamma_H, \delta)TO_\delta(f),$$

where q assigns to a connected reductive F-group the F-rank of its derived group, and m is the common value of the cardinalities of all the sets $\ker[H^1(F, T_H) \to H^1(F, I_{\gamma'_H})]$ and $\ker[H^1(F, T) \to H^1(F, J_\delta)]$. Of course we used sensible Haar measures and Rogawski's formula for the Shalika germ corresponding to the identity element, just as in the previously cited proofs. We also used that, when $t^d\gamma$ is G-regular, the θ-conjugacy classes having stable norm $t^d\gamma$ are represented by elements of the form $k'(t)\delta$, with δ again varying through twisted conjugacy classes of $\delta \in G(E)$ whose stable norm is γ, and (for fixed δ with stable norm γ) k' varying through a set of representatives for the stable conjugacy classes of embeddings $k' : T \hookrightarrow J_\delta$ of the kind appearing in the discussion leading up to proposition A.3.9. Dividing both sides of our equality by $m(-1)^{q(I_0)}$, where I_0 is a common quasi-split inner form of all the groups $I_{\gamma'_H}$ and J_δ, we obtain $SO_{\gamma_H}(f^H) = TO_{\gamma_H}(f)$, as desired.

When γ is not a stable norm from $G(E)$, we must show that $SO_{\gamma_H}(f^H) = 0$. Looking at the degree 0 part of the stable Shalika germ expansion for the maximal torus T_H in H (see the sentence just before proposition 1 in [K8], p. 639), we see that it is enough to show that $SO_{t_H}(f^H) = 0$ for all G-regular $t_H \in T_H(F)$ near γ_H. For this it is enough to show that elements t_H near γ_H, when viewed in $T(F)$, are not stable norms from $G(E)$, and this follows from lemma A.3.4. □

A.3.15 A correction to [K8]

In the course of looking through section 2 of [K8] I noticed an error in the definition of the Euler-Poincaré function f_{EP}. The sign character sgn_σ occurring in the definition of f_{EP} should be defined as follows: $\mathrm{sgn}_\sigma(g)$ is 1 if g preserves the orientation of the polysimplex σ, and it is -1 if g reverses that orientation. When σ is a simplex, $\mathrm{sgn}_\sigma(g)$ is just the sign of the permutation induced by g on the vertices of σ. When writing [K8] I carelessly assumed that the same is true for polysimplices, but this is in fact not the case even for the product of two copies of a 1-simplex. Then there is a reflection (obviously orientation reversing) that induces a permutation with cycle structure (12)(34) (obviously an even permutation) on the four vertices of the square. This situation actually arises for the Euler-Poincaré function on the group $PGL_2 \times PGL_2$.

With this corrected definition of the sign character, the formula $\mathrm{sgn}_\tau(\gamma) = (-1)^{\dim(\tau)-\dim(\tau(\gamma))}$ used in the proof of theorem 2 of [K8] becomes correct and so no change is needed in that proof.

Bibliography

[A1] J. Arthur, *A trace formula for reductive groups I. Terms associated to classes in* $\mathbf{G}(\mathbb{Q})$, Duke Math. J. 45 (1978), no. 4, 911–952

[A2] J. Arthur, *The invariant trace formula. I. Local theory*, J. Amer. Math. Soc. 1 (1988), no. 2, 323–383

[A3] J. Arthur, *The invariant trace formula. II. Global theory*, J. Amer. Math. Soc. 1 (1988), no. 3, 501–554

[A4] J. Arthur, *The local behaviour of weighted orbital integrals*, Duke Math. J. 56 (1988), no. 2, 223–293

[A5] J. Arthur, *Intertwining operators and residues. II. Invariant distributions*, Composition Math. 70 (1989), no. 1, 51–99

[A6] J. Arthur, *The L^2-Lefschetz numbers of Hecke operators*, Inv. Math. 97 (1989), 257–290

[AC] J. Arthur and L. Clozel, *Simple algebras, base change, and the advanced theory of the trace formula*, A. Math. Stud., 120, Princeton University Press (1989)

[Bo] A. Borel, *Automorphic L-functions*, in *Automorphic forms, representations, and L-functions* (Proc. Symposia in Pure Math. 1977), part 2, 26–61

[BC] A. Borel and W. Casselman, *L^2-cohomology of locally symmetric manifolds of finite volume*, Duke Math. J. 50 (1983), 625–647

[BJ] A. Borel and H. Jacquet, *Automorphic forms and automorphic representations*, in *Automorphic forms, representations, and L-functions* (Proc. Symposia in Pure Math. 33, 1977), part 1, 189–202

[BT] A. Borel and J. Tits, *Groupes réductifs*, Pub. Math. IHES 27 (1965), 55–151

[BW] A. Borel and N. Wallach, *Continuous cohomology, discrete subgroups, and representations of reductive groups, 2nd ed.*, Math. Surveys and Monographs 57, AMS (2000)

[BL] J.-L. Brylinski and J.-P. Labesse, *Cohomologie d'intersection et fonctions L de certaines variétés de Shimura*, Ann. Sci. École Norm. Sup., 4ième sér. 17 (1984), 361–412

[BuW] J. Burgos and J. Wildeshaus, *Hodge modules on Shimura varieties and their higher direct images in the Baily-Borel compactification*, Ann. Sci. École Norm. Sup. (4) 37 (2004), no. 3, 363–413

[Car] P. Cartier, *Representations of* $\mathfrak{p}$*-adic groups : a survey*, in *Automorphic forms, representations, and L-functions* (Proc. Symposia in Pure Math. 33, 1977), part 1, 111–156

[Cas] W. Casselman, *The unramified principal series of* $\mathfrak{p}$*-adic groups I. The spherical function*, Comp. Math. 40 (1980), no. 3, 387–406

[CCl] G. Chenevier and L. Clozel, *Corps de nombres peu ramifiés et formes automorphes autoduales*, http://www.math.polytechnique.fr/chenevier/articles/galoisQautodual2.pdf, J. Amer. Math. Soc., to appear

[C] V. I. Chernousov, *The Hasse principle for groups of type* E_8, Soviet Math. Dokl. 39 (1989), 592–596.

[Cl1] L. Clozel, *Orbital integrals on p-adic groups: a proof of the Howe conjecture*, Ann. of Math. 129 (1989), no. 2, 237–251

[Cl2] L. Clozel, *Changement de base pour les représentations tempérées des groupes réductifs réels*, Ann. Sci. École Norm. Sup. 15 (1982), no. 1, 45–115

[Cl3] L. Clozel, *The fundamental lemma for stable base change*, Duke Math. J. 61 (1990), no. 1, 255–302

[Cl4] L. Clozel, *Motifs et formes automorphes: application du principe de fonctorialité*, in *Automorphic forms, Shimura varieties, and L-functions*, Proc. Ann Arbor conference, ed. L. Clozel and J. Milne (1990), I, 77–159

[Cl5] L. Clozel, *Représentations galoisiennes associées aux représentations automorphes autoduales de* $\mathbf{GL}(n)$, Pub. Math. IHES 73 (1991), 97–145

[CD] L. Clozel and P. Delorme, *Pseudo-coefficients et cohomologie des groupes de Lie réductifs réels*, C.R. Acad. Sc. Paris 300, série I (1985), 385–387

[CL] L. Clozel and J.-P. Labesse, *Changement de base pour les représentations cohomologiques de certains groupes unitaires*, Astérisque 257 (1999), 119–133

[D1] P. Deligne, *Variétés de Shimura: interprétation modulaire, et techniques de construction de modèles canoniques*, in *Automorphic forms, representations and L-functions* (Proc. Sympos. Pure Math. 33, part 2, 1979), 247–290

[D2] P. Deligne, *La conjecture de Weil. II.*, Pub. Math. de l'IHES 52 (1981), 137–251

[DK] P. Deligne and W. Kuyk (ed.), *Modular functions of one variable II*, Lecture Notes in Mathematics 349, Springer (1973)

[De] P. Delorme, *Théorème de Paley-Wiener invariant tordu pour le changement de base* $\mathbb{C}/\mathbb{R}$, Compositio Math. 80 (1991), no. 2, 197–228

[DeM] P. Delorme and P. Mezo, *A twisted invariant Paley-Wiener theorem for real reductive groups*, Duke Math. J. 144 (2008), no. 2, 341–380

[Di] J. Dieudonné, *La géométrie des groupes classiques. 3rd ed.*, Ergebnisse der Mathematik und ihrer Grenzgebiete 5, Springer (1971)

[vD] G. van Dijk, *Computation of certain induced characters of p-adic groups*, Math. Ann. 199 (1972), 229–240

[Fa] L. Fargues, *Cohomologie des espaces de modules de groupes p-divisibles et correspondance de Langlands locale*, Astérisque 291 (2004), 1–200

[F] K. Fujiwara, *Rigid geometry, Lefschetz-Verdier trace formula and Deligne's conjecture*, Invent. Math. 127 (1997), 489–533

[GHM] M. Goresky, G. Harder and R. MacPherson, *Weighted cohomology*, Invent. Math. 166 (1994), 139–213

[GKM] M. Goresky, R. Kottwitz and R. MacPherson, *Discrete series characters and the Lefschetz formula for Hecke operators*, Duke Math. J. 89 (1997), 477–554 and Duke Math. J. 92 (1998), no. 3, 665–666

[Ha] T. Haines, *The base change fundamental lemma for central elements in parahoric Hecke algebras*, submitted, http://www.math.umd.edu/ tjh /fl29.pdf

[H1] T. Hales, *A simple definition of transfer factors for unramified groups*, in *Representation theory of groups and algebras*, eds. J. Adams, R. Herb, S. Kudla, J.-S. Li, R. Lipsman and J. Rosenberg, Contemp. Math. 145 (1993), 109–134

[H2] T. Hales, *On the fundamental lemma for standard endoscopy: reduction to unit elements*, Can. J. Math. 47 (1995), no. 5, 974–994

[HT] M. Harris and R. Taylor, *The geometry and cohomology of some simple Shimura varieties*, Ann. of Math. Stud. 151, Princeton University Press (2001)

[He] G. Henniart, *La conjecture de Langlands locale pour* **GL**(3), Mémoires de la SMF 2ème sér., tome 11–12 (1983), 1–186

[JS] H. Jacquet and J. A. Shalika, *On Euler products and the classification of automorphic forms II*, Amer. J. of Math. 103, no. 4 (1981), 777–815

[J] J. Johnson, *Stable base change* $\mathbb{C}/\mathbb{R}$ *of certain derived functor modules*, Math. Ann. 287 (1990), no. 3, 467–493

[Ka] D. Kazhdan, *Cuspidal geometry of p-adic groups*, J. Anal. Math. 47 (1986), 1–36

[Ke] D. Keys, *Reducibility of unramified unitary principal series representations of p-adic groups and class-*1 *representations*, Math. Ann. 260 (1982), 397–402

[K1] R. Kottwitz, *Rational conjugacy classes in reductive groups*, Duke Math. J. 49 (1982), 785–806

[K2] R. Kottwitz, *Sign changes in harmonic analysis on reductive groups*, Trans. A M S 278 (1983), 289–297

[K3] R. Kottwitz, *Shimura varieties and twisted orbital integrals*, Ann. of Math. 269 (1984), 287–300

[K4] R. Kottwitz, *Stable trace formula : cuspidal tempered terms*, Duke Math. J. 51 (1984), 611–650

[K5] R. Kottwitz, *Isocrystals with additional structure*, Compositio Math. 56 (1985), 365–399

[K6] R. Kottwitz, *Base change for units of Hecke algebras*, Compositio Math. 60 (1986), 237–250

[K7] R. Kottwitz, *Stable trace formula : elliptic singular terms*, Math. Ann. 275 (1986), 365–399

[K8] R. Kottwitz, *Tamagawa numbers*, Ann. of Math. 127 (1988), 629–646

[K9] R. Kottwitz, *Shimura varieties and λ-adic representations*, in *Automorphic forms, Shimura varieties and L-functions*, (Proc. of the Ann Arbor conference, eds. L. Clozel and J. Milne, 1990), vol. I, 161–209

[K10] R. Kottwitz, *On the λ-adic representations associated to some simple Shimura varieties*, Inv. Math. 108 (1992), 653–665

[K11] R. Kottwitz, *Points on some Shimura varieties over finite fields*, J. Amer. Math. Soc. 5, no. 2 (1992), 373–444

[K12] R. Kottwitz, *Isocrystals with additional structure II*, Comp. Math. 109 (1997), 255–339

[K13] R. Kottwitz, unpublished

[KRo] R. Kottwitz and J. Rogawski, *The distributions in the invariant trace formula are supported on characters*, Canad. J. Math. 52 (2000), no. 4, 804–814

[KS] R. Kottwitz and D. Shelstad, *Foundations of twisted endoscopy*, Astérisque 255 (1999)

[La1] J.-P. Labesse, *Fonctions élémentaires et lemme fondamental pour le changement de base stable*, Duke Math. J. 61 (1990), no. 2, 519–530

[La2] J.-P. Labesse, *Pseudo-coefficients très cuspidaux et K-théorie*, Math. Ann. 291 (1991), 607–616

[La3] J.-P. Labesse, *Cohomologie, stabilisation et changement de base*, Astérisque 257 (1999), 1–116

[Lan] K.-W. Lan, *Arithmetic compactifications of PEL-type Shimura varieties*, Ph.D. thesis, Harvard University (2008), http://www.math.princeton.edu/klan/articles/cpt-PEL-type-thesis.pdf

[L1] R. Langlands, *Modular forms and ℓ-adic representations*, in *Modular Functions of One Variable II*, Lecture Notes in Mathematics 349, Springer (1972), 361–500

[L2] R. Langlands, *Stable conjugacy: definitions and lemmas*, Can. J. Math. 31 (1979), 700–725

[L3] R. Langlands, *Les débuts d'une formule des traces stable*, Publ. Math. Univ. Paris VII 13, Paris (1983)

[LR] R. Langlands and D. Ramakrishnan (ed.), *The zeta function of Picard modular surfaces*, Publications du CRM Montréal (1992)

[LS1] R. Langlands and D. Shelstad, *On the definition of transfer factors*, Math. Ann. 278 (1987), 219–271

[LS2] R. Langlands and D. Shelstad, *Descent for transfer factors* (Grothendieck Festschrift, vol. II), Progr. Math. 87, Birkhäuser (1990), 485–563

[Lau1] G. Laumon, *Sur la cohomologie à supports compacts des variétés de Shimura pour* $\mathbf{GSp}(4)_{\mathbb{Q}}$, Compositio Math. 105 (1997), no. 3, 267–359

[Lau2] G. Laumon, *Fonctions zêta des variétés de Siegel de dimension trois*, dans *Formes automorphes II. Le cas du groupe* **GSp**(4), Astérisque 302 (2005), 1–66

[LN] G. Laumon and B.-C. Ngo, *Le lemme fondamental pour les groupes unitaires*, Ann. of Math. 168 (2008), no. 2, 477–573

[Lo] E. Looijenga, *L^2-cohomology of locally symmetric varieties*, Compositio Math. 67 (1988), no. 1, 3–20

[LoR] E. Looijenga and M. Rapoport, *Weights in the local cohomology of a Baily-Borel compactification*, in *Complex geometry and Lie theory*, Proc. Sympos. Pure Math. 53 (1991), 223–260

[MW] C. Moeglin and J.-L. Waldspurger, *Le spectre résiduel de* $\mathbf{GL}(n)$, Ann. Scient. É.N.S., 4ᵉ sér. 22 (1989), 605–674

[Ma1] E. Mantovan, *On certain unitary group Shimura varieties*, Astérisque 291 (2004), 201–331

[Ma2] E. Mantovan, *On the cohomology of certain PEL type Shimura varieties*, Duke Math. 129 (2005), 573–610.

[M1] S. Morel, *Complexes d'intersection des compactifications de Baily-Borel : le cas des groupes unitaires sur* $\mathbb{Q}$, Ph.D. thesis, Université Paris 11 (2005), http://www.math.ias.edu/ morel/complexes _d_intersection.pdf

[M2] S. Morel, *Complexes pondérés sur les compactifications de Baily-Borel. Le cas des variétés de Siegel*, Journal of the AMS 21 (2008), no. 1, 23–61

[M3] S. Morel, *Cohomologie d'intersection des variétés modulaires de Siegel, suite*, preprint, http://www.math.ias.edu/ morel/groupes _symplectiques.pdf

[Ne] J. Neukirch, *Algebraic number theory*, Grundlehren der mathematischen Wissenschaften, 322 (1999)

[Ng] B. C. Ngo, *Le lemme fondamental pour les algèbres de Lie*, submitted, arXiv:0801.0446

[O] T. Ono, *On Tamagawa numbers*, in *Algebraic groups and discontinuous subgroups*, editors A. Borel et G. Mostow, Proc. of Symposia in Pure Math. 9 (1966)

[P1] R. Pink, *Arithmetical compactification of mixed Shimura varieties*, Ph.D. thesis, Bonner Mathematische Schriften 209 (1989)

[P2] R. Pink, *On ℓ-adic sheaves on Shimura varieties and their higher direct images in the Baily-Borel compactification*, Math. Ann. 292 (1992), 197–240

[P3] R. Pink, *On the calculation of local terms in the Lefschetz-Verdier trace formula and its application to a conjecture of Deligne*, Ann. of Math. 135 (1992), 483–525

[Ra] M. Rapoport, *On the shape of the contribution of a fixed point on the boundary: the case of $\mathbb{Q}$-rank one*, in [LR], 479–488, with an appendix by L. Saper and M. Stern, 489–491

[Ro1] J. Rogawski, *Automorphic representations of unitary groups in three variables*, Ann. of Math. Stud. 123, Princeton University Press (1990)

[Ro2] J. Rogawski, *Analytic expression for the number of points mod p*, in [LR], 65–109

[SS] L. Saper and M. Stern, *L^2-cohomology of arithmetic varieties*, Ann. of Math. 132 (1990), no. 1, 1–69

[Sh1] D. Shelstad, *L-indistiguishability for real groups*, Math. Ann. 259 (1982), 385–430

[Sh2] D. Shelstad, *Endoscopic groups and base change $\mathbb{C}/\mathbb{R}$*, Pacific J. Math. 110 (1984), 397–416

[Shi1] S. W. Shin, *Counting points on Igusa varieties*, to appear in Duke Math. Journal

[Shi2] S. W. Shin, *Stable trace formula for Igusa varieties*, submitted

[Shi3] S. W. Shin, *Galois representations arising from some compact Shimura varieties*, preprint

[Sp] T. A. Springer, *Linear algebraic groups, 2nd ed.*, Progress in Math. 9, Birkhäuser (1998)

[V] Y. Varshavsky, *A proof of a generalization of Deligne's conjecture*, Electron. Res. Announc. Amer. Math. Soc. 11 (2005), 78–88

[Vi] M.-F. Vignéras, *Caractérisation des intégrales orbitales sur un groupe réductif p-adique*, J. Fac. Sci. Univ. Tokyo, Sect 1A Math. 28 (1982), 945–961

[Wa1] J.-L.Waldspurger, *Le lemme fondamental implique le transfert*, Comp. Math. 105 (1997), no. 2, 153–236

[Wa2] J.-L.Waldspurger, *Endoscopie et changement de caractéristique*, J. Inst. Math. Jussieu 5 (2006), no. 3, 423–525

[Wa3] J.-L.Waldspurger, *L'endoscopie tordue n'est pas si tordue*, Mem. Amer. Math. Soc. 194 (2008), no. 908

[W] J. Wildeshaus, *Mixed sheaves on Shimura varieties and their higher direct images in toroidal compactifications*, J. Algebraic Geom. 9, no. 2 (2000), 323–353

[SGA 4 1/2] P. Deligne and al., *Séminaire de géométrie algébrique 4 1/2. Cohomologie étale*, Lecture Notes in Mathematics 569, Springer (1977)

[SGA 5] A. Grothendieck and al., *Cohomologie ℓ-adique et fonctions L*, Lecture Notes in Math. 589, Springer (1977)

Index

www.ingramcontent.com/pod-product-compliance
Ingram Content Group UK Ltd.
Pitfield, Milton Keynes, MK11 3LW, UK
UKHW022227121224
452420UK00005B/288